Constants

1 torr = "1 mm Hg" = $\frac{1}{760}$ atm

1 cal (thermochemical) ≡ 4.184 J

1 cal (mean) = 4.190 02 J

1 cal (15°) 4.185 80 J

1 cal (20°) = 4.181 90 J

1 cal (IST) = 4.186 8 J [International Steam Table]

1 BTU = 1.05(5) $\times 10^3$ J

1 horsepower = 7.(4) $\times 10^2$ W

1 poise ≡ 0.1 Pa · s

1 pound ≡ 4.448 221 615 260 5 N

$c = 2.997\ 924\ 58 \times 10^8\ \frac{\text{m}}{\text{s}}$ Speed of light

$m_e = 9.109\ 534 \times 10^{-31}\text{kg}$ Electron rest mass

$g \equiv 9.806\ 65\ \frac{\text{N}}{\text{kg}}$ Standard gravitational field

Modern Thermodynamics with Statistical Mechanics

Robert P. Bauman
University of Alabama at Birmingham

Modern Thermodynamics with Statistical Mechanics

Macmillan Publishing Company
New York
Maxwell Macmillan Canada
Toronto
Maxwell Macmillan International
New York Oxford Singapore Sydney

Editor: Robert A. McConnin
Production Supervisor: Elaine W. Wetterau
Production Manager: Sandra Moore
Text Designer: Eileen Burke
Cover Designer: Jane Edelstein
Illustrations: Precision Graphics

This book was set in Times Roman by Publication Services,
printed and bound by Hamilton.
The cover was printed by Hamilton.

Printed in the United States of America

Macmillan Publishing Company
866 Third Avenue, New York, New York 10022

Macmillan Publishing Company is part
of the Maxwell Communication Group of Companies.

Maxwell Macmillan Canada, Inc.
1200 Eglinton Avenue East
Suite 200
Don Mills, Ontario M3C 3N1

Library of Congress Cataloging in Publication Data

Bauman, R. P. (Robert Poe)
Modern thermodynamics with statistical mechanics / Robert P.
Bauman.
p. cm.
Includes bibliographical references and index.
ISBN 0-02-306780-2
1. Thermodynamics. 2. Statistical mechanics. I. Title.
TJ265.B198 1992
621.402'1–dc20 91-22735
CIP

Printing: 1 2 3 4 5 6 7 8 Year: 2 3 4 5 6 7 8 9 0 1

To David

Preface

This introduction to thermodynamics and statistical mechanics is written for science students. It assumes some understanding of partial derivatives but does not require prior knowledge of thermodynamics. An introductory college-level physics course is a prerequisite. Understanding of chemistry at the freshman level is also assumed.

Thermodynamics is an essential preparation for many careers, including not only high school teachers, physicists, chemists, and engineers, but also those who will study the earth and planetary sciences, materials science, and astrophysics, among others.

Even as mature a science as thermodynamics undergoes significant development with time. New understandings in several areas justify a fresh presentation of classical thermodynamics and its younger sibling science, statistical mechanics. In particular, (1) recent reexamination of the definition of work has led to more precise statements of when work can and cannot be operationally defined and when the transfer of energy can be separated into Q and W terms; (2) consideration of the power output of heat engines has emphasized the distinction between maximum efficiency and optimum performance; and (3) previously unpublished methods of presentation of phase equilibria tie customary phase diagrams more clearly and explicitly to Gibbs' phase rule, simplifying their interpretation by beginners.

A new presentation of the subject has also provided an opportunity for minor modifications in notation and terminology. These newer forms serve as steady reminders of distinctions between concepts that experience has shown are often confused, even by textbook authors. Demonstration of the links between thermodynamics and other fields of physics is also aided by a notation that is consistent across fields. Without reteaching other subjects, it is shown that thermodynamic arguments are relevant to these other areas.

Thermodynamics was not discovered in linear form and it is never learned in linear form, despite the enthusiasm for teaching it in orderly fashion. A sincere effort has been made to avoid using concepts before they are explained, unless they are typically well covered in the prerequisite courses. Where exceptions seem necessary or appropriate, cross-references are given. Several topics are amenable

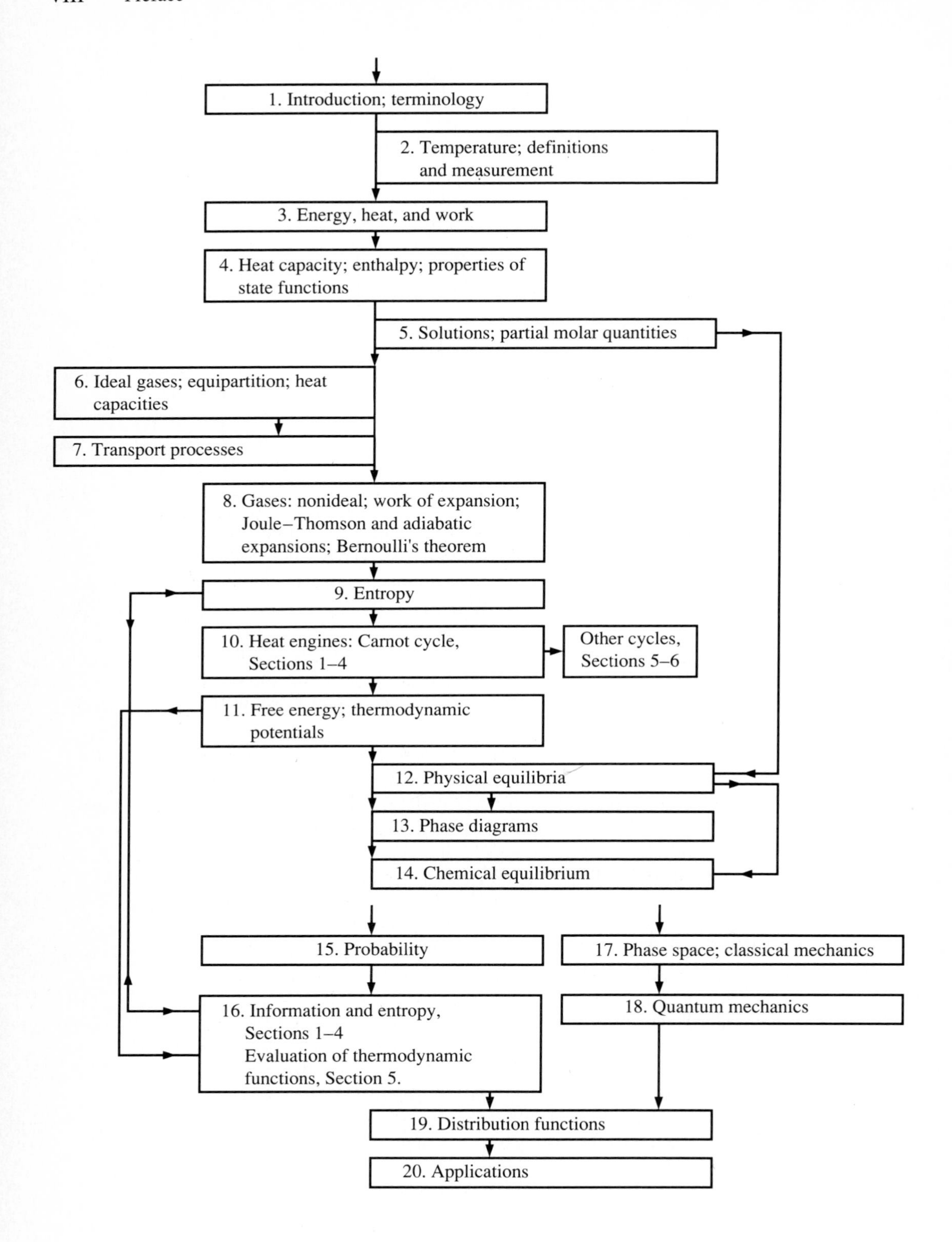

1. Introduction; terminology
2. Temperature; definitions and measurement
3. Energy, heat, and work
4. Heat capacity; enthalpy; properties of state functions
5. Solutions; partial molar quantities
6. Ideal gases; equipartition; heat capacities
7. Transport processes
8. Gases: nonideal; work of expansion; Joule–Thomson and adiabatic expansions; Bernoulli's theorem
9. Entropy
10. Heat engines: Carnot cycle, Sections 1–4
Other cycles, Sections 5–6
11. Free energy; thermodynamic potentials
12. Physical equilibria
13. Phase diagrams
14. Chemical equilibrium
15. Probability
17. Phase space; classical mechanics
16. Information and entropy, Sections 1–4 Evaluation of thermodynamic functions, Section 5.
18. Quantum mechanics
19. Distribution functions
20. Applications

to a spiral approach. An introduction to the physical concepts, with less than rigorous mathematical analysis, often helps students to gain perspective on the topic before becoming immersed in a later detailed development.

The number of variations in thermodynamics/statistical mechanics courses appears to be somewhat greater than the number of instructors. There is equal diversity in the background of students, particularly now as some traditional topics are being squeezed out of the first-year courses. Most, but not all, students will have completed a course devoted largely to modern physics. Some students put off thermodynamics until they have completed upper-level classical mechanics and introductory quantum mechanics.

The present textbook is intended to be sufficiently flexible to accommodate most needs of instructors and students. If students are well prepared, or if time constraints are severe, some chapters or sections may appropriately be omitted. Some reordering of chapters is practical, as represented in the accompanying figure.

Although the numbered order is preferable, when time permits, other sequences are practical. Chapters shown adjacent to the main line could be omitted without significantly hampering the study of later chapters, except where linkages are indicated, such as between Chapters 5 and 12 and between these and 13 and 14. Chapter 11 and its precursors are assumed for the discussion of thermodynamic functions in Chapter 16, but the functions could be introduced in Chapter 16.

Chapter 21 may be treated as an appendix, to be dipped into when the corresponding topics are first encountered. It is likely to be more effective, however, as a retrospective view, after the topics have been considered in more conventional fashion.

The most important message students should receive about thermodynamics is the universality of the theory. Although a variety of approximations are typically employed in specific applications, modern thermodynamics has the power to derive exact relationships for real materials, based on empirical measurements of properties of substances. When combined with the more probing techniques of statistical mechanics, based on specific atomic and molecular models of matter, thermodynamics can cut both ways, making predictions of macroscopic properties and behaviors and testing the detailed validity of the molecular-level models.

Thermodynamics is not an easy subject. Many of the concepts appear initially to be quite abstract. The greatest difficulty students have, however, is not in understanding, but in tying together seemingly unconnected ideas and remembering them when they are needed. Many of the problems are designed specifically to encourage development of this facility; to the instructor a problem may appear trivial, whereas it may appear insoluble to the novice. With some familiarity, however, the thermodynamic potentials and related functions will soon become good friends that may often be called upon in time of need. And as with other good friends, growing familiarity should provide real fun as, with them, new discoveries are made and new truths uncovered.

Acknowledgments

Any textbook author is in debt not only to the instructors and earlier authors who were primary sources of information, but also to colleagues who have raised questions and tested new ideas and, especially, to the students who probe the material they have been given with excruciating care for weaknesses in logic or clarity. The contributors to this book are accordingly too numerous to list here. Special mention is necessary, however, for Ernest Loebl, who aided some decades ago in the recognition that conventional wisdom must be tested as carefully as new ideas; Rolf Schwaneburg, who contributed to the pedagogy and to the analysis of Bernoulli flow; and to Celia C. Chow, who has contributed a set of demonstrations and laboratory exercises that appear in the companion *Instructor's Manual*.

Special thanks must go to Robert McConnin, who shepherded the book from manuscript to bound volume and to Mrs. Elaine Wetterau, who made the always difficult production processes relatively enjoyable. I also wish to acknowledge the true heroine of the project, my wife, Edith, who has cheerfully put up with the extended inconveniences and schedule distortions of the writing and publication process.

R. P. B.

Contents

PART II

Second Law of Thermodynamics: The Irreversibility of Time

PART III

Statistical Mechanics: Thermodynamics at the Molecular Level

Modern Thermodynamics with Statistical Mechanics

CHAPTER 1

The Language and Methods of Thermodynamics

Classical thermodynamics is rightfully considered one of the foundation stones of physics, along with mechanics, optics, and electricity and magnetism. It is also a fundamental part of chemistry, and much of the development of the field has been by those trained in chemistry. At the same time, thermodynamics has long been recognized as having a special importance in engineering.

1.1 Classical and Statistical Thermodynamics

The apex of classical thermodynamics came in the first quarter of the present century, before its sibling science, statistical thermodynamics, or statistical mechanics, began crowding it for attention. Thermodynamics has had a tradition of close ties to practical problems, from the experiments and inventions of Rumford to the monograph by Carnot on steam engines to modern applications in engineering processes. Although it now must share the spotlight with statistical mechanics, it remains a powerful tool in its own right.

Statistical mechanics has a strong base in the nineteenth-century efforts of Clerk Maxwell, Boltzmann, Clausius, and Gibbs, among others. But statistical mechanics could not mature until quantum mechanics developed in the 1920s. Indeed, one must sympathize with the frustration of Willard Gibbs, who found his beautiful theory and mathematics in conflict with the seemingly uncooperative behavior of molecules, acting under the quantum rules only hinted at by the work of Planck in 1900.

Writing in December of 1901, in the preface to his monograph titled *Elementary Principles of Statistical Mechanics Developed with Especial Reference to the Rational Foundation of Thermodynamics*, Gibbs observed that it was not yet possible to integrate thermodynamics with theories of light and electrical interactions of atoms, although "any theory is obviously inadequate which does not take account of all these phenomena." Citing one well-known example of disagreement between theory and experiment (the problem of heat capacities of gases, later resolved with the aid of quantum mechanics), he deprecates the validity of his own magnificent work. "Certainly, one is building on an insecure foundation, who rests his work on hypotheses concerning the constitution of matter."

Statistical mechanics and classical thermodynamics are complementary sciences, supporting and fulfilling each other, sharing terminology and functions, while maintaining distinct foundations. The strength of classical thermodynamics lies in its almost total lack of dependence on, or knowledge of, the atomic and molecular composition of matter.[1] By avoiding any reliance on molecular theory, classical thermodynamics is impervious to the uncertainties and peculiarities of the materials to which it is applied. This gives thermodynamics a unique robustness. It has been said, "If you can prove it by thermodynamics, you don't need to do the experiment," and there is some virtue in that sweeping generalization.

By contrast, statistical mechanics depends from the beginning on assumptions about the constitution of matter. Incorrect assumptions about the properties of the molecules very probably lead to incorrect predictions about the bulk properties of the substance. Therein lies the strength (as well as the weakness) of statistical mechanics, for having made provisional assumptions about the atoms and molecules and having then calculated the expected bulk properties, we have a check on those assumptions. If the calculated values agree with experiment, the assumptions regarding the molecules are probably correct.

Time is specifically *not* considered as a relevant variable in thermodynamics, in contrast to the importance of time in the topic of dynamics within mechanics. On the other hand, the processes by which change occurs are important, so "thermostatics" would certainly not be an appropriate label. We therefore retain the traditional title, *thermodynamics*.[2]

> *Thermodynamics deals with the flow of energy under conditions of equilibrium or near-equilibrium, and with the associated properties of the equilibrium states of matter.*

New fields of science typically begin with a collection of observations and empirical generalizations. As the subject develops, the theoretical base grows until eventually virtually the entire field can be described as arising from a small number of principles, which are accepted as postulates, or laws. Thermodynamics and statistical mechanics are both mature fields in this sense. Each is based on three assumptions. Although they are expressed differently, the same assumptions, or laws, underlie statistical mechanics as classical thermodynamics. These are

1. The *first law of thermodynamics* is the law of conservation of energy.
2. The *second law of thermodynamics* is an assumption about the direction of change, which requires that the universe move toward more probable states, or a condition of greater disorder.
3. The *third law of thermodynamics* describes the degree of disorder expected at the absolute zero of temperature, or the impossibility of reaching absolute zero.

[1]Although classical thermodynamics is nearly independent of the structure of matter, there are exceptions. For example, certain properties called *colligative* (i.e., collective) properties depend on the number of molecules or ions in a volume of solution, and some electrochemical potentials depend significantly on the nature of the ions present in solution (e.g., Hg_2^{2+} versus Hg^+). Other connections are more subtle, relying on Dalton's law of partial pressures and Avogadro's law, or utilizing as empirical constants values such as heat capacities that reveal molecular properties when interpreted by statistical mechanics. Statements of the third law of thermodynamics depend on values of heat capacities at low temperatures, and therefore on quantum predictions.

[2]*Thermology*, suggested in Fourier's *Analytical Theory of Heat* (1822), is also inappropriate, for it neglects the importance of work. The name *thermodynamics* reflects the early emphasis on heat (*thermo*) and work (*dynamics*) as means of transferring energy.

Some authors like to add an additional postulate, which amounts to an existence theorem for temperature. Accepting the familiar concept of temperature, however, the thermodynamic temperature scale is rigorously defined from the second law, as proposed by Kelvin. The "zeroth law" then follows from the second law.

There is also a practical law, for most processes, requiring that the number of particles of each type (i.e., the number of atoms or molecules) remains constant. This is not the same as the law of conservation of mass, although they are superficially similar.

1.2 System and Surroundings

An important step in applying thermodynamics or statistical mechanics to any problem is to select some part of the universe for consideration. This part of the universe is then called the *system*. A system may be something as obvious as a block of wood, or a solution contained in a flask, or a certain sample of a gas. It may also be more ephemeral: a set of gas molecules as they mix with another gas, or a volume defined by imaginary fixed boundaries within a flowing liquid, or a block of ice as it melts to liquid and vaporizes.

The remainder of the universe—everything but the system—is called the *surroundings*. In practice, to avoid involvement in cosmology, the surroundings are taken to be that part of the universe that might be affected by the process under consideration involving the system. Whether an adjacent laboratory is or is not included as part of the surroundings should not change any of the answers for a system located within our laboratory. Usually, when the system has been defined, the surroundings need not be specified. However, errors of interpretation often arise because part of the surroundings is inadvertently considered along with the system, or because a property of a remote part of the surroundings is interpreted as the property of the surroundings in contact with the system.

When a piece of ice is dropped into water, we may take the ice as the system and the water as the surroundings, provided that there is no significant interaction of the water or ice with the adjacent atmosphere or table. However, it is equally valid to exchange the labels. The water might be called the system and the ice called the surroundings. There is no implication that the surroundings must "surround" the system. Exchanging labels often helps ensure that we treat the system and surroundings properly and equivalently.

Careful definition of the system is important.

For example, given a gas in a cylinder, are the walls and the piston considered part of the system or part of the surroundings? When a beaker of solution is put into a desiccator, is the atmosphere above the solution considered part of the system? The cylinder and piston confining a gas will typically not be included as part of the system if the specific properties of the cylinder and piston do not affect the problem. Even when the mass of the piston, heat capacity of the walls, or friction between piston and cylinder are significant, it would usually be simpler to include these characteristics as part of the surroundings. Whether the atmosphere above the solution is included as part of the system changes the way in which the problem is analyzed but does not change predictions of measurable quantities.

A change from one state of the system to another is called a process (Fig. 1.1).

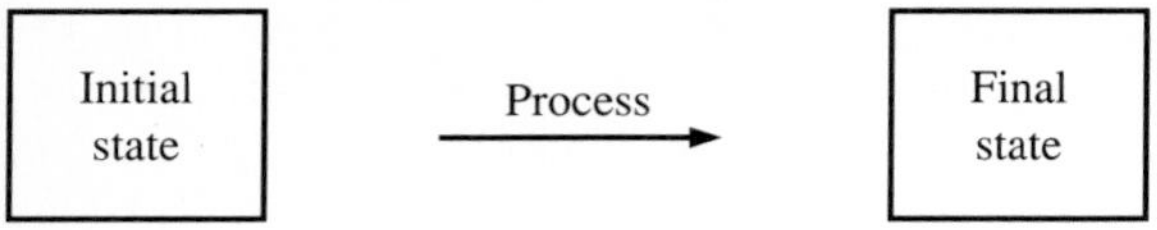

Initial state → Process → Final state

FIGURE 1.1 Any change is called a process.

1.3 State Functions

Every system has certain *properties*. For example, a certain block of ice has a mass, a volume, a temperature, a hardness, a density, and is under a certain pressure (typically, atmospheric pressure). Any set of values for all the properties defines a *state* of the system. In turn, if the state of a system is fixed, *all* of the properties of the system are determined.

A property *of a system that depends only on the state is called a* state function.

Typically, for a system of fixed composition and mass, in equilibrium with itself and the surroundings, only two variables must be specified to determine the state of the system. For example, a given mass of a specific ideal gas has related properties of pressure, volume, and temperature. Any two of these determine the third and also determine density and other properties. Similarly, two variables determine the equilibrium states of homogeneous nonideal gases, liquids, and solids.[3]

It is necessary to know more than two properties to specify the state fully and to determine all other properties of a system if there are additional external variables, such as electrical or magnetic fields, or additional internal variables, such as concentrations of components. Additional variables are also needed to describe inhomogeneities, such as mechanical strains or varying concentrations of solutes in different parts of the sample, or hysteresis effects or other nonequilibrium conditions. The mass of each component[4] of the system must be known to specify the system fully.

State functions depend *only* on the state of the system, so if the system changes in any fashion but eventually returns to the original state, all state functions return to their original values. Hence there is no change in a state function for any sequence of steps that returns the system to the original state. This is usually expressed as an integral.

The integral around any closed path of the change in any state function must be zero.

Often there is an equation, theoretical or empirical, that relates the state functions. The best known example is the ideal gas law, $PV = nRT$, relating pressure, volume, and temperature for a given gas sample. Any such equation relating the state functions is called an *equation of state*.

State functions may be *intensive* or *extensive*. An intensive variable is the same for one-half of a uniform system as for the entire system. Examples are pressure,

[3]The equilibrium state of any homogeneous pure substance is specified by two variables, in the absence of external fields, but it may be necessary to choose the proper variables to avoid ambiguity. For example, the temperature of water is not a single-valued function of density near the freezing point, and the dependence of density on pressure is too weak to be of practical value. Temperature and pressure specify the system adequately, but neither volume nor density, with a second variable, would suffice.

[4]Definition of the number of components of a system is discussed in Section 13.1.

Table 1.1 Examples of Properties and Nonproperties

Intensive	Extensive	Neither
Pressure	Volume	Resistance
Density	Mass	Absorbance
Temperature	Energy	Capacitance
Viscosity	Heat capacity	Surface area
Surface tension	Entropy	Surface energy

temperature, and density. Extensive variables are proportional to the amount of the system. Examples include mass, volume, and energy. Other examples are given in Table 1.1.

A ratio of two extensive properties is an intensive property. Therefore, an extensive property expressed per gram or per mole is an intensive property. Molar properties are represented by small capital letters. Hence C_v is the total heat capacity[5] of a system comprising n moles and C_v is a molar heat capacity; $C_v = n\mathrm{C}_v$.

Certain other characteristics are neither proportional to the mass of the system nor independent of the amount. Such quantities are not state functions, or properties, in the usual thermodynamic sense. For example, electrical resistance increases with the length of the sample but decreases in proportion to the cross-sectional area. By contrast, resistivity is the resistance for unit length and for unit cross-sectional area, $\rho = RA/l$, and is therefore an intensive state function, or property, independent of sample size and shape.

1.4 Equilibrium and Reversibility

Let the progress of a change be measured by a quantity λ. If water is evaporating, λ could be the mass of water vapor; if hydrogen is burning, λ could be the amount of water formed; if salt is dissolving, λ could be the concentration of salt in solution.[6]

Some processes can go only in one direction. When a protein is heated, molecular changes occur that cannot be reversed by cooling the sample or by any other changes in the macroscopic conditions. Molecular-level repairs would be required. A scrambled egg cannot be converted to a whole fresh egg.[7]

Other processes can go in either direction, given proper conditions; $d\lambda/dt$ may be positive or negative. If a small crystal of sodium chloride is added to a cup of water, the salt will dissolve and will not recrystallize, but if sufficient water is removed from the liquid phase, by evaporation or freezing, solid salt will form again. As conditions of temperature and humidity change, water may evaporate or condense from the atmosphere.

[5]We are often interested in the total heat capacity of a system, represented here by C. At other times, we prefer to know the molar heat capacity, represented by C. But for some purposes it is more convenient to know heat capacity per unit of mass, often called the *specific heat*. Where that is required, we will represent it by c. The meaning should be clear from the context. Later it will be necessary to distinguish between heat capacities measured at constant volume (C_v) and at constant pressure (C_p). For many substances these are nearly the same. The distinction may be ignored until Section 4.1.

[6]A definition of the progress variable for chemical reactions is developed in Section 14.1.

[7]Even feeding the egg to a hen would, at best, convert the egg to some mixture of meat, egg, carbon dioxide, and waste.

Table 1.2 Examples of Processes That May Be Reversible

Process	Measure of Process	Controlling Condition
Freezing water	Mass of ice	Temperature or pressure
$3H_2 + N_2 \rightarrow 2NH_3$	Moles NH_3	Temperature *or* Partial pressures
Expansion of a gas	Volume of gas	Temperature or pressure
Osmotic diffusion	Moles pure solvent *or* Mole fraction solute *or* Pressure	Pressure *or* Mole fraction solute

> *A process is said to be* thermodynamically reversible *if the direction in which the process moves, or the sign of* $d\lambda/dt$, *can be changed with an infinitesimal change in conditions.*

Dissolution of salt in water is thermodynamically reversible if and only if the solution is saturated with salt. Evaporation of water is thermodynamically reversible when the water vapor is at its condensation point. Examples of some reversible processes that may be carried out under thermodynamically reversible conditions, and appropriate conditions that could control the direction of change, are given in Table 1.2.

> *A thermodynamically* reversible *process passes through a sequence of thermodynamic* equilibrium *stages.*

Unless indicated otherwise, the term *reversible* will mean *thermodynamically reversible* and *equilibrium* will mean *thermodynamic equilibrium*.

A familiar example of a reversible process is the motion of a smooth cylinder, rolling on a frictionless, horizontal surface, as shown in Fig. 1.2a. A very, very small push from the left will make the cylinder move to the right or will increase the velocity to the right. A very, very small push to the left will increase the velocity to the left. This example is called *neutral equilibrium* because there is no unique equilibrium position (or velocity); the cylinder is in equilibrium wherever it may be on the horizontal surface or whatever its speed.

A more common type of equilibrium is shown in Fig. 1.2b. If the cylinder is given a small push from the left, it will move to the right, but not very far. If it is given a push from the right, it will move to the left, but not far. The preferred position is at the bottom of the trough. This is called *stable equilibrium*.

If the cylinder is at the exact top of a smooth hill, as in Fig. 1.2c, the cylinder is also in equilibrium, but the equilibrium is *unstable*. Such states are possible,

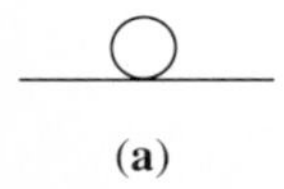

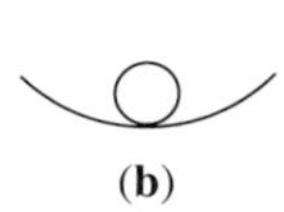

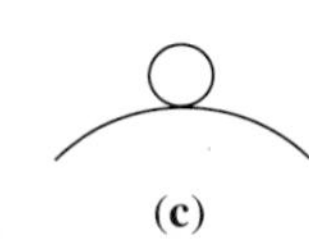

FIGURE 1.2 Equilibrium states: (a) Neutral; (b) stable; (c) unstable.

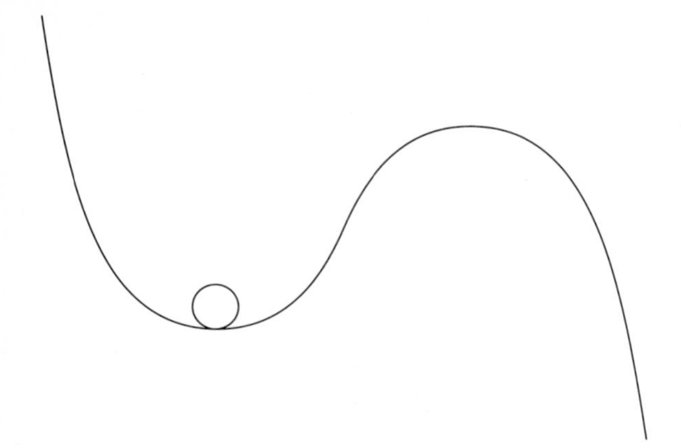

FIGURE 1.3 Metastable equilibrium.

but of little importance (other than theoretically) because they do not last very long.[8]

Another common type of equilibrium is shown in Fig. 1.3. For small displacements, the equilibrium is stable. If the system is disturbed from the equilibrium point somewhat further, however, it may go over the hump to the right, possibly after one or more oscillations to the left. When the system reaches the hump on the right, the equilibrium becomes unstable and the system may go on down the curve to the right, to a lower equilibrium state. A trough such as that shown is called *metastable equilibrium*. It is indistinguishable from stable equilibrium for small disturbances, but is unstable to large disturbances.

Every system can undergo more than one kind of change. For example, a sample of hydrogen gas may be compressed or expanded; it may be warmed or cooled; it may be raised or lowered in a gravitational field; or it may, at least in principle, undergo fusion. Any state of a system can be said to be metastable because there always exists a lower energy state somewhere. Unless there is some reasonable probability of the system being able to reach a lower energy state, however, we regard the state in which the system resides as stable.

A reversible process goes from one stable state to another through a series of equilibrium states. Such processes are sometimes called *quasi-static,* to emphasize that a truly reversible process would require an infinite amount of time (hence the states would be static). Real processes move through states that are only approximately equilibrium states. We will not, in general, attempt to distinguish between *equilibrium, reversible,* and *quasi-static* as labels for processes and the intervening states of a system.

EXAMPLE 1.1

Which of the following processes are reversible?

a. Air is pumped into a tire.
b. Air leaks out of a tire.
c. Water is frozen to ice cubes in a freezer. □

[Note: Solutions to examples are given at the top of the following page. The examples will be most helpful if you attempt to answer them before looking at the solutions.]

[8]The transition states of chemical kinetics are considered unstable, equally likely to go in either direction along the reaction coordinate. In practice they are likely to be metastable. That is, they are stable for very small displacements but unstable to larger displacements.

ANSWER 1.1

No real process can be truly reversible because it would require infinite time to occur. Questions about reversibility are therefore interpreted to mean, "Does, or can, the process approach reversibility?" (a) The process is approximately reversible, but not quite, because the pressure exerted on the pump must be greater than the pressure in the tire to overcome friction and viscosity. The air gets warm because of the compression (as well as friction), and the subsequent loss of thermal energy contributes to lack of reversibility. (b) The process is highly irreversible. More than a minor change of temperature/pressure conditions is required to get air back into a flat tire. (c) The process of freezing water at 0°C would be reversible, but at the temperature of the freezer (typically, −10°C) the process is irreversible.

Irreversible processes are also important. Classical thermodynamics, however, avoids the details of irreversible processes by considering the equilibrium states before and after the process. Often an idealized reversible path is considered between the initial and final states. Because the change in state functions is the same for any path, between fixed endpoints, the reversible path accurately predicts the changes for the irreversible process.

Nonequilibrium thermodynamics attempts to analyze processes that occur under certain types of nonequilibrium conditions. It tends to be more complex, mathematically, than equilibrium thermodynamics and is not nearly so well advanced in its theory.

1.5 Definitions and Notation

Although many of the terms required for thermodynamics will only be meaningful as they are subsequently defined by the developing theory, it may be helpful to provide a brief glossary here of some terms that will be defined, and explain some of the notation (see also Table 1.3).

A process at constant temperature is called *isothermal*, *iso* meaning essentially "the same" and *thermal* referring to heat and temperature. *Isobaric* means at the same pressure, a constant-pressure process. *Isochoric*,[9] a less common term, indicates a process at constant volume. Processes at constant energy, constant enthalpy, or constant entropy are known, respectively, as *isoergic, isenthalpic,* and *isentropic*.

Table 1.3 Prinicipal Symbols

T =	temperature	V =	volume
P =	pressure	N =	number of molecules
E =	energy	n =	number of moles or $= N/V$
Q =	thermal energy transfer to the system as a result of a temperature difference		
W =	work done on the system by an external agent		

[9]The origin of *isochoric* is obscure, but apparently it is based on a Greek word indicating locus, or place. Other labels are occasionally applied to indicate constant volume processes. Isobaric processes are sometimes called *isopiestic*.

Table 1.4 Process Labels

Term	Meaning	Symbols
Isothermal	Constant temperature	$\Delta T = 0$
Isobaric	Constant pressure	$\Delta P = 0$
Isochoric	Constant volume	$\Delta V = 0$
Isoergic	Constant energy	$\Delta E = 0$
Isenthalpic	Constant enthalpy	$\Delta H = 0$
Isentropic	Constant entropy	$\Delta S = 0$
Adiabatic	No thermal energy transfer	$Q = 0$

Every object contains energy that is randomly (and therefore more or less uniformly) distributed among the particles of the system. If the object is warmed, this randomized energy is increased; if the object is cooled, the randomized energy is decreased.

We call this randomized energy within the system thermal energy.

The definition of thermal energy will be explored further in later chapters.

If two systems have different temperatures, when they are brought into contact there will be a tendency for some of the thermal energy from the warmer body to pass to the cooler body.

We represent the transfer of thermal energy *from the surroundings to the system by the symbol Q.*

Note that a transfer of thermal energy to a system may, but need not, increase the thermal energy of the system; one must know what other processes are adding energy to, or removing energy from, the system. Adding thermal energy to a system may, but need not, increase the temperature of the system.

If an external force acts on a system and the point of application of the force moves in the direction of the force, energy is transferred to the system.

We represent the transfer of energy from the surroundings to the system by an external force acting on the system by the symbol W, which we call work.

A process in which no thermal energy is transferred between the system and the surroundings is called *adiabatic*.[10] An adiabatic process occurs within an insulated system, or occurs sufficiently rapidly that there is insufficient time for thermal energy transfer. A summary of common terms is given in Table 1.4.

In contrast to classical systems, for which rapid processes occur without thermal energy transfer and are therefore adiabatic, an adiabatic process in quantum mechanics is a slow, or gradual, change in the energy of the system. For example, a slow change in potential, such as a change in the size of a quantum mechanical box, alters the energy state but leaves the quantum numbers of the particle(s) unchanged. The change in energy of the system can be ascribed to the work done on the system. By contrast, absorption or emission of a photon causes a particle to jump from one energy level to another, a nonadiabatic process.

[10] *Diabatic* (= diabetic) indicates "passing through." *Adiabatic* thus means no flow, specifically no "heat" flow, or no thermal energy transfer across a boundary between system and surroundings.

In an adiabatic process, any change in energy of a closed[11] *system is equal to the work done on the system.*

Following conventional mathematical notation, the Greek capital delta, Δ, represents a change in the variable it precedes (e.g., $\Delta E = E_{\text{final}} - E_{\text{initial}}$). An infinitesimal change, or a *differential*, is denoted by a small d, as in dV. A *derivative,* such as dV/dT, may be regarded as a ratio of differentials.[12]

Note that a sum of changes is a change:

$$\sum \Delta X = \Delta X$$

and

$$\int dX = \Delta X$$

for any function X, whether or not X is a state function.

A system has many distinct properties, which are interrelated. We typically concern ourselves with only a few properties, and only two may be considered independent for a fixed quantity of a pure substance at equilibrium unless there are externally applied fields.

EXAMPLE 1.2

Atoms and molecules often have angular momentum and magnetic properties identified with nuclear and/or electronic "spins." Does this require additional variables (beyond two, such as temperature and pressure) to specify fully the state of the system? □

To call attention to the fact that one or more possible variables are being held constant, the constant quantities are placed next to the derivative as a right subscript, and the d symbol is replaced by the *partial derivative* symbol, ∂, as an additional reminder that it is important that some quantity is being held constant.[13]

A difficult notational problem arises for quantities that are *not* changes in any measurable function. It is often (but not always!) possible to divide energy transfers, between system and surroundings, into two types, called *thermal energy transfer*, Q (or, in less precise language, either *heat* or *heat transfer*) and *work*, W. Then, for closed systems (i.e., no transfer of material between system and surroundings),

[11] We describe a system for which no matter is added or removed as a *closed system*. Elsewhere the term may mean "isolated."

[12] There is an interesting history to this equivalence. Newton adopted the principle that dy/dx is the ratio of two quantities, each of which approaches zero. Later he became concerned about the indeterminacy of 0/0. He then reverted to an earlier argument that both numerator and denominator could be regarded as speeds (or *fluxes*) with regard to a "pure mathematical time." He recognized that it is experimentally meaningful to say that an object has a speed, dx/dt, at a point. Leaning on this observation, he interpreted the ratio of the fluxes as giving the derivative, with the mathematical time disappearing in the ratio. As the theory of limits has developed, the need for fluxes, or *fluxions*, to justify differentials and derivatives has disappeared.

[13] There is a deeper significance to this customary notation, which may be illustrated with the energy function(s), E. The energy of a system is often taken as a function of T and $V: E = E(T, V)$. At other times it is important to consider energy as a function of T and S (entropy): $E = E(T, S)$. Although both functions are equal in value to the property called energy, the two functions are mathematically distinct and should therefore have different symbols [i.e., $E = f(T, V) = g(T, S)$]; in particular, $\frac{\partial E(T,V)}{\partial T} \neq \frac{\partial E(T,S)}{\partial T}$. The notation $(\frac{\partial E}{\partial T})_V$ is the customary shorthand designation for the former; $(\frac{\partial E}{\partial T})_S$ designates the latter.

$$\Delta E = Q + W$$

or

$$\{W = 0\} \qquad \Delta E = Q$$

and

$$\{Q = 0\} \qquad \Delta E = W$$

Note in particular that Q and W are *amounts* of energy transferred to the system. They are *not* "changes" of any property of the system.

When the amount of energy transferred is infinitesimal, we encounter a notational difficulty. The sum of a large number of small amounts is still an amount of energy transferred.

$$\int Q = Q \qquad \text{and} \qquad \int W = W$$

Such notation is fundamentally sound (and was followed, for example, by G. N. Lewis in his classic 1923 textbook on thermodynamics). However, it does not look right and it omits an important distinction between infinitesimal and finite quantities.

Many authors attempt to adapt the mathematical notation of *change* to these amounts of energy transfer. For example, some will write

$$\{\text{Unsatisfactory notation}\} \qquad \Delta E = \Delta Q + \Delta W$$

or

$$\{\text{Unsatisfactory notation}\} \qquad dE = dQ + dW$$

but

$$\{W = 0\} \qquad \Delta Q = \Delta(\Delta E) \neq \Delta E$$

and

$$\{Q = 0\} \qquad \Delta W = \Delta(\Delta E) \neq \Delta E$$

The mathematical symbol for the differential has a clearly defined meaning as an infinitesimal *change* in the quantity it precedes. Thus

$$\{W = 0\} \qquad dQ = d(\Delta E) \neq dE$$

and

$$\{Q = 0\} \qquad dW = d(\Delta E) \neq dE$$

Others recognize that differential notation is not appropriate, so they choose a modification,[14] such as

[14]Some confusion is introduced because the work depends on the path. The product of P and the differential of V, dV, is an *inexact differential* (see Section 5.2), a change in PV arising from the change in V. The integral $\int P\,dV$ is called a *line integral*. Similarly, a change in resistance could be represented by ΔR or dR, even though electrical resistance is not a state function. In general, however, w is *not a change* in any definable function. Because w is not a change, it cannot be a differential; it can only be misleading to call it a differential of any form.

ANSWER 1.2

If an external magnetic field is applied, that field changes the spin state and also provides an additional variable that must be specified. In the absence of such applied fields, the equilibrium spin state is determined by the temperature, so no additional variables are necessary to determine the state of the system.

$$dE = đQ + đW$$

so that

$$\{W = 0\} \qquad đQ = đ(\Delta E)$$

and

$$\{Q = 0\} \qquad đW = đ(\Delta E)$$

but this, also, is misleading.

We choose the simplest notation consistent with the interpretation of Q and W. Let

$$\Delta E = Q + W \tag{1-1}$$

and for infinitesimal processes,

$$dE = q + w \tag{1-2}$$

and therefore

$$\Delta E = \int q + \int w = Q + W$$

emphasizing that q and w are infinitesimal *amounts* of energy transfer but not infinitesimal changes in any definable quantity.

All temperatures in thermodynamics must be on the thermodynamic temperature scale, called the Kelvin scale. However, the kelvin (symbol K) is exactly the same magnitude as a degree Celsius (symbol °C), so it is acceptable, and often far more convenient, to express temperature *differences* as differences in Celsius temperatures.

1.6 The Fundamental Concepts of Thermodynamics

Thermodynamics is constructed on two basic premises, known as the first and second laws. The first of these has its roots in the work of Leibnitz, in the seventeenth century, with contributions from Newton, Black, Rumford, and many nineteenth-century scientists. The second appears to be implicit in the work of Carnot in the first half of the nineteenth century, but was clarified only toward the end of that century. It is discussed in Part II.

Both laws depend strongly on an understanding of temperature and on "heat." Unfortunately, difficulties in understanding heat are deeply embedded in everyday language. The two most common meanings of the term refer to temperature and to energy within a body. Confusion exists between these two concepts and between these and the quantity that has been called "heat" in thermodynamics.

English terminology is strongly influenced by the *caloric theory* of Lavoisier in the eighteenth century. He considered temperature to be a measure of caloric in a body. Rumford and his successors replaced the caloric model with an interpretation based on energy.

We follow the usual path, for introductory courses, of assuming that temperature is a familiar concept from which we may develop other concepts, including energy and entropy. Later we will be able to give a more rigorous definition of temperature in terms of entropy, which can be defined as a statistical measure not directly dependent on the definition of temperature.

We recognize that every object, at equilibrium, has a well-defined property known as *temperature*. When two objects brought into contact have different temperatures, thermal energy tends to flow from the warmer to the cooler body (Section 9.3) until they acquire the same temperature. Because temperature is a property of each system, it follows that if system A has the same temperature as system B, and system A has the same temperature as system C, then system B has the same temperature as system C. This is often called the *zeroth law of thermodynamics*.

Does a thermometer measure energy? Certainly not. A large beaker of water contains more energy than does a small beaker of water if they are at the same temperature.

Does a thermometer measure energy density (e.g., energy per unit of mass or energy per unit of volume)? No. Equal volumes, or equal masses, of two gases that have very different amounts of energy may show the same temperature.

Does a thermometer measure energy flow? Certainly not. If temperature readings depended on the initial state of the thermometer, temperature would be of little value.

A thermometer measures the *average kinetic energy* of the molecules of an object.[15] The kinetic energy is usually a small part of the total internal energy. It is important to distinguish between (1) total energy, which includes kinetic energy of the center of mass of the system, rotational energy of the system about its center of mass, and gravitational potential energy attributed to the system; (2) total internal energy of the system; and (3) the molecular kinetic energy that determines temperature.

> *The internal energy of a system is the total energy of the system minus the energy associated with the position or motion of the center of mass of the system and the energy of rotation of the entire system about its center of mass.*

Although one must always beware of arbitrary reference levels for potential energy, most of the internal energy attributed to an object at room temperature is nuclear energy and chemical energy. Some is potential energy arising from the

[15]Even this statement is not rigorously true. It is correct according to classical physics, as is proved in Section 20.1, and nearly correct for most systems of interest. Corrections are required for systems in which quantum mechanical effects are important, especially at low temperatures, when temperatures are ascribed to systems not fully in equilibrium, and for relativistic systems. If there is doubt about how temperature should be defined or measured in a system, a system that obeys classical physics may be brought into thermal equilibrium with the nonclassical system. The temperature of the nonclassical system must then be the same as that of the classical system.

relative positions of the molecules. For example, in solids the atoms oscillate, with average potential energy equal to the average oscillational kinetic energy. The importance of potential energy in condensed phases is illustrated by water. More energy can be extracted from water vapor at 100°C by condensing it to water at 100°C than by cooling the water to freezing, and almost as much is released when the liquid water freezes to ice as is obtained by the cooling process. The heat of vaporization and heat of fusion represent changes in potential energy of the water molecules.

Internal energy may be regarded as consisting of three parts: the "fixed" nuclear and chemical energy, nonrandomized energy such as strains in a solid, and the potential and kinetic energies that increase or decrease when the temperature is raised or lowered. It is difficult to separate these rigorously, but the concept is nevertheless useful.

The potential and kinetic energies that change as a consequence of changes in temperature we call the thermal energy.

The definition must be adjusted to account accurately for energies associated with phase transitions (so-called "latent heat") if these transitions occur in the temperature range of interest.

When the temperature of an object is changed (by "heating" it), the average kinetic energy changes, but usually other forms of energy change also. For example, in most ideal gases, a change of temperature changes molecular rotational energy and vibrational energy as well as the translational kinetic energy.

It is helpful to consider the analogy of a series of wells with steep sides (Fig. 1.4). Each well has the same surface area (with exceptions explained below) and each is connected to the others. When water is added to any of the wells, it distributes itself equally among all of them. Three of these wells are designated as measures of "temperature." However, it is clear that it is not possible to predict, from a temperature change, how much water has been added without knowing just how many reservoirs are connected.

The problem is further complicated. First, some reservoirs (corresponding, for example, to nuclear and chemical energy) act as fully enclosed caverns that store a fixed amount of water regardless of the height of the water above or below them (i.e., the temperature). Most fixed-volume containers connect to the other reservoirs only at a fixed level; this is typical of the potential energy terms associated with phase changes. Certain other wells have bottoms that may be described as V-shaped. If the water level is low, very little water is stored in those reservoirs. As the overall level rises, more and more goes into them, until eventually they become full sharing partners. Such quantum effects are examined briefly in Chapter 6 and more fully in Part III.

Because we are often interested only in the portion of the energy that is added or removed for small changes in temperature, we call the "upper layer" of energy, or water in our analogy, the *thermal energy*.

An increase in temperature is an indication that there has been an increase in thermal energy of our system.

That increased thermal energy may have come from outside, or may be a result of restructuring of the system (through a chemical reaction, for example). An increase in thermal energy of the system does not require that thermal energy was added to the system.

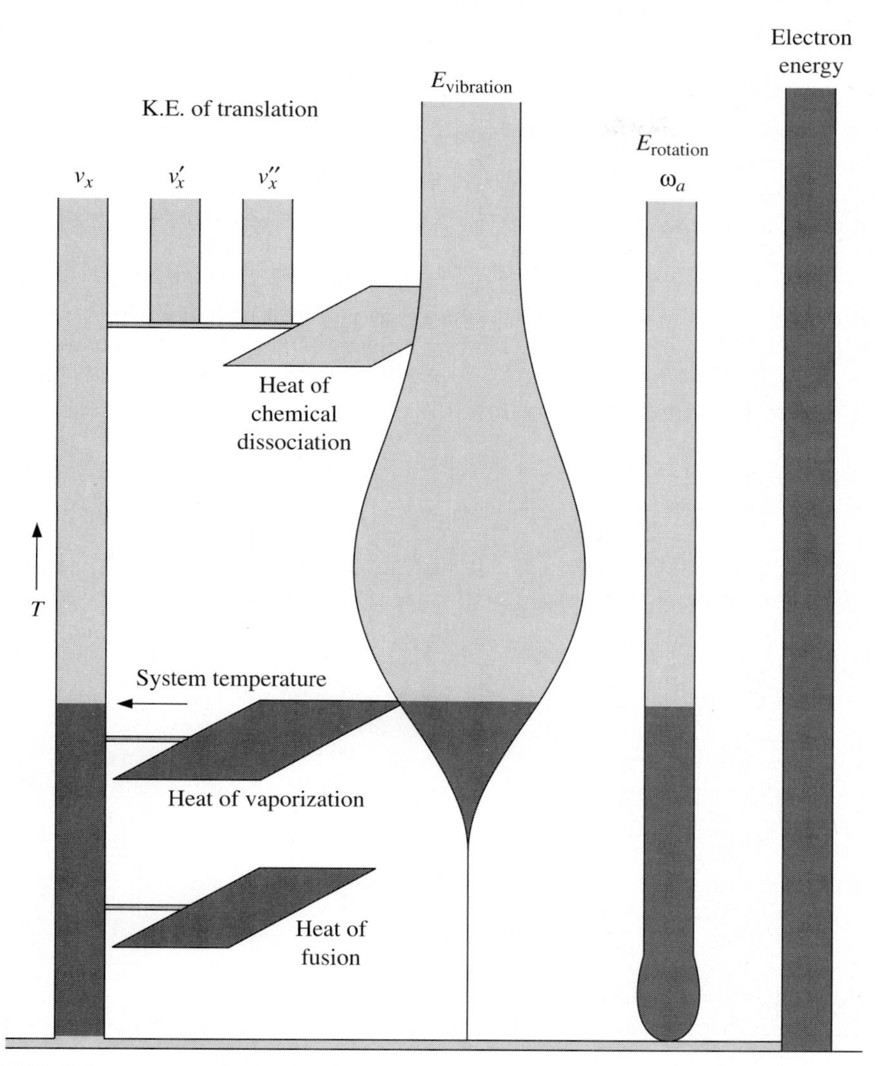

FIGURE 1.4 Conceptual model (not to scale) of thermal energy as water storage in an interconnected system of wells and caverns. The level of water in the v_x and ω_a wells (and $v_y, v_z, \omega_b, \omega_c$—not shown) represents the temperature. For solids, the electron energy is much greater than kT. At high temperatures, a compound may dissociate, giving new particles with new degrees of freedom (v'_x, v''_x etc.).

Although an increase in temperature requires that there was an increase in thermal energy (but not necessarily in total energy), the converse is not necessarily true. Adding thermal energy need not increase the thermal energy in the system. If the thermal energy may go into a new outlet, such as a phase transition or a chemical reaction, the level may be left unchanged.

At the risk of straining our analogy, we could add that water (thermal energy) may be added to or removed from the system either by streams (analogous to thermal energy transfer, caused by a temperature, or level, difference between the system and its surroundings) or by pumping (energy transfer as work). It is not possible to tell from a change in level how the energy was added, nor is it possible to predict, knowing only that energy was added (or removed) by one mechanism, what the final change in level will be unless we know the amount added (or removed) by other mechanisms.

EXAMPLE 1.3

Which of the following statements are valid?

a. An isothermal process is one for which there is no thermal energy.
b. An isothermal process is one for which $Q = 0$.
c. In an adiabatic process, $\Delta E = 0$.
d. In an adiabatic process, $\Delta T = 0$.
e. In an adiabatic process, there is no change in thermal energy. □

The associations that block understanding include confusion between thermal energy transfer to a system (Q), temperature of the system (T), and thermal energy in the system, and also between work done on a system (W) and mechanical energy (in the system and/or in the surroundings).

When thermal energy is put into a system, the temperature of the system may or may not rise.

When work is done on a system, the temperature of the system may or may not rise.

Thermal energy transfer, Q, does not require a temperature change of the system or of the surroundings. Work done on a system, W, does not require a change in mechanical energy of the system or of the surroundings.

It is the subtle interconnections between Q, W, temperature effects, and mechanical energy that distinguish modern thermodynamics from the caloric theory far more than considerations about the properties of the caloric "fluid." If the legacy of the caloric theory that is embedded in the vernacular language of today can be set aside, the concepts of thermodynamics discussed in the following chapters will be more easily grasped.

SUMMARY

The *first law of thermodynamics* is the law of conservation of energy. Energy of system and surroundings is constant.

The *second law of thermodynamics* gives the direction of change of allowed processes.

The *third law of thermodynamics* specifies the unattainability of absolute zero, or the lack of dependence of entropy on (equilibrium) state at 0 K.

The *zeroth law* describes equilibrium for equal temperatures (and can be derived from the second law if temperature is defined).

The *system* must be carefully specified; then everything else (or all else relevant to the process) is the *surroundings*.

State functions are properties of the system, depending only on the (equilibrium) state of the system (not its history). State functions may be classified as intensive and extensive.

A *reversible process* is one that passes through a series of equilibrium stages; such processes are often called quasi-static.

Iso indicates "constant"; for example, isothermal, isobaric, and isochoric specify constant T, P, and V. *Adiabatic* indicates no transfer of thermal energy between the system and the surroundings.

ANSWER 1.3

None of statements (a) to (e) are valid.

Notation: Q = *thermal energy transfer* to the system from the surroundings; W = energy transfer, as *work*, to the system from the surroundings; and q and w are infinitesimal amounts of energy transfer as thermal energy or work.

The energy of a stationary (and nonrotating) system is *internal* energy. Part of this is randomized and contributes to the temperature; this part is called *thermal energy*.

Temperature is a measure of the average kinetic energy of the particles within a system, for systems that obey classical mechanics.

QUESTIONS

1.1. An important distinction must be maintained between states of a system and processes that involve a change from one state to another. For each of the following symbols, is the description of a *state* or of a *process*?

a. E. **b.** ΔE.
c. W. **d.** w.
e. T. **f.** ΔT.

1.2. Do the following terms describe *states* or *processes*?

a. Isobaric. **b.** Adiabatic.
c. Isothermal. **d.** Isochoric.

1.3. Describe a process for which

a. $Q = 0$, $W = 0$, $\Delta E = 0$, but $\Delta T > 0$.
b. $Q > 0$, $\Delta E = 0$, $\Delta T = 0$.
c. $Q > 0$, $W \approx 0$, $\Delta E > 0$, $\Delta T = 0$.
d. $Q = 0$, $W > 0$, $\Delta E > 0$, $\Delta T > 0$.
e. $Q > 0$, $W \approx 0$, $\Delta E > 0$, $\Delta T > 0$.
f. $W > 0$, $\Delta E = 0$, $\Delta T = 0$.

[**Note:** For an ideal gas, the energy depends on the temperature only; it does not depend on pressure or volume.]

1.4. Would (or could) the following processes be thermodynamically reversible?

a. Combustion of wood.
b. Trickle discharge of a storage battery.
c. Boiling of water into a vacuum.
d. Compression of a spring.

1.5. Temperature and pressure, T and P, are variables that describe the state of a system, so they are clearly internal to the system. Why, then, are T and P often described as "external" variables?

1.6. Describe carefully a situation for which internal and external temperature and/or pressure would not be the same. How would you then define

a. T and P of the system?
b. T and P of the surroundings?

CHAPTER 2

Temperature

We are most familiar with temperature as a sensory perception. Quantitative measures of temperature, such as mercury-in-glass thermometers, more sharply define the property. The thermodynamic properties and definition of temperature are given in Sections 9.3 and 10.3. Understanding of temperature must rely on the methods of statistical mechanics, as explained in Section 6.3 and Chapter 20. It is most easily understood, however, from the model of classical mechanics.[1]

Temperature is a measure of the average kinetic energy of the molecules.[2]

At normal temperatures, $T = \frac{2}{3}(\overline{\text{K.E.}}/k)$, where k is the Boltzmann constant and the kinetic energy is *only* that arising from translational degrees of freedom of the molecule. The Boltzmann constant is the gas constant, R, divided by Avogadro's number (i.e., expressed per molecule): $k = R/N_A = 1.38 \times 10^{-23}$ J/K.

2.1 Temperature Sensing

Our first experiences with temperature measurement come from our own skin, which is a moderately sensitive thermometer. Nevertheless, our sensory response to temperature is one that can easily be fooled.

Even simple measurements with thermometers can be misleading. If you blow your breath on a thermometer bulb initially at room temperature, the temperature reading rises, showing that your breath is warmer than the ambient air (Fig. 2.1a). Blowing on a hot iron cools the iron.

Yet, when you blow on your hands in warm weather, your breath feels cool (Fig. 2.1b) because your hands are moist. Circulating air causes evaporation, which cools the skin. If the thermometer bulb is moistened by putting a damp

[1]As discussed previously, we will see that this description is rigorously true where classical mechanics is valid. It must be modified for relativistic effects and, especially at low temperatures, for quantum effects.

[2]The term *molecule* will include monatomic species such as helium.

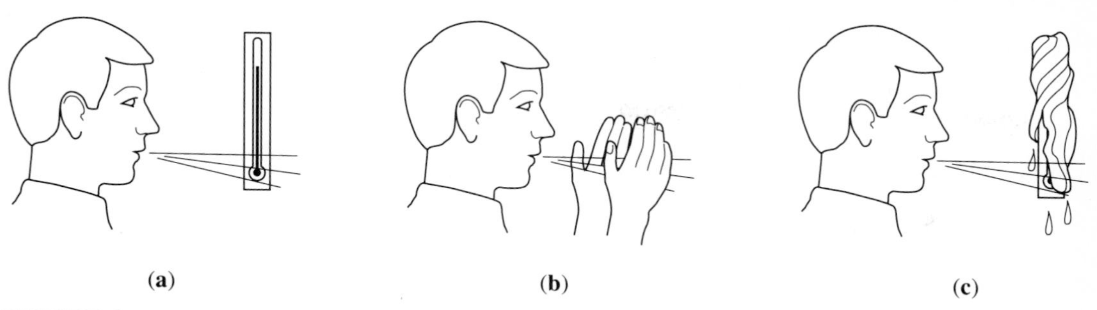

FIGURE 2.1 (a) Blowing on the thermometer raises the temperature reading. (b) Blowing on your hands cools them in warm weather (but warms them in cold weather). (c) If the thermometer bulb is damp, blowing on the bulb lowers the reading.

cloth around it, blowing across the bulb lowers the temperature reading (Fig. 2.1c and Table 2.1). Circulating air with a fan or swinging the thermometer bulb through the air lowers the temperature of the wet bulb even more. The amount of evaporation depends on the relative humidity of the air. Therefore, the drop in temperature serves as a measure of humidity, when compared with charts for relative humidity versus temperature and temperature difference.

A second peculiarity of temperature sensing will be noted if you put your hand on a piece of aluminum and a piece of wood, or step in bare feet from a rug or wooden floor onto a ceramic tile floor (Fig. 2.2). The objects have had ample time to reach equilibrium with their surroundings, so they must be at the same temperature, yet the aluminum and the tile feel much cooler than wood or cloth. The aluminum and ceramic have higher thermal conductivity, so they conduct thermal energy away from the skin and from the site where the skin is in contact, whereas thermal insulators allow the area in contact with the skin to increase in temperature until the local temperature is close to skin temperature.

Table 2.1 Relative Humidity from Wet- and Dry-Bulb Thermometer Readings[a]

Dry-Bulb Temperature	Wet-Bulb Temperature										
	15	**16**	**17**	**18**	**19**	**20**	**21**	**22**	**23**	**24**	**25**
15	100										
16	90	100									
17	81	90	100								
18	73	82	91	100							
19	65	74	82	91	100						
20	59	66	74	83	91	100					
21	52	60	67	75	83	91	100				
22	47	54	61	68	76	83	92	100			
23	42	48	55	61	69	76	84	92	100		
24	37	43	49	56	62	70	77	84	92	100	
25	33	38	44	50	57	63	70	77	84	92	100

[a] At the temperature shown in the left-hand column (dry bulb), a wet-bulb temperature reading shown in the top row indicates the relative humidity value (%) shown in the table.

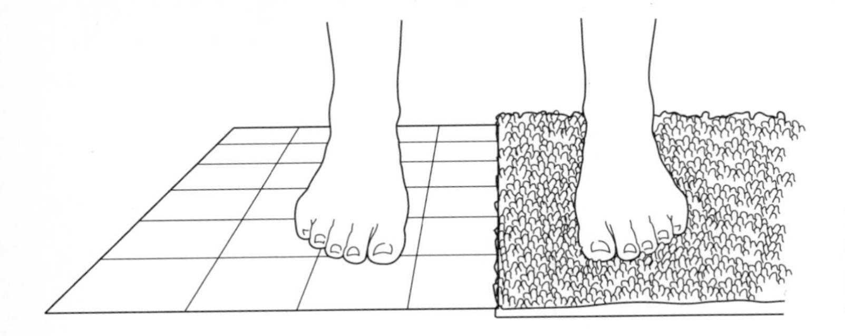

FIGURE 2.2 The foot on the tile floor is chilled compared to the foot on the adjacent rug.

One definition, or description, of temperature is that it is a measure of the tendency to transfer energy "as heat" (i.e., as thermal energy transfer). This is likely to be more misleading than helpful. The ceramic tile transfers more energy than the rug, although both are at the same temperature. The upper atmosphere has a high temperature but supplies negligible energy because the density of gas is very low. The definition also violates the rules for "negative" temperatures, discussed in Section 20.6. Temperature is a property of the system, independent of the properties of the surroundings.

A third effect is also well known. It is best demonstrated by a blindfolded subject who places one hand in a pan of warm water and the other in a pan of cool water, as shown in Fig. 2.3. The subject then moves each hand to a different pan, filled with water at room temperature, without seeing that both hands are now in a common pan. The hand previously in warm water will signal that the third pan of water is cool; the hand previously in cool water will signal that the third pan of water is warm. Preconditioning has a major influence on temperature perception.

Sensory perception is not reliable for comparison of temperatures, even to determine which of two objects, if either, is warmer. It is even less satisfactory for determining how much warmer or cooler one object is than another. Even a qualitative scale of temperature requires more objective techniques. There are many phenomena available for this purpose.

2.2 Expansion Thermometers

Almost all materials, whether solid, liquid, or gaseous, expand as they are warmed. The change in volume provides a reproducible measure of temperature change, provided that extraneous influences can be avoided (Fig. 2.4). Galileo devised a *thermoscope* that permitted comparison of the temperature of a patient with that of the physician, when they alternately put their mouths over the glass bulb at the top, an unacceptable procedure by today's standards of accuracy or sanitation.

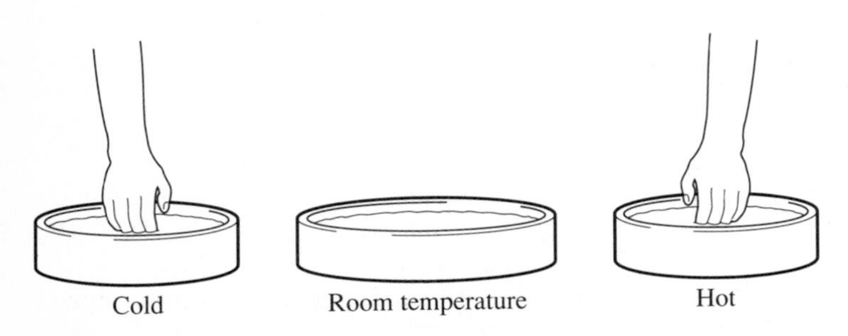

FIGURE 2.3 The center container of water feels warm to one hand and cold to the other.

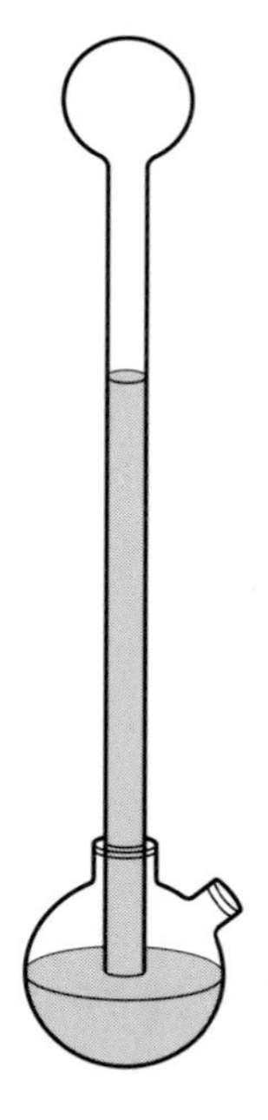

FIGURE 2.4 Sensitive but crude "thermoscope" built by Galileo. After warming the upper bulb with his hands, he inverted the tube in a vessel containing liquid. Changes in atmospheric pressure also influence the reading.

A gas expands much more, in comparison to its own volume, than any solid. Therefore, the temperature can be measured by the volume change of a gas, at a fixed pressure, ignoring the change in volume of the container. However, near room temperature a gas expands only about 0.3% of its own volume for each degree of temperature rise. To make this change visible the thermometer is designed so that gas from a large container expands into a narrow tube. Then even a small volume change, arising from a very small percentage change in the large volume, can produce a large linear displacement in the tube.

Gas thermometers may also operate at constant volume. The changing pressure is measured with an adjustable mercury column, as shown in Fig. 2.5, or with a pressure sensor.

To a first approximation there is no difference in sensitivity or accuracy between the constant-pressure gas thermometer and the constant-volume gas thermometer. The differences in practice arise because of minor differences in deviation of gas behavior from that expected for an ideal gas and from the slight differences in measurement techniques. Both types of gas thermometer are valuable for certain kinds of calibration measurements, but gas thermometers are too inconvenient for most routine measurements.

The common liquid-in-glass thermometer is convenient and sufficiently accurate for most temperature measurements, even though liquids expand less than gases (Table 2.2). Alcohol has a coefficient of thermal expansion only about one-third as great as that of typical gases, although its coefficient is relatively large for liquids. Mercury expands about one-sixth as much as alcohol, or about one-twentieth as much as air for the same temperature difference. Nevertheless, mercury has significant advantages over alcohol and most other liquids, particularly because it does not adhere to a clean glass surface.

The coefficient of thermal expansion of soft glass is about 20% as great as the coefficient for mercury, so the change in volume of the glass enclosure is far from negligible. Liquid-in-glass thermometers must therefore take advantage of the principle described above for the gas thermometer, employing a large reservoir of liquid in a glass container and a thin tube for measurement. The liquid expands

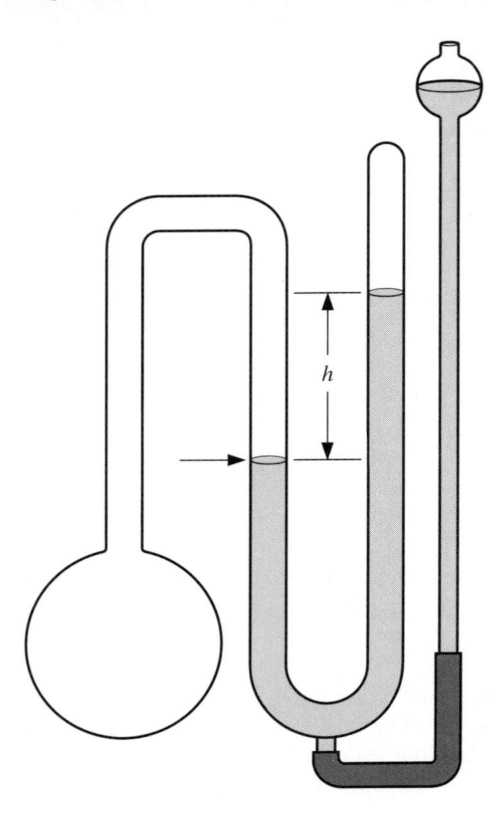

FIGURE 2.5 Constant-volume gas thermometer. The pressure is measured from the height difference, h, when the mercury is adjusted to the mark.

Table 2.2 Representative Coefficients of Thermal Expansion

$\frac{1}{L}\left(\frac{dL}{dT}\right)$		$\frac{1}{V}\left(\frac{dV}{dT}\right)^a$	
Aluminum	25×10^{-6}/K	Air	$1/T \approx 3.5 \times 10^{-3}$/K
Brass	19×10^{-6}/K	Ethanol	11×10^{-4}/K
Copper	17×10^{-6}/K	Mercury	1.8×10^{-4}/K
Iron		Oil	ca. 10×10^{-4}/K
Steel	$9\text{–}13 \times 10^{-6}$/K	Water	ca. $15\text{–}30 \times 10^{-4}$/K
Invar	0.9×10^{-6}/K		
Glass			
Soft	12×10^{-6}/K		
Pyrex	3×10^{-6}/K		
Vycor	0.8×10^{-6}/K		
Silica			
Fused	0.5×10^{-6}/K		
Quartz[b]	8×10^{-6}/K; 14×10^{-6}/K		
Wood[b]	$3\text{–}10 \times 10^{-6}$/K; $30\text{–}60 \times 10^{-6}$/K		

[a]The volume coefficient is approximately three times the linear coefficient of the same material.
[b]First value parallel to axis or grain; second value perpendicular.

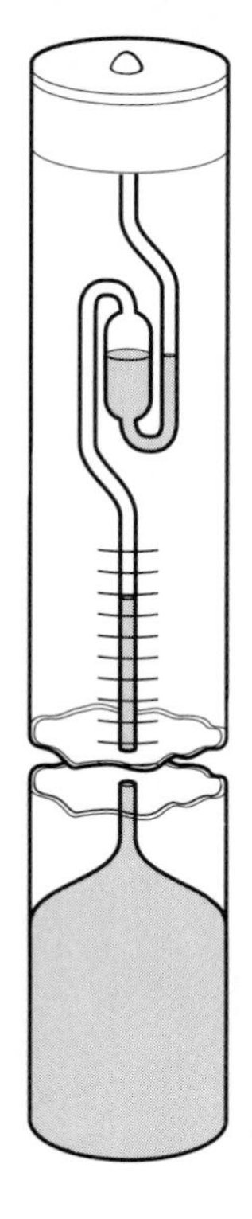

FIGURE 2.6 Beckmann thermometer. Although the range of temperature is only 5°C, that range may be set wherever desired by warming the thermometer until the mercury joins, then cooling and separating the column by tapping until the proper amount of mercury is in the lower portion alongside the scale. Temperatures can be read directly to 0.01°C and estimated to about 0.001°C. The sensitivity arises from the very large mercury reservoir at the bottom.

more than the glass, and the difference (80% of the volume change of the liquid) provides an "overflow" volume. This overflow volume passes into a thin capillary tube, so that a small volume change of the liquid gives a large change in length of the liquid column.

Accurate measurement with a liquid-in-glass thermometer requires careful attention to how long the thermometer has been in the medium to be measured and to the temperature of the stem above the reservoir. Glass thermometers typically require some hours to achieve equilibrium, but rough measurements are often made within a minute or less.

The Beckmann thermometer, common in organic chemistry laboratories for many years (Fig. 2.6), has a large reservoir (several cubic centimeters of liquid mercury) feeding into a very small capillary, permitting accurate measurement of small temperature differences over a small range. To extend the range, the capillary is provided with an overflow reservoir. The amount of mercury in the capillary tube is adjusted by the user to fit a selected temperature range, and the thermometer then reads temperature differences to 0.01°C or better. To reset the thermometer, the bulb is warmed until the mercury column reunites with the reservoir supply above. The thermometer is then tapped (carefully!) to break the mercury column at the top to give the necessary amount of mercury in the column.

More convenient, and certainly more rugged, devices are made entirely of metal. Coefficients of thermal expansion of metals are roughly comparable to that of soft glass, so a direct measurement of change of length of a metal rod is not very practical. Quite sensitive devices can, however, be constructed by combining sheets of two different metals of different coefficient of thermal expansion. As the temperature changes, such a strip bends. A relatively straight strip can bend itself away from an electrical contact in a relay at a predetermined temperature (Fig. 2.7a), or a bimetallic coil, because of its greater overall length, will give

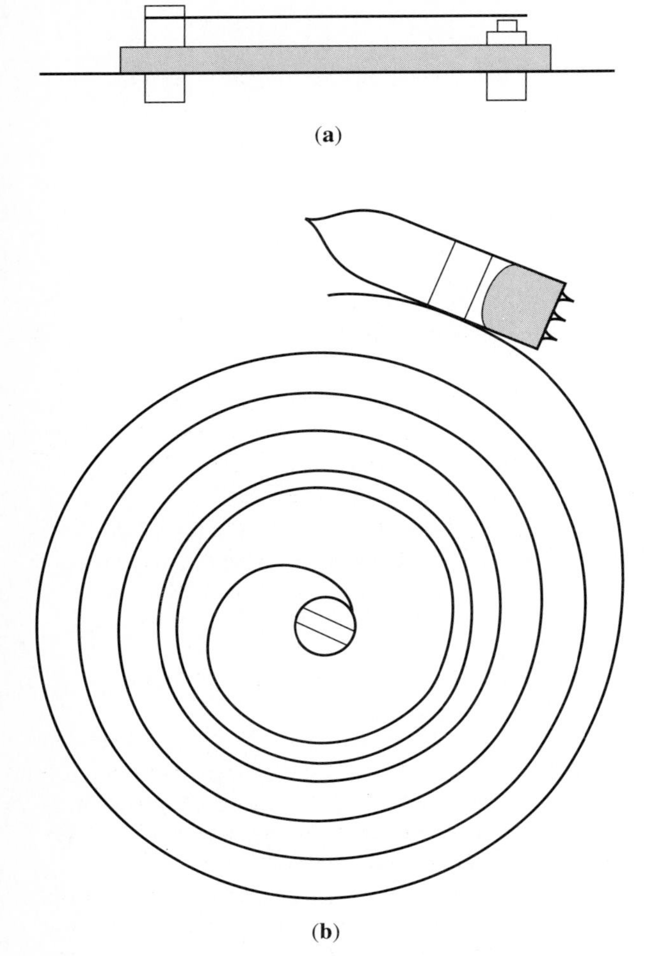

FIGURE 2.7 Thermostatic switches: (a) The bimetallic strip bends away and breaks the circuit as the temperature rises or falls; (b) the expanding coil tips the mercury switch to turn off a furnace or turn on air conditioning.

a sizable displacement of one end. This end may carry a pointer that moves across a scale or it may tip a switch, containing mercury, to open or close an electrical circuit (Fig. 2.7b).

2.3 Nonexpansion Thermometers

As important as thermal expansion may be for thermometry, it is far from the only property of matter with a useful temperature dependence. Among other important properties are the Seebeck effect, electrical resistance, and emission of radiation.

An electrical potential exists between any two dissimilar materials.[3] If dissimilar materials are brought into contact, electrons tend to flow in one direction or the other. Rubbing sometimes increases the amount of contact but otherwise has no effect. The potential difference between substances is a function of temperature, which gives rise to the *Seebeck effect*.

When two metal wires of different composition are twisted, soldered, or welded together at each end, forming a loop, electrons flow from one to the other until equilibrium is established. If the temperature of one of the junctions is raised

[3]See, for example, P. A. Schroeder, in R. M. Besançon, ed., *The Encyclopedia of Physics*, 3rd ed. (New York: Van Nostrand Reinhold Company, Inc., 1985), pp. 1236–1239, for analysis from a more detailed model.

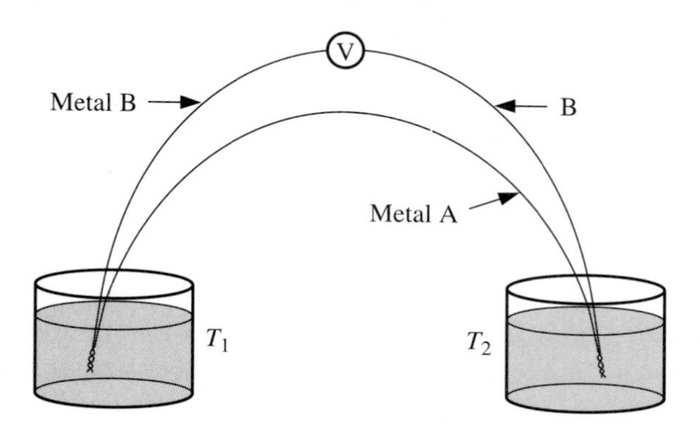

FIGURE 2.8 The voltmeter gives a reading that increases with temperature difference between left and right junctions. A and B are different metals.

(Fig. 2.8), the initial electrical potential difference is decreased at that junction. The original potential difference remains at the cooler junction, so electrons fall through the potential drop of the cooler junction, then flow through the conducting wire to the warmer junction, where they climb the lower barrier. A potential difference is established between the two junctions that depends on the temperature difference between the junctions. The wire loop is called a *thermocouple*.

EXAMPLE 2.1

What is the effect of the current flow on the temperature of each junction? □

If current is forced through the loop, which may initially be at a uniform temperature, by an externally applied potential, one junction (where electrons fall through a potential drop) gets warm and the other junction (where electrons must climb a barrier) gets cool. This provides a convenient refrigeration device for small thermal loads in the laboratory. It is called the *Peltier effect*.

The resistance of a metal wire increases with rising temperature (except below some very low temperature at which it may become superconducting). A *platinum resistance thermometer* consists of a platinum wire wrapped on a glass core and encased in glass. The thermometer is immersed in the sample whose temperature is to be measured, and connected into a Wheatstone bridge circuit. The resistance is not linear with temperature but is highly reproducible. Other resistors, designed for high sensitivity and rapid response to temperature changes, are called *bolometers*.

Not all materials increase in resistance as temperature increases. In particular, semiconductors show the reverse effect, often with very large decreases in resistance for small temperature increases. Semiconductor resistors employed for measuring temperature, called *thermistors,* are particularly convenient because the resistance change is large and the measurement circuitry can therefore be relatively simple. Direct digital reading thermistor devices have largely replaced mercury-in-glass thermometers in hospitals.

None of the thermometric devices considered above would be suitable for immersion in molten steel or for remote measurement of a flame or explosion. For such purposes, as well as special kinds of measurement near room temperature, radiation detectors are effective.

A perfect absorber of radiation must necessarily also be a perfect emitter. (Otherwise, it would accumulate energy from lower-temperature emitters in its vicinity.) Because it absorbs all frequencies, it is called a *blackbody*.

A convenient way to produce a blackbody is to provide a small opening to a cavity. If the area of the opening is negligible compared to the inner surface area, radiation entering the cavity will be multiply reflected within the cavity until it is

ANSWER 2.1

As electrons fall down the potential drop of the cooler junction they gain kinetic energy that is then lost to the nuclei, warming the junction. As they are pushed up the potential barrier, they lose kinetic energy, which is regained from the nuclei, cooling the warmer junction. Therefore, the passage of current, caused by the temperature difference, tends to destroy the temperature difference.

---

absorbed, for no surfaces are perfect reflectors. Within such a cavity, the surface is in equilibrium with its own radiation. Radiative power from a blackbody, or from a surface in equilibrium with its own radiation, is proportional to area, A, and to the fourth power of the temperature (see Sections 7.4 and 11.3):

$$\mathscr{P} = \sigma A T^4 \tag{2-1}$$

This T^4 law is called the *Stefan–Boltzmann law*. The constant σ, relating emitted power to temperature, is called the *Stefan–Boltzmann constant*. It is equal to 5.67×10^{-8} watt/meter$^2 \cdot$ K^4.

A blackbody appears black to the human eye only near or below room temperature. Because the emission follows Planck's equation for intensity versus frequency, the frequency distribution, and hence apparent color, of a blackbody is a function solely of temperature. As the temperature increases, the radiation peak moves from the infrared into the red end of the visible spectrum, giving a temperature called "red heat." With further temperature increase the intensity maximum moves into the middle of the visible region and the peak becomes wider, so the object appears white. At even higher temperatures an object appears bluish white.

A *radiation pyrometer* typically heats a platinum wire with an electric current until the wire color matches the color of the hot object in the background. The wire temperature, known from the electrical properties of the heating circuit, then matches the temperature of the remote object.

More specialized radiation detectors measure the radiated power from different parts of an object in selected frequency ranges. The detector may be a sensitive thermocouple, a thermistor, or a bolometer. Temperature distributions are measured over building surfaces to detect points of heat loss, warm objects are detected in military surveillance, and "hot spots" provide evidence of tumors beneath the skins of human patients.

Many other physical effects are applied to measurement of temperature.[4] Liquid crystals provide attractive desk thermometers that display numerical temperature or may be prepared in flat sheets that show small temperature differences. For example, when a steel block is struck with a hammer, a thin liquid-crystal film immediately placed over the block will display the "hot spot" produced. Vapor pressure is a function of temperature. Practical application of this property for thermometric measurements has been primarily at low temperatures, especially above liquid ^{4}He and ^{3}He.

Curie's law for magnetic susceptibility includes a temperature dependence, so measurement of magnetic susceptibility of weak paramagnets, at temperatures well above the critical temperature, serves as a thermometer. Similarly, nuclear spin orientation, at very low temperatures, is temperature dependent. The orientation

[4]See, for example, B. W. Mangum, "Temperature and Thermometry" in R. M. Besançon, ed., *The Encyclopedia of Physics*, 3rd ed. (New York: Van Nostrand Reinhold Company, Inc., 1985), pp. 1215–1223.

may be measured by γ-ray anisotropy. At high temperatures, vibrational populations may be measured by coherent anti-Stokes Raman spectroscopy (CARS).

Electrical devices with temperature dependence, applicable to thermometric measurements, include Johnson and Nyquist noise in resistors, dielectric constant of a gas (an indirect measurement of density), and properties of p-n junction diodes. Pyroelectric crystals such as triglycine sulfate (TGS) are good radiation detectors, as are charge-coupled devices (CCDs) and metal-oxide semiconductors (MOSs). Nuclear quadrupole resonance frequencies and the natural frequency of a quartz oscillator have served as thermometric scales.

Other devices combine principles. For example, a pulse of radiation absorbed in a gas gives an acoustic signal that may be measured as a pressure pulse, or the rise in temperature in the gas may be measured through the increase in pressure with a pressure-sensitive capacitor.

2.4 Thermodynamic Temperature

With so many different types of thermometers available we may naturally wonder how well they agree with each other. If the capillary stems of mercury-in-glass and alcohol-in-glass thermometers are calibrated at 0°C and 100°C and marked in uniform-length divisions between the freezing point and boiling point of water, will they agree for intermediate temperature values, or will either agree with temperatures measured with a gas thermometer? The best general answer is that none of the devices would agree with any other. That leaves wide open the question of which, if any, should serve as the reference scale against which others are to be calibrated. Is any one of the possible temperature scales fundamentally better than any other?

A *thermodynamic temperature*, T, may be defined from measurements on an arbitrary system. As will be shown in Section 10.3, the amount of thermal energy transferred, at each of two temperatures, defines the ratio of temperatures. This uniquely defines the thermodynamic temperature, apart from the size of the degree. Furthermore, the thermodynamic temperature scale can be related experimentally to any arbitrary temperature scale that increases as T increases but need not be a linear function of T. Practical thermometers therefore serve as adequate measures of temperatures in thermodynamic experiments.

The uniqueness of the thermodynamic temperature scale *and* the existence of a one-to-one relationship between the thermodynamic scale and an arbitrary scale of temperature (monotonic but not necessarily linear) is critically important, particularly in view of the difficulty of finding a single, conceptually simple definition of temperature that is universally valid.

The thermodynamic temperature scale was first defined by Lord Kelvin near the middle of the nineteenth century. It is a scale with a zero point approximately 273°C below the melting point of ice. It is therefore equivalent to the older *absolute temperature* scale, defined by reference to the behavior of ideal gases. In scientific literature it is universally accepted that the size of the temperature degree (now called a *kelvin*) should be the same as 1°C, which removes the only uncertainty in the thermodynamic temperature scale.

2.5 Temperature Scales

The kelvin scale, or thermodynamic scale, is defined in principle by calorimetric measurements related to heat engines (Chapter 10). The kelvin is defined as 1/273.16 of the temperature of the triple point (Chapter 13) of water. This makes

Table 2.3 Fixed Points of IPTS-68

Equilibrium	Substance	T_{68} (K)	t_{68} (°C)
tp	H_2	13.81	−259.34
bp	H_2	17.042	−256.108 $\frac{25}{76}$ atm
bp	H_2	20.28	−252.87
bp	Ne	27.402	−246.048
tp	O_2	54.361	−218.789
bp	O_2	90.188	−182.962
tp	H_2O	273.16	0.01
bp	H_2O	373.15	100.
[or fp	Sn	505.1181	231.9681]
fp	Zn	692.73	419.58
fp	Ag	1235.08	961.93
fp	Au	1337.58	1064.43

the melting point/freezing point of water, under 1 atm pressure, 273.15 K, and the boiling point of water, under 1 atm, 373.15 K. Temperatures on the Celsius scale are now defined as the temperature in kelvin minus 273.15 K.

The definitions of the preceding paragraph are sufficient in principle but do not tell the experimentalist how to measure a temperature or how to calibrate a thermometric system against the thermodynamic scale. To meet this need, the International Practical Temperature Scale has been defined, most recently in 1968.[5] Supplementary scales have been recommended for temperatures below 13.81 K.

The points shown in Table 2.3 are standard values on IPTS-68. The melting/freezing points and boiling points are all under 1 atm pressure, except for the boiling point of hydrogen at 250 torr (33,330.6 Pa), and the triple points (a melting/freezing point of a pure substance under its own vapor pressure). Because there can be some variation in isotopic composition of water from different sources, with consequent slight variations in equilibrium temperatures, the water is specified as having the isotopic composition of ocean water. The hydrogen gas is specified as equilibrium hydrogen (Section 18.2).

Even with the additional reference points, the temperature scale is not adequately specified without knowing how temperature should be interpolated between the fixed points. For this purpose, the scale is broken into four intervals. From 13.81 to 273.15 K, tables are supplied for converting the resistance readings of a platinum resistance thermometer into T_{68} values. A second empirical relationship is given, for the platinum resistance thermometer, for the interval of 273.15 to 903.90 K (630.75°C). The interval from 630.74 to 1064.43°C is specified in terms of voltages for a rhodium–platinum/platinum thermocouple. Above 1337.58 K (1064.43°C), temperatures are measured photometrically.

SUMMARY

For classical systems, temperature is a measure of the average kinetic energy of the molecules.

[5]The official English version is reported in *Metrologia*, **5**(2), 35 (1969). The scale is designated IPTS-68 and temperatures on the scale are often represented as T_{68}. See Mangum, loc. cit.

Temperature is measured indirectly, by measuring some property that varies with temperature. Most often this is volume, but it may be electrical resistance, radiated power, or many other properties.

A thermodynamic temperature scale may be uniquely defined (Chapters 9 and 10) and practical temperature measurements can be uniquely related to the thermodynamic temperature, in principle. In practice, convenient temperature points and scales are defined by the International Practical Temperature Scale of 1968.

PROBLEMS

2.1. Find the volume coefficient of thermal expansion of a gas

a. At 500°C.

b. At 300 K.

c. At 90 K.

2.2. The linear coefficient of thermal expansion of aluminum is $\alpha = 25 \times 10^{-6}/\text{K}$, so the change in length is $\alpha L\,\Delta T$. The change in volume of a cube of aluminum is $dV = (L + dL)^3 - L^3$. Find

a. The change in volume of a cube of aluminum for $L = 10$ cm and $\Delta T = 10$°C.

b. For this example, the error in the usual approximation that the volume coefficient of expansion, β, is equal to 3α.

2.3. Assume that object A absorbs radiation at frequency ν_1 (only) and radiates at frequency ν_2 (only), whereas object B absorbs radiation at frequency ν_2 (only) and radiates at frequency ν_1 (only). Show how you could construct a perpetual motion machine by moving filters in and out of the space between A and B. (As shown in Chapter 10, it is sufficient for the construction of a perpetual motion machine that one object become warmer and/or another cooler without doing work on either object or bringing up a warmer or cooler object to transfer thermal energy.)

2.4. Starting from the premise that 0°C = 273.15 K and 1°C = 1 K (size of degree), derive exact conversions between the Celsius scale and

a. The Fahrenheit scale (0°C = 32°F, 100°C = 212°F).

b. The Réaumur scale (0°C = 0°Re, 100°C = 80°Re).

c. The Rankine scale (0°R = 0 K, 1°R = 1°F, size of degree).

2.5. Body temperature is 37°C. Assume that you have a surface area of 2 m^2.

a. How much power would you radiate, as a 37°C blackbody, to the walls around you?

b. If the walls are at 25°C, how much power would you absorb from the walls (assuming blackbody radiation and absorption)?

c. Based on a diet of 2000 Cal/day, the average person is often described as the equivalent of a 100W light bulb. How can you explain the discrepancy between the difference of Parts (a) and (b) and this 100W estimate?

2.6. Assuming blackbody radiation and absorption, compare the power radiated (per m^2) by

a. A tungsten filament at 2000 K versus an ice cube at 0°C.

b. An ice cube at 0°C versus dry ice at −80°C.

2.7. When one junction of a copper-constantan thermocouple is at 0°C and the other at 100°C, a potential difference of 4.28 mV is produced.

a. Predict the potential difference if the hot junction is at 50°C, assuming linearity.

b. Observed potential differences are (for hot junction temperatures): 40°C: 1.61 mV; 70°C: 2.91 mV. Predict the value for 50°C assuming linearity over this smaller interval.

2.8. Using values from P2.7, predict the potential differences to be expected for junction temperatures of
a. 0 and 30°C.
b. 40 and 70°C.

2.9. The linear coefficient of thermal expansion of tungsten varies with temperature from approximately 4 to 6×10^{-6}. What type of glass (or silica) would most nearly match tungsten for wire-in-glass seals?

2.10. Two reservoirs are tested and found to give a ratio of thermal energy transfers, $Q_H/Q_C = T_H/T_C = 1.1677$. If the difference, $T_H - T_C$, is set equal to 50 K, what are T_C and T_H?

2.11. What ratio of $Q_H/Q_C = T_H/T_C$ should be found in an experiment that compares boiling water (at 1 atm) to the triple point of water? [In practice, there is a significant uncertainty in measurements of the boiling point.]

2.12. Find the average kinetic energy of a helium atom at 25°C.

2.13. Find the average kinetic energy of a sodium ion in NaCl at 25°C.

2.14. Find the resistance of a 100-W light bulb. Why will this not agree with the value measured by an ohmmeter for the same bulb?

EXERCISES

2.1. Place a sheet of aluminum against a piece of foam plastic. Grasping the two outer surfaces (aluminum and plastic), judge the (apparent) relative temperature of the two. [Both should be maintained at room temperature beforehand.]

2.2. Select four empty soft drink cans. Fill one with very hot water and a second with ice water. Fill the other two with tap water (room temperature). Let someone hold the hot and cold cans, one in each hand. Then the same person should hold the two room-temperature cans. After deciding which is (apparently) warmer, switch the two room-temperature cans between hands and judge again.

PART I

First Law of Thermodynamics: Energy Bookkeeping

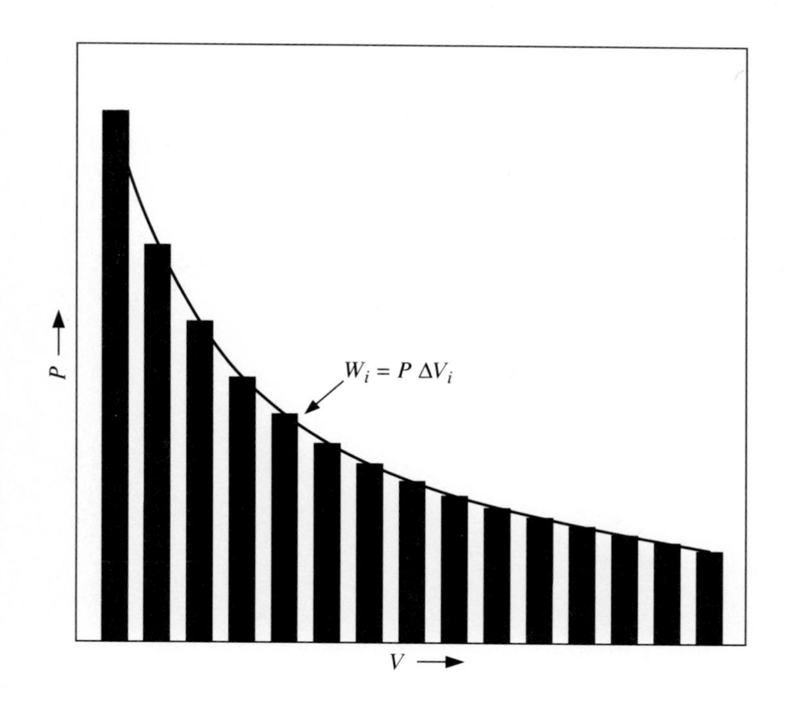

CHAPTER 3

Energy and the First Law

The first law of thermodynamics is the law of conservation of energy. Modern terminology for energy and its transfer is rooted in eighteenth-century science. Unfortunately, the misconceptions of that early time are embedded in both our technical and vernacular language, causing unnecessary difficulties for modern students. The terminology is particularly difficult to clarify because of the overlap with common language. To show where the terminology came from and how the modern theory differs from common perceptions, the topic is initially developed historically.

3.1 Heat and Calorimetry

Alchemists had proposed that combustion was accompanied by the escape of an elusive substance they called *phlogiston*. It was the phlogiston, released in burning, that was credited with producing the sensation of hotness. Newton suggested, about 1700, that the motion of particles within a body increased with temperature, but he never pursued this line of investigation.

Soon afterward, Joseph Black in Scotland, and later Antoine Lavoisier in France, suggested an alternative model. An intangible fluid, which Lavoisier named *caloric* (derived from *calorem*, the Latin word for heat), was believed to pervade all matter. Adding caloric to a substance would usually raise the temperature of the substance. *Calorimetry*, which takes its name from Lavoisier's model, is the measurement of temperature changes within an overall insulated container, from which energy changes can be determined. Lavoisier's model is generally adequate for the restricted set of processes involved in calorimetry, although our interpretation of the processes has changed.

Several refinements to the basic caloric theory were required.

1. The same amount of caloric added to different quantities of matter produced different temperature increases.

It was not the amount *of caloric, but rather its* concentration, *that determined the temperature effect.*

2. A given amount of caloric added to different substances would produce different temperature increases.

Each substance had a "chaleur specifique," or "calorique specifique," or "heat capacity" that determined how effective caloric would be in raising its temperature.

The amount of caloric required to raise 1 gram of water by 1 degree Celsius was called 1 *calorie*.

EXAMPLES 3.1

A. How many calories are required to warm 50 g of water from 20°C to 40°C?
B. Find the final temperature if 50 g of water at 20°C is added to 150 g of water at 95°C. □

In the early nineteenth century, Dulong and Petit surmised from the limited information available that every solid element had a heat capacity of 6 cal/mol · deg. Lead, for example, has a molar mass of 207 and a heat capacity of 0.0306 cal/g · deg.

EXAMPLES 3.2

A. How many calories would be required to raise the temperature of 50 g of lead from 20°C to 40°C?
B. What is the final temperature if 100 g of lead at 80°C is added to 100 g of water at 20°C? □

3. Caloric added to a substance was not always fully effective in raising the temperature. Caloric could bind to the substance and remain hidden. A substance such as brass contained caloric that only caused a temperature effect when it was released by cutting the brass. Some of the caloric added to ice would cause a phase transition (melting of the ice) but did not manifest itself through a temperature increase (Fig. 3.1).

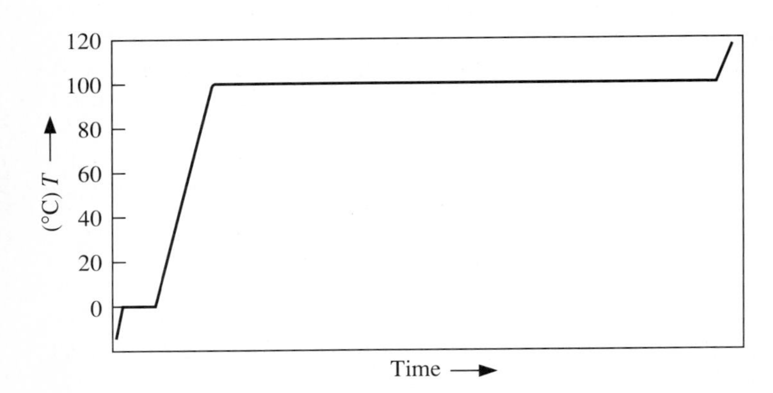

FIGURE 3.1 Temperature versus time for water as it is heated at a constant rate (idealized).

Caloric held within a substance but not influencing the temperature was called hidden *or* latent.

EXAMPLES 3.3

A. It requires 80 cal to melt 1 g of ice at 0°C. How many calories must be added to 10 g of ice to melt the ice and raise the water produced to 15°C?
B. Find the final temperature if 25 g of ice at 0°C is added to 150 g of water at 85°C.
C. Find the final temperature if 100 g of ice at 0°C is added to 100 g of water at 50°C. □

Lavoisier supported the new theory by his investigation of oxygen. He showed that combustion was a combination of the fuel with oxygen gas from the air. There was no evidence for release of phlogiston, the earlier explanation of combustion.

Benjamin Thompson, a contemporary of Lavoisier, was born in Woburn, Massachusetts, in 1753. At 19 he married a wealthy widow from Rumford (now, Concord) New Hampshire, who provided him with important political contacts. At the time of the American Revolution, he left the colonies for England, traveling later to Munich, where he distinguished himself in service to the Elector of Bavaria. For this he was named Graf, or Count, and he selected Rumford for his title.

Rumford's fascination with heat is shown by his published works, which deal with the nature of heat, practical applications of heat, and devices and techniques involving heat, in addition to smaller numbers of works on light, gunpowder, silk, coffee, and public institutions (care and feeding of the poor, army regulations, and the Royal Institution of Great Britain, which he founded in 1800).

Among Rumford's responsibilities in Munich was supervision of armament production. His published account of his research on heat explains his motivation as well as his procedures. "Being engaged lately in superintending the boring of cannon in the workshops of the military arsenal at Munich, I was struck with the very considerable degree of Heat which a brass gun acquires in a short time in being bored, and with the still more intense Heat (much greater than that of boiling water, as I found by experiment) of the metallic chips separated from it by the borer. The more I meditated on the phenomena, the more they appeared to me to be curious and interesting."[1]

The iron bit was initially at the same temperature as the brass, so there was no reason to believe that any caloric was being added by the bit. The "modern doctrines of latent Heat, and of caloric" attempted to explain the dramatic rise in temperature by the escape of latent caloric, trapped in the brass until it was released by shaving the brass. Rumford was skeptical of the explanation, and therefore performed several experiments, which added new requirements to be met by caloric.

4. If the brass cuttings had lost caloric previously trapped in the bulk metal, that should change the metal. The heat capacity of the caloric-poor cuttings should be greater than that of the bulk metal. However, Rumford's measurements showed no difference between the shavings and the original metal.

[1] Quotations from S. C. Brown, ed, *Collected Works of Count Rumford*, Vol. I (Cambridge, Mass.: Harvard University Press, 1968).

ANSWERS 3.1

A. 50 g × 1 cal/g · deg × (40 − 20) deg = 1000 cal.
B. The heat capacity of water is c_p = 1.0 cal/g · K, so that 50 g × 1 cal/g · K × (t − 20) K = heat gained by cooler water = heat lost by warmer water = 150 g × 1 cal/g · K × (95 − t) K. Solving for t gives $50t - 1000 = 14{,}250 - 150t$ and $200t = 15{,}250$, or t = 76.(25)°C.

Alternative 1:
$\sum(C_p \Delta T)_j$ = 50 g × 1 cal/g · K × (t − 20) K + 150 g × 1 cal/g · K × (t − 95) K = 0. This leads to the same arithmetic as the first solution.

Alternative 2:
The initial thermal energy, distributed among the substances present, is equal to the final thermal energy, all taken relative to an arbitrary reference temperature, t_0. $\sum_j C_{pj}(t_j - t_0) = \sum_j C_{pj}(t_f - t_0)$, where the t_j are the initial temperatures of the substances present, each with heat capacity C_{pj}; t_0 is an arbitrary temperature equal to or less than the lowest t_j; and t_f is the common final temperature. If t_0 is taken as 0°C, then 50 g × 1 cal/g · K × (20 − 0) K + 150 g × 1 cal/g · K × (95 − 0) K = 200 g × 1 cal/g · K × (t_f − 0). The method is attractively simple for this example but may become complex for other problems involving phase transitions and/or changes of heat capacity with temperature.

ANSWERS 3.2

A. 50 g × 0.0306 cal/g · deg × (40 − 20) deg = 30.6 cal.
B. 100 g (Pb) × 0.0306 cal/g · deg × (80 − t) deg = 100 g (H_2O) × 1.00 cal/g · deg × (t − 20) deg, or, dividing by 100 cal, 80 × 0.0306 − 0.0306t = t − 20, so that

$$t = \frac{2.448 + 20}{1 + 0.0306} \text{ deg}$$

and t = 21.(78)°C.

ANSWERS 3.3

A. 10 g × [80 cal/g + 1 cal/g · deg × (15 − 0) deg] = 950 cal.
B. Heat gained by ice = 25 g × [80 cal/g + 1 cal/g · deg × (t − 0) deg] = heat lost by warm water = 150 g × 1 cal/g · deg × (85 − t) deg. Solving for t gives $175t$ = 10,750, or t = 61.(4)°C.
C. Gain by ice = 100 g × [80 cal/g + 1 cal/g · deg × $t$deg] = loss by water = 100 g × 1 cal/g · deg × (50 − t) deg so that (100 + 100)t = (5000 − 8000), or t = −3000/200 = −15°C. This is impossible; one cannot add ice at 0°C and water at 50°C and obtain a temperature less than 0°C or greater than 50°C. The assumption that all the ice melted must therefore have been incorrect. If only m grams of ice melted, with $m < 100$, the final temperature must be 0°C and gain by ice = m grams × 80 cal/g = loss by water = 100g × 1 cal/g · deg × (50 − 0) deg. Solving for m gives m = 5000/80 = 62.5 g melted.

Losing caloric did not measurably change the capacity of the brass for caloric.

5. He found that as the bit grew dull, so that *less* metal was cut, the amount of caloric released appeared to *increase*. He found, when cutting with the dull bit, that even though only 54 g of brass was cut, "the very considerable quantity of Heat that was produced . . . would have been capable of melting $6\frac{1}{2}$ lb of ice," or enough to raise the temperature of $\frac{1}{4}$ million gram of water or 2 million gram of brass by 1°C.

More latent caloric was released when the metal was not cut than when it was cut.

Also,

An arbitrarily large amount of heat could be obtained from the brass.

Rumford interpreted this work as an "experiment . . . made in order to ascertain how much Heat was actually generated by friction."

6. To determine the properties of the hypothetical caloric, it was considered important to know its weight. Fordyce, in England, had attempted to measure the change in weight of water when it froze to ice, releasing caloric, and found a measurable difference. Rumford repeated the experiment, taking greater precautions to avoid condensation of moisture on the sample and air currents produced by temperature differences.

Rumford found no change in weight, to an accuracy of 1 part in a million.

EXAMPLES 3.4

A. A century later Einstein showed that there is a direct connection between energy and mass, and therefore weight. His equation was $E = mc^2$, where $c = 3 \times 10^8$ m/s is the speed of light. If 1 g of ice melts, should the liquid weigh more or less than the ice?
B. The energy added is 80 cal = 333 J. How much difference in mass would there be between the solid and liquid? □

Based on his experiments, Rumford supported Newton's theory of a century earlier that heat was associated with the motions of the particles within a body. It would seem that Rumford's experiments would have been sufficient to eliminate the caloric theory, but his ideas were considered "old fashioned" in contrast to the "modern" theory of Lavoisier, and were therefore largely ignored.

Rumford's interpretation was soon supported by Davy, in England; by Robert Mayer, a Bavarian physician; by Hermann von Helmholtz, professor of physiology and then professor of physics in Germany; and others. However, it was not until James Prescott Joule, a brewer and amateur scientist in Manchester, England, performed some quantitative experiments, that the scientific community recognized the validity of Rumford's arguments. Joule showed the quantitative relationship between mechanical energy input to a system and the temperature rise of the system.

In English, the term *caloric* has been replaced in common speech and understanding by the term *heat*, which is interpreted not as a fluid but as a form

---

ANSWERS 3.4

A. More. An increase of energy represents an increase in mass (c^2 is positive).
B. $\Delta m = \Delta E/c^2 = 3.7 \times 10^{-15}$ kg, or a weight change of 3.6×10^{-14} N.

---

of energy that can flow from one body to another. The *temperature* of a body is then explained as a measure of the concentration of heat, or heat energy, in the body, recognizing that different amounts of heat, or heat energy, are required by different substances for the same temperature increase.

This common perception leads to many of the same difficulties as Lavoisier's caloric model.

Rumford and Joule showed that it was not necessary to add caloric, or heat, to a substance to raise the temperature of the substance.

It is necessary to distinguish very carefully between "heat" added to a body, "heat" in a body, and the "degree of heat" or temperature of the body (as well as "heat" as a verb or, occasionally, a rate of heating).

In thermodynamics, the term *heat* is (nominally) reserved for an *amount of energy transferred as a consequence of a temperature difference*. According to this definition, there is *no* heat in a body. The amount of thermal energy transferred (i.e., the amount of heat) is represented by the symbol Q.

In practice, the vernacular meaning of heat is so embedded in our language and imagery that even thermodynamics textbooks use the term more often to mean a form of energy than a transfer of energy. Terms such as *heat capacity*, *latent heat*, *heat transfer*, and *heat flow* consolidate, in the listener's mind, the caloric model of heat as a fluid transferred to and then stored in a body.

The experiment by Joule, in which he stirred a tub of mercury and measured the temperature rise, is often called a measurement of the "mechanical equivalent of heat." Yet there is no "heat," in the modern technical sense, involved in his experiment. It is more accurate to say that Joule measured a conversion factor between two units for energy.

To avoid the confusion inherent in heat and its two meanings (in addition to a continuing confusion between "heat" and temperature), we limit the term *heat* to a general, nontechnical meaning. That part of the energy contained in a body that is responsible for the temperature of the body is called *thermal energy*. The exchange of energy between a system and its surroundings that occurs as a consequence of a temperature difference is called *thermal energy transfer*, represented by the symbol Q.

Thermal energy transfer occurs as a consequence of a temperature difference between systems in contact, or a system in contact with its surroundings (including, where appropriate, a radiation field). When there is no other form of energy transfer,

$$\left\{\begin{matrix}\text{no other}\\ \text{energy}\\ \text{transfer}\end{matrix}\right\} \qquad \Delta E = Q \qquad (3\text{-}1)$$

where E is the total energy of the system and therefore ΔE is the change in total energy.

Both thermal energy and the amount of thermal energy transferred may be measured in any energy units, such as calories or joules. The thermochemical

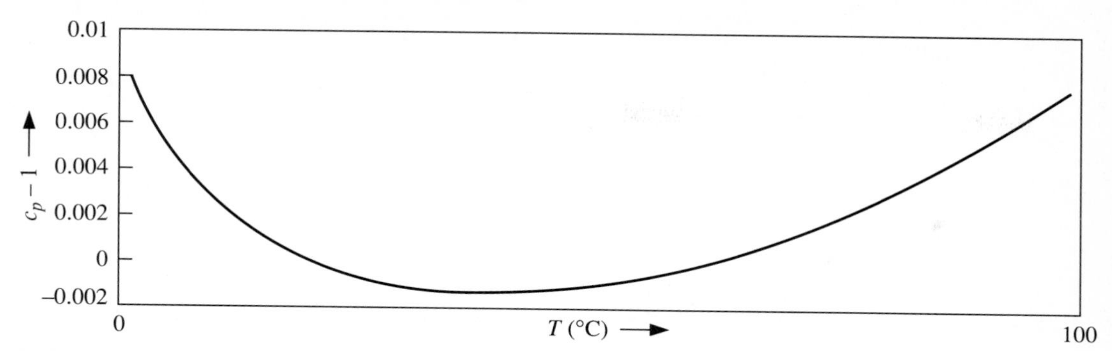

FIGURE 3.2 Heat capacity of liquid water between 0 and 100°C, expressed as $c_p - 1$ in thermochemical cal/g · K.

calorie is now defined by the relation

$$1 \text{ calorie} = 4.184 \text{ joule} \quad \text{(exactly)}$$

However, there are many other definitions of the calorie still in the literature (typically differing in the fourth or fifth place), so it is difficult to know which value an author has used unless it is stated explicitly. The dieters' Calorie is 1000 calorie, or 4 kJ. Values of the heat capacity of liquid water, steam, and ice at various temperatures are shown in Figs. 3.2 to 3.4.

3.2 Work

Energy may be transferred to a system in various ways. We add energy to our cars by filling the tanks with gasoline. We add energy to our homes in winter by burning oil or gas. Energy may also be added to a system by doing work on the system. Work has a central position in thermodynamics, so it is imperative that it be carefully defined and understood.

Work is often defined as force times distance,[2]

$$W = \int \mathbf{f} \cdot d\mathbf{s} \tag{3-2}$$

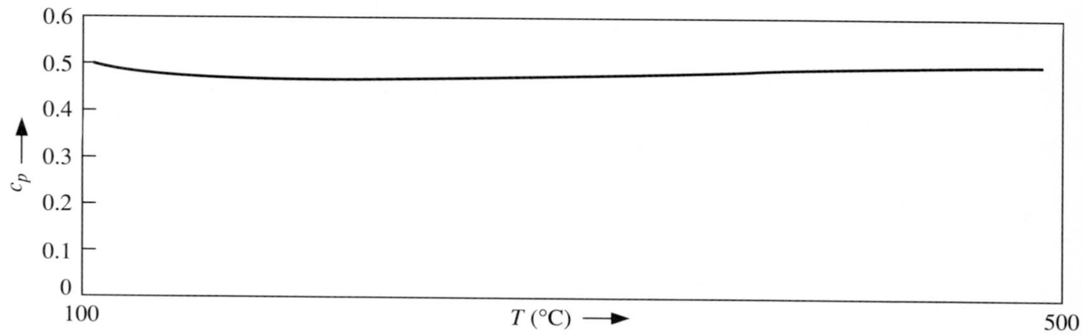

FIGURE 3.3 Heat capacity of steam at 1 atmosphere between 110 and 500°C, in cal/g · K.

[2] All integrals arising in physical problems are definite integrals. We will put explicit limits on the integrals only when the limits are relevant to the discussion.

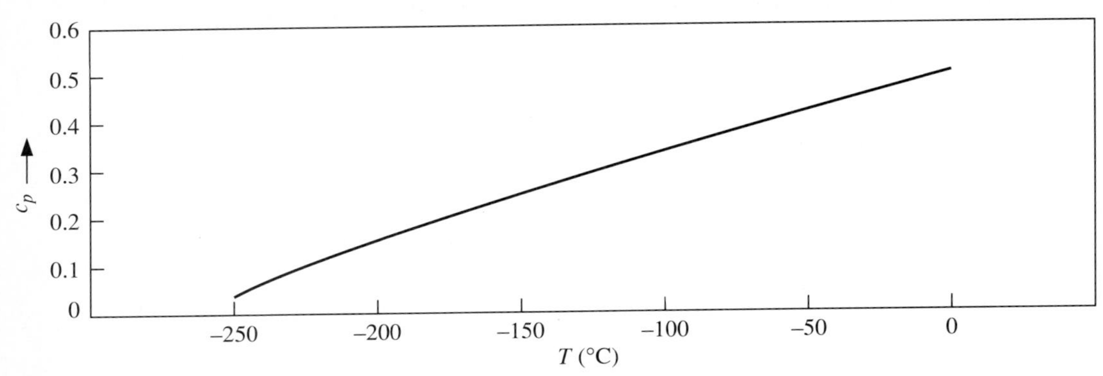

FIGURE 3.4 Heat capacity of ice between −250 and 0°C, in cal/g · K.

This statement is not wrong, but it is inadequate. What is exerting a force on what? (For every force there is an equal and opposite "reaction" force; which one are we to use?) What is the displacement? Is it, as in momentum problems, the displacement of the center of mass? Or is it the motion of the surface of the system? Most important, if incremental work is always expressible in the form $\mathbf{f} \cdot d\mathbf{s}$, does that require that $\mathbf{f} \cdot d\mathbf{s}$ is necessarily equal to an increment of work?

For the moment we will only partially answer these questions. We can return to them after the theory is somewhat better developed. The important ideas for right now are the following.

a. Work is a measure of *energy transferred* as directed energy from the surroundings to the system. Work done on the system increases the energy of the system. If there is no energy transfer other than as work,

$$\begin{Bmatrix}\text{no other}\\ \text{energy}\\ \text{transfer}\end{Bmatrix} \qquad \Delta E = W \qquad (3\text{-}3)$$

where ΔE is the change in the *total* energy of the system.

It follows that

A system cannot do work on itself.

Therefore, the force must be the *force exerted on the system by the surroundings.*

b. Work is the *scalar product*, or dot product, of the force exerted on the system by an external agent and the *displacement of the point of application* of that force. Note, in particular, that the displacement is often *not* the displacement of the center of mass of the system, nor of the surface of the system. When a single external force acts on the system, the work is

$$W = \int \mathbf{f}_{i,\text{ext}} \cdot d\mathbf{s}_i \qquad (3\text{-}4)$$

EXAMPLE 3.5

A centripetal force is a force toward the center of the motion. Force is a vector quantity, but work is not. What does it mean for a centripetal force to do negative work on an orbiting body? □

c. When more than one force is acting, the work done by each force must usually be found separately. Work terms (positive or negative, depending on whether the surroundings or the system is doing the work) can be added.

$$W = \sum_i \left[\int \mathbf{f}_{i,\text{ext}} \cdot d\mathbf{s}_i \right] \tag{3-5}$$

Note in particular that W is *not* equal, in general, to the *net force* times a displacement.

$$W \neq \int \left(\sum \mathbf{f}_{i,\text{ext}} \right) \cdot d\mathbf{s} = \int \mathbf{f}_{\text{net}} \cdot d\mathbf{s}$$

d. *Work can only be calculated, or measured, in certain idealized experiments.*[3] For general processes such as those involving friction, work cannot be known, because the motion of the point of application of the force is at the molecular level and unobservable.

Just as the term *heat* obscures the distinction between thermal energy and Q, thermal energy transfer, several terms involving the word *work* have been introduced that obscure the definition of work. Such terms as *internal work* and *pseudowork* or *center-of-mass work* have nothing to do with W, the thermodynamic work. They should be strenuously avoided.

A trick that is often helpful in analyzing energy transfers is to recognize that we can always interchange system and surroundings. Observable quantities must be the same for either choice of the system. There can be only a sign change when the system and surroundings are interchanged.

Four examples of work illustrate the meaning of the definition. These are work on a particle, on a body free to rotate, on a spring, and on an ideal gas.

1. Work on a Particle

The simplest example of work is that done on a particle. The term *particle* does not necessarily mean that the object is small. Rather, it means that we need not worry about the size or structure of the object. That is, in the processes considered, a particle cannot break apart, vibrate, get hot, rotate, or do anything else that might complicate the problem.

The internal energy of a particle is constant.

A rolling sphere cannot be considered a particle because the rotational energy is comparable in magnitude to the translational kinetic energy. On the other hand, the earth or other large body may be considered to be a particle for almost any problem in which rotational and tidal energies are negligible.

From the definition of work, applied to a particle,

$$W = \int \mathbf{f} \cdot d\mathbf{s}$$

it is clear that if the force and displacement are perpendicular, no work is done by the force. Thus motion in a circular orbit involves no work by the centripetal force. However, if the particle moves in an elliptical orbit, the displacement is no longer perpendicular to the force at all times, so the centripetal force does

[3]See B. A. Sherwood and W. H. Bernard, *Am. J. Phys.* **52**, 1001 (1984). Because $Q = \Delta E - W$, Q also is unknown for most real processes in which work may be performed.

ANSWER 3.5

A centripetal force is exerted by a central body through a field (e.g., electrical or gravitational) or by an elastic cord. If the field or cord does work on the orbiting body, it increases the kinetic energy of the moving body, transferring energy from the field or stretched cord. If the work is negative, the orbiting body slows down and transfers energy to the field or cord. In either case, work is done only to the extent that the motion has a component in the direction of (parallel or antiparallel to) the centripetal force.

---

work on the particle, which may be positive or negative, depending on the sign of $\mathbf{f} \cdot d\mathbf{s}$.

When a (net) force does a positive amount of work on a particle, the energy of the particle increases. The only form of energy a particle may have is kinetic energy. (Potential energy of a particle must reside in the field that exerts a force on the particle.) Therefore, when work is done on a particle, the kinetic energy of the particle increases. If the force opposes the motion (i.e., $\mathbf{f} \cdot d\mathbf{s} < 0$), the kinetic energy decreases.

$$\mathbf{f} \cdot d\mathbf{s} = m\mathbf{a} \cdot d\mathbf{s} = m\left(\frac{d\mathbf{v}}{dt}\right) \cdot d\mathbf{s} = m\left(\frac{d\mathbf{s}}{dt}\right) \cdot d\mathbf{v} = m\mathbf{v} \cdot d\mathbf{v} = d\left(\frac{1}{2}mv^2\right)$$

$$W = \int \mathbf{f} \cdot d\mathbf{s} = \int d\left(\frac{1}{2}mv^2\right) = \frac{1}{2}mv_2^2 - \frac{1}{2}mv_1^2 \tag{3-6}$$

Work done on a particle is equal to the change in kinetic energy of the particle. This result is called the *work-energy theorem*. It applies *only* to particles.

> *The work-energy theorem is the statement that when a system (a particle) can gain energy only as kinetic energy, the energy transferred to that system appears as kinetic energy.*

2. Work on an Object Free to Rotate

If a force is directed along a line that does not pass through the center of mass of the system, the system rotates and work is done on the system. Part of the energy goes into rotational energy of the system and the residue goes into translational kinetic energy if the body is rigid; otherwise it goes into translational kinetic energy and internal energy. As shown in Fig. 3.5, the displacement of the point where the force acts is greater than the displacement of the center of mass. The work done on the rigid body is equal to the sum of the change in translational kinetic energy and the change in rotational energy.

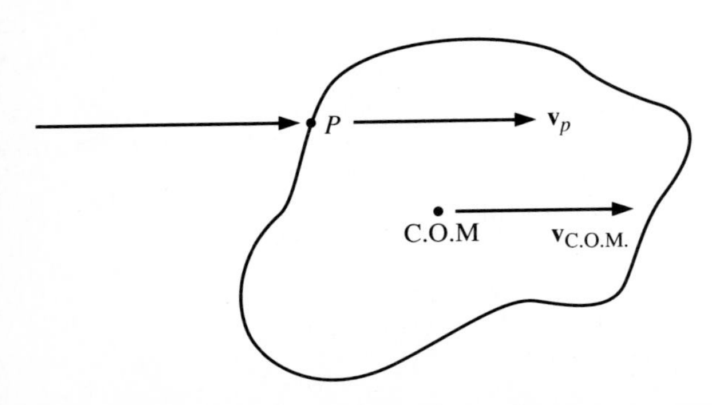

FIGURE 3.5 The velocity of one point (P) of a body may be very different from the velocity of other points (e.g., the center of mass), even if the body is rigid.

3. Work Done on a Spring

Consider a system consisting of a (perfect) spring, held in place by a rigid wall at one end and being pushed in at the free end. The work done on the spring adds energy to the spring, which is stored as potential energy of the spring. There is no kinetic energy imparted to the spring.

The potential energy of the spring is internal energy of the spring, but it is not thermal energy. It is not randomized; it can be quantitatively converted to kinetic energy or other mechanical energy at a constant temperature.

The force depressing the spring is not the only force acting (Fig. 3.6). The constraining force of the wall is as great, and in the opposite direction, so the net force on the spring is zero and the spring acquires no momentum. However, the constraining force does no work, because there is no motion of the boundary of the spring and wall. Therefore, the only work done is by the force acting on the movable end. That work is equal to the increase in potential energy of the spring.

4. Work Done on an Ideal Gas

An ideal gas is a substance that obeys the law

$$PV = nRT \tag{3-7}$$

where R is the gas constant ($R = 8.314$ J/mol · K) and n is the number of moles of substance. [An alternative definition, which can be shown to be equivalent, is that an ideal gas is a substance for which $(\partial E/\partial V)_T = 0$.] Although ideal gases are sometimes limited to monatomic gases, we choose the more general definition.

In gases and other isotropic fluids it is more convenient to deal with pressure than with total force. Pressure is force per unit area, $P = f/A$. Multiplying the displacement of a piston by the area of the piston gives the change in volume, $A\,dx = dV$. The pressure exerted by the fluid is

$$P_{\text{fluid}} = \frac{f_{\text{by fluid}}}{A} = \frac{-f_{\text{on fluid}}}{A}$$

so

{isotropic fluid} $$W = \int \mathbf{f} \cdot d\mathbf{x} = \int \frac{f}{A} \cdot (A\,dx) = -\int P \cdot dv \tag{3-8}$$

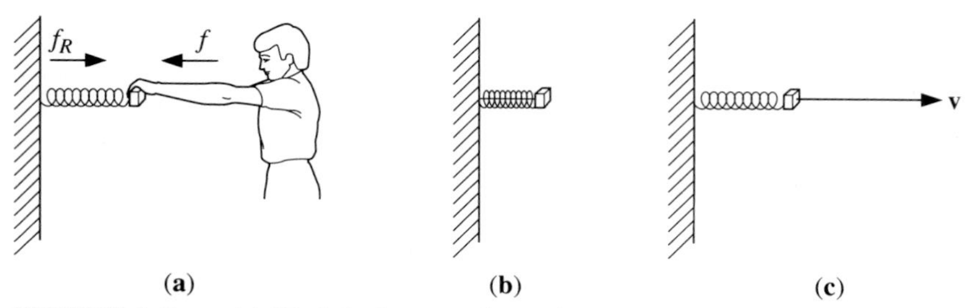

FIGURE 3.6 (a) Work is done on the spring when it is compressed, although the net force is zero. (b) The energy is stored in the spring. (c) If the constraining thread breaks, the stored energy is recovered as kinetic energy.

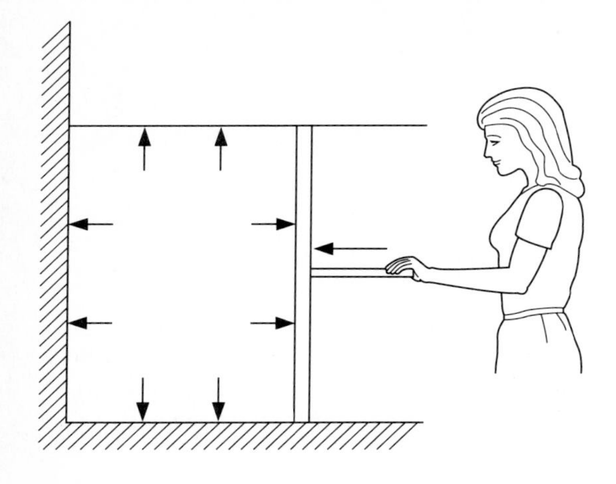

FIGURE 3.7 Work done on the gas increases the random kinetic energy of the molecules. No net momentum is given to the gas.

The pressure is the pressure of the fluid on the moving surface and dV is the change in volume of the fluid. Pressure exerted by a static fluid is always perpendicular to the boundary and opposite in direction to the external force, so $\mathbf{f} \cdot d\mathbf{x} = -P \cdot dV$; pressure behaves as a scalar quantity for isotropic materials.

If we assume that the gas is held in a cylinder at rest (Fig. 3.7), there is no net force on the gas while it is being compressed, no kinetic energy imparted to the center of mass of the system, no rotational energy given to the system, and no increase in potential energy of the system. Despite some "springiness" of the compressed gas, the energy of compression cannot be recovered as mechanical energy in an isothermal process.

Energy transferred to the gas as work appears within the gas as thermal energy. The molecules of the gas are moving faster as a consequence of work done on the gas, although $Q = 0$. No thermal energy transfer has occurred, but $\Delta T > 0$.

$$\{Q = 0\} \qquad \Delta E = Q + W = W$$

Work done on the system, W, has increased the internal energy of the system, and in particular, it has increased the random part of the internal energy, which we call the thermal energy.

Several standard types of processes are of particular importance. First, if the pressure is held constant during the expansion, either because the temperature is slowly increased or because the gas leaks slowly from its container to a cylinder at lower pressure (Fig. 3.8), the work done (Fig. 3.9) is simply

$$\{P_{\text{constant}}\} \qquad W = -\int P \cdot dV = -P \int dV = -P(V_2 - V_1) = -P\,\Delta V$$

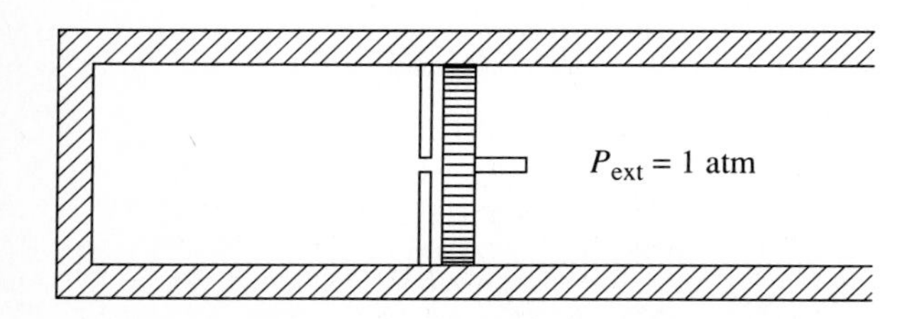

FIGURE 3.8 Gas escaping through the pinhole expands against a constant external pressure (1 atm).

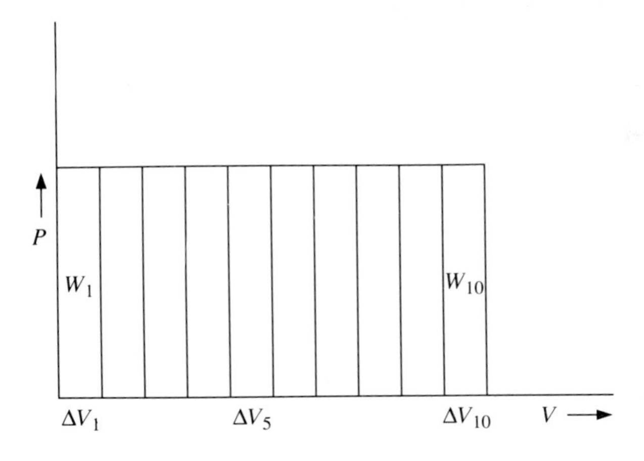

FIGURE 3.9 The incremental amount of work done is the constant pressure times ΔV_i.

This is typical of slow expansions of a gas or any other system against atmospheric pressure.[4]

EXAMPLES 3.6

A. Find the work done on 1 g of water as it freezes to ice (Fig. 3.10). The density of ice is approximately 0.9 g/cm^3, and atmospheric pressure is approximately 1×10^5 Pa (1 Pa = 1 newton/meter^2).
B. Find the work done on 1 g of water as it boils and forms vapor at 100°C, assuming that the vapor is an ideal gas.
C. 3 mol of nitrogen gas at 2 atm is expanded at constant pressure from a temperature of 27°C to a temperature of 77°C (Fig. 3.11). How much work is done on the gas? □

Work is often done on gases at constant temperature. By assuming the gas to be ideal and the pressure to be the equilibrium pressure of the gas (i.e., the process is reversible), the work done may be calculated to be

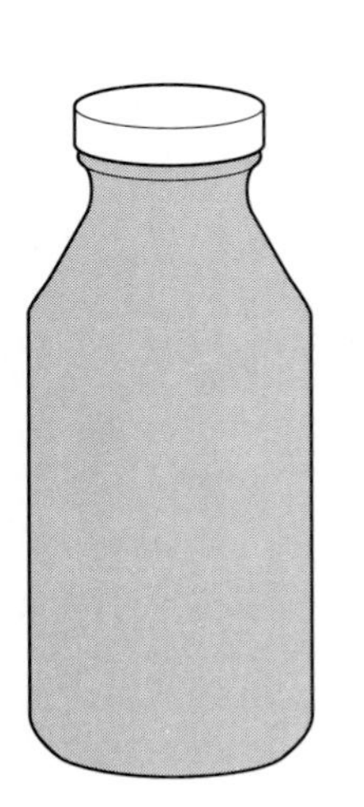

FIGURE 3.10 As milk freezes, it pushes the cap upward (or breaks the bottle). How much work is done by the atmosphere on the milk?

[4]The derivation assumed a gas. If a solid or liquid is expanding or contracting, the work calculation is less direct. The force against which the motion is occurring is exerted by the atmosphere, so atmospheric pressure is required, rather than an internal pressure of the solid or liquid. For the gas it was assumed that the gas pressure and the external pressure were equal.

ANSWERS 3.6

A. V(1 g water) $= 1.0$ cm^3, V(1 g ice) $= 1.1$ cm^3, $W = -P\,\Delta V = -1 \times 10^5$ Pa $\times$ 0.1 cm^3 $\times$ (1 m/100 cm)^3 $= -1 \times 10^{-2}$ J ($= -2 \times 10^{-3}$ cal). Compare with the heat of fusion of 0.33 kJ or 80 cal.
B. V(1 g vapor) $= nRT/P = \left(\frac{1}{18}\text{ mol}\right) \times 8.314$ J/mol·K $\times$ 373 K/1 $\times 10^5$ Pa $= 1.7 \times 10^{-3}$ m^3; V(1 g water) $= 1$ cm^3 $= 10^{-6}$ m^3. $\Delta V = 1.7 \times 10^{-3}$ m^3. $W = -P\Delta V = -10^5$ Pa $\times 1.7 \times 10^{-3}$ m^3 $= -0.17$ kJ ($= -41$ cal). Compare with the heat of vaporization of 2.3 kJ or 540 cal.
C. $V_i = nRT/P = 3 \times 8.314 \times 300/2 \times 10^5 = 3.74 \times 10^{-2}$ m^3; $V_f = 3 \times 8.314 \times 350/2 \times 10^5 = 4.36 \times 10^{-2}$ m^3. $\Delta V = 6.2 \times 10^{-3}$ m^3. $W = -$P $\Delta V = -2 \times 10^5 \times 6.2 \times 10^{-3} = -1.2(5)$ kJ ($= -0.3$ kcal).

{ideal gas; constant T}
$$W = -\int P\,dV = -\int nRT\,\frac{dV}{V} = -nRT\int \frac{dV}{V}$$

and therefore

{ideal gas; constant T}
$$W = -nRT\ln\frac{V_f}{V_i} \tag{3-9}$$

EXAMPLES 3.7

A. Find the work done by 3 mol of nitrogen if it expands reversibly at 27°C from 3 atm to 1 atm (Fig. 3.12).
B. Find the work done by 2 mol of nitrogen if it is compressed reversibly at 127°C from 1 atm to 3 atm. □

3.3 Other Forms of Work

The general definition of work is a force times a distance,

$$W = \int \mathbf{f}\cdot d\mathbf{s}$$

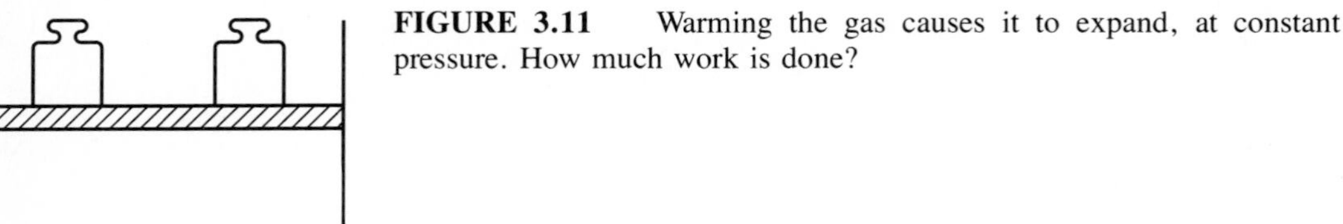

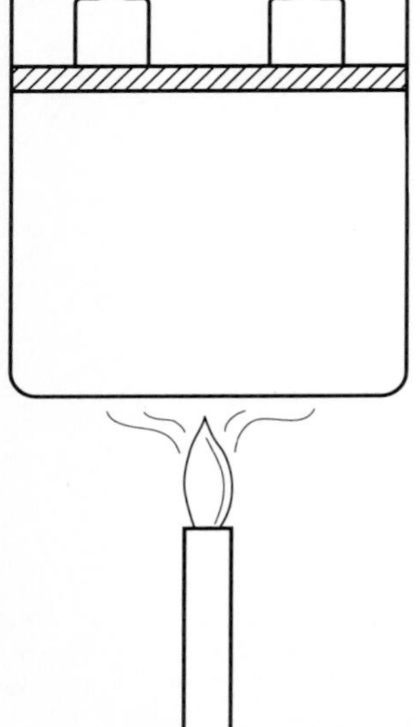

FIGURE 3.11 Warming the gas causes it to expand, at constant pressure. How much work is done?

where $d\mathbf{s}$ is the infinitesimal displacement of the point of action of the force. For expansion/compression work done on a fluid or by any body undergoing expansion/compression against a fluid, $f/A = -P$ and $A\,d\mathbf{s} = dV$, so the work can be written

$$W = -\int P\,dV$$

There are other types of interactions that are not conveniently expressed in terms of pressure and volume.

Energy of Force Fields

A particularly common example of a force acting on a body is an external force field, such as a static gravitational, electric, or magnetic field. Then the work done on the system is

$$W = \sum_i \left[\int \mathbf{f}_i \cdot d\mathbf{s}_i\right] = \sum_{\substack{\text{nonfield}\\ \text{forces}}} \left[\int \mathbf{f}_i \cdot d\mathbf{s}_i\right] + \sum \int \mathbf{f}_{\text{field}} \cdot d\mathbf{s} \qquad (3\text{-}10)$$

where $d\mathbf{s}$, in the last sum, is the displacement of the center of mass of the system, assuming that the field forces are *body forces*, causing motion of the center of mass of the system but no deformation of the system.

Stretched Wire

If a wire of length L is under tension $\mathcal{T}$ (Fig. 3.13), the surroundings must do work on the wire to increase its length. The displacement is dL and the force acting is equal to the tension, $\mathcal{T}$, so the work done on the system is

$$W = \int \mathcal{T}\,dL \qquad (3\text{-}11)$$

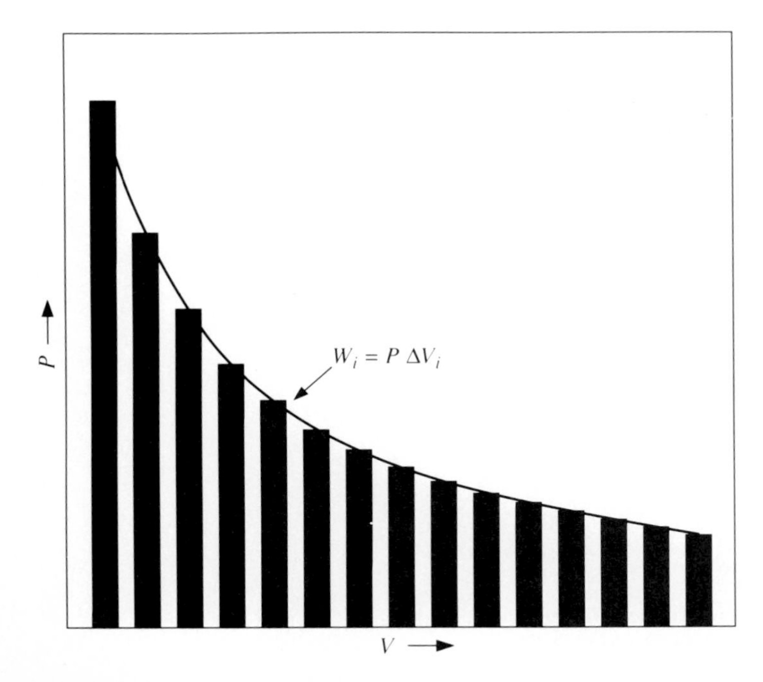

FIGURE 3.12 At constant temperature the pressure drops and the work done by the expanding gas is less than at constant pressure.

ANSWERS 3.7

A. $W = nRT \ln(V_2/V_1) = 3\,\text{mol} \times 8.314\,\text{J/mol}\cdot\text{K} \times 300\,\text{K} \times \ln[(nRT/1\,\text{atm})/(nRT/3\,\text{atm})] = 3 \times 8.314\,\text{J} \times 300 \times \ln 3 = 8.22\,\text{kJ}$.
B. $W = 2\,\text{mol} \times 8.314\,\text{J/mol}\cdot\text{K} \times 400\,\text{K} \times \ln \frac{1}{3} = -7.3\,\text{kJ}$.

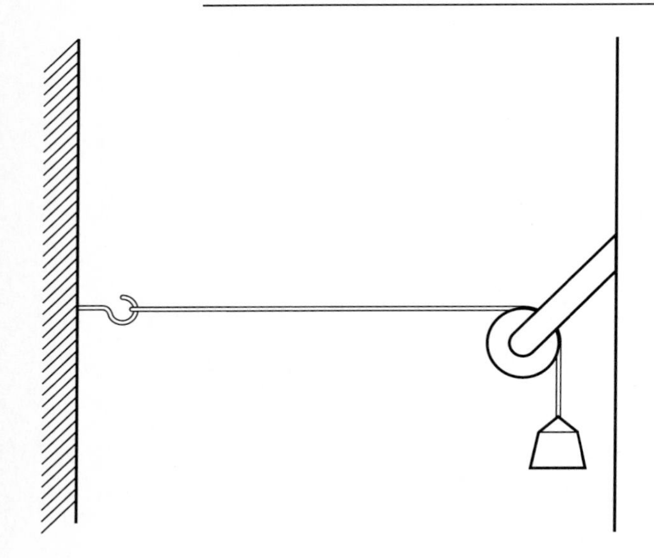

FIGURE 3.13 Work done as the wire stretches at constant tension $W = \mathcal{T}\Delta L$.

Consider two common approximations.

1. If the tension is constant, $W = \mathcal{T}\Delta L$.
2. If the tension is proportional to ΔL, as for a spring ($\mathcal{T} = k\Delta L$), the work is $W = \frac{1}{2}k(\Delta L)^2 = \frac{1}{2}\mathcal{T}\Delta L$.

The tension of a spring arises from torsion as the spring expands, in contrast to the constant tension expected from a stretching wire.

Electric Field

In an electrostatic field of potential V_E the work done by an external agent on a charge, q, to move it slowly through a change in field (Fig. 3.14) is

$$W = \int \mathbf{f} \cdot d\mathbf{x} = \int \left(q \frac{\partial V_E}{\partial \mathbf{x}} \right) \cdot d\mathbf{x} = \int q \, dV_E \tag{3-12}$$

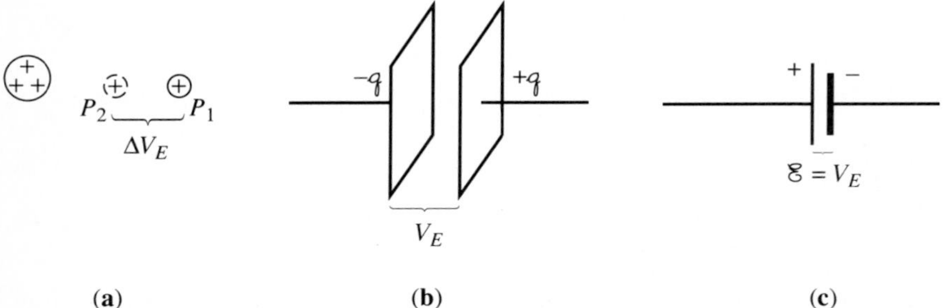

FIGURE 3.14 Work done by an electric field depends on the charge, q, and on the change in potential, V_E: (a) Coulombic field; (b) capacitor; (c) electrochemical cell.

The external force is required because the electric field is exerting an equal and opposite force on the charge. Hence the net work done *on the charge* is zero. Work is done on the electric field when the charge is moved against the field.

If the field is across a capacitor, the work done in charging the capacitor may be written in several equivalent forms:

$$W = \int q \, dV_E = \int C V_E \, dV_E = \frac{1}{2} C V_E^2 = \frac{1}{2} q V_E = \frac{1}{2} \frac{q^2}{C} \tag{3-13}$$

where C is the capacitance and $q = C V_E$. The work is done on the field within the capacitor; we would say that the energy of the capacitor is increased.

Similarly, the work done by an electrochemical cell in moving a charge q across the (fixed) potential difference of the cell, $\mathcal{E} = V_E$, is

$$W = \int q \, d\mathcal{E} = q\mathcal{E} \tag{3-14}$$

The energy is supplied by the chemical reaction. The energy is transferred, by pushing the charge against the electric field, to the external electric field (e.g., the electrical load in the circuit), where it is likely to be further converted.

A dielectric material, subjected to an electric field, becomes polarized. The forces exerted and the significant resultant motions at the atomic level are nonrandom, so that energy is transferred to the sample as work.

The energy density (E/V) of an electric field (**E**) is

$$\frac{E}{V} = \frac{1}{2} \mathbf{E} \cdot \mathbf{D}$$

where

$$\mathbf{D} = \epsilon_0 \mathbf{E} + \mathbf{P}$$

P is the electric polarization. When the electric field, **E**, changes, the change in energy of the sample is

$$dE = w = \frac{1}{2} V [d\mathbf{E} \cdot (\epsilon_0 \mathbf{E} + \mathbf{P}) + \mathbf{E} \cdot (\epsilon_0 \, d\mathbf{E} + d\mathbf{P})]$$

The work to produce an electric field in a vacuum is

$$dE = w = \epsilon_0 V \mathbf{E} \cdot d\mathbf{E}$$

Subtracting this energy attributable to the vacuum from the total gives the polarization energy of the sample,

$$\Delta E = W = \frac{1}{2} V \int (\mathbf{P} \cdot d\mathbf{E} + \mathbf{E} \cdot d\mathbf{P}) \tag{3-15}$$

The work is the same whether the sample is moved into a stationary electric field or the field is increased around a stationary sample by moving the external charges responsible for the field.

Magnetic Field

An unmagnetized sample ($\mathbf{M} = 0$), with nonzero magnetic susceptibility, becomes magnetized when it is subjected to a magnetic field, $\mathbf{H}$. The equations are similar to those for electric polarization.

The energy density of the magnetic field in the sample is

$$\frac{E}{V} = \frac{1}{2}\,\mathbf{B}\cdot\mathbf{H}$$

The local magnetic field is

$$\mathbf{B} = \mu_0(\mathbf{H} + \mathbf{M})$$

where $\mathbf{M}$ is the magnetization per unit volume.

When the field changes, the change in energy is

$$dE = w = \frac{1}{2}\,\mu_0 V\,[(\mathbf{H} + \mathbf{M})\cdot d\mathbf{H} + (d\mathbf{H} + d\mathbf{M})\cdot\mathbf{H}]$$

or

$$w = \mu_0 V\left[\mathbf{H}\cdot d\mathbf{H} + \frac{1}{2}(\mathbf{M}\cdot d\mathbf{H} + \mathbf{H}\cdot d\mathbf{M})\right]$$

Subtracting the change in energy of the vacuum field,

$$dE = w = \mu_0 V\mathbf{H}\cdot d\mathbf{H}$$

gives the magnetization energy of the sample and the work to produce that magnetization.

$$\Delta E = W = \frac{1}{2}\mu_0 V\int(\mathbf{H}\cdot d\mathbf{M} + \mathbf{M}\cdot d\mathbf{H}) \qquad (3\text{-}16)$$

The work required is the same whether a sample is moved into a magnetic field or a magnetic field is built up around a stationary sample.

Gravitational Field

Work done on a mass, m, by a gravitational field (not part of the system) depends on the change in the potential, Φ. The force on the mass per unit mass is

$$\frac{f}{m} = -\frac{d\Phi}{dx}$$

so the work done on the mass is

$$W = \int \mathbf{f}\cdot d\mathbf{x} = -\int m\,\frac{d\Phi}{d\mathbf{x}}\cdot d\mathbf{x} = -\int m\,d\Phi \qquad (3\text{-}17)$$

The gravitational potential may be written $\Phi = -\mathbf{g}\cdot\mathbf{h}$ near the surface of the earth or $\Phi = -GM/r$ at a distance r from the center of any spherically

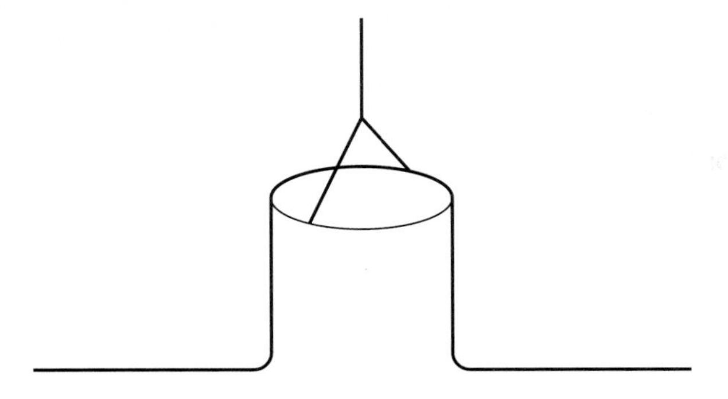

FIGURE 3.15 As the ring is lifted out of the liquid, the liquid follows because of surface tension. Work is done to increase the surface area of the liquid.

symmetric mass M. In either case, as h or r increases, the mass transfers energy to the field (e.g., a ball thrown upward slows down). Recognizing that $\mathbf{g}$ and $\mathbf{h}$ are antiparallel, we have

$$W = -\int m\,d\Phi = \int m\,\mathbf{g}\cdot d\mathbf{h}$$

or

$$W = -\int m\,d\Phi = -\int m\,d\left(\frac{-GM}{r}\right) = -\int \left(\frac{GMm\,dr}{r^2}\right)$$

If an external force acts on the mass, it performs work exactly equal and opposite to the work done on the mass by the gravitational field, provided that the kinetic, rotational, and internal energies of the mass are left unchanged.

Surface Tension

A liquid surface tends to contract, because the attractive cohesive forces from within are not balanced by molecules pulling outward. The effect depends on the surface tension, γ, which is the force per unit of length or the energy per unit of area. The value of γ depends on the substance and on the temperature. The work done on a liquid, for a change in area, is

$$W = \int \gamma\,dA \tag{3-18}$$

This is the increase in energy of the liquid if the surface of the liquid is extended (e.g., by a ring, as in Fig. 3.15). Note that a thin cylinder of liquid has an inside face and an outside face, so the area is $2\pi Dh$, for a ring of diameter D pulled to height h, and the work is twice as great as for a single sheet, leaving γ unaffected.

Energy is an extensive property. Therefore, any contributing term to the energy must be a product of an extensive property with one or more intensive properties. For example, mass is extensive and v^2 is intensive, so the product $\frac{1}{2}mv^2$ is extensive. A list of corresponding extensive and intensive properties is given in Table 3.1. Field properties ($\mathbf{g}$ or Φ, $\mathbf{E}$, and $\mathbf{H}$) are properties of the system or of the surroundings, depending on the definition of the system.

The generalized force may take many forms, but it always multiplies a geometric differential. A more general form of the first-law equation, therefore, would be

$$dE = q + w = q - P\,dV + \sum_i X_i\,dx_i \tag{3-19}$$

Table 3.1 Examples of Thermodynamic Work

Extensive	Intensive	w
$\mathbf{s}$, displacement	$\mathbf{f}$, force	$\mathbf{f} \cdot d\mathbf{s}$
x_i, "generalized displacement"	F_i, "generalized force"	$F_i\, dx_i$
L, length	$\mathcal{T}$, tension	$\mathcal{T}\, dL$
A, area	γ, surface tension	$\gamma\, dA$
V, volume	$-P$, pressure	$-P\, dV$
m, mass	Φ, gravitational potential	$-m\, d\Phi$
$\mathcal{q}$, charge	V_E, electric potential	$\mathcal{q}\, dV_E$
$V\mathbf{P}$, polarization	$\mathbf{E}$, electric field	$\frac{1}{2}V(\mathbf{P} \cdot d\mathbf{E} + \mathbf{E} \cdot d\mathbf{P})$
$V\mathbf{M}$, magnetization	$\mathbf{H}$, magnetic field	$\frac{1}{2}\mu_0 V(\mathbf{H} \cdot d\mathbf{M} + \mathbf{M} \cdot d\mathbf{H})$

where the X_i are called *generalized forces* and the x_i are *generalized displacements* (see Section 17.2). When both terms have vector properties (as in $\mathbf{P} \cdot \mathbf{E}$ and $\mathbf{M} \cdot \mathbf{H}$), two terms can contribute to the transfer of energy, as work, between system and surroundings.

Integration of a differential expression in Table 3.1 requires an assumption about what is held constant. For example, if V_E is constant, as in an electrochemical cell, $W = \mathcal{q}\, V_E$, but if V_E arises from $\mathcal{q}$, as in a capacitor, $W = \frac{1}{2}\mathcal{q}\, V_E$.

3.4 Relation of Energy, Heat, and Work

It is particularly important to keep in mind the distinction between how energy enters a system and how that energy is stored within the system. Q and W are only means of transferring energy, just as coins and bills are means of transferring money into a bank account. The account shows only the net dollars, with no record of how many coins or how many bills were deposited or withdrawn.

A system retains no memory of how much thermal energy was transferred in or out, or how much work was done.

Only the net energy change has any significance for the state of the system.

The amount of thermal energy transferred, Q, is often found by measurement of a temperature difference,

$$Q = \Delta E = C_v\, \Delta T$$

If the volume changes, or other work is done on the system, the measurement must be corrected for the work.

$$\left\{\begin{matrix}\text{constant mass} \\ \text{constant momentum}\end{matrix}\right\} \qquad Q = \Delta E - W = C_v\, \Delta T - W \qquad \text{(3-20)}$$

If the system is not well behaved (i.e., $\Delta E \neq C_v\, \Delta T$, perhaps because the system is undergoing a phase change or a chemical reaction), Q may be measured from ΔT of well-behaved surroundings. For example, a system consisting of ice can be placed in surroundings of warm water and the temperature change of the water measured as the ice melts.

Unfortunately, nature sometimes passes its money (ΔE) in a bag, so that we have no way of determining how much, if any, was passed as Q and how much as W. When the transaction is over, there is no trace of the mechanism, so we cannot apply the formalism of thermal energy transfer and work to such processes.

The division of energy transfers into Q and W has a deeper significance in reversible processes, for which W can be determined. This is discussed in Part II, where we discuss the second law of thermodynamics.

SUMMARY

For a large class of problems, $Q = C\,\Delta T$ or $Q = C\,\Delta T + Q_{\text{phase changes}}$, and therefore, in insulated systems,

$$\sum C_i\,\Delta T_i + \sum Q_{\text{phase changes}} = 0$$

or if there are two parts, A and B,

$$\sum_A (C\,\Delta T + Q_{\text{phase changes}}) = -\sum_B (C\,\Delta T + Q_{\text{phase changes}})$$

Work is a transfer of energy to the system from the surroundings; $W = \sum \int \mathbf{f}_i \cdot d\mathbf{s}_i$, where $d\mathbf{s}_i$ is the displacement of the point of application of the external force $\mathbf{f}_i$ to the system.

Q and W are not defined for most real processes (e.g., those involving friction), but may be satisfactorily approximated in the idealized processes usually considered.

For fluids, $\mathbf{f}_i \cdot d\mathbf{s}_i = -P\,dV$, where P is the pressure exerted by the system on the surroundings.

Special cases:

A. $W = \int X_i\,dx_i$ and X_i constant: $W = X_i\,\Delta x_i$
Example: Constant P, $W = -P\,\Delta V$

B. $W = \int X_i\,dx_i$ and X_i proportional to x_i : $W = \frac{1}{2}\Delta(X_i x_i)$
Example: Capacitor, $\mathcal{E} \sim q$: $W = \frac{1}{2}\,\mathcal{E} q$

C. $P = nRT/V$, T constant : $W = -nRT\ln(V_2/V_1)$

At constant volume, $Q = \Delta E = C_v\ \Delta T$ (C_v is measured at constant V).
Energy transfers can often be expressed in the form

$$dE = q + w = q - P\,dV + \sum_i X_i\ dx_i$$

(in the usual limit of idealized processes, so that w and q are defined).

QUESTIONS

3.1. Under what conditions (expressed in modern terminology) is the caloric theory of Lavoisier an adequate model for predicting the results of experiments?

3.2. Some writers have suggested that the second alternative method of solution for Example 3.1A should let $t_0 = 0$ K. What are the conceptual and practical advantages and disadvantages of such a choice?

3.3. Rumford's writings were apparently all initially in English, although some were subsequently translated into German. How would you translate, into modern technical English, his word *Heat* as it appears in the phrases quoted in the text:

a. Degree of Heat.
b. Intense Heat.
c. Latent Heat.
d. Quantity of Heat . . . produced.
e. How much Heat . . . was generated.

3.4. Objections have been raised to the word *heat* as a verb. What does the verb mean in the following examples? (Explain in terms of thermal energy, Q, and T or ΔT.)

a. A pan of water may be heated by placing it on a stove.
b. A gas may be heated by compressing it.
c. A glacier melts when it is heated by the sun.

3.5. Identify the primary form of energy transfer as Q, W, or other (or none) in each of the following.

a. A vat of mercury is stirred with a paddle.
b. An ice cube melts in a glass.
c. Hot water is poured into a beaker of cold water (with the system defined as the initial contents of the beaker).
d. Hot water is poured into a beaker of cold water (with the system defined as the final contents of the beaker).
e. Hot water is poured into a beaker of cold water (with the system defined as the instantaneous contents of the beaker).
f. A steel block is warmed by being struck by a hammer.
g. A hydrogen/oxygen explosion is contained within an insulated steel bomb.
h. A cup of coffee cools by evaporation.

PROBLEMS

3.1. How many calories or joules must be added to 40 g of water at 15°C to warm it to 75°C?

3.2. How many calories or joules must be added to 75 g of ice at −20°C to melt it (at 0°C) and warm it to 30°C?

3.3. What is the final state if 100 g of ice at −15°C is added to 1000 g of water at 25°C?

3.4. What is the final state if 200 g of water at 20°C is added to 50 g of steam at 110°C?

3.5. What is the final state if 50 g of ice at −20°C is added to 10 g of steam at 115°C?

3.6. What is the final temperature if 100 g of iron at 95°C is added to 20 g of water at 15°C? (Heat capacities are given in Tables 3.2 and 3.3.)

3.7. How much work is done on 2.0 mol of H_2 gas at 20°C if it is compressed isothermally from 0.50 atm to 2.50 atm?

3.8. How much work is done by 4.0 mol of O_2 gas at 40°C if it expands isothermally and reversibly from 100 kPa to 20 kPa? (1 Pa = 1 N/m^2; 1 atm = 1.01 bar = 101 kPa.)

Table 3.2 Heat Capacities

Substance	$\frac{\text{cal}}{\text{g} \cdot \text{deg}}$	$\frac{\text{J}}{\text{g} \cdot \text{deg}}$
$H_2O(l)$[a]	1.00	4.18
Ice[a]	0.50	2.1
Steam[a]	0.50	2.1
Al	0.214	0.896
Cu	0.0921	0.385
Fe	0.107	0.448
S(rh)	0.177	0.741
S(m)	0.193	0.809
Pb	0.0306	0.128

[a]For many purposes it is sufficient to set $c_p(\text{steam}) = c_p(\text{ice}) = \frac{1}{2}c_p(\text{liquid } H_2O) = \frac{1}{2}(1 \text{ cal/g} \cdot \text{K}) = \frac{1}{2}(4.2 \text{ J/g} \cdot \text{K})$. The heat of fusion of water (0°C) is 79.69 cal/g = 333.5 J/g. The heat of vaporization of water (100°C) is 539.42 cal/g = 2.258 kJ/g (80 and 540, or 333 and 2.26 are adequate for most purposes).

3.9. Young's modulus is the ratio of stress to strain, or the force per unit cross section area divided by the percentage change in length. The moduli for aluminum, brass, and steel are approximately 7, 9, and 20, respectively, each times 10^{10} N/m^2.

a. How much work is done on a brass wire if a mass of 10 kg is hung from a wire of length 2.0 m and diameter 1.0 mm? [**Note:** The tension is constant; the process is not reversible.]

b. What is the work if the wire is 4.0 m long?

c. What is the work if the wire is 0.50 mm in diameter and 2.0 m long?

d. What is the work if 20 kg is hung from the wire (1.0 mm and 2.0 m)?

3.10. The surface tension of water is 0.073 N/m, compared to 0.023 N/m for ethanol and 0.487 N/m for mercury. If a ring 2.5 cm in diameter is lifted 0.50 cm above the surface, pulling a cylindrical film of water with it, how much work is done on the water?

3.11. A spaceship and astronauts, with a combined mass of 5 metric ton, are to be lifted to an orbit 320 km above the surface of the earth.

a. How expensive should the operation be if the only cost were the energy,

Table 3.3 Heat Capacities of Gases

Substance	At Constant V		At Constant P	
	$\frac{\text{cal}}{\text{mol} \cdot \text{K}}$	$\frac{\text{J}}{\text{mol} \cdot \text{K}}$	$\frac{\text{cal}}{\text{mol} \cdot \text{K}}$	$\frac{\text{J}}{\text{mol} \cdot \text{K}}$
He	3.0	13	5.0	21
Ar	3.0	13	5.0	21
H_2	5.0	21	7.0	29
N_2	5.0	21	7.0	29
O_2	5.0	21	7.0	29
H_2O	6.6	28	8.7	36
CH_4	6.5	27	8.5	35
SO_2	10.0	41.8	10.0	41.8

at 10 cents/kWh? [The radius of the earth is about 6400 km, from which the product GM_E is available.]

b. The virial theorem states that if V is a homogeneous function of the coordinates of degree ρ, the average values of kinetic and potential energies are related by

$$\bar{T} = \frac{\rho}{2}\bar{V}$$

For example, harmonic oscillators have $V = \frac{1}{2} k x^2$, so $\rho = 2$ and $\bar{T} = \bar{V}$; inverse square forces have $V \sim 1/r$ so $\rho = -1$ and $\bar{T} = -\frac{1}{2}\bar{V}$. Find the additional cost (at 10 cents/kWh) for giving the space ship sufficient kinetic energy to maintain the orbit at 320 km above the surface. How do these answers compare with actual launch costs?

3.12. A typical large van de Graaff accelerator may have a terminal voltage of ± 20 MV.

a. How much work is done on an electron or proton by the electric field as the charge moves down the discharge tube?

b. How much work is done on a coulomb of charge?

3.13. Show that when $W = \oint P\,dV$ around a closed cycle, the work is represented by the area enclosed by the path on a pressure-volume diagram.

3.14. Two identical motors are each operating at 50 W. The first motor is running freely. The second motor is lifting a weight, and therefore turning more slowly. Do the two motors get equally hot? If not, which gets warmer? Explain your answer.

3.15. Some textbooks specify that work should be calculated as $\int \mathbf{f}_{\text{net}} \cdot d\mathbf{s}$ when more than one force is acting.

a. What is the net force acting on a gas as it is slowly compressed inside a cylinder? Is work being done on the gas?

b. What is the net force acting on an inverted bicycle wheel as it is spun in place? Is work being done on the wheel?

3.16. Explain why $\int \mathbf{f} \cdot d\mathbf{s}$ does not properly calculate work done by the floor on a ball as the ball bounces if $d\mathbf{s}$ is taken as the incremental displacement of the center of mass of the ball.

CHAPTER 4

Application of the First Law

In Chapter 3 we introduced the important principles for the first law and its applications. In this chapter we describe some of the mathematical tricks and the definitions that have been found helpful in applying the first law to real systems under laboratory conditions.

4.1 Heat Capacity

Heat capacity is defined as the thermal energy transferred to the system divided by the temperature rise of the system,

$$C = \frac{Q}{\Delta T} \tag{4-1}$$

Processes at Constant Volume

In the limit of infinitely small amounts of thermal energy transfer, q, the heat capacity at constant volume, C_v, is

$$\begin{Bmatrix} \text{constant } V \\ w = 0 \end{Bmatrix} \qquad C_v = \frac{q}{dT}$$

or because $dE = q$ when there are no other forms of energy transfer,

$$\begin{Bmatrix} \text{constant } V \\ w = 0 \end{Bmatrix} \qquad C_v = \frac{dE}{dT}$$

The constant volume condition is expressed in partial derivative notation by writing

$$C_v = \left(\frac{\partial E}{\partial T}\right)_v \tag{4-2}$$

That is, C_v gives the change in energy caused by a change in temperature only. The total differential, dE, may include other terms, involving change of volume or changes in variables other than temperature and volume.

The thermal energy transfer, Q, is the integral of C_v over temperature. If $W = 0$, then $\Delta E = Q = \int C_v\, dT$. Usually, C_v is approximately constant for moderate temperature intervals, so

$$\Delta E = \int dE = \int q = C_v \int dT = C_v\, \Delta T$$

For example, the value of c_v for helium is very nearly $\frac{3}{2}R = 12.5$ J/mol · K. If 0.5 mol of He is warmed from 20°C to 40°C at *constant volume,*

$$\Delta E = Q = \int C_v\, dT = nc_v \int dT = 0.5 \text{ mol} \times 12.5 \frac{\text{J}}{\text{mol} \cdot \text{K}} \times 20 \text{ K}$$
$$= 125 \text{ J}$$

Processes at Constant Pressure

When there is a significant change in volume, some energy will usually be transferred as work. Then the change in energy of the system, ΔE, is no longer equal to the thermal energy transfer, Q. For example, if 0.5 mol of helium is warmed from 20°C to 40°C at *constant pressure,* rather than at constant volume, thermal energy is transferred *to* the system but energy is transferred, as work, *from* the system.

$$W = -\int P\, dV = -P(V_f - V_i) = -(nRT_f - nRT_i) = -nR\,\Delta T$$
$$= -0.5 \text{ mol} \times 8.314 \frac{\text{J}}{\text{mol} \cdot \text{K}} \times 20 \text{ K} = -83 \text{ J}$$

The energy change is $\Delta E = Q + W$, but Q also differs from the previous calculation at constant volume.

The energy of an ideal gas depends only on the temperature, not on the pressure or volume. Therefore, in the constant-pressure process, $\Delta E = 125$ J, as calculated for constant volume. The thermal energy absorbed is

$$Q = \Delta E - W = 125 \text{ J} - (-83) \text{ J} = 208 \text{ J}$$

We must put 208 J of energy into the gas, by thermal energy transfer, to increase the thermal energy of the gas by 125 J. The rest of the energy is transferred back to the surroundings as work.

EXAMPLES 4.1

A. For hydrogen gas, $c_v = \frac{5}{2}R = 21$ J/mol · K. Find ΔE when 1 mol of hydrogen is warmed, at constant volume, from 15°C to 25°C.
B. Find W, ΔE, and Q when 1 mol of hydrogen is warmed at constant pressure from 15°C to 25°C. □

Similarly, the energy difference between liquid water and water vapor at 100°C involves both thermal energy transfer and work done against the atmosphere. The thermal energy transfer at constant pressure, Q (i.e., the *heat of vaporization*), is 2.26 kJ/g. The water expands as it evaporates, pushing against the atmosphere, so $\Delta E \neq Q$ for evaporation of water at constant pressure.

Sometimes the work done by expansion can be ignored. For example, the heat capacity of lead is approximately $3R = 25$ J/mol · K $= 0.12$ J/g · K. If 5 g of lead is warmed from 20°C to 30°C, the change in volume is small enough that $W = -\int P\, dV$ is much smaller than Q, so it is not particularly surprising that even at constant pressure,

$$\Delta E = Q = 5\text{g} \times 0.12 \text{ J/g} \cdot \text{K} \times 10 \text{ K} = 6 \text{ J}$$

EXAMPLES 4.2

A. From $Q_{\text{vaporization}} = 2260$ J/g for water at 100°C, find $\Delta E_{\text{vaporization}}$ of water at this temperature.
B. From $Q_{\text{fusion}} = 333$ J/g for ice at 0°C, find ΔE_{fusion} for ice at this temperature. [Density of ice $= 0.917$ g/cm^3.] □

Heat capacity was defined above for constant volume; $C_v = (q/dT)_V$. A *constant-pressure* heat capacity is defined as

$$C_p = \left(\frac{q}{dT}\right)_P \tag{4-3}$$

Heat capacity at constant pressure is greater than the heat capacity at constant volume, especially for gases, because some of the thermal energy added is returned to the surroundings as work, pushing back the atmosphere. In solids, expansion is related to anharmonicity of the lattice vibrations. The expansion has little effect on the energy, so $c_p - c_v$ is typically very small.

EXAMPLES 4.3

A. Show that when 1 mol of hydrogen is warmed from 15°C to 25°C, Q is found easily from the value of $c_p = \frac{7}{2}R = 29$ J/mol · K.
B. Show that for any ideal gas, $c_p - c_v = R$, the gas constant.
C. What is wrong with the following argument?
To find the energy change of lead when it is warmed at constant pressure, let

$$dE = \left(\frac{\partial E}{\partial V}\right)_T dV + \left(\frac{\partial E}{\partial T}\right)_V dT$$

The first term is $(w/dV)_T\, dV = (-P\, dV/dV)\, dV = -P\, dV$ and the second term is $C_v\, dT$. The volume changes very little, and P is only 1 atm, so $\int P\, dV$ is very small. Therefore, $dE = C_v\, dT$ is a good approximation for the warming of lead at constant pressure. □

4.2 Energy Transfers in Chemical Reactions

Chemical reactions are often carried out at constant pressure and may involve changes of volume. For example, if hydrogen burns in oxygen at approximately 100°C and 1 atm pressure to give water vapor, the reaction is

ANSWERS 4.1

A. $\Delta E = Q = nc_v \Delta T = 1.0 \text{ mol} \times 21 \text{ J/mol} \cdot \text{K} \times 10 \text{ K} = 210 \text{ J} = 50 \text{ cal.}$
B. $W = -P\Delta V = -nR\Delta T = -83$ J. $\Delta E = 210$ J (from A), so $Q = \Delta E - W = 210 \text{ J} - (-83 \text{ J}) = 290$ J.

ANSWERS 4.2

A. $W = -P\Delta V = -nRT$ (neglecting the volume of liquid compared to the vapor) $= -\frac{1}{18}$ mol $\times$ 8.314 J/mol $\cdot$ K $\times$373 K $= -172$ J. $\Delta E = Q + W = 2260$ J/g -172 J/g $= 2.09$ kJ/g.
B. $W = -P\Delta V = -1.01 \times 10^5 \text{ Pa} \times (1.00 \times 10^{-6} \text{ m}^3 - 1.09 \times 10^{-6} \text{ m}^3) = 9$ mJ/g. This is negligible compared to Q, so $\Delta E = Q = 333$ J/g.

ANSWERS 4.3

A. $Q = C_p \Delta T = 29$ J/mol $\cdot$ K $\cdot$ 1 mol$\cdot$ 10 K $= 290$ J, in agreement with the earlier answer (Example 4.1B).
B. $C_p \Delta T = Q$ (at constant pressure) $= \Delta E - W$ and $C_v \Delta T = Q$ (at constant volume) $= \Delta E$. Therefore, the difference, $(C_p - C_v)\Delta T = -W$, the work done by the system on the atmosphere at constant pressure. For an ideal gas, $W = -P\,\Delta V = -nR\,\Delta T$, so the difference per mole is $(c_p - c_v)\Delta T = R\,\Delta T$ per mole, and $c_p - c_v = R$.
C. Although the conclusion is correct (c_p and c_v are nearly the same for lead), the argument is fallacious. It assumes that the change in energy with volume is equal to the work done on the system divided by the change in volume. This neglects thermal energy transfers, and is therefore not generally true.
Try the same analysis on an ideal gas. This method would set

$$dE = -P dV + C_v\, dT = w + q$$

But we know that for an ideal gas, energy depends only on temperature, so $dE = C_v\, dT$, even at constant pressure, so $P dV = 0$ and therefore $P = 0$ for any ideal gas! This is clearly erroneous.

$$2H_2 + O_2 \rightarrow 2H_2O$$

There is 3 mol of gas on the left and only 2 mol on the right, so at constant temperature and pressure the volume decreases to two-thirds of the original value. The thermal energy absorbed by the reacting gases, called the *heat of reaction*, is -242.5 kJ/mol(H_2). Thermal energy is transferred from the system (the gases) to the surroundings.

The work done on the system is (Fig. 4.1)

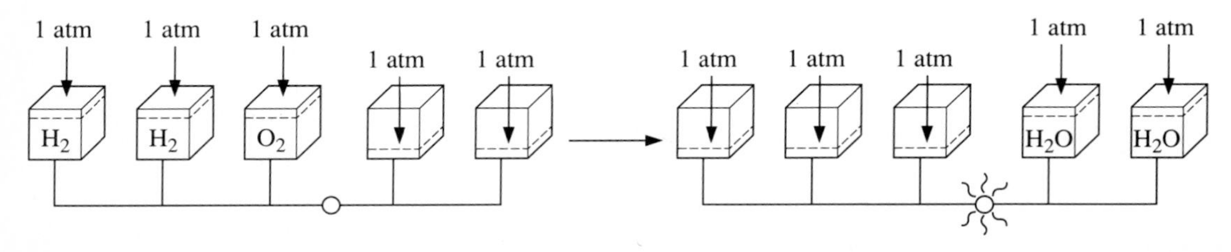

FIGURE 4.1 Three volumes of ($2H_2 + O_2$) go to 2 volumes of ($2H_2O$).

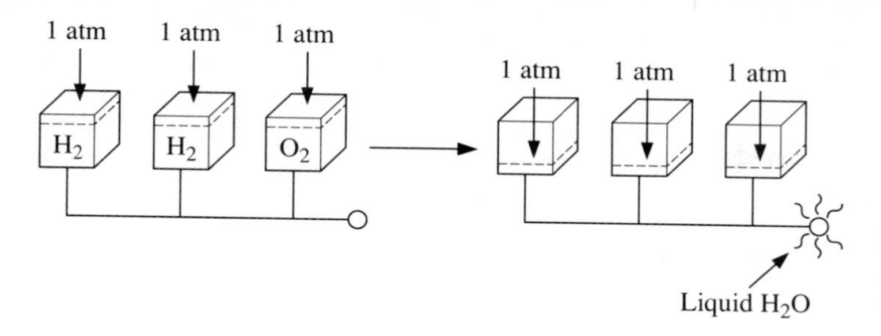

FIGURE 4.2 Three volumes of $(2H_2 + O_2)$ go to a negligible volume of liquid H_2O.

$$W = -\int P\,dV = -P(V_f - V_i) = -P\left(n_f\frac{RT}{P} - n_i\frac{RT}{P}\right) = -(n_f - n_i)RT$$

$$W = -(\Delta n)RT = -(-1\text{ mol}) \times 8.314\text{ J/mol}\cdot\text{K} \times 373\text{ K}$$

$$= 3.1\text{ kJ/}(2\text{ mol H}_2\text{O})$$

The energy change of the reacting system is therefore

$$\Delta E = Q + W = 2\text{ mol} \times (-242.5)\text{ kJ/mol} + 3.1\text{ kJ}$$

$$= -481.9\text{ kJ/}(2\text{ mol H}_2\text{O}) \quad \text{or} \quad -241\text{ kJ/mol of water}$$

A different amount of work is done on the system if hydrogen and oxygen react, at approximately the same temperature and pressure, producing liquid water instead of water vapor (Fig. 4.2). $W = -\int P\,dV = -P\Delta V$. The volume of a liquid is smaller by about three orders of magnitude than the volume of its vapor, near room temperature. Therefore, the volume of the liquid may be neglected and ΔV set equal to the change in volume of the vapor,[1] which is

$$\Delta V = V_f - V_i = \frac{n_f RT}{P} - \frac{n_i RT}{P} = (\Delta n)\frac{RT}{P}$$

The work done on the system is

$$W = -P\Delta V = -(\Delta n)RT = (3\text{ mol}) \times 8.314\text{ J/mol}\cdot\text{K} \times 373\text{ K}$$

$$= 9.3\text{ kJ (per 2 mol H}_2\text{O).}$$

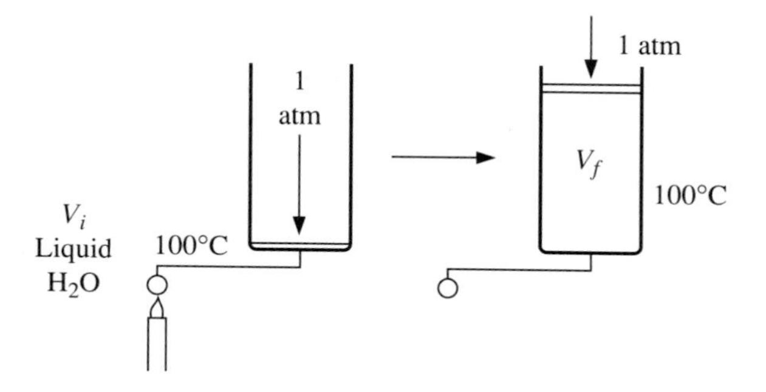

FIGURE 4.3 The difference in heat of reaction between forming water vapor and forming liquid water is equal to the heat of vaporization of the liquid.

[1]In the reaction $2H_2 + O_2 \rightarrow 2H_2O$, $\Delta V = V(2H_2O) - V(2H_2) - V(O_2)$, but the volume of 2 mol of liquid water is only 36 cm^3, compared to about 25 L for the oxygen and 50 L for the hydrogen. Thus it is quite adequate to neglect the volume of liquid and take ΔV as the change in volume of the gases present, which makes Δn the change in number of moles of gas (-3 mol in this reaction). Similarly, solids have negligible volume compared to gases, so in the reaction $C + CO_2 \rightarrow 2CO_2$, $\Delta V = \Delta nRT = RT$ because the number of moles of gas increases from one to two.

Table 4.1 Heats of Reaction for Producing Water

Reaction	Temperature	Q = Heat of Reaction = $\Delta H_{\text{reaction}}$
$H_2 + \frac{1}{2} O_2 \rightarrow H_2O(g)$	100°C	−242.5 kJ
$H_2 + \frac{1}{2} O_2 \rightarrow H_2O(l)$	100°C	−283.2 kJ

The thermal energy transfer, Q, is greater in magnitude when liquid water is formed, by the heat of vaporization of the water (Fig. 4.3 and Table 4.1).

$$
\begin{aligned}
Q &= 2 \text{ mol}(H_2O) \times (-242.5 \text{ kJ/mol}(H_2O) - 2 \text{ mol} \times 2.26 \text{ kJ/g} \times 18 \text{ g/mol} \\
&= -566 \text{ kJ/}(2 \text{ mol } H_2O)
\end{aligned}
$$

For the reaction to produce liquid water at 100°C, the energy change is

$$
\begin{aligned}
\Delta E &= Q + W = -566 \text{ kJ/}(2 \text{ mol } H_2O) + 9.3 \text{ kJ/}(2 \text{ mol } H_2O) \\
&= -557 \text{ kJ/}(2 \text{ mol } H_2O) \quad \text{or} \quad -278.5 \text{ kJ/mol of water}
\end{aligned}
$$

This differs from the value of ΔE calculated previously (forming water vapor) by $278.5 - 241 = 37.(5)$ kJ/mol, which is just 18 times the energy difference per gram between liquid water and water vapor calculated in Example 4.2; 18 g/mol × 2.09 kJ/g = 37.(6) kJ/mol.

4.3 Enthalpy

Making corrections for work done against atmospheric pressure can be a serious inconvenience. It would be helpful to have a new function, which could be called a "heat function," H, defined in such a way that the change in this function would be equal to Q, the thermal energy transfer, at least when the pressure is constant.

To obtain such a function, define H as

$$H \equiv E + PV \tag{4-4}$$

Then, for constant pressure,

$$\{\text{constant } P\} \qquad dH = dE + P\,dV$$

or

$$\{\text{constant } P\} \qquad \Delta H = \Delta E + P\Delta V$$

If we assume that $\Delta E = Q + W$, *and* that no work is done other than $P\,dV$ work,

$$\Delta H = Q + W + P\Delta V = Q + W - W$$

or

$$\left\{\begin{matrix} \text{constant P} \\ W = P\,\Delta V \end{matrix}\right\} \qquad \Delta H = Q \tag{4-5}$$

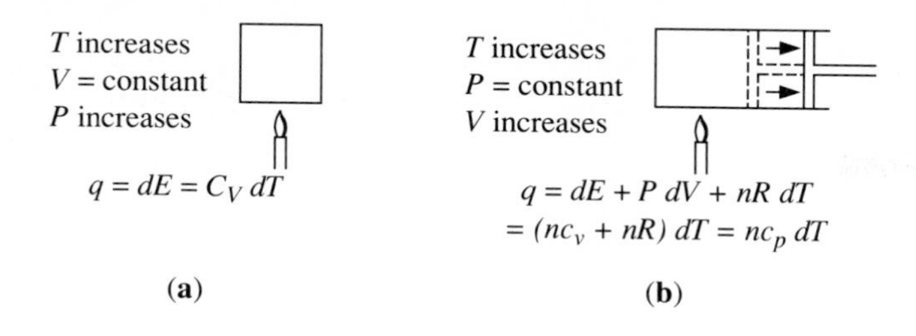

FIGURE 4.4 More thermal energy is required to expand a gas against constant pressure (b) than to warm it by the same amount at constant volume (a).

which shows that the new function accomplishes its purpose.

The new function, $H \equiv E + PV$ *is called* enthalpy.

For constant pressure,

$$C_p = \frac{q}{dT}$$

Because $dH = q$ for constant pressure,

$$C_p = \left(\frac{\partial H}{\partial T}\right)_P \tag{4-6}$$

It is the heat capacity at constant pressure, C_p, rather than C_v that is most often tabulated for liquids and solids, although they are typically nearly the same[2] (Fig. 4.4).

For an ideal gas, $c_p - c_v = R$; for a real gas, $c_p - c_v$ is approximately equal to R, but may differ slightly. Thus water vapor, at 100°C, has $c_v = 6.6$ cal/mol · K = 27.5 J/mol · K, and $c_p = 8.7$ cal/mol · K = 36.4 J/mol · K, for a difference, $c_p - c_v$, of 2.1 cal/mol · K = 8.9 J/mol · K, slightly greater than R.

A review of the calculations on water will illustrate the advantages of enthalpy over energy for many routine calculations. First, for the transition from liquid water to water vapor at 100°C, at constant pressure (with no other forms of work), the enthalpy change is

$$\Delta H = Q = 2.258 \text{ kJ/g} \times 18 \text{ g/mol} = 40.6 \text{ kJ/mol}$$

The *heat of vaporization* is equal to $\Delta H_{\text{vaporization}}$.

Similarly, the *heat of reaction* is equal to $\Delta H_{\text{reaction}}$. For the reaction at approximately 100°C and 1 atm (Fig. 4.5),

$$2H_2 + O_2 \rightarrow 2H_2O(\text{vapor})$$

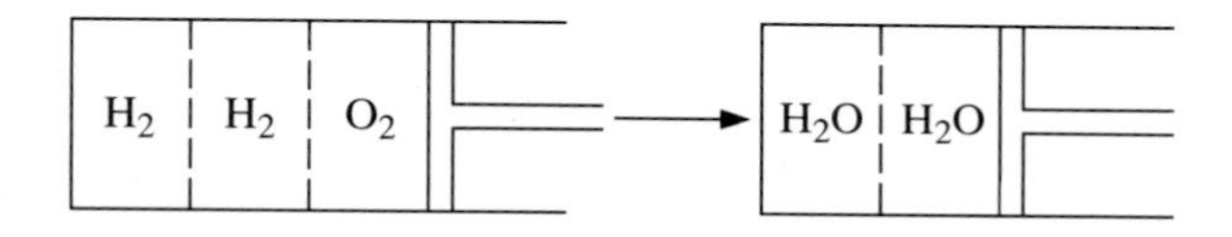

FIGURE 4.5 Three volumes of ($2H_2 + O_2$) go to 2 volumes of ($2H_2O$) vapor.

[2] It can be shown (e.g., from the Bridgman relations) that $C_p - C_v = TV\beta^2/\kappa$, which has an average value for solids of about 0.4 cal/mol·K = 1.7 J/mol · K. $\beta = (1/V)(\partial V/\partial T)_P$ and $\kappa = -(1/V)(\partial V/\partial P)_T$. See ref. 26, p. 142.

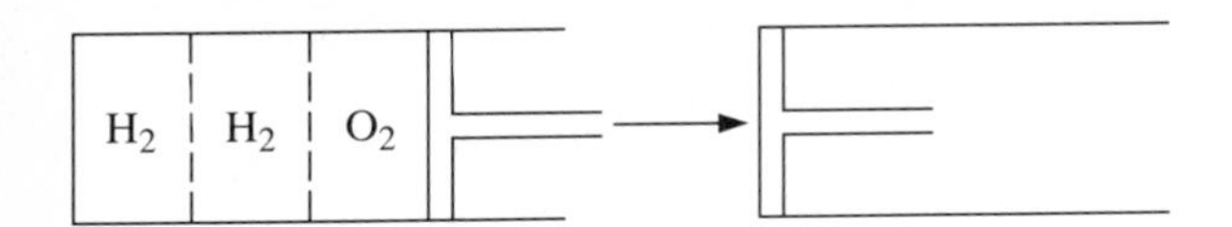

FIGURE 4.6 Three volumes of ($2H_2$ + O_2) give a negligible volume of liquid.

the enthalpy change is therefore −242.5 kJ/mol(H_2O) (from Table 4.1). For the reaction of hydrogen and oxygen to form liquid water, at the same temperature and pressure, ΔH is the difference of the heat of reaction to form the vapor and the heat of vaporization.

$$\Delta H(\text{to form vapor}) = \Delta H(\text{to form liquid}) + \Delta H(\text{vaporization})$$

so

$$\begin{aligned}\Delta H(\text{to form liquid}) &= -242.5\ \text{kJ/mol(H}_2\text{O)} - 40.7\ \text{kJ/mol} \\ &= -283.2\ \text{kJ/mol(H}_2\text{O)}\end{aligned}$$

as shown in Fig. 4.6.

For an ideal gas, $H = E + PV = E + nRT$, so enthalpy, like energy, depends only on the temperature; $H = H(T)$ for an ideal gas.

4.4 Total Differentials and Process Paths

We often employ C_v or C_p in processes for which volume and/or pressure are not constant, especially for gases. The justification for this is found in an elementary property of differentials.

Let Z be any function of two variables, x and y:

$$Z = Z(x, y)$$

Then the total differential of Z is defined as

$$dZ = \left(\frac{\partial Z}{\partial x}\right)_y dx + \left(\frac{\partial Z}{\partial y}\right)_x dy$$

That is, for sufficiently small changes, the total change in Z is the change due to the change in x plus the change due to the change in y. And the change due to the change in x, for example, is the rate of change of Z with x, multiplied by the actual (very small) change in x.

For larger changes in independent variables, the change in the function, or dependent variable, is not a sum of arbitrary changes due to each independent variable. For example, the dependence of V on P and T, for an ideal gas, is represented in Fig. 4.7. A decrease in pressure, ΔP, followed by an increase in temperature, ΔT, gives a larger volume change than the same increase in temperature followed by the same decrease in pressure. However, in the limit, as $\Delta P \to dP \approx 0$ and $\Delta T \to dT \approx 0$, the change in volume, $dV \approx 0$, becomes independent of the order of change in P and T. The path differences depend only to second order on ΔP and ΔT.

The energy is, in general, a function of (at least) two state functions. It is often appropriate to choose the variables as volume and temperature; $E = E(V, T)$.

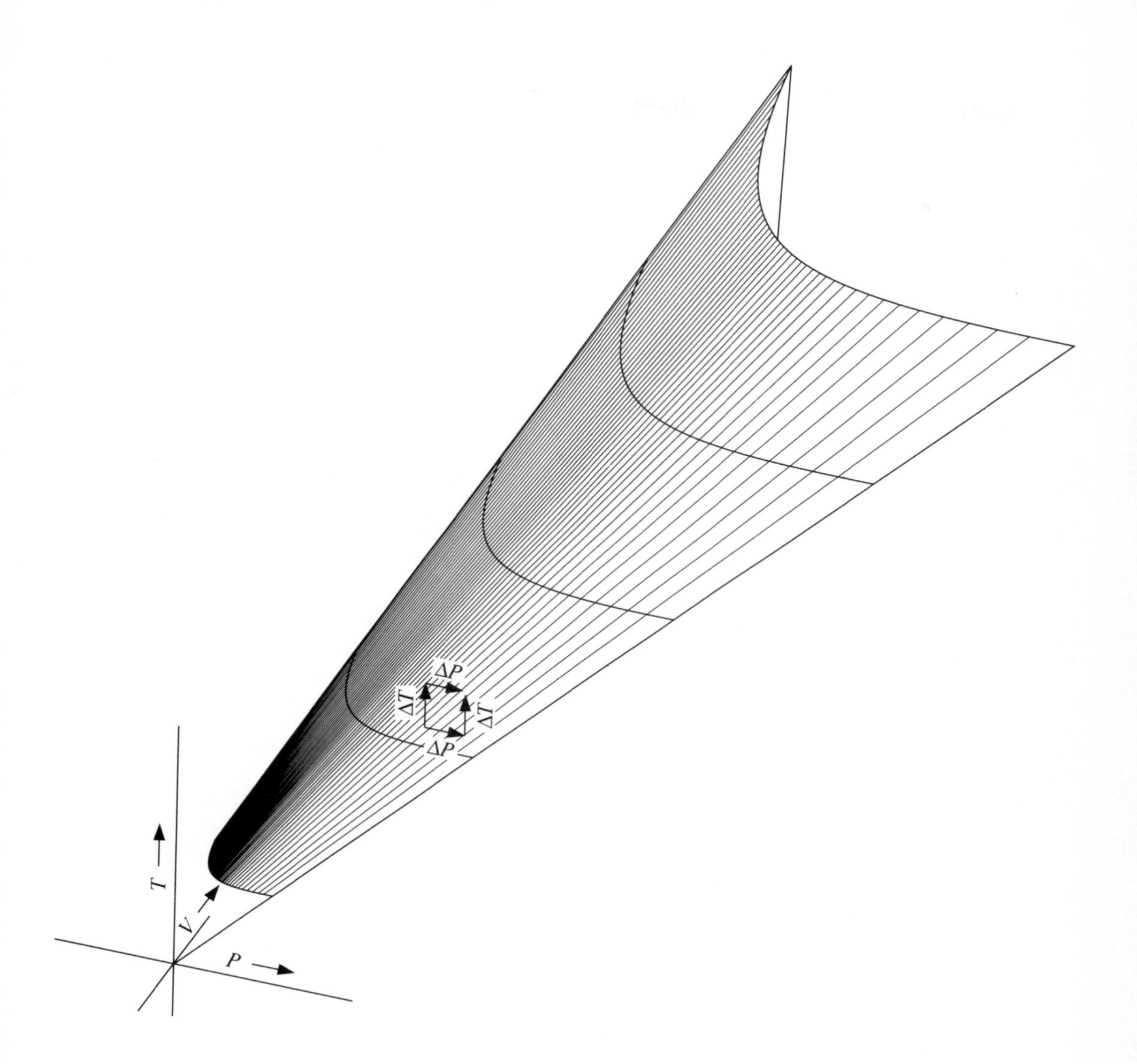

FIGURE 4.7 $\Delta V \neq (\partial V/\partial T)_P \, \Delta T + (\partial V/\partial P)_T \, \Delta P$, but $dV = (\partial V/\partial T)_P \, dT + (\partial V/\partial P)_T \, dP$, independent of the order of steps.

Then

$$dE = \left(\frac{\partial E}{\partial V}\right)_T dV + \left(\frac{\partial E}{\partial T}\right)_V dT$$

The last term may be recognized as $C_v \, dT$. For an ideal gas, the first term is zero, because the energy of an ideal gas is independent of volume. Therefore, for an ideal gas,

{ideal gas} $$dE = C_v \, dT \tag{4-7}$$

for any process, whether it occurs at constant volume or not. Similarly, writing enthalpy as a function of pressure and temperature,

$$dH = \left(\frac{\partial H}{\partial P}\right)_T dP + \left(\frac{\partial H}{\partial T}\right)_P dT$$

The last term is $C_p\, dT$, and the first term is zero because enthalpy is independent of pressure for an ideal gas. Therefore,

{ideal gas} $$dH = C_p\, dT \tag{4-8}$$

for the ideal gas, regardless of changes in pressure or volume.

For any ideal gas,

{ideal gas} $$dE = C_v\, dT \qquad \text{and} \qquad dH = C_p\, dT$$

whether or not volume or pressure are held constant.

This does *not* mean that $dE = q$ or that $dH = q$ for an ideal gas regardless of conditions. Only if $w = 0$ is $q = dE = C_v\, dT$ and only if $w = -d(PV)$ is $q = C_p\, dT$. A constant-volume restriction usually satisfies the first condition; a constant-pressure restriction usually satisfies the second condition.

EXAMPLES 4.4

A. Find Q, W, ΔE, and ΔH when 2 mol of hydrogen gas is warmed at constant volume from 25°C to 50°C.
B. Find Q, W, ΔE, and ΔH when 2 mol of hydrogen gas is warmed at constant pressure from 25°C to 50°C. □

Calculations of enthalpy changes due to changes in temperature are essentially identical to the calorimetric calculations discussed previously, because most changes occur at constant pressure, for which $\Delta H = Q$. To find ΔH for warming 10 g of water from 15°C to vapor at 120°C, we write

$$\Delta H = Q = 10\text{ g} \times [4.2\text{ J/g}\cdot\text{K} \times (100 - 15)\text{ K} + 2.26\text{ kJ/g} + 2.1\text{ J/g}\cdot\text{K} \times (120 - 100)\text{ K}] = 26.6\text{ kJ}$$

Values of heats of fusion, heats of vaporization, and heats of reaction in tables are typically labeled as values of ΔH_{fusion}, $\Delta H_{\text{vaporization}}$, and $\Delta H_{\text{reaction}}$. (The significant exception is the "heat of combustion," which is taken as $-\Delta H_{\text{combustion}}$ so that it will be positive.)

A *path* is the set of states that a system goes through between the initial state and the final state. The path therefore reflects the conditions under which a process is carried out. If a system undergoes an arbitrary series of changes, then returns to the original state, every property, or state function, of that system must return to the original value. Therefore, the integral of the change in any state function, around *any* closed path, must be zero. This is true for such functions, already considered, as energy, enthalpy, pressure, volume, temperature, C_v, and C_p.

State functions depend only on the state of the system. Therefore, the change in such a function depends only on the initial and final states, not on the path taken. For example,

$$\Delta E = E_2 - E_1 = \int dE$$

Integrands for which the integral depends only on the endpoints are called *exact differentials*, or *total differentials*. Some familiar examples of exact differentials are

$$\int x\, dx = \frac{1}{2}x_2^2 - \frac{1}{2}x_1^2$$

and

$$\int (x\, dy + y\, dx) = \int d(xy) = x_2y_2 - x_1y_1$$

As you know, the way to evaluate an integral (analytically) is to recognize the integrand as being an exact differential, or to manipulate the integral (e.g., integrating by parts or making substitutions) until the terms can be so recognized.

EXAMPLES 4.5

A. Can you recognize the integrand and thus evaluate the following integral? (Be sure to test your result by differentiation.)

$$\int (x\, dy - y\, dx)$$

B. Can you evaluate the integral $\int y\, dx$? □

The infinitesimal change in any function $\mathcal{F}(x, y)$ can be written as a sum of the change in the function caused by the change in x and the change in the function caused by the change in y. That is,

$$d\mathcal{F} = \left(\frac{\partial \mathcal{F}}{\partial x}\right)_y dx + \left(\frac{\partial \mathcal{F}}{\partial y}\right)_x dy$$

This is the *total differential*, or *exact differential*. (It is *exactly* the differential of $\mathcal{F}$ and it is the *total* differential change in $\mathcal{F}$, arising from changes in all variables.)

The mathematical test of an exact differential requires that the partial derivative, with respect to the second variable, of the coefficient of the differential of the first variable must be equal to the partial derivative, with respect to the first variable, of the coefficient of the differential of the second variable. That is,

$$\frac{\partial}{\partial y}\left(\frac{\partial \mathcal{F}}{\partial x}\right) = \frac{\partial^2 \mathcal{F}}{\partial y\, \partial x} = \frac{\partial^2 \mathcal{F}}{\partial x\, \partial y} = \frac{\partial}{\partial x}\left(\frac{\partial \mathcal{F}}{\partial y}\right)$$

because the order of differentiation does not change the value of a second derivative. Or, more simply, if

ANSWERS 4.4

A. The work done at constant volume is zero, so $\Delta E = Q = C_v \, \Delta T = 2 \text{ mol} \times 21 \text{ J/mol} \cdot \text{K} \times 25 \text{ deg} = 1.0 \text{ kJ}$ and $\Delta H = C_p \, \Delta T = 2 \text{ mol} \times 29 \; J/\text{mol} \cdot K \times 25 \; K = 1.5 \text{ kJ}$. Note that $\Delta H \neq Q$ at constant volume. The enthalpy change is greater than the energy change, because $\Delta H = \Delta E + \Delta(PV)$, and the increase in temperature causes an increase in P, and hence in the product PV.
B. At constant pressure, $\Delta E = C_v \, \Delta T = 1.0 \text{ kJ}$ and $\Delta H = Q = C_p \, \Delta T = 1.5$ kJ. Note that $\Delta E \neq Q$ at constant pressure.

ANSWERS 4.5

A. The integrand is not an exact differential, so there is no way of carrying out the integration as indicated. It would be necessary to have additional information about how y varies with x.
B. Again, because the integrand is not an exact differential, the integral cannot be evaluated without knowing the function $y = y(x)$.

$$d\mathscr{F} = \mathscr{F}_1 \, dx + \mathscr{F}_2 \, dy$$

then

$$\left(\frac{\partial \mathscr{F}_1}{\partial y}\right)_x = \left(\frac{\partial \mathscr{F}_2}{\partial x}\right)_y \tag{4-9}$$

EXAMPLES 4.6

A. Apply this test to the integrand $x \; dy - y \; dx$, in Example 4.5A.
B. Apply the test to the integrand $y \; dx$ in Example 4.5B. □

When the integrand is not an exact differential, it is necessary to supply additional information about how one variable changes with changes in the other. This is equivalent to specifying a function, or specifying a line, on a graph, along which the integral will be evaluated. To integrate $\int y \; dx$, it is necessary to specify the function, or line, $y = y(x)$, along which to integrate. Such integrals are therefore called *line integrals*.

4.5 State Functions and Alternative Paths

Whereas energy and enthalpy are state functions, so dE and dH are always exact differentials, or total differentials, the expression for work is $-\int P \; dV$, which is not the integral of an exact differential. Thus, in general, the work depends on the path and can only be calculated when a path has been specified. That is, the path is represented mathematically by a line on a graph giving P and V at each point during the process. Only when we know P as a function of V can we integrate $\int P \; dV$. Similarly, we can only find the thermal energy transfer, Q, when we know the path taken between initial and final states. Only the sum, $Q + W$, is independent of the path.

Kirchhoff's Law

Because energy and enthalpy are state functions, the changes depend only on the endpoints. Changes in these functions can be calculated by alternate paths, chosen for convenience. Consider, for example, the following problem. The heat of vaporization of water at 100°C has been measured to be 2.26 kJ/g. What is the heat of vaporization of water at room temperature, say 25°C?

An analogy here would be that if you want to cross the river, and there is no bridge close at hand, it will probably pay to go to the bridge, cross there, and come back on the opposite side (Fig. 4.8). Although there is no simple way to calculate the heat of vaporization at 25°C directly, ΔH can be calculated for the process of warming the water to 100°C, vaporizing the water at 100°C, and cooling the vapor back to 25°C. For 1 g of water,

$$\begin{aligned}\Delta H &= 1\text{ g} \times [4.2\text{ J/g}\cdot\text{K} \times (100 - 25)\text{ K} \\ &\quad + 2.26\text{ kJ/g} + 2.1\text{ J/g}\cdot\text{K} \times (25 - 100)\text{ K}] \\ &= 2.42\text{ kJ/g for vaporization at 25°C}\end{aligned}$$

EXAMPLE 4.7

Find the heat of fusion, ΔH_{fusion}, for ice at −20°C. (This is $-\Delta H$ for the spontaneous process of liquid freezing to ice at −20°C.) □

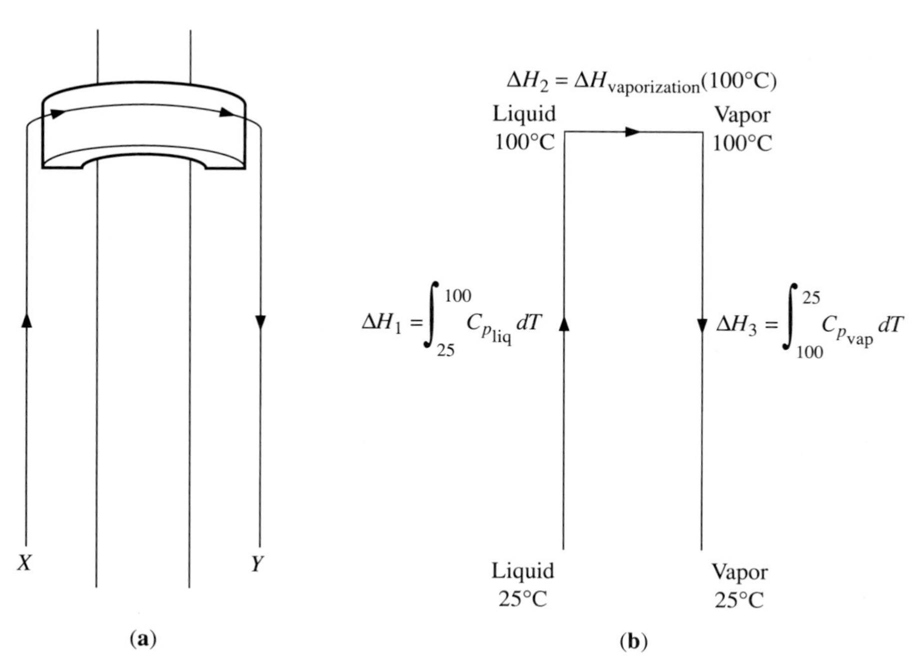

FIGURE 4.8 (a) The easiest way to get across the river from X to Y is usually to go to the bridge, cross the river, and return on the opposite side. (b) $\Delta H_{\text{vap}}(25°\text{C}) = \Delta H_1 + \Delta H_2 + \Delta H_3 = \int_{25}^{100} C_P(\text{liquid})\, dT + \Delta H_{\text{vap}}(100°\text{C}) + \int_{100}^{25} C_P(\text{vapor})\, dT$.

ANSWERS 4.6

A. $(\partial x/\partial x)_y = 1$ and $(\partial(-y)/\partial y)_x = -1$. These are not equal. (Note that for the integrand $x\ dy + y\ dx$ the partial derivatives are both equal to $+1$, and it is possible to write down the integral.)
B. $(\partial y/\partial y)_x = 1$, but the coefficient of $dy = 0$, so the partial derivative is zero. The two partial derivatives are not equal, showing that the expression is not an exact differential.

ANSWER 4.7

The steps are warming ice to 0°C, melting the ice, and cooling the water to −20°C. $\Delta H(-20) = 2.1\ \text{J/g} \cdot \text{K} \times 20\ \text{K} + 333\ \text{J/g} + 4.2\ \text{J/g} \cdot \text{K} \times (-20\ \text{K}) = 290\ \text{J/g}$.

The argument employed here is known as *Kirchhoff's law*. It says that the change in any state function is independent of path, and therefore the change is zero around any complete cycle.[3]

Hess's Law

A very similar principle, related to chemical reactions, is known as *Hess's law*. It allows us to determine heats of reaction and changes of other thermodynamic functions by adding together the effects of other chemical reactions.

Methane, CH_4, is the smallest of the hydrocarbons. Methane is produced by organic matter as it decays, so it is known as marsh gas. It is also a major component of natural gas and may be produced from water and coal as a component of "water gas," or artificial gas.

$$2H_2O + 3C \rightarrow CH_4 + 2CO$$

The energetics of the formation of methane,

$$C + 2H_2 \rightarrow CH_4$$

are important, but the reaction cannot be carried out in this simple way to obtain the direct measure of the heat of reaction. On the other hand, carbon, hydrogen, and methane will all burn.

$$C + O_2 \rightarrow CO_2 \qquad \Delta H_1$$

$$2H_2 + O_2 \rightarrow 2H_2O \qquad \Delta H_2$$

$$CH_4 + 2O_2 \rightarrow CO_2 + 2H_2O \qquad \Delta H_3$$

The sum of the first two reactions is

$$C + 2H_2 + 2O_2 \rightarrow CO_2 + 2H_2O \qquad \Delta H = \Delta H_1 + \Delta H_2$$

The negative of the third reaction is

$$CO_2 + 2H_2O \rightarrow CH_4 + 2O_2 \qquad \Delta H = -\Delta H_3$$

[3]Kirchhoff's name is also applied to a radiation/absorption law and to the two basic laws of electrical circuits, involving voltage and current. One of these, known also as the loop theorem, is closely related to the concept considered here; it says that the sum of the changes in potential is zero for any complete circuit loop.

Adding the last two equations (i.e., 1 + 2 − 3) gives

$$C + 2H_2 + (2O_2 + CO_2 + 2H_2O) \rightarrow$$

$$CH_4 + (2O_2 + CO_2 + 2H_2O) \qquad \Delta H = \Delta H_1 + \Delta H_2 - \Delta H_3$$

The identical terms on the two sides, shown in parentheses, may be dropped, showing that the heat of reaction for the formation of methane from carbon and hydrogen is $\Delta H_1 + \Delta H_2 - \Delta H_3$.

$$C + 2H_2 \rightarrow CH_4 \qquad \Delta H = \Delta H_1 + \Delta H_2 - \Delta H_3$$

EXAMPLES 4.8

A. Aragonite and calcite are two different forms of calcium carbonate, $CaCO_3$, found in nature. Most animals produce the aragonite form; chemical deposition produces calcite. They behave quite similarly chemically (e.g., each dissolves in hydrochloric acid, HCl, to give $CaCl_2$ and CO_2) but they have different crystal structures. Neither form converts readily to the other. How could you find $\Delta H_{\text{reaction}}$ for the process of converting aragonite to calcite?

B. Conversion of graphite to diamond has become commercially feasible, producing industrial grade diamonds required for cutting and grinding. In developing such a process, it is important to know the energetics of the reaction

$$C_{\text{graphite}} \rightarrow C_{\text{diamond}}$$

but the reaction does not go under conditions susceptible to measurement. How could you measure ΔH for this reaction? □

SUMMARY

Heat capacity $= C = Q/\Delta T$ or q/dT.
Define $H \equiv E + PV$.
Define $C_v = (\partial E/\partial T)_V$ and $C_p = (\partial H/\partial T)_P$.
At constant pressure (and $w = -P\,dV$), $q = dH$.
For chemical reactions, H^0 is tabulated (per mole, in defined state) at 25°C. Then heat of reaction = thermal energy absorbed $= Q = \Delta H$ for reactions at constant pressure, and $\Delta E = Q - P\Delta V$.
For phase changes, $Q_{\text{phase change}} = \Delta H_{\text{phase change}}$.
Mathematical properties: $Z = Z(u, v, w, \ldots) = Z(x_i)$.

$$dZ = \sum_i \frac{\partial Z}{\partial x_i}\,dx_i \qquad \text{and} \qquad \frac{\partial}{\partial x_j}\left(\frac{\partial Z}{\partial x_i}\right) = \frac{\partial}{\partial x_i}\left(\frac{\partial Z}{\partial x_j}\right)$$

dZ is called the *total*, or *exact*, *differential* of Z. $\int X_i\,dx_i$ is independent of path (dependent only on end points) if and only if $X_i\,dx_i$ is a total differential (differential of a state function). Note that q and w are not differentials, although under specified conditions they may be equal to differentials (e.g., $q = dE$ or $q = dH$).

Kirchhoff's law: Because the change in any state function is independent of path, changes may be calculated by any convenient path.

ANSWERS 4.8

A. Dissolve aragonite in HCl and measure ΔH_1. Dissolve calcite in HCl and measure ΔH_2. Then, because the products are the same, $\Delta H_1 - \Delta H_2$ is ΔH for the conversion of aragonite to calcite.
B. Burn graphite and diamond, separately, in oxygen and measure the heat of combustion. Then ΔH for graphite to diamond is the difference of the heats of reaction of graphite and diamond with oxygen.

[**Note:** "Heat of combustion" = $-\Delta H$ of combustion, by convention, making the heat of combustion a positive quantity.]

Hess's law: Because changes in state functions depend only on end points, changes in state functions may be calculated by adding and subtracting chemical reactions, and the corresponding changes in state functions (e.g., ΔE or ΔH).

QUESTIONS

4.1. Give alternative symbols for the following: $(\partial E/\partial T)_V$ and $(\partial H/\partial T)_P$.
4.2. Under what special conditions is $Q = \Delta E$? Why does $\Delta V = 0$ not necessarily ensure that this condition is met?
4.3. Under what special conditions is $Q = \Delta H$? Why is it necessary to impose more than one restriction?
4.4. The definition of enthalpy is $H = E + PV$. Is it always, sometimes, or never true that $\Delta H = \Delta E + P\Delta V + V\Delta P$? Explain.
4.5. The integral $\int P\, dV$ is an example of a *line integral*. Does that mean that $W = \int dW$ and dW is an inexact differential? Explain.
4.6. In finding the work done by hydrogen + oxygen as they react to form water, the expression $W = -\Delta nRT$ was obtained. What is the meaning of Δn, and why does it appear in the expression for work? Why does it have a value that depends on the final state of the system?
4.7. Justify the terminology "exact differential" = "total differential." Could the terminology, and a similar test, be extended to the differential of $\mathscr{F}(x, y, z)$? If $d\mathscr{F} = A\,dx + B\,dy + C\,dz$, what conditions must exist linking A, B, and C?
4.8. Occasionally, one encounters the statement that the first law of thermodynamics is equivalent to the condition that $\oint dE = 0$; that is, the integral of dE about any closed loop vanishes. Is this a valid statement of the conservation law? Explain.
4.9. Show that the Bridgman relations give the expression for $C_p - C_v$ given in the footnote on page 63.

PROBLEMS

4.1. 4.0 mol of He expands reversibly at 27°C from 1 m^3 to 3 m^3. Find
a. Q. **b.** W. **c.** ΔE. **d.** ΔH.
4.2. 2.0 mol of O_2 at 27°C expands reversibly against a constant pressure of 300 kPa from 0.0166 m^3 to 0.0222 m^3. Find
a. Q. **b.** W. **c.** ΔE. **d.** ΔH.

4.3. 3.0 mol of N_2 is expanded into a vacuum (which does not change its temperature) from 0.050 m^3 to 0.100 m^3 at 27°C. Find

a. Q. **b.** W. **c.** ΔE. **d.** ΔH.

4.4. 2.5 mol of N_2 at 27°C is warmed at constant pressure from 0.50 m^3 to 0.75 m^3. Find

a. Q. **b.** W. **c.** ΔE. **d.** ΔH.

4.5. 0.75 mol He is warmed from 0 to 100°C at a constant pressure of 1 atm. Find

a. Q. **b.** W. **c.** ΔE. **d.** ΔH.

4.6. For water at 60°C, find

a. ΔE_{vap}. **b.** ΔH_{vap}.

4.7. The heat of vaporization of ethanol is 204 cal/g (at 78.3°C). Assuming average values of 0.40 cal/g · K (liquid) and 0.70 cal/g · K (vapor) for specific heats, find ΔH_{vap} of ethanol at 25°C.

4.8. For the conversion of liquid water to ice at −30°C (neglecting changes of density with temperature), find

a. ΔE. **b.** ΔH.

4.9. One-tenth mole of an ideal gas at 1 atm and 27°C is expanded against a constant pressure of $\frac{1}{2}$ atm at constant temperature. The temperature is raised to 35°C, then the gas is compressed to a volume of 1 L at that temperature. It is then expanded reversibly and isothermally to a pressure of 0.8 atm, warmed to 40°C, and expanded into a vacuum at constant temperature to a final pressure of 0.5 atm. For the overall process, find ΔV.

4.10. 200 g of ice at −15°C is added to 1000 g of water at 35°C in a suitably insulated container. Taking the system as ice + water, find

a. The final temperature. **b.** ΔH.

4.11. Work is defined as the integral of force times displacement of the point of action of the force, $W = \int \mathbf{f}_i \cdot d\mathbf{x}_i$, which for a gas expansion can be written as (minus) the pressure on a piston times the volume displacement of the piston, $W = -\int P dV$.

Justify the inequality for an expansion,

$$-w_{\text{exp}} \leq P dV ,$$

where P is interpreted as the pressure measured for the gas as it expands. For an ideal gas, $P = nRT/V$.

4.12. Describe carefully a system and surroundings (of your choice) for which the pressure of the system is not equal to the external pressure. For your system, answer the following.

a. Is there a meaningful single definition of an external pressure?

b. Does the external pressure affect in any way the system or changes in the system? If so, does the external pressure determine the effect, or is there another parameter (e.g., stiffness of a wall or mass of a piston) that mediates the effect of the external pressure on the system?

CHAPTER 5

Mixed Substances and Chemical Reactions

Properties of materials often depend strongly on the presence of impurities and on the distribution of the impurities within the host substance. Trace elements change copper from soft and malleable to hard and rigid. The properties of iron vary markedly with carbon content, depending also on the form in which the carbon is present.

Two or more materials brought together may react chemically to form one or more new substances. Gasoline and air combine, when ignited, to form carbon dioxide and water; silicon exposed to oxygen forms a surface layer of SiO_2; sulfuric acid reacts with sugar or fabrics; ammonia vapor combines with hydrochloric acid vapor to form a white powder, ammonium chloride.

If the distinct substances brought together retain their separate physical and chemical properties, with negligible interaction except at the boundaries between regions of one substance and regions of the other, the combination is called a *mixture*. Familiar examples are sugar and salt crystals, or oil and water.

Very often, however, materials mix thoroughly, so that each loses its bulk properties but retains its principal chemical properties. Silver dissolves in gold, salt dissolves in water, water and alcohol mix completely, and all gases dissolve in each other. In these cases the mixing is at the molecular level and, unless conditions are changed, the materials will never separate again. Such *homogeneous mixtures* are called *solutions*. They may be gas in gas, gas in liquid, liquid in liquid, solid in liquid, solid in solid, or even liquid in solid, liquid in gas, solid in gas, or gas in solid.

> A solution *is a mixture that is locally homogeneous to approximately the molecular level.*

5.1 Solutions

The major component of a solution is usually called the *solvent* and other components are called *solutes*. In seawater, the water is the solvent and the sodium chloride and other impurities are solutes. However, this is a convenience, rather than a fundamental distinction. From time to time, we alter our point of view as to which component of a solution is to be called *solvent* and which is to be called *solute*.

> *The* solvent *(usually the major component) will be designated by the subscript* 1*; the* solutes *(any other components) by subscripts* 2, 3, *and so on.*

The relative amounts, or concentrations, of substances in a solution may be expressed in different ways, such as grams of solute per liter of solution or per liter of solvent, or grams of solute per kilogram of solution or per kilogram of solvent. However, the number of molecules present, typically expressed as the number of moles, is usually the most convenient measure of the amount of each substance present.

> *The number of moles of the* ith *substance,* n_i*, is called the* mole number.

The number of particles is Avogadro's number (6×10^{23}) times the mole number.

Three concentration measures based on the mole as the quantity of solute are mole fraction, molarity, and molality.

Mole Fraction

The fraction of the total number of molecules represented by the ith component is called the *mole fraction*, represented by X_i or N_i.

> *The* mole fraction, X_i*, is the number of moles of the* ith *substance divided by the total number of moles.*

$$X_i = \frac{n_i}{n_1 + n_2 + n_3 + \cdots} \qquad (5\text{-}1)$$

The sum of the mole fractions is necessarily equal to 1.

$$\sum_i X_i = \sum_i \frac{n_i}{n_1 + n_2 + n_3 + \cdots} = 1$$

Molarity

The most common measure of concentration is *molarity*, represented by the symbol M.

> *The* molarity *of a solution is the number of moles of solute per liter of solution.*

For example, if 2.92 g of NaCl is placed in a volumetric flask and water is added to bring the volume of the solution to 100 cm^3, the number of moles of NaCl is

$$n_2 = \frac{2.92 \text{ g}}{58.44 \text{ g/mol}} = 0.0500 \text{ mol}$$

and the concentration is

$$C_2 = \frac{0.0500 \text{ mol}}{0.100 \text{ L}} = 0.500\ M \text{ NaCl}$$

Molarity has two disadvantages for studies of solution properties. The amount of solvent is not known, and the molarity changes if the temperature is changed, because of expansion or contraction of the solution.

Molality

For precise work on properties of solutions, or studies of solutions over a range of temperatures, the molality is often a better measure of concentration than molarity.

Molality *is the number of moles of solute per kilogram of solvent.*

It is represented by the symbol m. One liter of water (at 4°C) has a mass of 1 kg, so the molarity and molality of dilute aqueous solutions are practically indistinguishable near room temperature. For other solvents, these measures may be quite different.

EXAMPLES 5.1

A. The density of water at 20°C is 0.99823 g/cm^3. The density of a 0.0500 M NaCl solution at the same temperature is 1.01866 g/cm^3. Find the molality of the 0.0500 M NaCl solution at this temperature. (The molar mass of NaCl is 58.44 g.)
B. The density of carbon tetrachloride, CCl_4, at 20°C is 1.595 g/cm^3. Neglecting any change in density or volume, find
 a. The molarity.
 b. The molality.
if 2g of benzene (molar mass = 78.1 g) is added to 100 cm^3 of CCl_4. □

Volume is a familiar and easily visualized property, as well as an important thermodynamic function. We will therefore look first at volume properties, then apply the same methods of mathematical analysis to other state functions.

Volume is an extensive property; half as much solution will have half as great a volume. In general, however, volumes are not additive, except in *ideal solutions*.

Volumes in Ideal Solutions

Any substance forms an ideal solution with itself. One of the properties of an ideal solution is that the total volume of the solution is the sum of volumes of the components.

The volume per mole, or the *molar volume*, or *molal volume*, of the *i*th component,[1] is v_i and n_i is the mole number of the component. The total volume of the solution, V, is then

$$\{\text{ideal solution}\} \qquad V = n_1\mathrm{v}_1 + n_2\mathrm{v}_2 + \cdots = \sum n_i\mathrm{v}_i \qquad (5\text{-}2)$$

[1]The terms *molar* and *molal* carry the same meaning of a value per mole. The choice of one term or the other is a combination of custom and taste, for each application, except for concentration where molarity and molality are distinguishable, as discussed above.

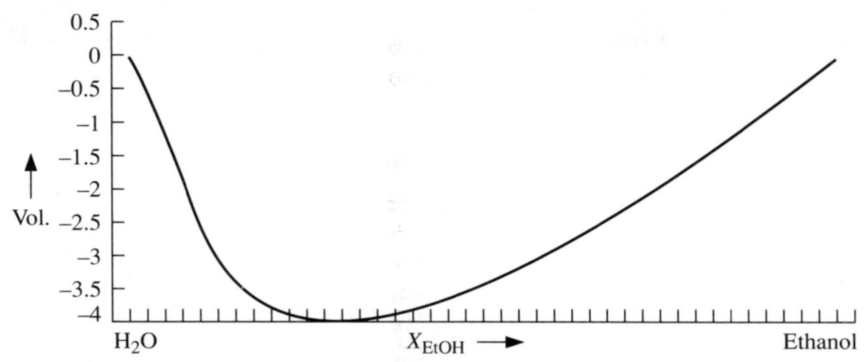

FIGURE 5.1 When the sum of the volumes of water and alcohol is 100 cm^3, the volume of the solution may be less by several cm^3. $n_1 v_1 + n_2 v_2 = 100$ cm^3; $V \leq 100$ cm^3.

Apparent Molar Volume

Solutions of alcohol and water are not ideal. The volume of the solution is not equal to the sum of the volumes of the constituents. The discrepancy, volume of solution minus sum of volumes of pure components, or $V - n_1 v_1 - n_2 v_2$, is plotted against mole fraction of alcohol in Fig. 5.1.

One way of describing the actual solution is to consider that the water retains its own volume and the difference between that volume and the volume of the solution is ascribed to the alcohol. Expressed per mole, this is called the *apparent molar volume* of the alcohol, regarded as the solute.

Let

V = total volume of solution
n_1 = number of moles of solvent
v_1 = molar volume of solvent
n_2 = number of moles of solute

The apparent molar volume, ϕ_2, of the solute is given by

$$V = n_1 v_1 + n_2 \phi_2$$

or

$$\phi_2 = \frac{V - n_1 v_1}{n_2} = \frac{\Delta V}{n_2} \tag{5-3}$$

The advantage of apparent molar volume is that it is simply related to direct experimental observations. When one substance is added to the other, the increase in volume, divided by the number of moles added, is equal to the apparent molar volume (Fig. 5.2).

Apparent molar volumes are almost always positive, even for solutions such as alcohol and water in which the total volume is less than the sum of the volumes of the water and alcohol. However, only if the volumes are additive will the apparent molar volume of the second substance in the solution be equal to the volume per mole of the pure second substance. In fact, negative values are possible.

ANSWERS 5.1

A. 1000 cm^3 of solution = 1018.66 g of solution = (1018.66 − 2.92) g of H_2O = 1015.74 g of H_2O. Therefore, the molality is

$$C_2 = \frac{0.0500 \text{ mol NaCl}}{1.015746 \text{ kg } H_2O} = 0.0492_3\ m$$

B. a. The molarity of benzene (ΦH) is

$$\frac{0.0256 \text{ mol ΦH}}{0.100 \text{ L soln.}} = 0.256\ M$$

b. The molality is

$$\frac{0.0256 \text{ mol ΦH}}{0.1595 \text{ kg } CCl_4} = 0.161\ m$$

If magnesium sulfate, $MgSO_4$, is added to water, the apparent molar volume of the magnesium sulfate is negative for dilute solutions. The magnesium and sulfate ions attract the water molecules very strongly, causing the water structure to collapse around the ions, giving a volume of solution less than the volume of the pure solvent. Magnesium sulfate crystals are a good drying agent because of this attraction.

5.2 Partial Molar Properties and Chemical Potential

The disadvantage of apparent molar volume is that the components of the solution are treated quite unequally. If alcohol is added to water, the water is considered to retain its own volume and the alcohol gets the difference, but if the water,

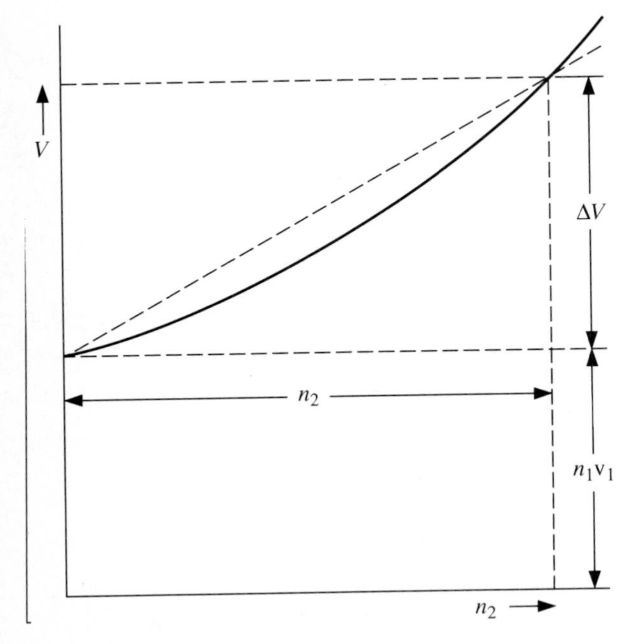

FIGURE 5.2 The apparent molar volume of the solute is the difference between the volume of solution and volume of solvent, divided by the number of moles of solute.

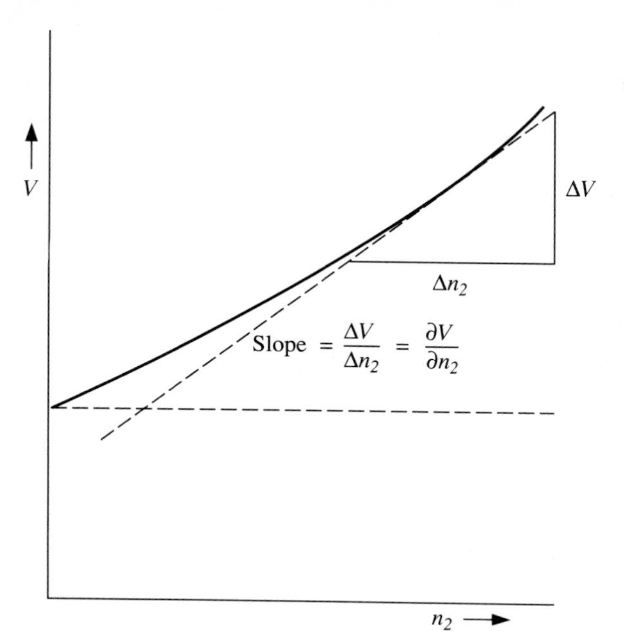

FIGURE 5.3 The slope of the curve of volume of solution versus moles of solute is the partial molar volume of the solute at that concentration.

as solute, is added to alcohol, as solvent, the alcohol has its own molar volume and the water gets the difference. Quite clearly this is not an appropriate way of understanding the properties of either substance in the solution.

A better measure of the effective value of a volume in solution is the partial molar volume (Fig. 5.3), which is independent of labeling choices for solvent and solute.

Partial molar properties are the effective values of the property per mole for the given substance.

Partial Molar Volume

Let V be the total volume of a solution and let $n_i \bar{v}_i$ be the effective contribution of the ith component to the total volume. Then $\bar{v}_i$ is called the *partial molar volume* of the ith component and

$$V = \bar{v}_1 n_1 + \bar{v}_2 n_2 + \bar{v}_3 n_3 + \cdots \tag{5-4}$$

Taking the total differential of V, the total volume of the solution at a given concentration, and hence for a given set of values of $\bar{v}_1$, $\bar{v}_2$, $\bar{v}_3$, and so on, we see that

$$dV = \bar{v}_1\, dn_1 + \bar{v}_2\, dn_2 + \bar{v}_3\, dn_3 + \cdots$$

The total differential of V is

$$dV = \left(\frac{\partial V}{\partial n_1}\right)_{n_2, n_3, \ldots} dn_1 + \left(\frac{\partial V}{\partial n_2}\right)_{n_1, n_3, \ldots} dn_2 + \cdots$$

Comparison of the last two equations shows that the effective volume per mole, or the partial molar volume, of the ith component is the partial derivative of the total volume with respect to the mole number of the *i*th component, or the slope of the curve of volume versus number of moles of solute (Fig. 5.3).

The partial molar volume *of the* ith *component in solution is*

$$\bar{v}_i = \left(\frac{\partial V}{\partial n_i}\right)_{n_j, n_k, \ldots, T, P} \tag{5-5}$$

It is the change in total volume of the solution per mole of the ith *component added, measured at a particular concentration, temperature, and pressure.*

All mole numbers except n_i are held constant, and the partial derivative is taken at constant temperature and pressure. It is a *volume* per *mole* defined by a *partial* derivative and is therefore called the *partial molar volume* or partial molal volume.

For example, suppose that a solution contains 2000 mol of water and 1000 mol of alcohol. If 1 mol of water is added, the total volume of the solution increases. The concentration of the solution is practically unchanged, so it is reasonable to assume that the mole of water added last occupies as great a volume, in the solution, as each of the 2000 moles already there. The change in the total volume, ΔV, therefore gives the volume per mole of water in the solution. Similarly, adding 1 mol of alcohol gives a new V, from which the effective volume per mole of alcohol at that concentration can be found.

In an ideal solution, the partial molar volume of each component is equal to the molar volume of the pure substance.

It is understood that partial molar quantities are evaluated at constant temperature and pressure, and all mole numbers but one are being held constant, so the symbol is sometimes simplified to

$$\bar{v}_i = \left(\frac{\partial V}{\partial n_i}\right)_{n_j}$$

Other Partial Molar Quantities

The technique of defining partial molar quantities as the effective value per mole may be extended to other properties, such as energy, enthalpy, heat capacity, and so on. These are values per mole, a ratio of an extensive property (volume, energy, etc.) to another extensive property (the mole number), and therefore are intensive properties.

The heat capacity at constant pressure of a solution is affected by all substances present in the solution. For the total solution,

$$C_P = \bar{c}_{P_1} n_1 + \bar{c}_{P_2} n_2 + \cdots$$

where

$$\bar{c}_{P_1} = \left(\frac{\partial C_P}{\partial n_1}\right)_{n_2, n_3, \ldots, T, P} \tag{5-6}$$

Similarly, define a partial molar energy such that

$$E = \overline{E}_1 n_1 + \overline{E}_2 n_2 + \cdots$$

and a partial molar enthalpy such that

$$H = \overline{H}_1 n_1 + \overline{H}_2 n_2 + \cdots$$

Note, however, that neither the partial molar energy nor partial molar enthalpy can be determined, as there is no way of knowing the total energy or total enthalpy. Energy and enthalpy *changes* can be measured, however, so we can find the *difference* in enthalpy for a substance in solution and for the same substance in some *standard reference state*.[2]

Heat of Solution

An important practical measure of whether two (or more) substances form an ideal solution is the *heat of solution*. If the solution tends to get warm when the solute is added, it gives off thermal energy in returning to the original temperature, Q is negative, and the heat of solution, ΔH, is negative. Many examples are known of negative heats of solution, such as sulfuric acid added to water.

On the other hand, when added to a solvent, some solutes cause the solution to cool. The solution must absorb thermal energy to maintain constant temperature. The heat of solution is positive. A familiar example of this is the addition of "hypo," sodium hyposulfite, more properly called sodium thiosulfate, $Na_2S_2O_3$, to water. This is the material commonly employed by photographers as a "fixer" solution to stop the process of development. When hypo is added to hot water, the solution becomes noticeably cool.

Heat of solution is related to the partial molar enthalpy, which is the partial derivative of the total enthalpy of the solution, with respect to mole number of the ith component, at constant temperature and pressure. The partial molar enthalpy is

$$\overline{H}_i = \left(\frac{\partial H}{\partial n_i}\right)_{n_j} \tag{5-7}$$

If solute is added to a solution, the heat of solution is

$$\Delta H = H_f - H_i = (\overline{H}_1 n_1 + \overline{H}_2 n_2)_f - (\overline{H}_1 n_1 + \overline{H}_2 n_2)_i$$

Heat of solution is a *change* in enthalpy, the difference between the enthalpy after the addition and the enthalpy before the addition. This difference is experimentally measurable; at constant pressure, $\Delta H = Q$.

Two measures of heat of solution are often encountered. The *integral* heat of solution is ΔH, per mole of solute, between the final solution and the constituents—pure solvent and pure solute. The *differential* heat of solution is ΔH, per mole of solute, when the amount of solute is increased slightly at a given concentration. These two values need not be the same and, in fact, are often of different sign.

[2] See the discussion of standard states below.

Chemical Potential

For reasons that will be explained later, when we consider Legendre transforms, it is particularly convenient to define a quantity almost identical to the partial molar energy, but with variables other than temperature and pressure held constant. With this slight caveat, we write the equation

$$dE = \left(\frac{\partial E}{\partial n_1}\right)_{n_2,n_3,\ldots} dn_1 + \left(\frac{\partial E}{\partial n_2}\right)_{n_1,n_3,\ldots} dn_2 + \cdots$$

As a reminder that we have varied the definition slightly, but even more because of its extreme importance, we then define a new symbol,

$$\mu_i = \left(\frac{\partial E}{\partial n_i}\right)_{n_j}$$

and give it a new name, the *chemical potential*. The total energy apart from an unknown additive constant is the sum

$$E = \mu_1 n_1 + \mu_2 n_2 + \cdots$$

The chemical potential gives us a more general form of the first law equation.

$$dE = q + w + \sum \mu_i \, dn_i$$

The importance of the chemical potential is suggested by the fact that when we similarly define a slightly modified partial molar enthalpy (i.e., keeping other variables than temperature and pressure constant), we find the same μ_i ,

$$\mu_i = \left(\frac{\partial H}{\partial n_i}\right)_{n_j}$$

and

$$H = \mu_1 n_1 + \mu_2 n_2 + \cdots$$

The chemical potential is defined further in Chapter 11 and exploited in later chapters.

5.3 Thermochemistry

We have already seen that it is possible to calculate ΔE or ΔH for a reaction, and that it is possible to add and subtract reactions to find ΔH for the exact reaction desired. Also, if we know ΔH at one temperature, ΔH may be calculated at any other temperature if we know the heat capacities of reactants and products. In this section we expand somewhat on the techniques involved.

Standard States

Energy is not known on an absolute scale.[3] Therefore, it is not possible to know the energy or enthalpy of any reactant or product. Enthalpies and other thermodynamic

[3] Measurement of mass provides the only "absolute" energy value, but the uncertainties in m and resultant uncertainties in $E = mc^2$ are much larger than the energy differences considered here.

potential functions are therefore always given as differences from a specified standard state. Although there are some variations among workers in the field in the choice of states, number of states selected, nomenclature, and symbols, the principles are the same and a single choice will suffice to illustrate the methods.

For each element a standard reference state[4] is chosen, which is then represented by a superscript zero, °, on the thermodynamic properties.

The standard state is generally the most stable, or most common, form of the element, under 1 *atmosphere pressure, at* 25°C.

Similarly, a standard state is chosen for each compound, or in some cases for each phase of a compound that is stable or metastable at the chosen temperature. Then all enthalpies are evaluated relative to the respective standard states.

The standard state of a solvent is taken as the pure solvent.

Pure solute is not easily related to solute in solution, so pure solute is not a satisfactory standard state. However, as solutions become increasingly dilute, the properties of solutes approach a well-defined, experimentally ascertainable limiting value.

The standard state for a solute is usually taken as the solute in an infinitely dilute solution.

Heats of Formation

The heat of reaction, $\Delta H_{\text{reaction}}$, to form any substance in its standard state from the elements in their standard states is called the (standard) *heat of formation*, or standard enthalpy of formation, represented as ΔH_f° . It follows from the definition that:

The enthalpy of formation of any element in its standard state is zero.

Expression of enthalpies relative to the elements illustrates an important principle, applicable to the other thermodynamic functions. (Entropy is the exception, for reasons considered in Section 10.7.) Consider the following (simplified) reaction for removing silicon dioxide by etching.

$$SiO_2 + 4HF \rightarrow SiF_4 + 2H_2O$$

The enthalpy change is

$$\Delta H_{\text{reaction}} = 2H_{H_2O} + H_{SiF_4} - H_{SiO_2} - 4H_{HF}$$

These molar enthalpy values are not known, but the heats of formation from the elements are measurable quantities.

[4]Some authors distinguish between *standard states* and *reference states*. To minimize confusion, we will use only the term *standard state*. Standard states are not determined once and for all. It is sometimes convenient to modify the definition to fit a specific application. Standard states based on the working temperature or on the absolute zero of temperature are discussed in later chapters.

$$O_2 + Si \rightarrow SiO_2 \qquad \Delta H_f(SiO_2) = H_{SiO_2} - H_{O_2} - H_{Si}$$

$$2H_2 + 2F_2 \rightarrow 4HF \qquad 4\,\Delta H_f(HF) = 4H_{HF} - 2H_{H_2} - 2H_{F_2}$$

$$Si + 2F_2 \rightarrow SiF_4 \qquad \Delta H_f(SiF_4) = H_{SiF_4} - H_{Si} - 2H_{F_2}$$

$$2H_2 + O_2 \rightarrow 2H_2O \qquad 2\,\Delta H_f(H_2O) = 2H_{H_2O} - 2H_{H_2} - H_{O_2}$$

By adding and subtracting equal quantities, we may then write

$$\begin{aligned} \Delta H_{\text{reaction}} &= (2H_{H_2O} - 2H_{H_2} - H_{O_2}) + (H_{SiF_4} - H_{Si} - 2H_{F_2}) \\ &\quad - (H_{SiO_2} - H_{O_2} - H_{Si}) - (4H_{HF} - 2H_{H_2} - 2H_{F_2}) \\ &= 2\,\Delta H_f(H_2O) + \Delta H_f(SiF_4) - \Delta H_f(SiO_2) - 4\,\Delta H_f(HF) \end{aligned}$$

The molar enthalpy values of the elements do not appear explicitly in the final expressions. It is convenient, therefore, and represents no loss of generality, to treat the molar enthalpies of the elements as if they were zero. The standard state is usually selected as a stable form of the pure element at 25°C and at atmospheric pressure.

For example, the standard state of sulfur is the common yellow rhombic solid and the standard state of oxygen is the gas at 1 atm and 25°C. From calorimetric measurements it is found that under standard conditions at 25°C,

$$S_{rh} \rightarrow S_{monoclinic} \qquad \Delta H = 0.3 \text{ kJ}$$

and

$$S_{rh} + O_2 \rightarrow SO_2 \qquad \Delta H = -296.9 \text{ kJ}$$

Because ΔH_f° is zero for rhombic sulfur and for oxygen, thermodynamic tables show that ΔH_f° of monoclinic sulfur is 0.3 kJ and ΔH_f° of gaseous SO_2 is -296.9 kJ/mol. Some additional representative values are shown in Table 5.1.

The enthalpy may be determined for an arbitrary temperature, T, relative to the standard state at T° if the heat capacity is known. (Some heat capacities are given in Table 3.2.)

$$\Delta H_T = \Delta H^\circ + \int_{T^\circ}^{T} C_p\, dT \tag{5-8}$$

The enthalpy of a gas is independent of pressure in the ideal gas approximation. The correction for nonideality is calculated from C_p and from the Joule–Thomson coefficient (Section 8.3).

Heats of Reaction

Heats of reaction may be calculated directly from tabulated heats of formation. For example, the standard heat of formation of gaseous SO_3, from Table 5.1, is -395.2 kJ/mol. Apply Hess's law to the pair of reactions

$$S_{rh} + \tfrac{3}{2}O_2 \rightarrow SO_3(g) \qquad \Delta H_f^\circ = -395.2 \text{ kJ} = \Delta H_1$$

$$S_{rh} + O_2 \rightarrow SO_2(g) \qquad \Delta H_f^\circ = -296.9 \text{ kJ} = \Delta H_2$$

Table 5.1 Standard Enthalpies of Formation[a]

Compound	State	ΔH_f°	Compound	State	ΔH_f°
AgBr	c	−99.50	HCl	g	−92.30
AgCl	c	−127.03		aq	−167.46
AgI	c	−62.38	HI	g	26.
Ag_2O	c	−30.57		aq	−55.94
Al_2O_3	c	−1675.3	H_2O	g	−241.83
Br_2	g	30.9		liq	−285.84
C	diamond	1.90	H_2S	g	−20.15
	g	−922.6	I_2	g	62.3
CF_4	g	−679.1	NH_3	g	−46.19
CCl_4	liq	−139.		aq	−80.83
CO	g	−110.53	NH_4Cl	c	−315.4
CO_2	g	−393.51	N_2O	g	82.05
	aq	−412.92	NO	g	90.29
$CaCO_3$	calcite	−1206.87	NO_2	g	33.
	aragonite	−1207.04	N_2O_4	g	9.08
$CaCl_2$	c	−795.0	NaCl	c	−411.1
	aq	−877.88		aq	−407.11
CaF_2	c	−1214.6	NaOH	c	−427.77
	aq	−1201.18		aq	−469.60
CaO	c	−635.5	SO_2	g	−296.9
$Ca(OH)_2$	c	−986.6	SO_3	g	−395.2
	aq	−1002.8	C_2H_2	g	226.7
FeO	c	−266.5	C_2H_4	g	52.3
Fe_2O_3	c	−822.2	C_2H_6	g	−84.667
Fe_3O_4	c	−1117.1	C_2H_5OH	g	−235.3
HBr	g	−36.2		liq	−277.63

[a]Values are in kJ/mol for 25°C and 1 atm, from Circular 500, National Bureau of Standards, "Selected Values of Chemical Thermodynamic Properties," or from the JANAF Tables. Standard states for liquids (liq) and crystalline solids (c) are pure materials at 1 atm; for gases (g) the pure gas at 1 atm in the "ideal gas state" (extrapolated from low pressure). "aq" refers to the (hypothetical ideal) 1 *m* solution in water. 1 thermochemical cal ≡ 4.1840 J.

Then, subtracting the second reaction from the first,[5] we have

$$\tfrac{1}{2}O_2 = SO_3(g) - SO_2(g) \qquad \Delta H = \Delta H_1 - \Delta H_2$$

so for

[5]Chemical reactions are often called *chemical equations* because the atoms on the left are equal to (and, in fact, identical to) the atoms on the right. Similarly, when considering the enthalpies or other state functions, it is convenient to think of the chemical reaction as an ordinary mathematical equation, to which materials may be added or subtracted. Adding a substance to both sides adds the enthalpy of that substance to both sides and does not change ΔH.

$$SO_2(g) + \tfrac{1}{2}O_2 \rightarrow SO_3(g)$$

$$\Delta H^\circ_{\text{reaction}} = -395.2 - (-296.9) = -98.3 \text{ kJ/mol}(SO_3)$$

SUMMARY

Definitions: A *solution* is a mixture that is locally homogeneous to approximately the molecular level.

The *mole number*, n_i, is the number of moles of the ith substance.

Mole fraction, X_i (or N_i) $= \dfrac{n_i}{n_1 + n_2 + \cdots}$; $\sum X_i = 1$.

Molarity is the number of moles of solute per liter of solution.

Molality is the number of moles of solute per kilogram of solvent.

Apparent molar property is the difference in the property caused by adding the solute, calculated per mole of solute; for example,

$$\phi_2 = \frac{V - n_1 \text{v}_1}{n_2} = \frac{\text{volume of solution} - \text{volume of solvent}}{\text{number of moles of solute (2)}}$$

Partial molar property is the rate of change of property of a solution with addition of one component; it is the effective value of the property per mole of the added component; for example,

$$\bar{\text{v}}_i = \left(\frac{\partial V}{\partial n_i}\right)_{n_j,T,P} \quad \text{and} \quad V = \sum_i n_i \bar{\text{v}}_i$$

The *chemical potential* of the ith substance $= \mu_i$ gives the dependence of the thermodynamic "potentials" (E, H, and others to be defined) on the mole numbers of the components of a solution.

Thermochemistry: *heat of formation* $= \Delta H_{\text{form}} = \Delta H_{\text{reaction}}$ in the reaction of elements to form a compound. In general,

$$\textit{heat of reaction} = \Delta H_{\text{reaction}} = \sum_{\text{products}} \Delta H_{\text{form}} - \sum_{\text{reactants}} \Delta H_{\text{form}}$$

QUESTIONS

5.1. Define each of the following quantities, distinguishing between them as clearly and carefully as you can. Give the standard symbol for each.

a. Molarity. **b.** Molality.
c. Mole fraction. **d.** Molar volume.
e. Apparent molar volume. **f.** Partial molar volume.

5.2. Volumes are additive for ideal solutions, and $\bar{\text{v}}_i = \text{v}_i$. Would you also expect enthalpies to be additive for ideal solutions? Would $\bar{\text{H}}_i = \text{H}_i$? Would you expect heat capacities to be additive? Would $\bar{\text{C}}_p = \text{C}_p$?

5.3. Which of the following could have negative values?

a. Molar volume. **b.** Apparent molar volume.
c. Partial molar volume. **d.** Total volume.

5.4. Partial molar enthalpy is the *change* in enthalpy of a solution when one component is added. Although H cannot be known, we measure ΔH by holding pressure constant and setting $\Delta H = Q$. Will this enable us to measure $\bar{\text{H}}_i$? If not, why not?

PROBLEMS

5.1. Find ΔH for the following reactions.
a. $2FeO + \frac{1}{2}O_2 \rightarrow Fe_2O_3$.
b. $2Fe_2O_3 + 3C \rightarrow 4Fe + 3CO_2$.

5.2. Find $\Delta H_{\text{reaction}}$ at 150°C for

$$H_2(g, 1 \text{ atm}) + \frac{1}{2}O_2(g, 1 \text{ atm}) \rightarrow H_2O(g, 1 \text{ atm})$$

5.3. When coal or charcoal burns, CO is often produced initially. This gas then reacts with oxygen to form CO_2. Which step releases more thermal energy? Give a quantitative comparison.

5.4. Calcium metal is quite reactive and not generally found outside the laboratory. Quick lime, CaO, and slaked lime, $Ca(OH)_2$, are familiar industrial materials. Slaked lime can also react with CO_2 from the atmosphere to form $CaCO_3$. Compare the thermal energy released in each step for these solid compounds.
a. $Ca \rightarrow CaO$.
b. $CaO \rightarrow Ca(OH)_2$.
c. $Ca(OH)_2 \rightarrow CaCO_3$.

5.5. Find ΔH at 25°C for the reaction

$$H_2S + \tfrac{3}{2}O_2 \rightarrow SO_2 + H_2O(g)$$

5.6. Manganese can be prepared by a "thermite" process,

$$3Mn_3O_4 + 8Al \rightarrow 9Mn + 4Al_2O_3$$

The standard heats of formation of Mn_3O_4 and Al_2O_3 are -1387 and -1675 kJ/mol, respectively. Find the amount of thermal energy given off by the reaction as written, with reactants starting at room temperature and products ending at room temperature.

5.7. If 73.57 g of Al_2Cl_6 is added to water to give 1 L of solution, the density of the solution will be 1.0510 g/cm^3. Find
a. M. **b.** m.
c. % Al_2Cl_6 (by weight).

5.8. If 4.699 g of $MgCl_2$ is added to water to give 1 L of solution, the density of the solution will be 1.0033, compared to the density of water at the same temperature of 0.99936 g/cm^3. For this solution, find
a. M. **b.** m.
c. Weight % $MgCl_2$. **d.** Apparent molar volume, ϕ_2.

5.9. A 2% solution of $MgSO_4$ in water has a density of 1.0186 g/cm^3, compared to 0.998203 for water.
a. What is the concentration of $MgSO_4$ in g/L?
b. What is the apparent molar volume of the $MgSO_4$?

5.10. The volume of certain acetic acid solutions (in 1 kg of water) fits the equation $V = 1002.935 + 51.832\,m + 0.1394\,m^2$ as a function of molality. The density of water is 0.997044 at the same temperature. For a 2.00-m solution, find
a. $\overline{v}_2$ (acetic acid). **b.** $\overline{v}_1$ (water).
c. ϕ_2.

CHAPTER 6

Molecular Processes in Gases and Condensed Phases

Thermodynamics need not be concerned with molecular properties. However, it is the behavior of molecules that determines the experimental values of state functions such as pressure, volume, temperature, and heat capacity. Many thermodynamic properties are more readily understood when there is some insight into molecular processes. Therefore, in this and the following chapter we leave the classical thermodynamic approach temporarily to examine some molecular processes, without yet invoking the full power of statistical thermodynamics.

6.1 Pressure of an Ideal Gas

A gas is conveniently described as having no definite shape and no definite volume, expanding to fill all volume available to it. A liquid has a definite volume but no definite shape. A solid has a definite volume and a definite shape. These descriptions are usually adequate to distinguish between the phases, but there are supercooled liquids (such as glass and tar) that have the properties of solids, and it is often not meaningful to distinguish between liquid and gas phases for fluids under pressure.

The dependence of gas volume on the pressure at a given temperature was investigated in 1662 by Robert Boyle, a contemporary of Newton. The dependence of the product PV on temperature was measured by a series of workers during the eighteenth century, including Charles, who found the proper equation in 1789 but did not publish his results, and Gay–Lussac, who published his findings in 1802. The result that V is proportional to temperature, at constant pressure, is called Gay–Lussac's law or Charles' law. Boyle's law, $PV =$ constant at a given temperature, when combined with Gay–Lussac's law, gives the relationship $PV/T =$ constant.

Charles and others recognized from the beginning that temperatures must be converted from the common room-temperature scales to a scale with zero at $-273°C$. It was not until 1802, however, that Dalton introduced the name *absolute zero* for this temperature.

At the beginning of the nineteenth century, Dalton suggested that all matter consists of indivisible atoms, of elements, that combine to form molecules of compounds. Avogadro subsequently proposed, in 1811, that the same volume, temperature, and pressure of any gas would contain the same number of molecules. Most of Avogadro's contemporaries rejected this idea, but by 1834 Clapeyron had introduced the symbol R for the gas constant and had written a form equivalent to

$$PV = nRT \tag{6-1}$$

known now as the *ideal gas law*.

The temperature scale based on the ideal gas law was called the *absolute temperature scale*, represented by the unit °A. This scale has now been superseded by the scale proposed by Kelvin in the middle of the nineteenth century, which is based on a thermodynamic definition and a single defined temperature (and absolute zero). The unit is called the kelvin, with symbol K (*not* °K), with 1 K= 1°A = 1°C.

From the molecular point of view, a gas is a collection of small particles, undergoing collisions with each other and with the walls. For equilibrium conditions, the energy of translational motion neither increases nor decreases with time. On the average, therefore, the collisions are necessarily elastic, The primary qualities of gases can be derived from this simple model. Refinements are necessary for precise quantitative predictions of properties, especially as the gas approaches the conditions at which it will condense to a liquid.

An *ideal gas* may be defined either as one that follows the ideal gas law, $PV = nRT$, or as a gas that consists of physical point masses undergoing elastic collisions with each other and with the walls. These definitions may be shown to be equivalent, and equivalent to the condition that $(\partial E/\partial V)_T = 0$.

Point masses are specified because it is important that the molecules are very small, compared to the container, so that they neither occupy significant volume nor involve significant time variation in flight because of collisions. Physical point masses are specified so that they can collide, exchanging energy and momentum. Mathematical points have no extension and therefore cannot collide.

Let a gas of N particles be enclosed in a cubic box of sides a. Consider initially only the components of momentum and velocity along one axis, say the x axis. Although the molecules constantly undergo collisions with each other and with the other walls, the p_x component is unchanged, on the average, by such collisions, by the law of conservation of momentum. Therefore, all collisions except those with the end walls can be ignored.

Each time a molecule undergoes a reflection at a wall it experiences a change in momentum component, perpendicular to the wall, of magnitude $\Delta\mathbf{p} = m\mathbf{v}_f - m\mathbf{v}_i = m\mathbf{v}_f - (-m\mathbf{v}_f) = 2m\mathbf{v}_f$. When a molecule with momentum p_{ix} strikes an end wall (at $x = 0$ or $x = a$), the change in momentum is $\Delta p_{ix} = 2m_i v_{ix}$. The time between end-wall collisions is a/v_{ix} . Therefore, the force exerted on the end walls is

$$f_i = \frac{dp_i}{dt} = \frac{2m_i v_{ix}}{a/v_{ix}} = \frac{2m_i v_{ix}^2}{a}$$

Pressure is force per unit area, and the total area of the walls of the box is $6a^2$. Summing over all the molecules and over the three directions (x, y, and z axes), and dividing by the area of the walls, gives the pressure.

$$P = \sum_{i=1}^{N} \frac{2m_i(v_{ix}^2 + v_{iy}^2 + v_{iz}^2)}{6a^2 \cdot a} \tag{6-2}$$

The volume is

$$V = a^3$$

Velocities add by Pythagorean addition; that is,

$$v_{ix}^2 + v_{iy}^2 + v_{iz}^2 = v_i^2$$

The sum over all N molecules is equal to N times the average value, which may be represented with angular brackets, $\langle\ \rangle$.

$$PV = N\left\langle \frac{1}{3}mv^2 \right\rangle = \frac{2}{3}N\left\langle \frac{1}{2}mv^2 \right\rangle \tag{6-3}$$

The product of pressure and volume is equal to two-thirds of the total translational kinetic energy of the molecules.[1]

The ideal gas law is

$$PV = nRT = NkT$$

where R is the gas constant in units of energy/mol · kelvin and k is the same constant in units of energy/kelvin, or energy/molecule · kelvin, known as the Boltzmann constant. The constants are related by

$$N = nN_A \qquad \text{and} \qquad R = N_A k$$

where N_A is Avogadro's number, 6.02×10^{23}. The Boltzmann constant is $k = R/N_A = (8.314/6.02 \times 10^{23})\ \text{J}/K = 1.38 \times 10^{-23}$ J/K.

Combining the two equations for PV gives

$$NkT = \frac{2}{3}N\left\langle \frac{1}{2}mv^2 \right\rangle$$

or

$$\left\langle \frac{1}{2}mv^2 \right\rangle = \frac{3}{2}kT \tag{6-4}$$

The average kinetic energy of the molecules of an ideal gas is equal to $\frac{3}{2}kT$.

The average kinetic energy does not depend on the mass of the molecules. Less massive molecules move faster; more massive molecules move slower. For all molecules the average kinetic energy depends *only* on the temperature.

[1] Although derived for an ideal, classical gas, the equation

$$PV = \frac{2}{3}E$$

will be found to be applicable to quantum statistics as well, provided that the kinetic energy is K.E. $= \sum_{i=1}^{3} p_i^2/2m$.

6.2 Forms of Energy Storage in Free Molecules

The preceding discussion has considered only translational energy of gas molecules.

$$\langle \text{K.E.} \rangle = \left\langle \frac{1}{2} m v^2 \right\rangle = \frac{3}{2} kT$$

Ideal gases, as well as real gases, may have other forms of energy storage, including molecular rotations and vibrations.

Rotations

Rotational energy may be written as a sum of rotations about three *principal* axes (see Section 17.3). The rotational energy then has the form

$$E_{\text{rot}} = \frac{1}{2} I_a \omega_a^2 + \frac{1}{2} I_b \omega_b^2 + \frac{1}{2} I_c \omega_c^2 \tag{6-5}$$

Quantum mechanics restricts the possible values of the angular momenta of small rotors, such as molecules. For a linear molecule,

$$E_{\text{rot}} = \frac{1}{2} I \omega^2 = \frac{(I\omega)^2}{2I} \equiv \frac{\mathscr{L}^2}{2I} \tag{6-6}$$

where $\mathscr{L} = I\omega$ is the angular momentum, limited to discrete values determined by an integer quantum number, J.

$$\mathscr{L} = \sqrt{\text{J}(\text{J}+1)}\hbar \tag{6-7}$$

where $\hbar \equiv h/2\pi$. The same equation is applicable to spherical rotors—molecules such as CH_4 or UF_6 with very high symmetry—or to symmetric rotors (threefold or higher axis of symmetry) to the approximation that rotational energy about the symmetry axis can be ignored. Representing the moment of inertia about a (nonunique) principal axis perpendicular to the symmetry axis by I_b , the rotational energy is

$$E_{\text{rot}} = \frac{\mathscr{L}^2}{2I_b} = \frac{\text{J}(\text{J}+1)\hbar^2}{2I_b} = \frac{\text{J}(\text{J}+1)h^2}{8\pi^2 I_b} \tag{6-8}$$

Spacing of rotational energy levels in terms of the higher value of J (i.e., J$'$, when J$'' \to$ J$'$ = J$''$ + 1) is

$$\Delta E_{\text{rot}} = 2\frac{h}{8\pi^2 c I_b} hc\text{J} \equiv 2Bhc\text{J} \tag{6-9}$$

The constant, B, and corresponding constants for other rotation axes, provide a convenient measure of rotational energy spacings. They are sometimes given as frequencies, in hertz, but more often are expressed as wavenumbers,[2] $\sigma = \nu/c$,

[2]Spectroscopic measurements yield accurate wavelengths, but the energies of interest are proportional to frequencies, which are not as easily measured. The wavenumber, or number of wave oscillations per unit of length, $\sigma = 1/\lambda = \nu/c$, has the advantage of being measurable (as its reciprocal, λ) and being theoretically significant (because it is proportional to $E = h\nu$). The angular wavenumber, $\kappa = 2\pi/\lambda$, is sometimes substituted for the wavenumber.

usually with units of cm^{-1}. Values of rotational constants for some molecules are given in Table 6.1. At room temperature, $hc\sigma = kT$ gives $\sigma \approx 200\ cm^{-1}$.

Vibrations

A diatomic molecule, regarded as a harmonic oscillator, vibrates at a single frequency,

$$\nu = \frac{1}{2\pi}\sqrt{\frac{k}{\mu}} \qquad \frac{1}{\mu} = \frac{1}{m_1} + \frac{1}{m_2} \tag{6-10}$$

The *force constant*, k, is a measure of the "stiffness" of the bond as a spring, and μ is the *reduced mass*. Quantum mechanics restricts the energies of vibrating systems to values determined by an integral quantum number, n.

$$E_{\text{vib}} = \left(n + \frac{1}{2}\right)h\nu_o = \left(n + \frac{1}{2}\right)\hbar\sqrt{\frac{k}{\mu}} \tag{6-11}$$

Adjacent energy levels therefore differ by

$$\Delta E_{\text{vib}} = h\nu_o \tag{6-12}$$

Polyatomic molecules follow similar rules, although the calculation of each of the vibration frequencies [corresponding to the *normal modes*, or independent modes of vibration (Fig. 6.1)] is more complex, involving all the masses and bond force constants and the bond lengths and angles (Section 17.4). Examples of normal mode frequencies (actually wavenumbers, ν/c, in units of cm^{-1}) or ranges are given in Table 6.2.

Each atom of a molecule requires three position coordinates, so a molecule with N atoms requires $3N$ coordinates to specify locations of all its atoms. However, the center of mass requires three position coordinates. The positions of N atoms are given relative to the center of mass by the remaining $3N-3$ coordinates.

For a molecule of low or moderate symmetry, three coordinates are required to describe the orientation of the molecule in space. Subtracting these three rotation coordinates from the $3N - 3$ position coordinates leaves $3N - 6$ coordinates required to describe the "internal" positions of atoms in the molecule. These $3N-6$ coordinates represent the *number of internal degrees of freedom* and therefore the number of modes of vibration of the molecule.

Table 6.1 Values of Molecular Rotational Constants[a]

Molecule	$A = \frac{h}{8\pi^2 cI_a}$	$B = \frac{h}{8\pi^2 cI_b}$	$C = \frac{h}{8\pi^2 cI_c}$
He	$\approx 4.7 \times 10^{12}$	$\approx 4.7 \times 10^{12}$	$\approx 4.7 \times 10^{12}$
N_2	$\approx 2.8 \times 10^{11}$	2.0	2.0
CO_2	$\approx 1.7 \times 10^{11}$	0.39	0.39
H_2O	26.64	14.40	9.16
C_2H_6	2.54	0.66	0.66

[a]All values in cm^{-1}. $I_a \le I_b \le I_c$ *or* a is the unique axis.

Table 6.2 Wavenumber Values for Vibrational Modes

Molecule	Wavenumber, $\sigma = \nu/c$ (cm^{-1})
N_2	2360
CO_2	667, ≈1334, 2350
H_2O	1595, 3652, 3756
C_2H_6	≈275–3000

EXAMPLES 6.1

A. How many vibrational modes does the water molecule have?
B. How many vibrational modes does ethane, C_2H_6, have? □

Linear molecules require only two rotational orientation coordinates, so there are $3N-5$ internal coordinates and $3N-5$ vibrational modes. Monatomic particles have no rotational orientation coordinates and $3N-3=0$ vibrational modes. The apparent lack of coordinates to describe the rotation of a linear molecule about its axis or the rotation of a single atom is discussed more fully below.

EXAMPLES 6.2

A. How many vibrational modes does the linear molecule CO_2 have?
B. How many vibrational modes does acetylene, C_2H_2, a linear tetratomic molecule, have?
C. How many vibrational modes does O_2 have? □

Comparison of the vibrational modes of water and CO_2 will reveal that the motion of the two hydrogen atoms of water out of the plane of the paper (Fig. 6.1)

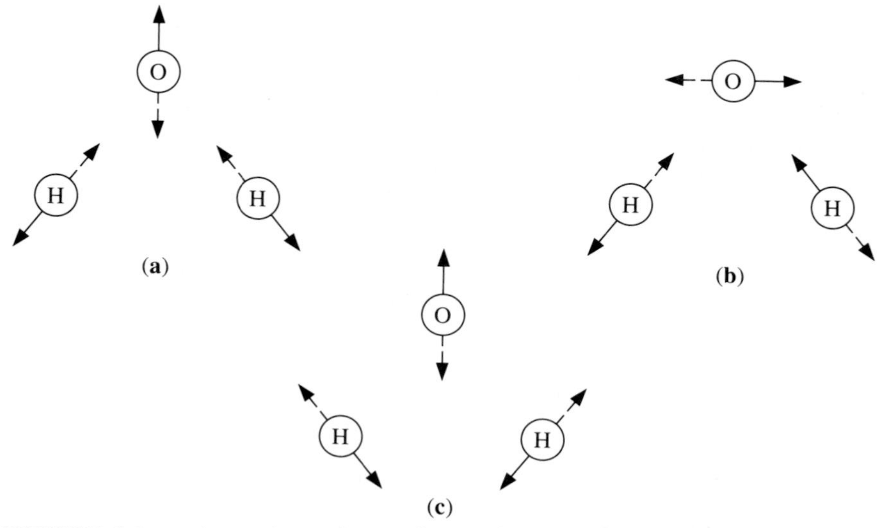

FIGURE 6.1 Approximate forms of normal modes of water. Solid arrows represent one phase of the motion; dashed arrows show the opposite phase.

ANSWERS 6.1

A. For water, $N = 3$ and $3N - 6 = 9 - 6 = 3$ There are three vibrational modes.
B. For ethane, $N = 8$ and $3N - 6 = 24 - 6 = 18$. There are 18 vibrational modes, but a smaller number of distinct frequencies of vibration because of the symmetry of ethane.

ANSWERS 6.2

A. For CO_2, $N = 3$ and $3N - 5 = 9 - 5 = 4$ vibrational modes. Two are vibrations that retain linearity (Fig. 6.2a and b). The other two are bending modes, one in the plane of the paper and an equivalent mode out of the plane of the paper (Fig. 6.2c).
B. For acetylene, $N = 4$ and $3N - 5 = 7$ modes of vibration, not all at different frequencies.
C. A diatomic molecule has only one vibrational mode. $3N - 5 = 6 - 5 = 1$.

would constitute a rotation, whereas the corresponding motion of the two oxygen atoms of CO_2 out of the plane (equivalent to Fig. 6.2c, rotated 90°) constitutes a vibrational mode of CO_2, changing the shape of the linear molecule. It is this conversion of a rotational mode into a vibrational mode that gives linear molecules one more vibrational mode than nonlinear molecules ($3N - 5$ as compared with $3N - 6$ vibrations).

6.3 Equipartition of Energy

The *law of equipartition of energy*,[3] in simplified form, states that if the energy depends on the square of a coordinate, and certain other conditions are satisfied, the average amount of energy stored in that degree of freedom will be $\frac{1}{2}kT$.

The criteria for a coordinate to fit the equipartition law are

1. The degree of freedom, described by the coordinate ξ_i, gives rise to an energy contribution of the form

$$E_i = A_i \xi_i^2$$

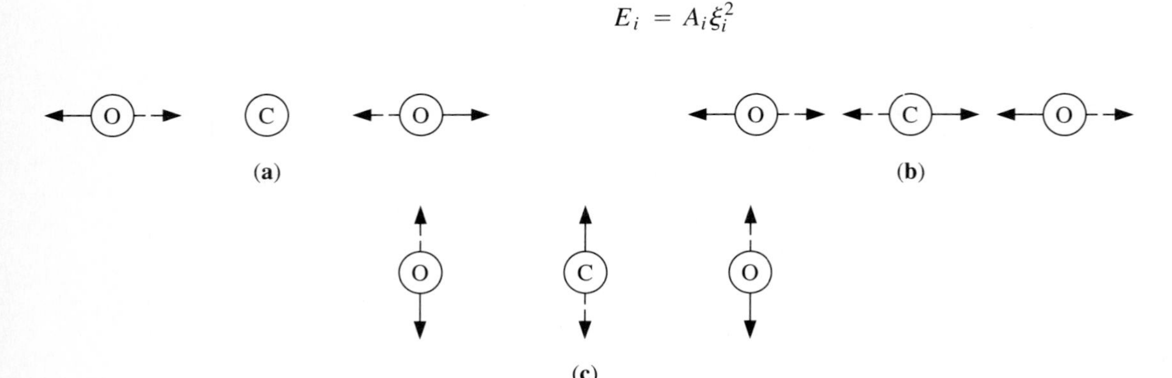

FIGURE 6.2 Normal modes of carbon dioxide. A bending mode equivalent to c would move the atoms out of the plane of the paper.

[3]See also Section 20.1.

2. The energy behaves properly for limiting values of ξ_i (which it will almost always do).
3. The system has enough energy to populate the next-to-lowest energy state.

Call these the *equipartition criteria*. If these criteria are satisfied, then:

The average energy in the degree of freedom will be

$$\langle E_i \rangle = \frac{1}{2}kT \tag{6-13}$$

This is surely one of the most remarkable results of classical statistical mechanics. Gravity is too weak to have a significant effect, so the equivalence in directions is to be expected.

$$\left\langle \frac{1}{2}mv_x^2 \right\rangle = \left\langle \frac{1}{2}mv_y^2 \right\rangle = \left\langle \frac{1}{2}mv_z^2 \right\rangle$$

Also, it is not particularly surprising that

$$\left\langle \frac{1}{2}mv_x^2 \right\rangle = \frac{1}{2}kT$$

for any gas, although intuition might suggest that the average kinetic energy would be different for H_2 than for Hg_2, because of the differences in mass and therefore in speed.

On the other hand, who would have predicted that

$$\left\langle \frac{1}{2}I_a\omega_a^2 \right\rangle = \left\langle \frac{1}{2}I_b\omega_b^2 \right\rangle$$

for the rotational modes of a molecule such as ethane, or that

$$\left\langle \frac{1}{2}I_b\omega_b^2 \right\rangle = \left\langle \frac{1}{2}mv_x^2 \right\rangle$$

for ethane and every other diatomic or polyatomic gas?

An atom within a crystal is constrained to its lattice site by the chemical bonds or lattice forces. It vibrates about this equilibrium location. An arbitrary motion can be resolved into motions along the x, y, and z axes, so the total energy of a particle within a lattice is given classically by

$$E = \frac{1}{2}m_x v_x^2 + \frac{1}{2}m_y v_y^2 + \frac{1}{2}m_z v_z^2 + \frac{1}{2}\mathscr{k}_x \Delta x^2 + \frac{1}{2}\mathscr{k}_y \Delta y^2 + \frac{1}{2}\mathscr{k}_z \Delta z^2$$

where m_x, m_y, and m_z are effective mass values (obtained by solving classical equations of motion for the system) and $\mathscr{k}_x$, $\mathscr{k}_y$, and $\mathscr{k}_z$ are classical harmonic force constants for the motions.

In a liquid, the molecules are neither free, as in an ideal gas, nor bound tightly, as in a perfect crystal. However, regardless of the binding, classical kinetic energy depends on the square of a speed. Therefore the average kinetic energy of the molecules in a liquid must be $\frac{3}{2}kT$, which is the same as the average kinetic energy in a gas.

It follows, then, from the law of equipartition of energy, that when water boils, the molecules have the same average (i.e., rms) speed in the vapor as in the liquid. The energy difference (approximately the heat of vaporization) reflects the difference in potential energy between molecules in the vapor phase and the liquid phase.

Because there is no potential energy contribution to the energy of an ideal gas, the energy of the ideal gas depends only on the temperature. Molecules in a condensed phase have the same average kinetic energy as the gas phase, for the same temperature, but condensed phases have potential energy contributions as well. The potential energy depends on the volume and therefore on the pressure applied to the condensed phase.

6.4 Heat Capacities of Gases

The law of equipartition of energy tells us the total (temperature-dependent) energy of any ideal gas of N molecules is

$$\langle E \rangle = N \sum_{i=1}^{\mathscr{S}} A_i \xi_i^2 = N \sum_{i=1}^{\mathscr{S}} \frac{1}{2} kT = \frac{\mathscr{S}}{2} NkT \qquad \text{(6-14)}$$

where $\mathscr{S}$ is the number of energy modes that satisfy the equipartition criteria. Expressed per mole, we have

$$\langle \mathrm{E} \rangle = \frac{\mathscr{S}}{2} RT$$

Heat capacity, at constant volume, is

$$C_v = \left(\frac{\partial E}{\partial T} \right)_V$$

or

$$\mathrm{C}_v = \left(\frac{\partial \mathrm{E}}{\partial T} \right)_V$$

Therefore,

$$\mathrm{C}_v = \frac{\partial}{\partial T} \left(\frac{\mathscr{S}}{2} RT \right) = \frac{\mathscr{S}}{2} R \qquad \text{(6-15)}$$

Thus if we know $\mathscr{S}$, the number of energy modes that satisfy the equipartition criteria, we will know the heat capacity.

The value of RT, or kT, at room temperature is approximately 8 J/mol · K × 300 K = 2500 J/mol or 4×10^{-21} J/molecule. For purposes of comparison, it is convenient to write

$$kT \equiv hc\sigma_T$$

or

$$\sigma_T \equiv \frac{kT}{hc}$$

Then σ_T is the wavenumber that would correspond to an energy of kT. For $T = 25°C = 298$ K, $\sigma_T = 207$ cm^{-1} $= 2.07 \times 10^4$ m^{-1}.

EXAMPLES 6.3

A. How does kT compare with the spacing of rotational levels, which may be taken to be $2Bhc$ (see Table 6.1):
a. For CO_2, about the axes of larger moment of inertia?
b. For N_2 about its symmetry axis?
B. How does kT at room temperature compare with the spacings of vibrational frequencies, $hc\sigma$, using the wavenumbers from Table 6.2? □

For helium or any other monatomic gas, the *only* energy modes of concern are the translations, $\frac{1}{2}mv_x^2$, $\frac{1}{2}mv_y^2$, and $\frac{1}{2}mv_z^2$. Therefore, $s = 3$ and $c_v = \frac{3}{2}R = 12.5$ J/mol · K. Experimental measurements agree very well with this prediction.

$$\left\{\begin{matrix}\text{any monatomic}\\ \text{ideal gas}\end{matrix}\right\} \qquad c_v = \tfrac{3}{2}R \qquad (6\text{-}16)$$

Diatomic gases, such as H_2, O_2, and N_2, have additional energy modes, corresponding to two rotations and one vibration. Classically, the energy is

$$E = N\left(\frac{1}{2}mv_x^2 + \frac{1}{2}mv_y^2 + \frac{1}{2}mv_z^2 + \frac{1}{2}I_a\omega_a^2 + \frac{1}{2}I_b\omega_b^2 + \frac{1}{2}I_c\omega_c^2 + \frac{1}{2}k\,\Delta r^2 + \frac{1}{2}\mu v_r^2\right)$$

The translations satisfy the equipartition criteria, as do the two "tumbling" rotations. However, neither the vibrations nor the rotations about the molecular axis satisfy the equipartition criteria, because the energy levels are spaced significantly farther apart than temperature energy, kT. Therefore, vibrations contribute little, if anything, to the heat capacity of diatomic molecules at room temperature and below.

If we count only the three translations and two rotations, the total number of contributing modes is 5. With $s = 5$, $c_v = \frac{5}{2}R = 21$ J/mol · K. This agrees well with observed values of c_v for diatomic molecules near room temperature.

The remaining coordinates, $3N - 6$ or $3N - 5$ (or $3N - 3 = 0$ for an atom), describe the relative positions of the atoms within the molecule, as the molecule undergoes vibrations (Figs. 6.1 and 6.2). For example, water has $3 \times 3 - 6 = 3$ vibrational modes. Carbon dioxide, because it is linear, has $3N - 5 = 9 - 5 = 4$ vibrational modes. Ethane, with eight atoms, has $3N - 6 = 18$ vibrational modes. Nitrogen, of course, has $3N - 5 = 6 - 5 = 1$ vibration. Each vibration has kinetic energy and potential energy.

Polyatomic molecules are not quite as simple. If the molecule is linear, we might expect three translational degrees of freedom, only two rotations, and no vibrations. The prediction for a linear polyatomic molecule would therefore be $s = 5$ and $c_v = \frac{5}{2}R$. However, any linear molecule containing three or more atoms has low-frequency bending modes, for which the vibrational frequency is

---

ANSWERS 6.3

A. a. $B = 0.39\ \text{cm}^{-1} = 39\ \text{m}^{-1}$, so $2Bhc = 2 \times 39 \times 6.6 \times 10^{-34} \times 3 \times 10^8$ J $= 1.55 \times 10^{-23}$ J. This is much smaller than $kT \approx 4 \times 10^{-21}$ J, so there is ample thermal energy available at room temperature for the CO_2 molecule to undergo its tumbling rotation.
b. For N_2, $A = 2.8 \times 10^{11}\ \text{cm}^{-1}$, so $2Ahc = 2 \times 2.8 \times 10^{13} \times 6.63 \times 10^{-34} \times 3 \times 10^8$ J $= 1.1 \times 10^{-11}$ J, which is more than a billion times larger than kT, so there is not enough thermal energy available at room temperature to make nitrogen rotate about its axis of symmetry.
B. σ is on the order of $10^3\ \text{cm}^{-1} = 10^5\ \text{m}^{-1}$, so $hc\sigma$ is roughly $6.6 \times 10^{-34} \times 3 \times 10^8 \times 10^5 \approx 2 \times 10^{-20}$ J. This is enough larger than kT at room temperature that it is a reasonable approximation to say that molecules do not vibrate at room temperature. (A more quantitative discussion is given in Part III.)

---

comparable to kT. Such bending modes therefore contribute, at least in part, to the heat capacity.

For nonlinear polyatomic molecules, there are three translational degrees of freedom and three rotations, so the number of modes contributing to the heat capacity is at least six, and with $s = 6$, $c_v = \frac{6}{2}R = 3R$. However, except for very small molecules, with only light atoms, there will be many vibrational modes, and some of those, especially bending modes, will have quite low frequencies. All we can say at this point, therefore, is that for nonlinear polyatomic molecules, $s \geq 6$ and

$$c_v \geq 3R = 25\ \text{J/mol} \cdot \text{K}$$

It is possible to calculate heat capacities accurately when the vibrational frequencies are known. Discussion of that theory is postponed to Part III. For the present we have adequate information for estimation of the importance of heat capacities in calculations involving temperature changes. For any ideal gas, $c_p - c_v = R$. Letting $\gamma \equiv C_p/C_v = c_p/c_v$, we can summarize the results as shown in Tables 6.3 and 6.4.

There is a small but finite moment of inertia for an individual atom, and therefore a calculable value of the rotational constant, A, for rotations of monatomic particles and of linear molecules about their axes of highest symmetry. Because I_a is small, A is large and the lowest rotational state of an atom such as helium is a high-energy state. On the E/k temperature scale it would correspond to a temperature of about 10^{11} K or an energy of about 10 MeV. This is about 1 billion times more energy than the average temperature energy, kT, so it is an extremely good approximation to say that single atoms do not rotate. Nuclear rotation energy levels are observed, but only in higher-energy processes such as nuclear decays. Rotations of monatomic particles and rotations of linear molecules about the axis of high symmetry are "frozen out" at room temperature.

Atoms and molecules also have electrons that can absorb energy as they change states. The corresponding absorption bands are usually in the visible or ultraviolet region, showing that the energy required to excite an electron to an empty level is equivalent to 1 to $5 \times 10^{14}\ \text{cm}^{-1}$. This is many times kT, so there is negligible probability of the electrons being thermally excited, in most molecules, and the electrons contribute nothing to the heat capacity near room temperature.

Table 6.3 Modes of Thermal Energy Storage in Molecules

Type of Molecule	Modes of Thermal Energy Storage	Number, s
Monatomic	Translational (v_x, v_y, v_z)	3
Diatomic	Translational (v_x, v_y, v_z) + rotational (ω_a, ω_b) (+ possible small vibrational contribution)	5
Polyatomic		
Linear	Translational (v_x, v_y, v_z) + rotational (ω_a, ω_b) (+ possible small vibrational contribution)	5
Nonlinear	Translational (v_x, v_y, v_z) + rotational (ω_a, ω_b, ω_c) (+ possible small vibrational contribution)	6

6.5 Heat Capacities of Condensed Phases

Electrons in solids occupy a wide range of energy levels compared with room temperature. All the lowest-energy levels are filled to capacity, so the electronic energy of the crystal can be increased only if the energy of the most energetic electrons is increased. There are relatively few of these electrons with enough energy to be near empty energy levels, and these already have energies in excess of 10,000 K. It is unlikely that collisions with molecules at 300 K will be able to excite these few electrons to even higher energies. Electrons therefore contribute little to the heat capacities of most solids at room temperature.

A crystalline element is well represented as a collection of identical point particles oscillating about equilibrium positions. The energy is the sum of three kinetic energy terms and three potential energy terms. The number of degrees of freedom is 6, and with $s = 6$,

$$c_v = \frac{6}{2}R = 25 \text{ J/mol}\cdot\text{K} \tag{6-17}$$

Without benefit of such a theoretical prediction, Dulong and Petit, in 1819, conjectured from scanty experimental evidence on metals that the molar heat capacity of all solid elements is 6 cal/mol·K = 25 J/mol·K. The generalization was helpful in following decades as attempts were made to determine molar masses for elements and compounds.

Table 6.4 Heat Capacities of Ideal Gases

Type of Molecule	s	c_v	c_p	$\gamma = \frac{c_p}{c_v}$
Monatomic	3	$\frac{3}{2}R$	$\frac{5}{2}R$	$\frac{5}{3} = 1.67$
Diatomic	5	$\frac{5}{2}R$	$\frac{7}{2}R$	$\frac{7}{5} = 1.40$
Polyatomic				
Linear	≥ 5	$\geq \frac{5}{2}R$	$\geq \frac{7}{2}R$	$\leq \frac{7}{5} = 1.4$
Nonlinear	≥ 6	$\geq 3R$	$\geq 4R$	$\leq \frac{4}{3} = 1.3$

Table 6.5 Heat Capacities of Solid Elements

Element	c (J/mol · K)	Element	c (J/mol · K)
Aluminum	24.1	Phosphorus	24.6
Cadmium	25.9	Silicon	20.
Carbon (*d*)	5.9	Silver	25.2
Copper	24.5	Sodium	28.4
Gold	25.8	Tin	26.9
Lead	26.5	Uranium	27.8
Magnesium	25.0	Zinc	25.3

EXAMPLE 6.4

A. Phosphorus and hydrogen form a compound in which 10.24 g of P combines with 1 g of H. Assuming a molar mass for H of 1, the molar mass of P can be calculated, depending on the formula assumed for the compound.

If formula is:	PH_3	PH_6
Molar mass of P is:	31.0	61.9

Which of these values would be most consistent with the observed value of the heat capacity of phosphorus of 0.79 J/g · K ?

B. Berzelius was among those who rejected the ideas of Avogadro and of Dulong and Petit. Consequently, he missed the molar mass of silver by a factor of 2. What value would he have calculated solely from the observed (modern) heat capacity value for silver of 0.233 J/g · K ? □

The law of Dulong and Petit is valid for *crystalline elements*, except that very light elements have lower heat capacities because of the same quantum limitations that limit the effects of molecular vibrations in gases. Table 6.5 shows the dependence of molar heat capacity on molar mass for some light elements.

The quantum mechanical spacing of vibrational levels is $\Delta E = h\nu$, where ν is the classical vibrational frequency. For a diatomic molecule, $\nu = (1/2\pi)\sqrt{k/\mu}$, where k is the Hooke's law force constant, a measure of the stiffness of the bond to stretching, and μ is the reduced mass. The reduced mass is equal to

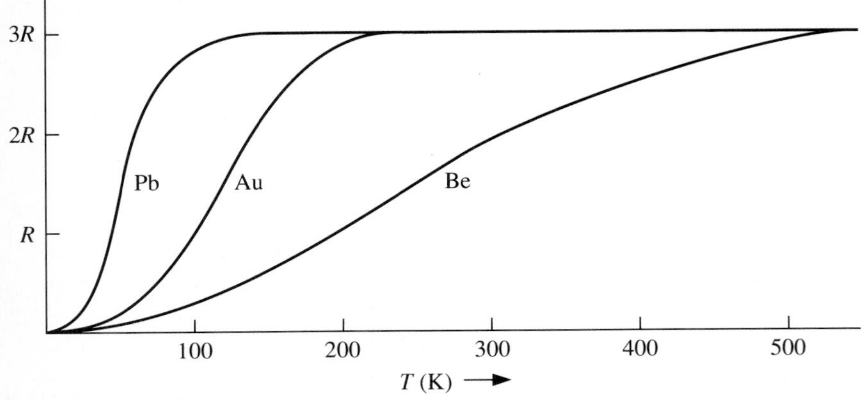

FIGURE 6.3 Approximate form of c_v versus T for some solid elements.

the lighter mass if they are very different, or equal to half the mass at each end if the masses are the same.

The force constant, k, is typically about the same for all bonds (within approximately a factor of 3). Thus we can expect that if μ is small (e.g., carbon or magnesium) the vibrational frequency will be high, the energy levels will be far apart, and the vibrations may be still partially frozen out at room temperature. Figure 6.3 shows a curve of heat capacity of several solid elements as a function of temperature. Generally, solid elements heavier than titanium (molar mass = 47.9 g/mol) follow the law of Dulong and Petit quite well at room temperature and above. Lighter atoms typically have lower values of heat capacity at room temperature.

Dulong and Petit's law may be extended, with caution, to some solid compounds. For example, a crystal of KBr contains two kinds of particles, and *each* kind contributes $3k$ per atom to the heat capacity. It is only the number of particles, not the chemical nature, that determines the heat capacity in the classical limit. If some of the particles are light, or the chemical bonding is especially strong, quantum effects may be important. The generalization that the heat capacity will be $3R$ times the number of atoms per mole provides an approximate upper bound.

EXAMPLES 6.5

A. What specific heat (J/g·K) would you predict for KBr (molar mass = 119.0 g/mol)?
B. What specific heat (upper limit) would you predict for calcium carbonate, $CaCO_3$ (molar mass = 100.1 g/mol)?
C. What specific heat (upper limit) would you predict for benzene, C_6H_6 (molar mass = 78 g/mol)? □

Kopp's rule is an empirical generalization that allows for variations in relative masses and binding strengths of elements in solid compounds. It is not intended to be highly accurate, but provides a good first guess when experimental values are not available. Each atom is assumed to give a contribution of 6.2 cal/mol·K (= 26 J/mol·K) except as follows: C = 1.8 (7.5), H = 2.3 (9.6), O = 4.0 (17), S = 5.4 (23), P = 5.4 (23), F = 5.0 (21), Si = 3.8 (16), and B = 2.7 (11), all in cal/mol·K (J/mol·K). For liquids, the values are C = 2.8 (12), H = 4.8 (20), O = 6.0 (25), S = 7.4 (31), P = 7.4 (31), F = 7.0 (29), Si = 5.8 (24), B = 4.7 (20), and all others 8.0 (33).

EXAMPLES 6.6

A. What specific heat would Kopp's rule predict for KBr?
B. For $CaCO_3$?
C. For benzene (liq), C_6H_6? □

Liquids lack the simplicity of gases, for which interactions can be neglected, or of solids, for which the interactions are strong but largely predictable. The kinetic energy contribution to heat capacity will be $\frac{3}{2}R$ per atom, except when strong chemical bondings and low mass causes quantum limitations. Molecules in a liquid are almost confined to oscillate in place, so the potential energy contribution to the heat capacity should also be about $\frac{3}{2}R$. Another mode of energy storage may be

---

ANSWERS 6.4

A. $\dfrac{25 \text{ J/mol} \cdot \text{K}}{0.79 \text{ cal/g} \cdot \text{K}} = 31.6$ g/mol.

B. $\dfrac{25 \text{ J/mol} \cdot \text{K}}{0.233 \text{ J/g} \cdot \text{K}} = 107.3$ g/mol, very close to 107.870.

ANSWERS 6.5

A. Two particles (K and Br), so that the predicted c_v would be $2 \times 3R =$ 50 J/mol · K = 0.42 J/g · K. The observed value of C_p is 0.439 J/g · K.
B. $CaCO_3$ has five atoms per molecule and $5 \times 3R = 125$ J/mol · K = 1.25 $J/g \cdot K$. The observed value is 0.836 J/g · K.
C. C_6H_6 has 12 atoms per molecule and $12 \times 3R = 300$ J/mol · K = 3.85 $J/g \cdot K$. The observed value is about 1.26 J/g · K.

ANSWERS 6.6

A. 0.437 (versus 0.439) J/g · K.
B. 0.84 (versus 0.86) J/g · K.
C. 2.46 (versus 1.7) J/g · K.

---

important, however, in many liquids, especially in hydrogen-bonding liquids such as water.

Water molecules interact through O···H links between adjacent molecules, called *hydrogen bonds*. Between the melting point and boiling point of water, many of these bonds are broken. To go from liquid water at 0°C to vapor at 100°C requires 640 cal/g = 2.7 kJ/g, compared to 50 cal/g = 200 J/g to heat water vapor over the same temperature interval. The difference, 10.6 kcal/mol = 44.4 J/mol, is the sum of changes in potential energy and hydrogen bonding energy of the liquid. A "normal" liquid would require about 35 kJ/mol for vaporization at 100°C (based on a Trouton constant of 90 to 95 J/mol·K; Section 9.4). The excess required by water, 44.4 − 35 = 9.4 kJ/mol, is comparable to the hydrogen-bonding energy in water. It is therefore reasonable to explain the high heat capacity of water as arising from the energy required to break hydrogen bonds, as the temperature increases.

Exact calculations of heat capacities are possible if all the vibrational frequencies are known. In fact, the calculations are sufficiently sensitive and reliable that being able to predict heat capacities, as a function of temperature, is an important check on the accuracy of assignments of vibrational frequencies, as discussed in Part III.

SUMMARY

For *ideal gases* (except where otherwise specified):

$$PV = nRT = NkT$$

$$P = \frac{\sum \text{momentum transfer to walls}}{\text{time} \times \text{area of walls}} = \frac{1}{V}\frac{2}{3}N\left\langle \frac{1}{2}mv^2 \right\rangle$$

Therefore, average kinetic energy $= \langle \frac{1}{2}mv^2 \rangle = \overline{\text{K.E.}} = \frac{3}{2}kT$.

$E_{\text{rotation}} = \dfrac{J(J+1)h^2}{8\pi^2 I_b} \equiv hcBJ(J+1)$ for a molecule with moment of inertia I_b (about an axis perpendicular to a 3-fold or higher symmetry axis) and with quantum number J.

$\Delta E = 2BhcJ$ and ΔE is typically small, compared to thermal energies (i.e., kT).

A molecule of N atoms has $3N - 6$ internal, or vibrational, modes (or $3N - 5$ if the molecule is linear). The energy increments are $\Delta E = h\nu$, where ν is the classical frequency of vibration. Typically, $\Delta E_{\text{vibration}} >$ thermal energy.

Equipartition principle: For classical degrees of freedom in a molecule, each mode contributes an average energy of $\langle E_i \rangle = \frac{1}{2}kT$.

Heat capacities: For $\mathcal{S}$ classical degrees of freedom,

$$c_v = \frac{\mathcal{S}}{2}R; \qquad c_p = \frac{\mathcal{S}+2}{2}R; \qquad \gamma \equiv c_p/c_v.$$

$\mathcal{S} = 3$ (for translations) + 3 (for rotations of polyatomic molecules; or 2 if molecule is linear) + possible contributions of low-frequency vibrational modes.

Solids in which each atom is held by lattice forces have $c_v = 3R$ (for elements) or $c_v = 3NR$ (N atoms per "molecule") unless the atoms are very light or very tightly bound, as in diamond or molecular crystals.

Kopp's rule is an empirical approximation for c_p of solids and liquids.

PROBLEMS

6.1. The measured value of c_p of SO_2 at room temperature is 41 J/mol · K. Based on this information, would you conclude that SO_2 is linear or nonlinear? Explain.

6.2. Predict the value of c_v for Pd (molar mass of 106.4 g).

6.3. Predict the value of c_v for BaO (molar mass of 137.34 g).

6.4. Predict an approximate value of c_p for NH_3 vapor. Explain the basis of the prediction.

6.5. A meter may be made arbitrarily sensitive. For example, assume that a small electric coil and mirror, with moment of inertia $I = 10^{-8}$ kg · m^2, is hung from a very long quartz fiber so that the torsional constant, κ, is 10^{-5} dyn · cm $= 10^{-14}$ N · m. A light beam is reflected from the mirror to a scale 3.0 m away. This gives a deflection of the light beam by 2ϑ when the coil rotates through the angle ϑ. Find the value expected, at room temperature, for

a. $\bar{\bar{\vartheta}}$, the rms deviation of the mirror direction from its equilibrium value.

b. $\bar{\bar{x}}$, the rms error in position on the scale.

c. $\bar{\bar{\omega}}$, the rms angular speed of the mirror.

d. $\bar{\bar{v}}$, the resultant rms speed of the light beam at the scale.

6.6. How would the quantities calculated in Problem 6.5 differ

a. If the space around the coil and mirror were evacuated to 10^{-12} torr?

b. If the temperature were lowered to 4 K?

6.7. A quartz fiber spring forms a very sensitive balance. Is there a natural lower limit to the mass that can be measured, at room temperature, with such a balance? Justify your answer with calculations. Could you, in principle, weigh 1 pg $= 10^{-12}$ g? Could you, in principle, weigh one electron?

6.8. Using the approximate value of the rotational constant B from Table 6.1 for He, find

a. The rotational energy for He in the first excited rotational state (J = 1).

b. The temperature equivalent of this energy ($T = E/k$).

6.9. **a.** Find the vibrational energy in joules of the N_2 molecule in its first vibrational state ($\sigma = 2360\ cm^{-1}$).

b. Find the equivalent temperature ($T = E/k$).

CHAPTER 7

Transport Processes

Many practical applications of thermodynamics are intimately tied to rates of cooling and warming, through conduction, convection, evaporation, and radiation. Fuel costs are often determined more by unwanted energy losses than by the desired effects of the energy at the point of use. In this chapter we examine some basic principles of transport processes, including macroscopic equations of diffusion processes and energy transport as well as analyses that depend directly on molecular speeds and mean free paths.

7.1 Maxwell–Boltzmann Distribution

The atomic model of matter, first seriously considered at the beginning of the nineteenth century, gained favor only slowly during that century, as the caloric theory was simultaneously being replaced by the idea that temperature is a measure of the kinetic energy of molecules. Many scientists, well trained in mathematics as well as chemistry and physics, contributed to this flood of scientific discovery and development of theory. Among the active contributors to the theory of gases were James Clerk Maxwell and Ludwig Boltzmann.

Gaussian Curve

It is not possible to follow in detail the motions of individual molecules within a gas. The gas must be described by a *probability distribution*. It will be helpful to begin by examining a more familiar distribution function, known as the error curve, normal curve, or Gaussian distribution,

$$\rho(x) = Ae^{-a(x-x_0)^2} \qquad (7\text{-}1)$$

which has the shape depicted in Fig. 7.1. The constant coefficient, A, in the distribution function *normalizes* the curve, so the area under the curve is equal to 1 (or some other selected value).

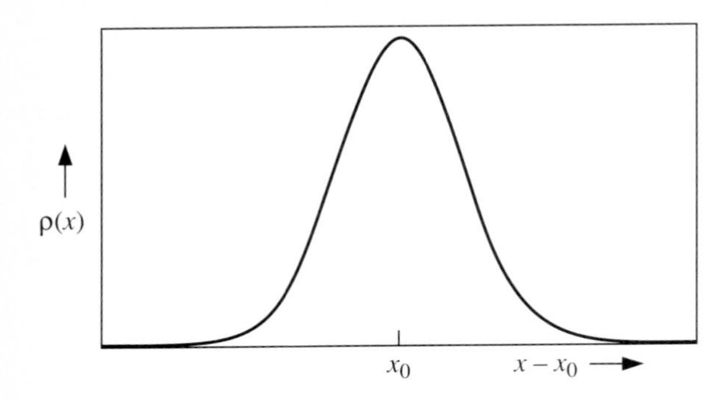

FIGURE 7.1 Gaussian, normal, or error curve.

The error curve is expected to arise when the number of measurements is large and each is subject to random error (see Section 15.7). Because we assume that the errors are random, not systematic, values to the right or to the left of x_o are equally likely. The value x_o, at the center, is considered the "true" value; the "error," in any measurement that yields the value x, is $x - x_o$. Any value of x is considered a possible result of a measurement, although the probability of getting any value of x must decrease as we get farther from the true value, x_o.

Elementary probability theory tells us that when N items are randomly distributed in M boxes, the probability of any particular item going in a specified box should be $1/M$. The probable number of items per box would be N/M. As M becomes large, the probability of finding a box occupied becomes small. If M becomes infinite while N is finite, the average occupancy level of the boxes goes to zero.

Now consider the following two questions about the error curve.

1. What is the most probable error for any measurement?
2. What is the probability of getting that error?

The first question is easily answered by looking at the curve. The greatest probability occurs right at the center, at $x = x_o$. The error is $x - x_o$, so the greatest probability is for $x - x_o = 0$. That is:

1. The most probable error is zero.[1]

The second question is only slightly more involved. If we make N measurements, where N may be any finite number, the probability of getting any particular value is N/M, where M is the number of possible values. For a continuous distribution, M is infinite, so the probability of getting any specific value is $N/M = N/\infty = 0$. The value $x = x_o$ is one such specific value, which must therefore have zero probability. Hence

2. The probability of getting zero error is zero.

Some range of values must be specified to have an experimental result fall in the "box." Probabilities are expressed as the chance that a measured value will fall within a specified range. That is equivalent to finding the area under some

[1]The *median error* is the value such that half of a large set of measurements will show errors smaller and half larger than this value. Statisticians usually call this the "probable error," although it is unrelated to the most probable error.

segment of the curve. The area under any mathematical point (a specific value of x) is identically zero.

In general, the probability of finding an experimental value in the interval x_1 to x_2 is finite, given by the integral of the distribution function, ρ, over that interval.

$$\text{Probability} = \int_{x_1}^{x_2} \rho(x)\,dx \tag{7-2}$$

where $\rho(x) = Ae^{-a(x-x_o)^2}$ if the errors are random.

Velocity Distribution

Molecular velocities may, in principle, have magnitudes from zero to infinity (or at least to some very high value such that all of the energy of the gas is invested in one molecule). Maxwell showed in 1859 that relative populations or probabilities of velocity states are proportional to $e^{-m(v_x^2+v_y^2+v_z^2)/2kT}$. In 1868, and over the following two decades, Boltzmann generalized this result to include potential energy and showed that it applied to polyatomic molecules. We will follow the usual custom of referring to the Maxwell distribution and Boltzmann's extension of it as the Maxwell–Boltzmann distribution.[2]

Velocities along the positive axis are equally as likely as velocities along the negative axis, so the distribution is symmetric. The distribution function for molecular velocities is therefore a normal, or Gaussian, curve, as shown in Fig. 7.2. A factor $(m/2\pi kT)^{1/2}$ is required to *normalize* the curve such that the integral will be equal to N, the number of molecules. The distribution function (Fig. 7.2) is

$$dN = N\left(\frac{m}{2\pi kT}\right)^{1/2} e^{-mv_x^2/2kT}\,dv_x \tag{7-3}$$

The velocity distribution in two dimensions would be

$$dN = N\left(\frac{m}{2\pi kT}\right) e^{-m(v_x^2+v_y^2)/2kT}\,dv_x\,dv_y \tag{7-4}$$

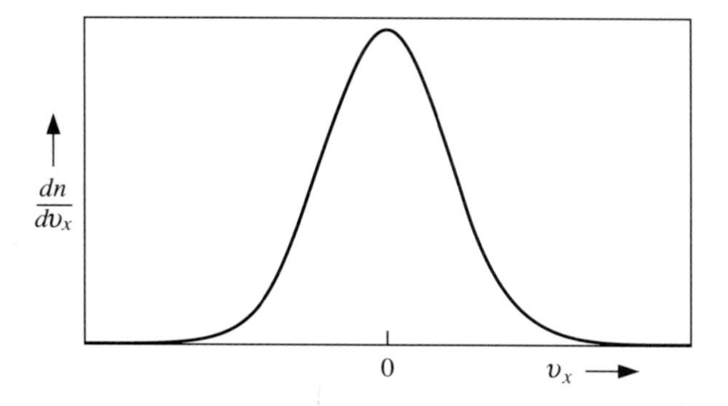

FIGURE 7.2 Velocity distribution in one dimension: $\rho(v_x)$.

[2]A familiar example of such a distribution is the atmospheric equation. The pressure in any fluid column is proportional to depth, h, and to density, ρ, so the change in pressure with height in the atmosphere is $dP = -\rho g\,dh$. Setting $\rho = M/V$ and $V = RT/P$ gives $dP/P = -Mg\,dh/RT$. Integration yields $P = P_o e^{-(Mgh/RT)}$, where M is the molar mass of air. This is of the form $e^{-(\Delta E/kT)}$ with $\Delta E = mgh$.

and in three dimensions,

$$dN = N\left(\frac{m}{2\pi kT}\right)^{3/2} e^{-m(v_x^2+v_y^2+v_z^2)/2kT}\, dv_x\, dv_y\, dv_z \qquad (7\text{-}5)$$

Speed Distribution

We are now prepared to look at a distribution function for molecular speeds. First we ask, what is the probability for finding $v = 0$? Clearly, $v = 0$ is one specific value of v, from infinitely many possible values, so the probability must be zero. More explicitly, the probability of obtaining $v = 0$ must be infinitesimal, and as there is only one possible vector of length zero, the total probability for speed zero remains infinitesimal.[3] The distribution function for molecular speeds must go through the origin; $\rho(v) = 0$ at $v = 0$.

Because not all the energy of the gas will be found in just one or a few molecules, the probability asymptotically approaches zero as the speed, v, becomes very large. Hence the curve starts at the origin, rises to a maximum, and falls asymptotically to zero. The shape is shown in Fig. 7.3. We only need to find the precise mathematical description.

The *speed distribution* differs markedly from the *velocity distribution*, because there can be many different velocities corresponding to the same speed. To find the speed distribution, we multiply the velocity distribution by the number of states for each velocity.

If there is more than one state of the same energy, the states are said to be *degenerate*.[4] Each of the degenerate states has the same probability, so the total probability for that energy value is increased in proportion to the number of

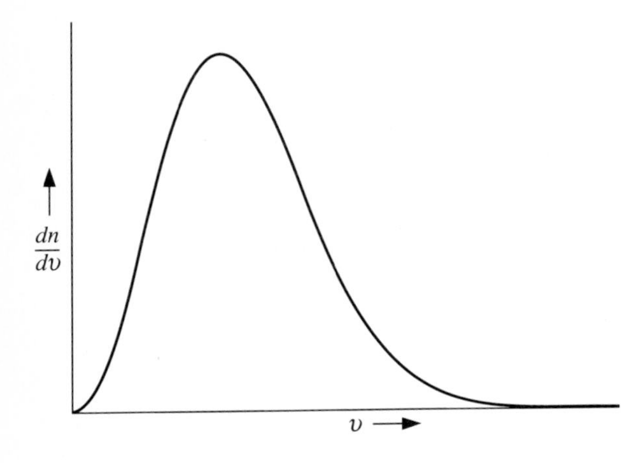

FIGURE 7.3 Maxwell–Boltzmann curve for distribution of molecular speeds in an ideal gas.

[3] By contrast, there are many vectors that have a projection, on the x axis, of zero, so the distribution function for $\mathbf{v}_x$ (Fig. 7.2) need not go to zero at $\mathbf{v}_x = 0$.

[4] The degeneracy considered here is the classical degeneracy arising from the symmetry of space. The probability of two molecules having exactly the same velocity is vanishingly small, but they may have the same speed. If the molecules are treated by quantum mechanics (Section 18.1)—the familiar particle-in-a-box problem—two kinds of degeneracy must be considered. Spatial degeneracy is no longer exact, or *necessary*; it occurs only in the unlikely event that the ratio of quantum number to length of container is the same in two directions, so that, for example, $n_a^2/a^2 = n_b^2/b^2$. A new type of degeneracy arises because the molecules can have only integral values of n_a, n_b, and n_c, so the probability of two molecules having the same value of all three quantum numbers does not vanish. For boxes large enough to hold a sample of gas, quantum mechanics gives the same predictions as classical mechanics.

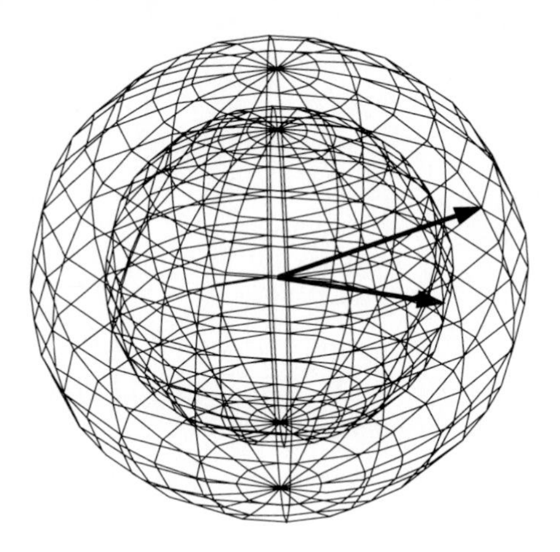

FIGURE 7.4 A spherical shell of radius v (representing the speeds, or magnitudes of the velocities) defines an energy level. The number of velocity states contributing to that level is proportional to $4\pi v^2$.

states of that energy. The number (or relative number) of states that have the same magnitude of a continuously variable vector, such as velocity, is counted with the help of a geometric model of the vector distribution.

The possible velocities, corresponding to a single speed, point in all directions. Represent these states by arrows, of length proportional to the speed, drawn from the origin (Fig. 7.4), terminating with some arbitrary arrow density on the surface of a sphere of radius R, with $R \sim v$. The number of such vectors is proportional to the area of that sphere, $4\pi R^2$. Hence the number of states of a given speed, v, increases as v^2, the square of the speed.

For small values of v, the exponential term is nearly constant, equal to 1. Therefore, the initial part of the distribution curve, at low speeds, should be parabolic; the probability is proportional to speed squared. At high speeds the curve is lowered by the exponential factor, $\exp(-\frac{1}{2}mv^2/kT)$, which can overcome any power of v to return the curve to the horizontal axis.

We obtain the speed distribution from the velocity distribution by integrating over a sphere. This replaces $dv_x\, dv_y\, dv_z$ with $4\pi v^2\, dv$. The total distribution function for molecular speeds is

$$dN = \left[4\pi N\left(\frac{m}{2\pi kT}\right)^{3/2}\right] v^2 e^{-mv^2/2kT}\, dv \tag{7-6}$$

The area under some segment of the curve gives the number of molecules with speed within that range. The area under the whole curve gives the number of molecules with all possible speeds, which must include all the molecules. Therefore, the total area under the curve must be constant, equal to N, the number of molecules in the gas.

$$\int \left(\frac{dn}{dv}\right) dv = \int dn = N$$

The combination of factors inside the square brackets is required to normalize the expression (i.e., to ensure that the sum of the molecules having each speed adds up to the total number of molecules, no more and no less). This normalizing factor is of no particular interest to us at present.

At low temperatures there is a small spread of speeds and a high maximum. The distribution becomes broader and the peak height falls as temperature increases, as represented in Fig. 7.5.

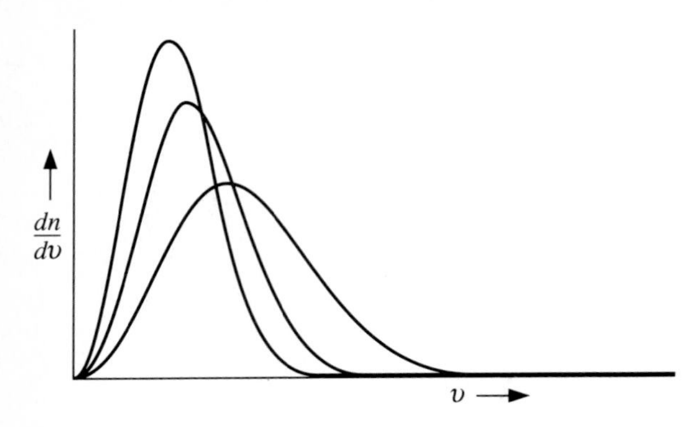

FIGURE 7.5 Maxwell–Boltzmann distribution for three temperatures.

The most probable speed is obtained by taking the derivative of the curve, setting

$$\frac{d}{dv}\left(\frac{dN}{dv}\right) = 0$$

which gives

$$v_{\text{prob}} = \sqrt{\frac{2kT}{m}} \tag{7-7}$$

The average speed is

$$\overline{v} = \frac{\int v\, dN}{\int dN} = 4\left(\frac{a^3}{\pi}\right)^{1/2} \int_0^\infty v^3 e^{-av^2}\, dv$$

with $a = m/2kT$. Integration by parts gives

$$\overline{v} = \sqrt{\frac{8kT}{\pi m}} \tag{7-8}$$

The average kinetic energy is

$$\overline{\text{K.E.}} \equiv \left\langle \frac{1}{2}mv^2 \right\rangle = \frac{\int \frac{1}{2}mv^2\, dN}{\int dN}$$

We easily anticipate the result of this integration if we recall that $\langle \frac{1}{2}mv^2 \rangle = \frac{3}{2}kT$. Then

$$\overline{\overline{v}} \equiv \sqrt{\langle v^2 \rangle} = \sqrt{\frac{3kT}{m}} \tag{7-9}$$

where we have introduced the customary notation of $\overline{\overline{v}}$ for the square *root* of the *mean* of the *square* of the velocity, called simply the *root-mean-square* velocity, and have represented average values by the angular brackets, $\langle\ \rangle$.

The values for v_{prob}, $\overline{v}$, and $\overline{\overline{v}}$ are shown in Fig. 7.6. The largest of these is $\overline{\overline{v}}$, because when an average is taken of the squares of a set of values, each large value is multiplied by a large value. Thus the large values are more heavily weighted.

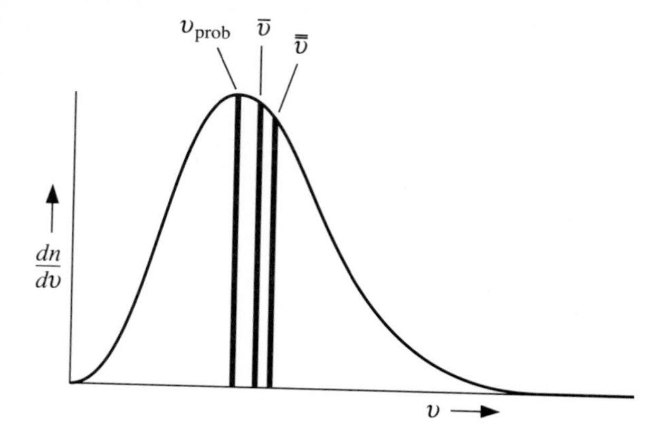

FIGURE 7.6 Most probable, average, and rms speeds.

7.2 Macroscopic Diffusion Processes

Before looking at the molecular equations for diffusive processes, we develop the macroscopic equations and introduce the symbols and nomenclature that will be required in the following sections.

Gaseous Diffusion

Assume an experimental arrangement equivalent to Fig. 7.7, where pure gas 1 is maintained at plane 1 (perhaps by flowing pure 1 past a highly permeable membrane) and pure gas 2 is maintained at plane 2, with the gas between 1 and 2 at constant temperature and pressure.

The number of 1 molecules crossing an arbitrary plane per unit of time, dN_1/dt, is proportional to the area and to the gradient of concentration (expressed as mole fraction), dX_1/dz, across the plane.

$$\frac{1}{A}\frac{dN_1}{dt} = -D_1\frac{dX_1}{dz} \tag{7-10}$$

The minus sign shows that the diffusion is *against* the gradient (i.e., from high concentration to low) and D_1 is called the *diffusion constant*.

After some time, the gas between planes 1 and 2 will reach a steady state. The concentration then varies with position, but not with time. This requires that the flow rate is the same across any plane between 1 and 2.

$$\text{flow rate} = \frac{dN_1}{dt} = \text{constant} \neq f(z)$$

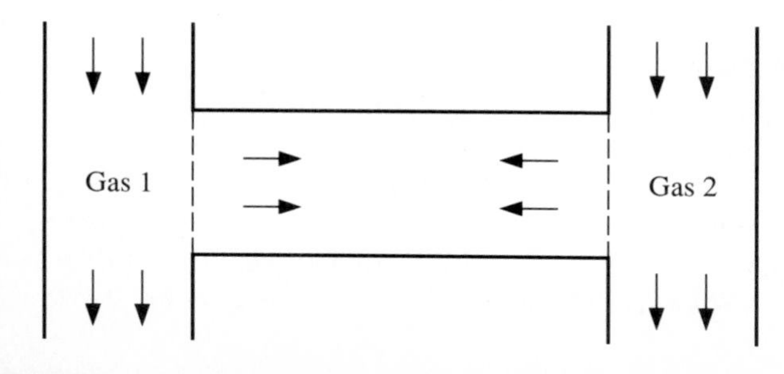

FIGURE 7.7 Two pure gases, on the left and right, diffuse toward each other in the intermediate zone at constant temperature and pressure.

Then

$$dX_1 = -\frac{1}{AD_1}\frac{dN_1}{dt}\,dz = \theta_1\,dz$$

where θ_1 is a constant, independent of X_1 and z. Therefore, X_1 is proportional to z.

Similarly,

$$dX_2 = -\frac{1}{AD_2}\frac{dN_2}{dt}\,dz = \theta_2\,dz$$

But the sum of mole fractions is 1, $\Sigma X_i = 1$, so

$$dX_1 = -dX_2 \qquad \text{and} \qquad dN_1 = -dN_2$$

Therefore, the diffusion constants are the same.

$$D_1 = D_2 = D_{12}$$

The diffusion of 1 is exactly counterbalanced by diffusion of 2. Often, mole fraction is replaced by mole number, as the measure of concentration, and the diffusion equation is written

$$-\frac{1}{A}\frac{dN_i}{dt} = D_i\,\frac{dn_i}{dz}$$

Viscous Flow

Newton observed that the viscous force, for laminar, or streamline, flow, may be represented as

$$f_z = \eta A_s\,\frac{dv}{dx} \tag{7-11}$$

Here dv/dx is the velocity gradient, perpendicular to the flow, A_s is the surface area of contact between adjacent layers for which the viscous force is measured, and η (eta), is the constant usually called the *viscosity*. Viscosity has traditionally been measured in units of dyn · s/cm^2, called a *poise* (after Poiseuille, who studied viscous flow), or in *centipoise*. The SI unit is 1 N · s/m^2 = 1 Pa · s, which is equal to 10 poise.

We are most familiar with viscosity in liquids. Water has a high viscosity compared to other molecules of comparable size, because of the strong attraction between adjacent molecules, through hydrogen bonding. As intermolecular attractions decrease, viscosity decreases.

We also have come to expect that viscosity will decrease with increasing temperature. Warm syrup, or warm oil, flows more freely than the cold liquid. However, viscosity in gases is quite different. Although it follows the same macroscopic equation, it does not depend on intermolecular attractions. Even ideal gases exhibit viscosity, by transport of momentum between adjacent layers, as shown in the following section. As the temperature increases, the viscosity of a gas increases.

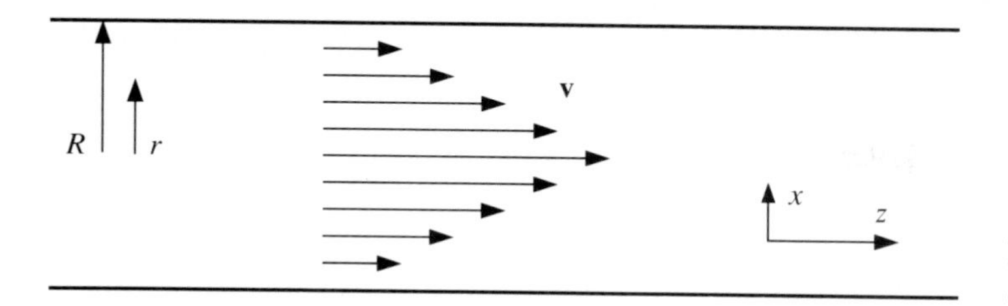

FIGURE 7.8 Viscous flow in a tube of radius R.

An example of viscous flow is flow of a liquid or gas through a tube (Fig. 7.8). Let the radius be R, the length L, and the speed of the flow $v = v(r)$, a function of the radial distance from the center of the tube. At the wall, $r = R$ and $v(R) = 0$.

The viscous drag on the surface of a solid cylinder of fluid of radius r (Fig. 7.9) is

$$f_{\text{drag}}(r) = -\eta \frac{dv}{dr} A_s = -\eta \frac{dv}{dr} 2\pi r L$$

with $A_s = 2\pi r L$ the area of contact between the moving cylinder of length L and the fluid adjacent (at larger r). The driving force on the cylinder of fluid is equal to the pressure difference, across the ends, times the cross-sectional area of the cylinder, A_c.

$$f_{\text{drive}}(r) = \Delta P A_c = (P_1 - P_2)\pi r^2$$

For steady flow, $f_{\text{drag}}(r) = f_{\text{drive}}(r)$.

$$-\eta \frac{dv}{dr} 2\pi r L = (P_1 - P_2)\pi r^2$$

Therefore,

$$-dv = \frac{(P_1 - P_2) r^2 \, dr}{2\eta r L}$$

and

$$-\int_{v=v}^{v=0} dv = \int_{r=r}^{r=R} \frac{(P_1 - P_2) r \, dr}{2\eta L}$$

This gives the speed at any radial distance.

$$v(r) - v(R) = v(r) = \frac{P_1 - P_2}{4\eta L}(R^2 - r^2) \tag{7-12}$$

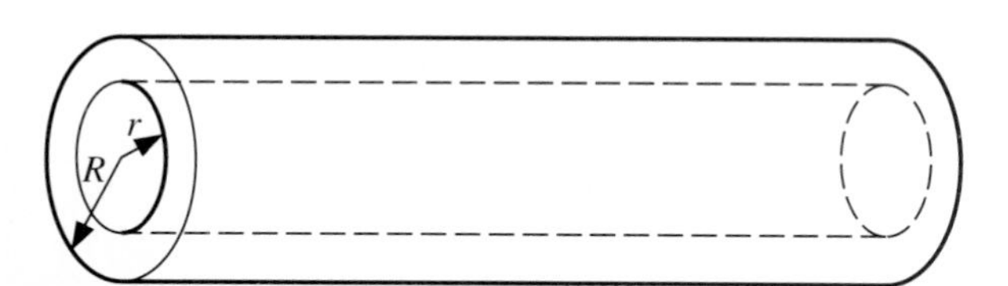

FIGURE 7.9 The cylinder of fluid of radius r is driven by the pressure difference $P_1 - P_2$ between the ends.

If the fluid is incompressible, the flow rate is

$$\frac{dV}{dt} = \int_{r=0}^{r=R} v(r)\, dA_c(r) = \int_{r=0}^{r=R} \frac{P_1 - P_2}{4\eta L}(R^2 - r^2)2\pi r\, dr$$

$$= \frac{\pi}{8}\frac{(P_1 - P_2)R^4}{\eta L} \tag{7-13}$$

For a gas the volume varies with pressure along the tube, but if the temperature is constant, PV is constant along the flow. Multiply each side by pressure and average the pressure, on the right side, along the tube, which gives $(P_1 + P_2)/2$. The pressure on the left, P_0, may be selected arbitrarily as the pressure at which dV is to be measured. Then

$$P_o \frac{dV}{dt} = \frac{P_1 + P_2}{2}\frac{\pi}{8}\frac{(P_1 - P_2)R^4}{\eta L}$$

or

$$\frac{dV}{dt} = \frac{\pi}{8}\frac{(P_1 - P_2)R^4}{\eta L}\frac{P_1 + P_2}{2P_o} = \frac{\pi(P_1^2 - P_2^2)R^4}{16\eta L P_o} \tag{7-14}$$

EXAMPLES 7.1

A. The viscosity of water is taken as 1.002 cP or 1.002×10^{-3} Pa · s, as a reference, at 20°C. If water is pushed through a 10-m pipe of diameter 2.0 cm with 1 atm gauge pressure (i.e., $P_1 = 2 \times 10^5$ Pa and $P_2 = 1 \times 10^5$ Pa), what is the flow rate?

B. What would be the flow rate of air ($\eta = 184\mu P = 1.84 \times 10^{-5}$ Pa · s at 20°C) through the same tube with the same pressure difference? □

7.3 Molecular Effusion and Diffusion

The ideal gas model is sufficiently simple that many properties can be calculated for such a gas. The values calculated agree well with experiments on most real gases, at least at moderate or low pressures.

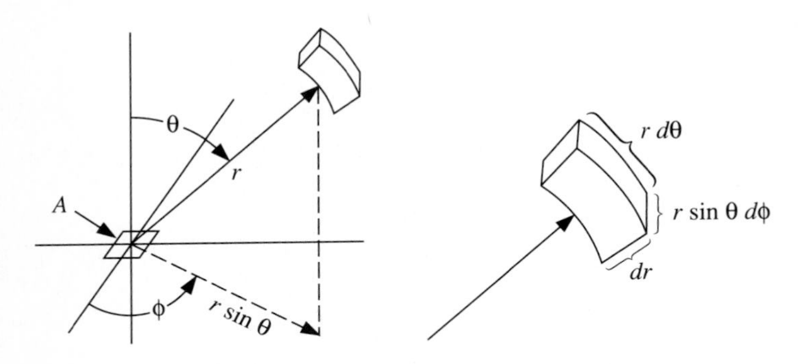

FIGURE 7.10 The volume element, dV, is located with respect to the origin by coordinates r, ϑ, and ϕ.

Effusion

An important application of elementary kinetic theory of gases is the problem of *effusion,* the rate at which a gas will escape (into vacuum) through a small hole. The hole should be sufficiently small that the velocity distribution in the gas is not appreciably disturbed and no collisions occur within the gas during passage through the two-dimensional opening (so viscosity is not a factor), but sufficiently large that any molecule that reaches the opening will pass through. The problem is simply to find the number of molecules striking an opening of area A in unit time from the side on which the gas is located.

Let A be a small horizontal area (Fig. 7.10a) centered at the origin of a reference frame described by the polar coordinates r, ϑ, and ϕ, with ϑ measured from the perpendicular to A and ϕ measured about that perpendicular. Define an increment of volume (Fig. 7.10b) about a point, $dV = dV(r, \vartheta, \phi) = r^2 \sin\vartheta \, d\vartheta \, d\phi \, dr$. If the density of molecules is represented by $n \equiv N/V$, the number of molecules in the increment of volume, dV, is

$$dN = \frac{N}{V}\, dV = n\, dV = nr^2 \sin\vartheta \, d\vartheta \, d\phi \, dr$$

Select the volume element, dV, at the distance $r = \overline{v}\,\Delta t$ from A, where $\overline{v}$ is the average speed of the molecules and Δt is some small increment of time. Then, on the average, the molecules in dV can just reach the area A in this time, Δt. Draw a sphere about dV, passing through the origin and therefore intersecting A at an angle (Fig. 7.11a). The angle between the surface of the sphere, which is perpendicular to r, and A is equal to ϑ, the angle between r and the perpendicular to A.

The probability of a molecule in dV striking A is proportional to the effective ratio of the area of A to the total area of the sphere about dV. The projection of A on a plane perpendicular to r is $A\cos\vartheta$, regardless of the shape of A. From dV, therefore, A looks like the area $A\cos\vartheta$ (Fig. 7.11b), and the probability is proportional to

$$\frac{A\cos\vartheta}{4\pi r^2}$$

The incremental number, with average speed $\overline{v}$, striking A in a time $\Delta t = r/\overline{v}$ is therefore

$$\delta N = \frac{N}{V}\, dV\, A\frac{\cos\vartheta}{4\pi r^2} = \frac{nA\cos\vartheta}{4\pi r^2} r^2 \sin\vartheta \, d\vartheta \, d\phi \, dr$$

This must be integrated over all the volume elements that are within the distance $r = \overline{v}\,\Delta t$, so the molecules can reach A in the time Δt *and* are on the proper side of A. We count only those striking a single face of A.

The total number striking A and therefore passing through if A is an opening is

$$\Delta N = \int_{r=0}^{r=\overline{v}\Delta t} \int_{\vartheta=0}^{\pi/2} \int_{\phi=0}^{2\pi} n\left(A\frac{\cos\vartheta}{4\pi}\right) \sin\vartheta \, d\vartheta \, d\phi \, dr$$

ANSWERS 7.1

A. $\dfrac{dV}{dt} = \dfrac{\pi}{8}\dfrac{(P_1 - P_2)R^4}{\eta L} = \dfrac{\pi}{8}\dfrac{10^5 \times (10^{-2})^4}{10^{-3} \times 10} = \dfrac{\pi}{80}\ \text{m}^3 = 39\dfrac{\text{L}}{\text{s}}(\approx 10\ \text{gal/s}).$

B. $\dfrac{dV}{dt} = \dfrac{\pi}{8}\dfrac{(P_1 - P_2)R^4}{\eta L}\dfrac{P_1 + P_2}{2P_o} = \dfrac{\pi}{8}\dfrac{10^5 \times (10^{-2})^4}{1.84 \times 10^{-5} \times 10}\dfrac{3 \times 10^5}{2 \times 10^5} = 3.2\ \text{m}^{3/5}$ air measured at 1 atm.

$$= n\bar{v}\ \Delta t\ \frac{A}{4\pi} 2\pi \int_{\vartheta=0}^{\pi/2} \sin\vartheta \cos\vartheta\ d\vartheta$$

The last integral may be written $\int \sin\vartheta\ d(\sin\vartheta)$, which is $\sin^2\vartheta/2$ evaluated between $\vartheta = 0$ and $\vartheta = \pi/2$, equal to $\frac{1}{2}$. The effusion rate, $\Delta N/\Delta t$ or dN/dt, is accordingly

$$\frac{dN}{dt} = \frac{1}{4} nA\bar{v} \tag{7-15}$$

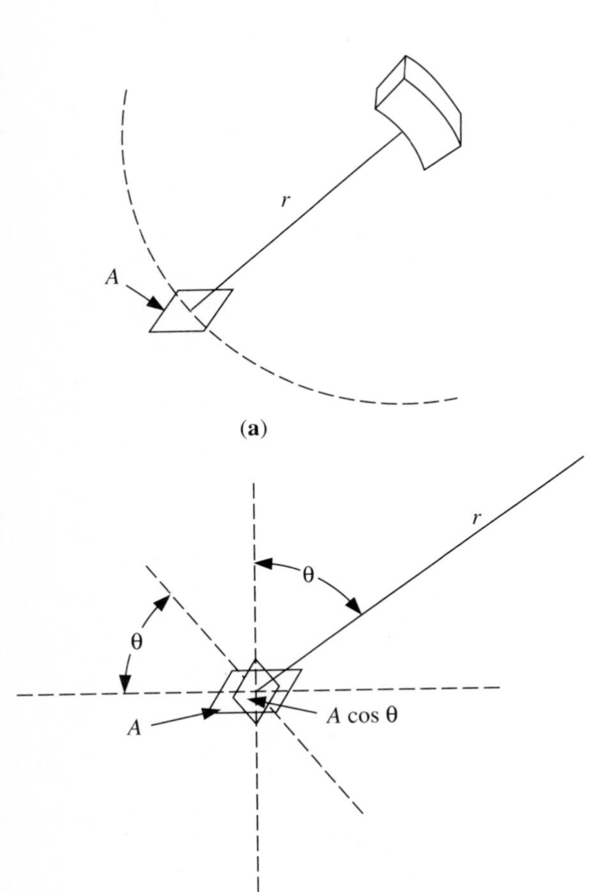

FIGURE 7.11 The aperture of area A at a distance r from the volume element presents a target area $A\cos\vartheta$.

It is[5] one-fourth of the density, $n = N/V$, times the area of the hole, A, times the average speed, $\overline{v}$.

EXAMPLE 7.2

Estimate the number of molecules of N_2 that will enter a vacuum system ($P = 0$) per second from air (let P of $N_2 = 1$ atm) through a pinhole $1\,\mu m$ in diameter at room temperature. □

The calculation of effusion rate does not allow for collisions of molecules between the source volume, dV, and the target area, A. For an equilibrium distribution it can be assumed that for each molecule starting toward A but deflected by collision, another will take its place. The final expression tells only how many molecules reach A per unit time, not which molecules reach A, where they (most recently) came from, or from what direction they came.

To find the average distance above the plane from which a molecule has come since its last collision, we introduce the average distance traveled by a molecule between collisions, λ, called the *mean free path* (see below). Then the average distance above the plane (perpendicular to A) from which the molecules come is

$$\overline{d}_{\perp} = \frac{\lambda \int\int\int \cos\vartheta\, \delta N}{\int\int\int \delta N} = \lambda \frac{\int_{\vartheta=0}^{\vartheta=\pi/2} \cos^2\vartheta \sin\vartheta\, d\vartheta}{\int_{\vartheta=0}^{\vartheta=\pi/2} \cos\vartheta \sin\vartheta\, d\vartheta} = \lambda \frac{\frac{1}{3}}{\frac{1}{2}} = \frac{2}{3}\lambda \qquad (7\text{-}16)$$

Effusion must be distinguished from normal *flow* and from *diffusion*. In normal flow the removal of gas at the opening is sufficiently rapid to cause a change in the velocity distribution within the gas. Diffusion may be fast or slow but is along a longer path, such as a tube, in which multiple collisions occur during the motion. For true effusion, the pinhole must have a thickness small enough that no collisions occur during transit.

Mean Free Path

Within a gas consisting of mathematical point particles, no collisions would occur. Any actual gas consists of molecules of finite size. To estimate the number or frequency of collisions experienced by a molecule, assume each molecule to be a sphere of radius r (Fig. 7.12a).

A collision is counted if the path of the projectile molecule brings its center to within r of the surface of a target molecule, or within $2r$ of the center of the target molecule. It is often more convenient to represent the problem as shown in Fig. 7.12b. Assign the radius $2r$ to the projectile molecule. Then a collision is counted if the cylinder swept out includes the location of (the center of) any other (target) molecule. Hence the number of collisions of the projectile molecule per unit of time is

[5] A pseudoderivation, or mnemonic, is the following. The number passing through A per unit of time must be proportional to the density of molecules, their average speed, and the area, A. Multiply by one-half because only molecules passing through in one direction are to be counted. Multiply by an azimuthal angle factor of $\cos\vartheta$ in the x-z plane and by a similar factor for the angle in the y-z plane. The average value of $\cos^2\vartheta$ is $\frac{1}{2}$. Therefore,

$$\frac{dN}{dt} = \left(\frac{1}{2}\right)\left(\frac{1}{2}\right) nA\overline{v}$$

ANSWER 7.2

An adequate approximation for the area is $\pi D^2/4 = D^2 = 10^{-12}\ \text{m}^2$.

$$\frac{N}{V} = \frac{N_A P}{RT} = \frac{6.02 \times 10^{23}\ \text{molecule/mol} \times 1\ \text{atm} \times 1.01 \times 10^5\ \text{N/m}^2 \cdot \text{atm}}{8.314\ \text{J/mol} \cdot \text{K} \times 300\ \text{K}}$$

$$= \frac{1}{4} \times 10^{26}\ \text{molecule/m}^3$$

The average speed is obtained approximately from $\frac{1}{2}mv^2 = \frac{3}{2}kT$, giving

$$v = \sqrt{\frac{3kT}{m}} = \sqrt{\frac{3RT}{M}} = \sqrt{\frac{3 \times 8.314 \times 300\ \text{J/mol}}{0.028\ \text{kg/mol}}}$$

$$= \sqrt{25 \times 10^4\ \text{J/kg}} = 5 \times 10^2\ \text{m/s}$$

The number per second passing through the opening is then

$$\frac{1}{4}\frac{N}{V}vA = \frac{1}{4} \times \left(\frac{1}{4} \times 10^{26}\text{molecule/m}^3\right)(5 \times 10^2\ \text{m/s}) \times 10^{-12}\text{m}^2$$

$$= \frac{1}{3} \times 10^{16}\ \text{molecule/s} = 3 \times 10^{15}\text{s}^{-1}$$

[Effusion is proportional to the average speed, whereas the kinetic energy expression gives us the root-mean-square (rms) speed. A more careful calculation, with $\overline{v} = \sqrt{8kT/\pi m}$, would give 2.3×10^{15}/s, or about 1.8×10^{15} nitrogen molecule/s and 0.4×10^{15} oxygen molecule/s.]

$$\nu = \frac{\text{collisions}}{\text{time}} \approx \pi d^2 v n$$

where v is the speed of the projectile molecule, $n = N/V$ is the density of (target) molecules, and d is the molecular diameter.

There is no difference between projectile and target molecules; each is in motion. Therefore, a better approximation to the collision frequency is obtained by replacing v with v_{rel}, the relative speed of projectile and target molecules.

Let $\rho(v)$ be some function that describes the probability of finding any given molecule with the speed v, or with a speed in the interval $v + dv$. Now, if the

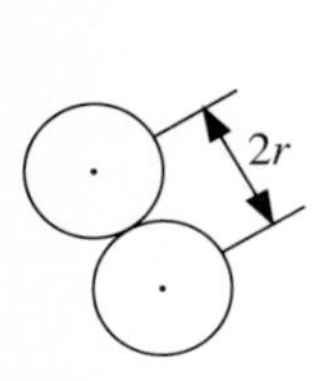

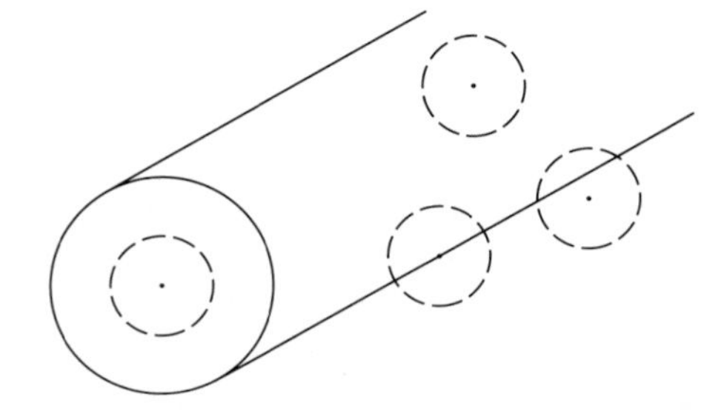

FIGURE 7.12 (a) The condition for a collision is that the surfaces of two molecules come in contact, or that the distance between centers is less than or equal to $2r = d$. (b) Another representation is to let a projectile molecule of radius d pass among point target molecules.

projectile and target molecules have velocity distributions $\rho_p(\mathbf{v}_p) = \rho_t(\mathbf{v}_t)$, then because $\mathbf{v}_p$ and $\mathbf{v}_t$ can be considered uncorrelated,

$$\rho(\mathbf{v}_p, \mathbf{v}_t) = \rho(\mathbf{v}_p)\rho(\mathbf{v}_t)$$

The velocity distribution is known for an ideal gas. All velocities are equally probable, except for the Boltzmann factor, which gives a velocity of magnitude v a relative probability of

$$e^{-mv^2/2kT}$$

The average value, $\overline{|\mathbf{v}_P - \mathbf{v}_t|}$, is found[6] to be equal to $\sqrt{2}\bar{v}$.

The number of collisions of a projectile molecule with target molecules per unit of time is therefore

$$\{\text{one projectile}\} \qquad \nu = \frac{\text{collisions}}{\text{time}} = \sqrt{2}\pi\, d^2 n \bar{v} \tag{7-17}$$

The total number of collisions in the gas per unit of time and per unit of volume is

$$\{\text{all molecules}\} \qquad \frac{1}{2}\nu n = \frac{1}{\sqrt{2}}\pi\, d^2 n^2 \bar{v} \tag{7-18}$$

The additional factor of n represents the number of projectile molecules per unit of volume, but these are the same molecules as the target molecules, so every collision would be counted twice—in the collision of a, as projectile, with b, as target, and for the same collision, with b as projectile and a as target. The factor of $\frac{1}{2}$ corrects for this double counting.

The time between collisions for any one molecule is

$$\tau = \frac{1}{\nu} = \frac{1}{\sqrt{2}\pi\, d^2 n \bar{v}} \tag{7-19}$$

The average distance traveled between collisions is $\tau\bar{v}$, called the *mean free path*, λ.

$$\lambda = \frac{1}{\sqrt{2}\pi\, d^2 n} \tag{7-20}$$

Molecular Diffusion

Equations were derived above for diffusion and viscous flow involving macroscopic parameters, such as temperature, pressure, and mole fraction. Equivalent equations may be obtained based on a molecular model.

[6]The average for the difference must be between the value $0 \cdot v$, for zero angle between $\mathbf{v}_p$ and $\mathbf{v}_t$, and the value $2 \cdot v$, for 180° between them; it turns out to correspond to 90°. An adequate demonstration for the present is to show that the root-mean-square value is $\sqrt{2}$ of the root-mean-square value of either velocity. For arbitrary angle between the velocities, $\mathbf{v}_p$ and $\mathbf{v}_t$, the vector sum of $\mathbf{v}_p$ and $(-\mathbf{v}_t)$ is $\overline{v_s^2} = (1/N)\sum[v_p^2 + v_t^2 + 2v_p v_t \cos(\mathbf{v}_p, \mathbf{v}_t)]$. Because $\mathbf{v}_p$ and $\mathbf{v}_t$ are uncorrelated, the cross term averages to zero. Thus $\overline{v_s^2} = \overline{v_p^2} + \overline{v_t^2} = 2\overline{v^2}$ and taking the square root of each side gives the factor $\sqrt{2}$. $\overline{(\mathbf{v}_1 - \mathbf{v}_2)^2} = 2\overline{\mathbf{v}_1^2} = 2\overline{\mathbf{v}_2^2}$.

The number of molecules crossing unit area in unit time (from either direction) was found above, in discussing effusion, to be

$$\frac{1}{A}\frac{dN}{dt} = \frac{1}{2}n\overline{v}$$

with $n = N/V$. Each molecule travels an average distance of λ, but this path is not in general perpendicular to the plane of A. The average path perpendicular to A was shown above to be $\frac{2}{3}$ of λ, so the net transport of molecules across the plane per unit time and unit area is

$$-\frac{1}{A}\frac{dN_i}{dt} = \frac{1}{2}\frac{dn_i}{dz}\left(\frac{2}{3}\lambda\right)\overline{v} = \frac{1}{3}\frac{dn_i}{dz}\lambda\overline{v} \tag{7-21}$$

The average speed, $\overline{v}$, is inversely proportional to the square root of the mass of the molecule and therefore inversely proportional to square root of density. The result is therefore consistent with *Graham's law of diffusion*, stated in 1829, that the diffusion rate is inversely proportional to the square root of density of the gas. It gives a diffusion constant $D_i = \frac{1}{3}\lambda\overline{v}$.

Viscosity

Viscosity appears in an ideal gas because adjacent layers of gas have different average momenta in the direction of flow. As molecules diffuse in a direction perpendicular to the flow, between layers moving at different speeds, they carry positive or negative momentum contributions with them. When a molecule moves a distance $\frac{2}{3}\lambda$ in the x direction, it transports an increment of momentum along the z axis given by

$$\Delta p_z = m\frac{dv_z}{dx}\Delta x = m\frac{dv_z}{dx}\frac{2}{3}\lambda$$

The number crossing per unit time is $dN/dt = \frac{1}{2}nA\overline{v}$, so the time rate of momentum transport is the rate of crossing times the momentum transport per crossing, or

$$\frac{dp_z}{dt} = f_z = \frac{dN}{dt}\Delta p_z = \frac{1}{2}nA\overline{v}m\frac{dv_z}{dx}\frac{2}{3}\lambda = \frac{1}{3}nm\lambda\overline{v}\frac{dv_z}{dx}A = \eta\frac{dv_z}{dx}A$$

This predicts that the viscosity, η, should be given by

$$\eta = \frac{1}{3}nm\lambda\overline{v} = \frac{1}{3}\rho\lambda\overline{v} = \frac{m\overline{v}}{3\sqrt{2}\pi d^2} \tag{7-22}$$

which contains the temperature dependence of $\overline{v} \sim \sqrt{T}$, in qualitative agreement with experiment. The equation also shows that because λ is inversely proportional to n, or ρ, the viscosity is, surprisingly, independent of density, ρ.

7.4 Transport of Energy

Newton examined the question of how fast a warm body cools. His conclusion may well be considered self-evident. The temperature of the warmer body decreases with time, at a rate that depends on the temperature difference between the object and its surroundings. If an object at T_1 immersed in a medium at T_2 cools at a rate of δT per second, the same object, at the temperature halfway between T_2 and T_1 immersed in the medium at T_2 will cool at the rate of $\frac{1}{2}\delta T$ per second. This is *Newton's law of cooling*: The rate of cooling is proportional to the temperature difference between an object and its surroundings.

Newton's law is correct as a limiting law for small temperature differences. Unfortunately, it does not accurately describe most real cooling problems, which involve not only conduction but also radiation and, often, evaporation.

Thermal conduction is a transport of thermal energy by and through matter, without net motion of the matter. Convection is a transport of thermal energy by motion of fluid matter that is at a higher or lower temperature than the surroundings. Radiation is the transport of energy (as electromagnetic waves) independently of matter.

Conduction

The mathematical theory of thermal conduction was developed in the first quarter of the nineteenth century by Joseph Fourier. His model for heat was closer to that of Lavoisier than of Rumford or Joule, but for the range of problems of interest, that was no hindrance.

Fourier proposed, and solved, problems such as the following. Given a flat ring, heated at one point, the temperature of each point in the ring will change with time until a *steady state* is reached. The rate of heat input is then equal to the rate of loss from the surface. In the steady state, each point of the ring will have a temperature, $T(x)$, where x measures position around the ring from the heating point. What will that temperature distribution be?

If the source of heat is subsequently removed from the ring, the temperatures will fall with time until eventually the entire ring again assumes a uniform temperature equal to the ambient temperature. How does $T(x)$ depend on time?

The most general problems of the type proposed by Fourier are not easily solved. The equations are differential equations, with coefficients that are not necessarily constant. With some reasonable simplifying assumptions, however, the equations may be written down and at least some properties of the solutions determined.

The amount of thermal energy transferred along the x direction, q, is proportional to the temperature gradient along the x direction, to the cross-sectional area, and to the time. The gradient of temperature is dT/dx.

$$q = dE = -\kappa \frac{dT}{dx} \, dA \, dt$$

The constant of proportionality, κ (kappa), called the *thermal conductivity*, depends strongly on the nature of the substance, but also depends somewhat on the temperature. We neglect the temperature dependence of κ here. The negative sign shows that the flow is from higher to lower temperature, or against the temperature gradient.[7]

For steady-state thermal flow, the flux is easily evaluated at any point for which the conductivity and gradient are known. The temperature is a function of the energy.

[7]Note another peculiarity of customary mathematical notation. If we integrate over time, area, and distance, dx, on the right, the left side is still $dE \sim dT$, just as we can write a mass increment as $dm = \rho \, dx \, dy \, dz$ or as $dm = \rho A \, dz$.

$$C_v = \frac{q}{dT} = \frac{dE}{dT}$$

and therefore

$$dT = \frac{1}{C_v}\, dE$$

so

$$\frac{dT}{dx} = \frac{1}{C_v}\frac{dE}{dx}$$

The equation then becomes

$$\frac{dE}{dt} = -\frac{\kappa A}{C_v}\frac{dE}{dx} \tag{7-23}$$

This is an equation for the diffusion of energy. The time rate of flow of energy is proportional to the gradient of energy, equivalent to Newton's law of cooling. Evaluation to find $E(x, t)$ requires *boundary conditions*, such as the distribution of energy at $t = 0$.

To the approximation that κ and C_v are independent of temperature, dE/dt is independent of x. The same amount of energy crosses each plane perpendicular to the energy gradient per unit of time, so the energy density varies linearly with distance.

Thermal conductivity varies widely with substance. Metals generally have high thermal conductivities (as well as high electrical conductivities). Liquids are intermediate, with κ typically much lower than for metals.

Gases (except helium) are extremely poor thermal conductors. The best forms of insulation usually consist of nonmetallic solids (e.g., glass or textile) plus layers or pockets of air. The gas should be in thin enough layers that convection within the gas does not contribute to conduction. The primary function of the solid is to encapsulate the gas, which is the principal thermal barrier.

Radiation

Newton's law of cooling fails in most practical situations because thermal conduction is only one of the mechanisms for energy transport. Unless the temperature differences are small, radiative transfer is likely to be more important than conduction.

Radiation from a surface is governed by *Kirchhoff's law of radiation,* which says that an object can radiate only those frequencies that it can absorb, and by the Stefan–Boltzmann law,

$$\mathcal{P} = \frac{dE}{dt} = \epsilon\sigma A T^4 \tag{7-24}$$

The power radiated is proportional to the area, to a quantity ϵ characteristic of the surface that is, in general, a function of frequency, and to the fourth power of the temperature. The *emissivity*, ϵ, is also an absorptivity. A good emitter must be a good absorber, at the same frequencies. A perfect absorber, called a *blackbody*, has $\epsilon = 1$ for all frequencies. The *Stefan–Boltzmann* constant is

$$\sigma = 5.67 \times 10^{-8} \mathrm{W/m^2 \cdot K^4}$$

When temperature differences between a sample and its environment are small, the difference between energy emitted and energy absorbed is

$$\frac{dE}{dt} = \sigma A(T_1^4 - T_2^4) = \sigma A(T_1 - T_2)(T_1 + T_2)(T_1^2 + T_2^2)$$

so the rate of energy transfer is approximately proportional to ΔT:

$$\frac{dE}{dt} \mathrel{\dot{\sim}} (T_1 - T_2)$$

For small temperature differences, Newton's law of cooling holds approximately, even when radiation transfer is included.

Because radiative transfer is so important for substantial temperature differences, a reflecting surface provides an effective barrier. A sheet of aluminum foil is roughly as effective in the attic of a home as the very much thicker bats of rock wool or fiberglass insulation. Dewar flasks, for holding liquid nitrogen and other low-temperature materials, have a reflecting surface on the inside of the glass, with the space between glass surfaces evacuated.

Convection

A liquid cools a warm object rapidly because there is much better thermal contact between the liquid and the immersed object than for solids or gases. After the initial cooling, which warms the liquid, the rate of cooling would be very slow except that the liquid flows, bringing cool liquid against the object and exposing the warmed liquid to cooler regions. This kind of flow is known as convection. It relies on the expansion of a liquid or gas as it warms, decreasing the density of the warmer fluid. Cooler fluid, being more dense, is pulled downward in the gravitational field, displacing the warm fluid.

Often this convective flow carries other materials with it. Convection carries water vapor from the oceans to land areas, where it is dropped as rain or snow. Other materials are also carried by the wind, which is driven primarily by temperature differences.

Convection disappears in a gravity-free environment. For example, a candle is rapidly extinguished if it is enclosed in a glass jar and dropped. During the second or so that it is in free fall there is no convection inside the jar. The waste products of combustion accumulate around the wick and no fresh oxygen is brought to the flame. The flame goes out, even though there is still ample oxygen left within the jar.

Calculation of energy transfer by convection is difficult because the flow depends on details of scale and local resistance to flow. Furthermore, although the moving fluid carries energy with it, transfer from the fluid to stationary objects nearly always involves a conduction step across a layer of immobile fluid at the surface.

For example, measurements of energy losses through windows show that aluminum frames contribute little more to heat transfer than do wooden frames. In each case the limiting step is the transfer, by conduction, between the solid material of the window frame or glass and the fluid moving past. That conduction is through a boundary layer of gas at the window surface.

Evaporation

Evaporation of a liquid can be one of the most effective mechanisms for cooling a hot object. Human beings are cooled by evaporation of sweat from the skin; dogs are cooled by evaporation of water from the mouth. A fast way to reduce a fever is with alcohol applied to the skin. The rapid evaporation of the alcohol extracts thermal energy from the surface.

Evaporation cools a liquid because only the most energetic molecules can escape from the potential well of the liquid (or solid). The molecules remaining have less energy, on the average, and therefore are cooler. The amount of cooling is the heat of evaporation times the amount evaporating. The rate of evaporation depends exponentially on temperature, so as the temperature increases, the rate of cooling by evaporation increases rapidly.

Energy lost to a liquid by evaporation may be recovered elsewhere when the vapor condenses. Moisture evaporating inside a house releases its energy to the surface of a cold window as it condenses on the pane. A rainstorm in summer is cooling, as the water evaporates and carries off energy. A rainstorm in cool weather is often accompanied by warming because the condensation of moisture has released the heat of evaporation.

Thermal Conductivity of Gases

The equation for thermal conductivity in an ideal gas is easily obtained from the similar equations for diffusion of molecules and momentum. Energy diffuses from one plane to another. The rate of energy diffusion in the x direction is

$$\frac{dE}{dt} = -\frac{1}{3}n\bar{v}\lambda \frac{dE}{dx} A$$

Comparing this equation with the general expression for thermal diffusion,

$$\frac{dE}{dt} = -\frac{\kappa A}{C_v}\frac{dE}{dx}$$

shows that the thermal conductivity, κ, is

$$\kappa = \frac{1}{3}n\lambda\bar{v}C_v \qquad (7\text{-}25)$$

with C_v the heat capacity per molecule (molar heat capacity$/N_A$) and $n = N/V$. Alternatively, the equation may be written in terms of heat capacity per unit of mass times the mass per molecule.

$$\kappa = \frac{1}{3}n\lambda\bar{v}mc_v$$

The last equation shows that the thermal conductivity should be equal to the product of viscosity and heat capacity

$$\kappa = \eta c_v \qquad (7\text{-}26)$$

In the analyses above, no distinction has been made between the various types of collision processes that occur in the gas. In practice, the mean free path between scattering processes is measurably different from the mean free path for momentum transfers or energy transfer. The equations derived, therefore, are correct in general form, but factors as great as 2 or more must be applied if constants are to be carried from one type of experiment to another.[8] For example, because energy is not totally exchanged in each collision, and for other minor reasons, such as the dependence of molecular speeds on the temperature, the experimental values show that the ratio of $\eta C_v/\kappa$ is typically in the range of 0.5 to 0.75, rather than 1.0.

SUMMARY

A *probability distribution* gives the (relative) probability of a value falling in a certain interval, for possible values of that interval.

The distribution function for molecular speeds contains three factors: v^2, because the number of possible states is proportional to v^2; $e^{-mv^2/2kT}$, because the probability for any state falls with energy as $e^{-\Delta E/kT}$; and a normalization constant.

Maxwell–Boltzmann distribution: $\dfrac{dN}{dv} = 4\pi N\left(\dfrac{m}{2\pi kT}\right)^{3/2} v^2 e^{-mv^2/2kT}$.

The most probable speed is $\sqrt{\dfrac{2kT}{m}}$; the average speed is $\sqrt{\dfrac{8kT}{\pi m}}$; the rms speed is $\sqrt{\dfrac{3kT}{m}}$.

Gaseous *diffusion* is proportional to the concentration gradient:

$$\frac{1}{A}\frac{dN_i}{dt} = -D_i\frac{dX_i}{dz}$$

In viscous flow, the viscous (drag) force $= f_z = \eta A_s \dfrac{dv}{dx}$.

Flow rate of an incompressible fluid through a tube is given by Poiseuille's equation,

$$\frac{dV}{dt} = \frac{\pi}{8}\frac{(P_1 - P_2)R^4}{\eta L}$$

or for a gas measured at P_o,

$$\frac{dV}{dt} = \frac{\pi(P_1^2 - P_2^2)R^4}{16\eta L P_o}$$

Rate of *effusion* is $\dfrac{dN}{dt} = \dfrac{1}{4}nA\bar{v}$.

Mean free path $= \lambda = \dfrac{1}{\sqrt{2}\pi d^2 n}$.

[8]See, for example, S. Chapman and T. G. Cowling, *The Mathematical Theory of Non-uniform Gases* (Cambridge: Cambridge University Press, 1958), for a more exact treatment.

Average transport distance $= \overline{d}_{\perp} = \frac{2}{3}\lambda$.

Mean time between collisions of a molecule $= \tau = \dfrac{\lambda}{\overline{v}} = \dfrac{1}{\sqrt{2}\pi d^2 n\overline{v}}$.

Rate of collisions is $\dfrac{1}{V}\dfrac{dN}{dt} = \dfrac{1}{\sqrt{2}}\pi d^2 n\overline{v}$.

Diffusion constant is $D_i = \frac{1}{3}\lambda\overline{v}$.

Viscosity of an ideal gas $= \eta = \dfrac{1}{3}nm\lambda\overline{v} = \dfrac{1}{3}\rho\lambda\overline{v} = \dfrac{m\overline{v}}{3\sqrt{2}\pi d^2}$.

Thermal conduction: $\dfrac{dE}{dt} = -A\dfrac{\kappa}{C_v}\dfrac{dE}{dx}$.

Power radiated $= \mathscr{P} = \epsilon\sigma AT^4$.

Thermal conductivity of a gas: $\kappa \approx \frac{1}{3}n\lambda\overline{v}C_v = \eta c_v$.

QUESTIONS

7.1. Sketch the form of the Maxwell–Boltzmann distribution curve.
  a. Label axes properly and show carefully the shape of the curve at each end.
  b. Show the approximate positions of the average speed, the most probable speed, and the rms speed.
  c. How many molecules have the most probable speed?
  d. How does the area under the curve change with temperature?
  e. Write the equation (representing the constants in front by A, if you wish). Specify the meaning of the symbols.

7.2. How does mean free path depend on the speed of the molecules? Can you justify this relationship with physical arguments? (Consider taking a movie of moving molecules and varying the speed of playback.)

7.3. Does the mean free path give the distance traveled since the last collision, or does it give the distance traveled before the next collision? Are these different? Are they additive? (Consider a similar question for throwing dice. What is the probability of getting 12 in future throws versus the probability of having gotten 12 in past throws, assuming no information is available on past throws? Should you add the average number forward and backward to find the number between getting 12's?) How do these questions relate to "preparation" of a system?

7.4. Effusion is proportional to the average speed of the molecules. Diffusion depends on viscosity, which is also a function of $\overline{v}$. For two molecules of the same size, how does the relative rate of diffusion depend on relative average speeds?

DERIVATIONS AND PROOFS

7.1. Show that $dv_x\, dv_y\, dv_z$ can be replaced by $v^2 \sin\vartheta\, d\vartheta\, d\phi\, dr$, and that this integrates to $4\pi v^2$, to be evaluated between $v = 0$ and v.

7.2. **a.** Show that if $\rho(v)$ is the Maxwell-Boltzmann distribution function, setting $d\rho/dv = 0$ yields $v_{\text{prob}} = \sqrt{2kT/m}$.
  b. Show how $4\sqrt{a^3/\pi}\int_0^\infty v^3 e^{-\alpha v^2}\, dv = \overline{v}$ can be integrated by parts to give $\overline{v} = \sqrt{8kT/\pi m}$.

c. Show that $\overline{\overline{v}} = \sqrt{3kT/m}$ can be obtained directly by integration of $\int_0^\infty v^2 \rho(v)\, dv$.

7.3. Find the drag force exerted on a plane surface, of area A, when another surface (of greater length) moves past it, a distance d away, at a (relative) speed v, if the space between is filled with a gas of viscosity η. Neglect edge effects.

7.4. Find the drag (expressed as a torque, τ) on a cylinder, of radius R and length L, if a second (longer) cylinder of radius $R + d$ rotates concentrically about it with angular speed ω if the space between the cylinders is filled with a gas of viscosity η.

7.5. Show that the energy transferred by an elastic collision of a molecule of mass m and speed v with a plane piston face moving at speed u through a gas is $W = 2mu(v \cos\vartheta - u)$, which is nonzero if and only if the speed of the piston is nonzero.

7.6. Consider a gas, at P and T, confined in a cylinder by a piston of mass M. The external pressure is $P_{ext} \ll P$.

a. Show that if the piston is released, the work done *by* the gas is $W = P_{int}\, dV \gg P_{ext}\, dV$, provided that $M \neq 0$ and therefore the speed of the piston is $u \ll v_{sound}$.

b. Show that P_{int}, the pressure exerted on the piston in part a, is less than P, the equilibrium pressure of the gas for any V and T.

c. Show, with a numerical calculation for reasonable values of M, ΔP, and distance traveled, that when a large force is applied to a piston externally, $u \ll v_{sound}$ is a good assumption and therefore the compression will be nearly reversible.

7.7. Show that for the conditions of Derivation 7.6 the pressure exerted by a gas on a moving piston is $P = (mN/V)(\frac{1}{3}\overline{\overline{v}} - uv + u^2)$. Show that for a Maxwell–Boltzmann distribution (hence small u), the pressure may therefore be written as $P/P_{eq} = 1 - (8/\pi)\alpha + (8/\pi)\alpha^2$, where $\alpha = u/\overline{v}$ (which is approximately the Mach number).

7.8. Show that if the incremental probability of a collision from a certain state in a time interval dt is $P(0)\nu\, dt$, where $P(0)$ is the probability (between 0 and 1) of the molecule being in the state at $t = 0$, then

$$P(t) = P(0)e^{-\nu t}$$

is the probability of the molecule still being in the state (i.e., not having suffered a collision) after a time t.

7.9. Show, by geometric reasoning, the dependence of temperature on radius when a hot wire, at temperature T_1, lies on the axis of a cool cylinder, at temperature T_2, if energy transfer through the intervening solid is by conduction only.

7.10. Find the fraction of the molecules that will pass through an opening, A, per second and be removed from a gas (or equivalently, will strike a cold surface of area A and be frozen out of the gas, if the sticking coefficient is taken to be unity) if there are originally N molecules in a volume V.

7.11. Show that the requirement for Newton's law of cooling to be followed in the presence of radiation between bodies at T_1 and T_2 ($T_1 > T_2$) is that $\Delta T/T_2 \ll 1$. What is the dependence of rate of energy transfer on T_2?

PROBLEMS

7.1. A 6:1 mixture, by volume, of neon and krypton (molar masses 20.183 and 83.80 g) is allowed to effuse through a small hole into an evacuated space.
  a. What is the composition of the mixture that first passes through, expressed as weight percentage?
  b. What is the composition as mole percentage?

7.2. Two identical containers hold equal numbers of molecules of He (molar mass 4.0 g) and Kr, respectively, at the same temperature.
  a. What is the ratio of the average momentum of a He molecule to the average momentum of a Kr molecule?
  b. What is the ratio of the number of wall collisions per second for He molecules compared to Kr molecules?
  c. What is the ratio of the product of average momentum and wall collisions per second for He compared to Kr? Why is this result expected?

7.3. Dry air is typically 78 mol % N_2, 21 mol % O_2, and 0.9 mol % Ar, plus smaller amounts of other gases. The molar masses are 14.01, 16.00, and 39.95 g for the atoms N, O, and Ar. What is the composition of the gas that will leak into a vacuum system through a pinhole?

7.4. Separation of ^{235}U from ^{238}U was undertaken on a large scale at Oak Ridge during the 1940s by passing UF_6 gas through membranes. If the fraction of $^{235}UF_6$ is initially θ_o, and the process can be treated as equivalent to effusion, what is the fraction of $^{235}UF_6$ after one separation step at room temperature?

7.5. For O_2 at 25°C, find
  a. The most probable speed.
  b. The average speed.
  c. The rms speed.

7.6. Calculate the mean free path for an oxygen molecule at 25°C at 1.0 atm, assuming an effective molecular diameter of 3.6Å.

7.7. The measured value of the mean free path of N_2 at 0°C and 1.0 atm is 60 nm.
  a. What is the effective molecular diameter of the N_2 molecule?
  b. What mean free path should be expected at 25°C?
  c. What mean free path should be expected at 77 K?

7.8. What effective molecular diameter would be calculated for N_2 from the measured viscosity of 1.67×10^{-3} Pa·s, at 0°C and 1.0 atm? (This is a less accurate approach to molecular diameter than is the mean free path.)

7.9. What is the maximum pumping rate, in g/s, on a vacuum system at 10^{-5} torr if the pumping line has a diameter of 1.0 cm and a length of 40 cm? Assume that the gas is N_2 at 25°C and $\eta = 1.75 \times 10^{-3}$ Pa·s.

7.10. Calculate the rate of evaporation of water molecules from the surface of ice at 0°C. The equilibrium vapor pressure of ice is 4.579 torr. Express the value in molecules/cm^2 · s and in g/cm^2 · s.

7.11. Molecular diameters have been measured, from viscosities, as 2.0 Å for He, 2.9 Å for Ar, and 3.15 Å for N_2. Calculate approximate values for the thermal conductivities, κ, for each of these gases.

7.12. The sun (radius 7.0×10^8 m) has an apparent surface temperature of 6000 K. It is 1.5×10^{11} m from the earth (radius 6.4×10^6 m), which has an average surface temperature of about 270 K.
  a. Find the power radiated by the sun if it acts as a blackbody.
  b. Find the power incident on the earth from the sun.

c. Find the solar power per square meter on the earth. (The value of 1 kW/m^2 is often accepted for such purposes as estimating the available energy for solar conversion. Can you explain the difference between your figure and 1 kW/m^2?)

d. Estimate the power radiated by the earth (using the average temperature).

7.13. Evaluate the drag on a cylinder, of radius 2.0 cm and length 10 cm, if a second (longer) cylinder of radius 3.0 cm rotates concentrically about it at 3450 rpm, if the gas is air ($\eta = 185 \mu P = 18.5 \mu Pa \cdot s$). [See Derivarion 7.4.]

7.14. Find the rate of thermal energy transport by conduction, in $J/m^2 \cdot s$, between two parallel conducting surfaces, one at 25°C, the other at 0°C, separated by 2.5 cm of

a. Air.

b. Helium.

c. Argon.

7.15. If a molecule takes N "steps," each of rms length λ and each in a random direction, the average distance from the starting point will be $\lambda \sqrt{N}$. Find the mean time for a molecule of ^{14}N-^{15}N to diffuse 1.0 m through air (approximating the air as pure $^{14}N_2$) at room temperature and 1.0 atm.

CHAPTER 8

Gases

Early thermodynamics emphasized the properties of gases, because gases are comparatively simple, especially when they approximate the behavior of ideal gases. Even nonideal gases are described quite well by the ideal gas equations, with correction terms that are typically small.

In this chapter we consider the equations of state for gases that deviate from the ideal gas law and will look at one of the key experiments that demonstrated the differences in thermodynamic behavior of ideal and nonideal gases. The chapter concludes with consideration of adiabatic expansions and application of the theories to moving fluids.

8.1 Nonideal Gases

An ideal gas follows the equation $PV = nRT$. For any given temperature, a plot of P versus V will be a hyperbola, as shown in Fig. 8.1. Different temperatures give different hyperbolas. Isothermal lines are called *isotherms*.

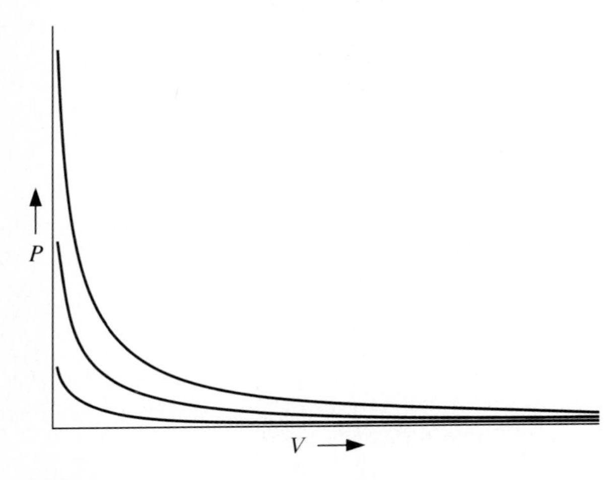

FIGURE 8.1 Pressure versus volume isotherms for an ideal gas.

A reasonable first-approximation correction to the ideal gas law was suggested by J. D. van der Waals, in 1867. First, the volume available to the gas molecules is not the total volume of the container, but that volume less the volume actually occupied by the gas molecules. Therefore, V is replaced by $V - nb$, where b is a constant that must be determined for each gas. A very rough approximation of this molecular volume would be the volume occupied by the liquid.[1]

The second consideration is that real gas molecules tend to stick to each other. That is why gases condense at low temperatures. The cohesiveness of the molecules decreases the pressure of the gas by a term that may be written n^2a/V^2.

The pressure exerted by gas molecules is proportional to the impulse per collision and to the number of collisions per second, which depends on the number of molecules per unit volume. A molecule within a gas is pulled equally in all directions, but a molecule approaching the wall is pulled unevenly, with a force proportional to the density of molecules pulling on it. The effect per molecule is proportional to N/V, and the number of molecules affected is proportional to N/V. Therefore, the correction to the pressure is $N/V(aN/V)$, where a describes the strength of the pull by one molecule on another.

The actual pressure of a gas should accordingly be

$$P = \frac{nRT}{V - nb} - \frac{n^2a}{V^2}$$

The constants a and b are called the *van der Waals constants* of the gas. The equation is usually written in the form

$$\left(P + \frac{n^2a}{V^2}\right)(V - nb) = nRT \tag{8-1}$$

or if V is taken as the volume per mole (i.e., $n = 1$, or $V/n \rightarrow V$),

$$\left(P + \frac{a}{V^2}\right)(V - b) = RT$$

Multiplying out the first equation gives a cubic equation in the volume, V.

$$PV^3 - n(Pb + RT)V^2 + n^2aV - n^3ab = 0$$

A plot of P versus V for this cubic equation gives the family of curves shown in Fig. 8.2. At high temperatures, the curve closely approximates the hyperbola of an ideal gas. At low temperatures the curve displays an intermediate maximum. One peculiarity of the curve is particularly striking. It indicates that for a given pressure there will be three different possible values of the volume.

[1]Analysis suggests that the value of b should be greater than the sum of the volumes of the molecules. The centers of two molecules cannot approach closer than one molecular diameter (cf. Fig. 7.11). Approximating the molecules as cubes, the volume excluded per molecule is $(2r)^3 = 8r^3$, compared to the volume of a (spherical) molecule of $\frac{4}{3}\pi r^3 \approx 4r^3$. This factor of $\frac{8}{4} = 2$ is close to values observed for loosely packed marbles. If the excluded volume is calculated as $\frac{4}{3}\pi(2r)^3 \approx 32r^3$, and this volume is attributed to two molecules, the resultant factor is $\frac{1}{2} \times \frac{32}{4} = 4$. The actual (maximum) volume of the liquid (at the critical point, discussed below) is approximately $3b$, giving a value of $b \approx \frac{1}{3}(2V_{mol})$, for cubes, or $\frac{1}{3}(4V_{mol})$ for spheres. In practice, b is considered an empirical parameter that is chosen for a best fit for the particular need.

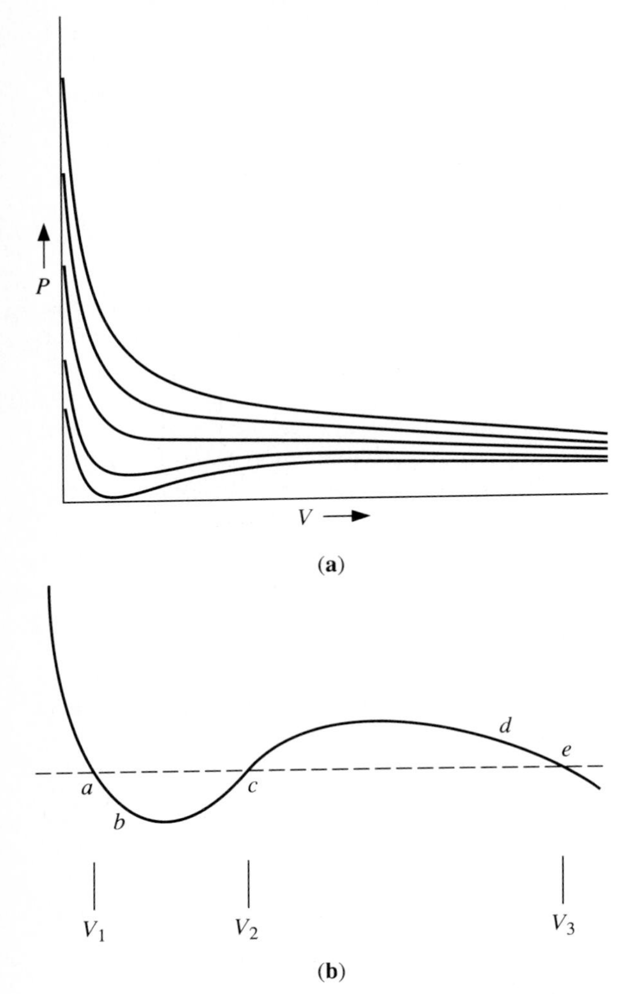

FIGURE 8.2 (a) Isotherms of a van der Waals gas. (b) Expanded segment (schematic) below the critical point, showing three volumes for the same pressure along a single isotherm.

To interpret this low-temperature van der Waals plot, consider first what will be present under limiting conditions. At very low pressure, the gas approaches an ideal gas, occupying a large volume. This must be the portion of the curve at the right. At the low-volume end to the left, on the other hand, the figure shows that a substantial increase in pressure will cause very little change in volume. The fluid is nearly incompressible. That is the behavior expected for a liquid.

If we interpret the left-hand segment of the curve as describing the liquid phase and the right-hand segment as gas, the two extreme values of the volume, V_1 and V_3, can be interpreted as the volume of the liquid and the volume of the vapor. These exist in equilibrium with each other. The intermediate volume may be considered to be a mathematical artifact, which has no physical significance.

It is easy to show that even if the intermediate volume, V_2, represented a real volume of the fluid, it could not be observed experimentally. It is possible to superheat liquid or supercool vapor and thus trace out portions of the P–V curve representing a–b and e–d. But suppose that we have a fluid (gas or liquid) with pressure, volume, and temperature corresponding to point c.

If the volume of a small volume element of fluid were to *increase* very slightly, as might be expected to occur from statistical fluctuations, the curve shows that the pressure of that clump should *increase*. If the pressure increases, that would *compress* the remainder of the fluid, which will *lower* the pressure of the bulk. This, in turn, allows the expanding clump to expand further and causes the bulk to be further compressed, until the portions go beyond the maximum or minimum of the curve and settle into the stable equilibrium of liquid plus vapor. The intermediate volume state could not persist for more than a small fraction of a second, and its properties could not be measured without destroying the state.

Maxwell pointed out that the two areas between the horizontal line and the two portions of the isotherm must be equal, if they are to represent equilibrium states of the gas. Otherwise, it would be possible to go through a reversible isothermal cycle around the enclosed areas with the system performing net work.

Interpretation of the low-temperature van der Waals curve as representing liquid-vapor equilibrium allows us to determine the constants, a and b, experimentally. The highest temperature at which the liquid-vapor equilibrium can occur is the curve T_c. This is known as the *critical point*.[2] The critical constants, T_c, V_c, and P_c, satisfy van der Waals' equation. The critical point is a horizontal inflection point; the slope is zero, $(\partial P/\partial V)_T = 0$, and the curvature is zero, so $(\partial^2 P/\partial V^2)_T = 0$.

We can also solve for a and b by recognizing that any polynomial can be written as a product of its factors, which can be expressed in terms of the roots. If the roots are x_1, x_2, and so on, then $(x - x_1)(x - x_2)\cdots = 0$. At the critical point the three roots of the cubic equation in V are identical, and the volume is equal to V_c, so the equation can be written

$$(V - V_c)^3 = 0$$

After expanding this product, the coefficients of powers of V may be equated with those in the original equation for V that are in terms of a and b. Solution of the equations for a and b gives

$$a = \frac{27R^2T_c^2}{64P_c} \qquad b = \frac{RT_c}{8P_c}$$

Van der Waals' equation is generally regarded as semiempirical. In practice, the van der Waals constants (Table 8.1) are often adjusted to give the best fit for a particular temperature region. Theoretical treatments of condensation phenomena show that the van der Waals equation cannot be correct. At the same time, it is clear that the equation does give a good approximate fit in the region of condensation, much better than the ideal gas equation and adequate for many purposes.

Other equations of state have been suggested for nonideal gases. Among them are the equations of Dieterici and of Berthelot.

$$\text{Dieterici:} \qquad P = \frac{RT}{V - nb}e^{-a/nVRT} \tag{8-2}$$

$$\text{Berthelot:} \qquad PV = nRT\left[1 + \left(1 - \frac{6T_c^2}{T^2}\right)\frac{9PT_c}{128P_cT}\right] \tag{8-3}$$

[2]The term *point* sometimes represents a specific state, or set of values of external variables, but the term is also applied to indicate a temperature, as in *boiling point*, *critical point*, *Boyle point*, and so on.

Table 8.1 Critical and van der Waals Constants for Selected Gases

	Critical Constants			van der Waals Constants	
Gas	T_c **(K)**	P_c **(MPa)**	$V_c \left(\frac{cm^3}{mol}\right)$	$a \left(\frac{Pa \cdot m^6}{mol^2}\right)$	$b \left(\frac{cm^3}{mol}\right)$
Acetone	508.2	4.8	217	1.409	99.4
Ammonia, NH_3	405.6	11.29	72.5	0.4225	37.07
Argon, Ar	151	4.9	75.2	0.1363	32.19
Benzene, C_6H_6	561.6	4.8	257	1.824	115.4
CO_2	304.2	73.9	95.7	0.3640	42.67
CO	134	3.5	90.1	0.1505	39.85
Chlorine, Cl_2	417.2	7.71	124	0.6579	56.22
Ethane, C_2H_6	305.2	4.94	140	0.5562	63.80
Helium, He	5.2	0.229	57.8	0.003457	23.70
Hydrogen, H_2	33.2	1.30	65.0	0.02476	26.61
Krypton, Kr	210	5.5	110	0.2349	39.78
Methane, CH_4	190.6	4.64	99.0	0.2283	42.78
Neon, Ne	44.4	2.62	41.7	0.02135	17.09
Nitrogen, N_2	126.0	3.39	84.65	0.1408	39.13
Oxygen, O_2	154.4	5.03	74.4	0.1378	31.83
Water, H_2O	647.2	22.06	45	0.5536	30.49
Xenon, Xe	331.4	5.90	113.7	0.4250	51.05

Source: Based on data from the *Handbook of Chemistry and Physics* and the Landolt–Bornstein Physical Chemical Tables. 1 MPa = 10^6 N/m^2; 1 cm^3 = 10^{-6}m^3.

These two fit experimental curves about as well as the van der Waals equation; all three have two adjustable parameters.

More accurate equations include that of Keyes:

$$\text{Keyes:} \qquad P = \frac{RT}{V - \beta e^{-\alpha/V}} - \frac{A}{(V - f)^2} \tag{8-4}$$

with empirical constants α, β, A, and f, and the equation of J. A. Beattie and O. C. Bridgeman,

$$P = \frac{RT}{V^2}\left(1 - \frac{C}{VT^3}\right)\left(V + B - \frac{bB}{V}\right) - \frac{A}{V^2}\left(1 - \frac{a}{V}\right) \tag{8-5}$$

with five adjustable parameters, a, b, A, B, and C. Sometimes it is preferable to employ a power-series expansion, such as

$$\frac{PV}{RT} = \alpha + \beta P + \gamma P^2 + \delta P^3 + \cdots$$

or

$$PV = nRT[1 + B(T)V^{-1} + C(T)V^{-2} + \cdots] \tag{8-6}$$

The coefficients $B(T)$, $C(T)$, and so on, are called the second, third, and so on, *virial coefficients*.

8.2 The Law of Corresponding States

A powerful conceptual argument may be derived from an equation such as that of van der Waals. Write each of the variables as a ratio to its critical value using molar volumes.

$$P' = \frac{P}{P_c} \qquad V' = \frac{V}{V_c} \qquad T' = \frac{T}{T_c}$$

Then the equation

$$(P + \frac{a}{V^2})(V - b) = RT$$

becomes

$$\left(P'P_c + \frac{a}{V'^2V_c^2}\right)(V'V_c - b) = RT'T_c$$

which reduces to the simple form

$$\left(P' + \frac{3}{V'^2}\right)(3V' - 1) = 8T' \tag{8-7}$$

The remarkable thing about this equation, first obtained by van der Waals in 1881, is that there are no longer any adjustable parameters to distinguish one gas from another. It says that insofar as gases obey van der Waals' equation, every gas behaves exactly like every other when described in terms of the reduced variables, P', V', and T'.

The consequences of this result are often expressed as the *law of corresponding states:*

> *Every substance will tend to act like every other substance for a proper scaling of temperature, pressure, and volume variables.*

Even though the law of corresponding states is too broad—it is easy to find exceptions—it is an extremely valuable method of evaluating properties and predicting effects of temperature.

A good example of the law of corresponding states is the family of plots of PV/nRT versus P of real gases for various temperatures, shown in Fig. 8.3. At low temperatures the curve initially has a negative slope, passing through a minimum before rising above the value one. At high temperatures, however, the slope is positive from the beginning. For some intermediate temperature, called the *Boyle point,* or *Boyle temperature,* the gas obeys the ideal gas law over a substantial range of pressure. The Boyle point is $T_B = a/bR$, in terms of

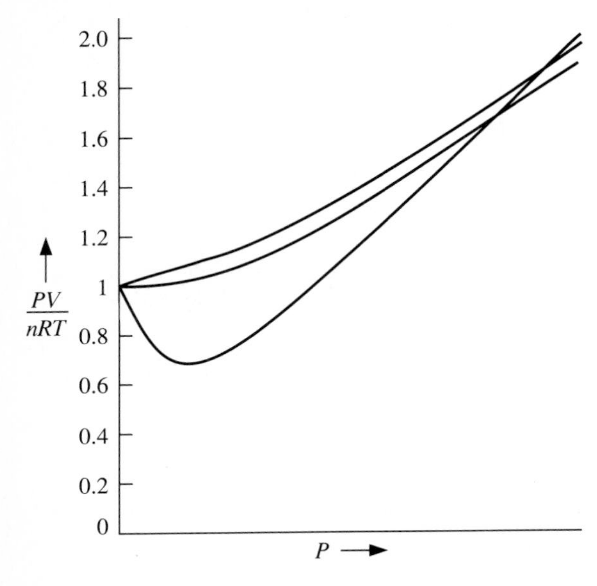

FIGURE 8.3 For real gases, PV is a function of P, which depends on the temperature. Upper curve: $T = 8.22T_C$ (H_2 at 273 K); middle curve: $T = 3T_C$ (N_2 at 378 K); lower curve: $T = 1.03T_C$ (CO_2 at 313 K). Pressure 0 to 1000 atm for upper curves; 0 to 10,000 atm for lower curve.

the van der Waals constants a and b. (Similarly, the Joule–Thomson inversion temperature, where μ changes sign, is equal to $2a/bR = 2T_B$; see Section 8.3.)

The behavior shown in Fig. 8.3 may be explained in terms of the forces between atoms or molecules of the gas. At large separations there is no significant force acting; the curve is flat. When atoms or molecules approach each other, however, there is an attractive force, known as the *van der Waals force*, completely apart from any ionic, polar, or bonding attractions between the molecules.

The weak van der Waals attraction is sufficient to cause condensation at low temperatures. A curve of potential energy against separation distance will

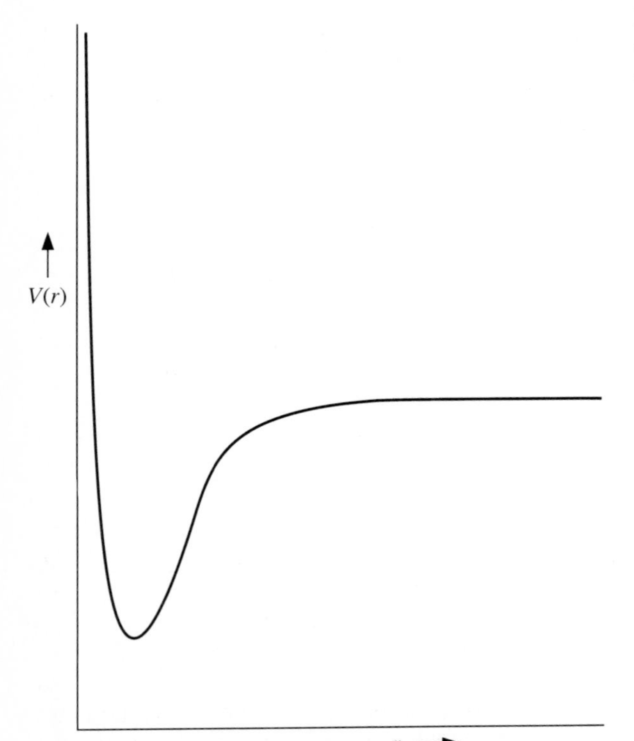

FIGURE 8.4 Potential energy versus intermolecular distance shows a minimum at small r.

be somewhat like that shown in Fig. 8.4. The force of attraction, equal to the slope of the curve, is positive (i.e., attractive) to the right of the equilibrium distance. To the left of the equilibrium distance nuclear–nuclear repulsions (as well as electron–electron repulsions) make the slope negative.

An appropriate way of visualizing the source of the van der Waals force is to consider two hydrogen atoms near each other. As the electron of one atom moves about the nucleus, there is, at any instant, a dipole moment formed by the positive nucleus and negative electron. This dipole moment induces a dipole moment in the second atom, so that the second atom will have its electron in such a position that the two dipole moments attract each other. Although this instantaneous dipole-induced dipole attraction is quite weak, it is sufficient to cause the atoms or molecules to stick together to form a liquid or solid.

The quantum mechanical solution of the problem, showing that dipole-induced dipole forces can explain the van der Waals attraction, was first obtained by London, in 1930. The forces, also called "dispersion" forces, depend on the polarizability, α, of the molecules, which is approximately the ratio of the induced dipole to the electric dipole field causing the induced dipole. The potential is of the form

$$U = -\frac{3h\nu_o\alpha^2}{4r^6} \tag{8-8}$$

where ν_o is a frequency characteristic of the molecule.

When the kinetic energy of an approaching gas particle is small, the attractive force makes the molecules stick together, reducing the pressure. At low temperatures, PV/nRT is less than 1.

On the other hand, if the energy of an approaching particle is large, the potential well has very little effect. The incoming particle sails across the top of the well, far enough above it that there is little change in the motion. The particle encounters the repulsive force at the inner side of the well. Because the high-energy particle experiences primarily this repulsive force, the pressure is greater than for an ideal gas, so PV/nRT is greater than 1.

Repulsion terms also dominate at very high pressure. If the volume is sufficiently small that the molecules are likely to approach closer than the equilibrium intermolecular distance, they encounter the repulsive slope of the potential well. This gives rise to the additive term, $-nb$, on the volume in van der Waals' equation. At high pressures, the a/V^2 term will be negligible compared to the pressure, but because $V - nb$ is smaller than V, the pressure must be higher than for an ideal gas at the same volume and temperature. Although b is predicted to be one-third of the volume at the critical point, experimental values typically give slightly larger values of b/V_c for a best fit.

8.3 Joule–Thomson Expansion

Soon after the dependence of pressure and volume on temperature was determined, Gay–Lussac attempted, in 1807, to measure the change in temperature of a gas as it expanded freely into a vacuum. His efforts were repeated by Joule in 1844, but both found only that there was no measurable temperature change.

The apparatus is shown schematically in Fig. 8.5. A gas at pressure P in the left container is connected, through a valve, to an identical, evacuated container on the right. The entire assembly is surrounded by water, at a known temperature.

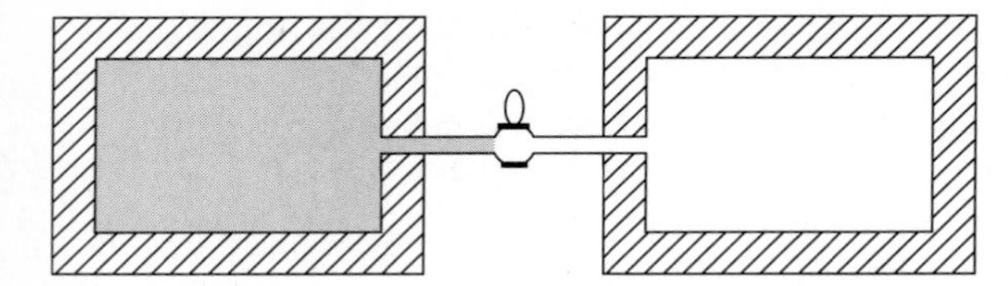

FIGURE 8.5 The gas initially in the left-hand container, of volume V_1, at a pressure P_1, expands to fill the evacuated right-hand container, giving a final pressure P_2 in the total volume V_2.

When the valve is opened, the gas expands to fill each container at half the original pressure. The gas cannot do work on itself, or on the rigid container, and initially, at least, $Q = 0$. The change in energy is

$$dE = \left(\frac{\partial E}{\partial T}\right)_V dT + \left(\frac{\partial E}{\partial V}\right)_T dV$$

or

$$dE = C_V\, dT + \left(\frac{\partial E}{\partial V}\right)_T dV$$

Before temperature equilibration with the surroundings, $Q = 0$, $W = 0$, and therefore, $\Delta E = 0$. Hence

$$\left(\frac{\partial E}{\partial V}\right)_T = -C_v \frac{\Delta T}{\Delta V}$$

The early experiments showed no temperature change of the gas as it expanded. More sensitive measurements will show a slight change in temperature for most gases, which may be either a temperature rise or drop, depending on the gas and its initial temperature and pressure. The temperature change is identically zero for an ideal gas and small (positive or negative) for a real gas. For an ideal gas, the temperature depends only on the energy, and the energy did not change in this experiment.

William Thomson (later Lord Kelvin) proposed a more sensitive test of the dependence of the energy of a gas on the volume, and he and Joule carried out a

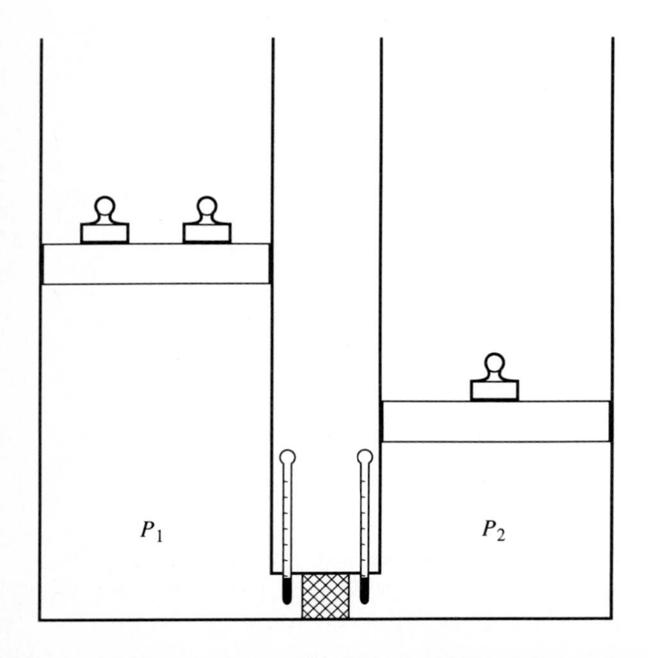

FIGURE 8.6 Joule–Thomson type of experiment. Each pressure is constant. A porous plug allows gas to pass from left to right.

series of measurements, in Joule's brewery, between 1852 and 1862. A schematic representation of the apparatus is shown in Fig. 8.6. The piston on the left and the piston on the right are each exerting constant pressure on the gas, with the pressure, P_1, on the left greater than the pressure, P_2, on the right.

Consider a certain volume of gas or liquid, V_1, initially on the left. The piston on the left does work on the fluid equal to $-\int_{V_1}^{0} P_1\,dV$, if the volume decreases to zero.[3] The fluid, as it emerges on the right, does work on the right-hand piston equal to $\int_0^{V_2} P_2\,dV$, if the volume increases from zero to V_2. The net work done on the fluid is

$$W = -\int_{V_1}^{0} P_1\,dV + -\int_{0}^{V_2} P_2\,dV = P_1V_1 - P_2V_2 = -\Delta(PV) \qquad (8\text{-}9)$$

There is no thermal energy transfer during this expansion, so $Q = 0$. The energy change of the fluid is

$$\Delta E = Q + W = 0 - \Delta(PV)$$

and the enthalpy change is

$$\Delta H = \Delta E + \Delta(PV) = -\Delta(PV) + \Delta(PV) = 0 \qquad (8\text{-}10)$$

Thus the Joule–Thomson expansion is an expansion at constant enthalpy.

The quantity directly measured in this experiment is called the Joule–Thomson coefficient. It measures how the temperature of the fluid changes as the pressure changes, from one constant pressure to another.

$$\mu_{\text{JT}} = \left(\frac{\partial T}{\partial P}\right)_H \qquad (8\text{-}11)$$

For the constant enthalpy condition of this expansion,

$$\{\Delta H = 0\} \qquad \Delta T = \mu_{\text{JT}}\,\Delta P$$

The Joule–Thomson coefficient is zero for an ideal gas. For an ideal gas, the energy and the enthalpy depend only on the temperature. $(\partial E/\partial V)_T = 0$ and $(\partial H/\partial P)_T = 0$. These conditions are equivalent to the statement that $PV = nRT$, or $PV =$ constant for a constant temperature. At constant temperature for an ideal gas, $\Delta(PV) = 0$ and no net work is done in a Joule–Thomson expansion. The temperature remains constant and

$$\Delta E = Q + W = 0 - \Delta(PV) = 0$$

For a real gas or a liquid, the dependence of enthalpy on pressure may be found in terms of the Joule–Thomson coefficient. Let $H = H(P,T)$. Then

$$dH = \left(\frac{\partial H}{\partial T}\right)_P dT + \left(\frac{\partial H}{\partial P}\right)_T dP$$

$$= C_p\,dT + \left(\frac{\partial H}{\partial P}\right)_T dP$$

[3]Although it is not experimentally feasible to reduce the volume to zero, the equations obtained deal only with rates of change of temperature with pressure and are thus independent of this assumption.

Set $dH = 0$ and divide through by dP to obtain

{constant H} $$\left(\frac{\partial H}{\partial P}\right)_T = -C_p \frac{dT}{dP}$$

The last derivative (at constant H) is the Joule–Thomson coefficient, so

$$\left(\frac{\partial H}{\partial P}\right)_T = -C_p \mu_{JT}$$

or

$$\Delta H = -\int_{P_1}^{P_2} \mu_{JT} C_p \, dP \tag{8-12}$$

A real gas may show a temperature increase or decrease in a Joule–Thomson expansion. Most gases show a drop in temperature in the vicinity of room temperature. Negative ΔT divided by a negative ΔP gives a positive μ_{JT}. The drop in temperature is easily observed, for example, if a CO_2 fire extinguisher is opened to the atmosphere. The gas expands from the high internal pressure of the tank to 1 atm external pressure. The chilling effect is sufficient to reduce a significant portion of the expanding gas to solid dry ice, at $-78.5°C$. Some commercial producers of dry ice employ this method of refrigeration.

Oxygen, nitrogen, and the other so-called "permanent" gases cannot be liquefied by compression, because they are well above their critical temperatures. They must be cooled as well as compressed. The primary method of cooling gases to liquefy them is by a series of Joule–Thomson expansions.

The gas is compressed, which raises its temperature. It is then cooled to the temperature of its surroundings and allowed to expand against an external pressure. The expansion step (for which μ_{JT} is positive) cools the gas and the cooled gas is brought in contact, through a heat exchanger, with incoming compressed gas. The incoming gas is thereby initially cooled. When it, in turn, is allowed to undergo a Joule–Thomson expansion, it reaches a lower final temperature, which permits incoming compressed gas to be cooled to a lower temperature. A succession of such steps gives compressed gas below its critical point, which therefore liquefies.

An example of liquefaction of oxygen will illustrate the magnitudes of energy terms involved. The enthalpy change necessary to liquefy 1 mol of oxygen, at 90 K, is approximately $c_p \Delta T = \frac{7}{2}R \times (300 - 90)\text{K} = 12$ kJ/mol, plus the heat of vaporization, at 90 K, which is approximately 6.7 kJ/mol. A total of about 19 kJ/mol must therefore be removed.

When oxygen expands from 200 atm to 1 atm in a Joule–Thomson expansion it cools about 50°C, representing an enthalpy change of approximately $c_p \Delta T = \frac{7}{2}R \times 50\,\text{K} = 1.5$ kJ/mol. This is roughly 10% of the energy extraction required to liquefy the mole of oxygen, so with heat exchangers about 10% of the oxygen can be liquefied.

The cost of liquefying the gas is the cost of the work to compress the oxygen gas initially, which is approximately $W = RT \ln(200/1) = R \times 298 \times 5.3 = 13$ kJ/mol, or 13 kJ/(0.1 mol condensed) $= 130$ kJ/mol of oxygen liquefied. Although it is theoretically possible to liquefy oxygen with about one-fourth of this energy expenditure (Section 8.4), there are advantages in the simplicity of the Joule–Thomson apparatus and process.

The calculations here are quite approximate. However, the principal correction terms involve the decrease in μ_{JT} and in c_p with temperature, and the lack of ideal thermodynamic efficiency in the compression and thermal energy exchange. For example, the escaping gas may be below room temperature, so some cooling effect is lost. The errors tend to compensate.

If hydrogen or helium escapes from a high-pressure cylinder, the gas will not cool, but rather, will get warm. The Joule–Thomson coefficient of these gases is negative.

$$\mu_{JT} = \left(\frac{\partial T}{\partial P}\right)_H \qquad \text{and} \qquad \Delta T = \mu_{JT}\,\Delta P > 0 \text{ when } \Delta P < 0$$

The Joule–Thomson coefficient depends somewhat on the pressure, but it does not go to zero as the pressure goes to zero, away from the inversion temperature.

From the law of corresponding states we may generalize:

$$\begin{array}{lll} \{\text{high temperatures}\} & \mu_{JT} < 0 & \Delta T > 0 \text{ when } \Delta P < 0 \\ \left\{\begin{array}{l}\text{inversion temperature} \\ \text{or ideal gas}\end{array}\right\} & \mu_{JT} = 0 & \Delta T = 0 \\ \{\text{low temperatures}\} & \mu_{JT} > 0 & \Delta T < 0 \text{ when } \Delta P < 0 \end{array} \tag{8-13}$$

The existence of an inversion temperature may be predicted from van der Waals' equation. The equation predicts an inversion temperature of approximately

$$T_{\mathrm{I}} = \frac{2a}{Rb} \text{ compared with } T_c = \frac{8a}{27Rb}$$

in terms of the van der Waals constants, a and b. This is in qualitative agreement with experiment. A more complete analysis, retaining all terms in the equation, gives a quadratic equation,

$$T_{\mathrm{I}}^2 - \frac{2a}{Rb}T_{\mathrm{I}} + \frac{3aP}{R^2} = 0$$

which predicts two inversion temperatures for each pressure. The Joule–Thomson coefficient should become negative for very low temperatures as well as high temperatures. Although there is some experimental confirmation of two inversion temperatures, the lower one may occur in the liquid phase, for which the Joule–Thomson equations are valid but van der Waals' equation is not.

Above a pressure of $P = a/3b^2$, the equation predicts that the Joule–Thomson coefficient should be negative at all temperatures, indicating that repulsive forces between molecules are more important at these very high pressures than are the attractive forces.

8.4 Adiabatic Expansion

In the first expansion process considered above (a Joule or Gay–Lussac expansion), a gas expanded into a vacuum, doing no work on the surroundings. In the limit of an ideal gas, there is no temperature change, and even for nonideal gases, the temperature changes (which may be positive or negative) are small.

The second expansion process (Joule–Thomson) differed in that the gas expanded against an opposing pressure. A Joule–Thomson expansion is also irreversible and for an ideal gas the temperature change is still zero. For real gases, however, the temperature change can be substantial. It is usually negative, but is positive for gases at high temperatures (relative to their critical points).

A third type of expansion is at least equally important, as we will see in later chapters. It is an adiabatic, reversible expansion.[4] Because thermal energy transfer is relatively slow, most gas expansions approximate adiabatic conditions, even without insulation of the gas from the surroundings.

If an ideal gas expands, or is compressed,

$$\{\text{ideal gas}\} \qquad dE = \left(\frac{\partial E}{\partial T}\right)_V dT + \left(\frac{\partial E}{\partial V}\right)_T dV = C_v\, dT = n c_v\, dT$$

If the process is adiabatic and reversible, the P appearing in the expression for work is equal to an equilibrium pressure of the gas and the work done on the gas is

$$\begin{Bmatrix}\text{reversible}\\ \text{ideal gas}\end{Bmatrix} \qquad w = -P\, dV = -nRT\,\frac{dV}{V}$$

To integrate the equation for dE we would have to know the final temperature, and the equation for work cannot be integrated because T changes as V changes. The problem appears quite hopeless, *unless we remember the key.*

Because $q = 0$, the two equations may be combined.

$$dE = w$$

$$n c_v\, dT = -nRT\,\frac{dV}{V} \tag{8-14}$$

Now separate variables by dividing both sides of the equation by nT.

$$c_v\,\frac{dT}{T} = -R\,\frac{dV}{V}$$

Integrate each side, between the limits T_1 and T_2 or V_1 and V_2, to give

$$c_v \ln\frac{T_2}{T_1} = -R\ln\frac{V_2}{V_1}$$

or

$$c_v \ln\frac{T_2}{T_1} = R\ln\frac{V_1}{V_2} \tag{8-15}$$

[4]Adiabatic and reversible are almost mutually contradictory terms for a gas expansion or compression. To be reversible the expansion must be infinitely slow. To be adiabatic the expansion must be rapid; otherwise, thermal energy will be exchanged with the walls of the container, even if they are insulating walls. In practice, an expansion is nearly reversible if the speed of the piston is small compared to the speed of sound, and nearly adiabatic if the time of expansion is even moderately short.

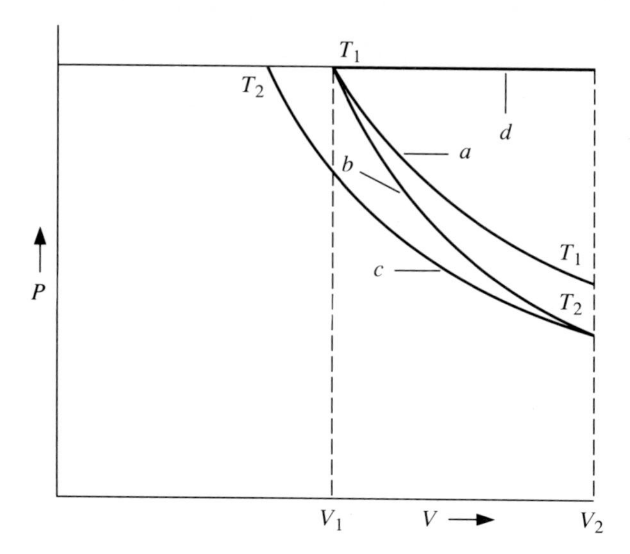

FIGURE 8.7 The work done by a gas in a reversible expansion is the area under the curve. The adiabatic expansion (b) does less work than an isothermal expansion (a) at the same initial temperature but more than an isothermal expansion (c) at the same final temperature. Expansion at the constant initial pressure (d) does the greatest amount of work.

In an expansion, V_2 is greater than V_1, and therefore T_1 is greater than T_2. That is, the gas cools as it expands because it is performing work on the surroundings.

Having found T_2 from the last equation, ΔE and W are easily calculated.

$$W = \Delta E = n\text{c}_v \, \Delta T = n\text{c}_v(T_2 - T_1)$$

Work done by a gas ($-W$, area under the PV curve) in an adiabatic reversible expansion is less than in an isothermal reversible expansion between the same initial and final volume for the same initial temperature, because the temperature decreases during the adiabatic expansion, and therefore the pressure decreases more rapidly as the volume increases. By a similar argument, less work is done by a gas in an isothermal reversible compression than in an adiabatic reversible compression between the same initial and final volumes (V_2 and V_1, respectively, in Fig. 8.7), if the final temperature is the same. Maximum work is done by the gas, for a given change in volume, if the pressure is held constant (e.g., by warming the gas).

Adiabatic processes are of sufficient importance that we should examine alternative expressions relating the initial and final states.

Starting with $dE = w$, or

$$n\text{c}_v \, dT = -P \, dV$$

with $V = nRT/P$, the change in volume is

$$dV = d\left(\frac{nRT}{P}\right) = nR\,\frac{dT}{P} - nRT\,\frac{dP}{P^2}$$

and therefore

$$n\text{c}_v \, dT = -P \, dV = -nR \, dT + nRT\,\frac{dP}{P}$$

Separating variables gives

Table 8.2 Equations for Adiabatic, Reversible Expansion or Compression of an Ideal Gas

$$c_v \ln \frac{T_2}{T_1} = R \ln \frac{V_1}{V_2}$$

$$c_p \ln \frac{T_2}{T_1} = R \ln \frac{P_2}{P_1}$$

$$\frac{dP}{P} = -\gamma \frac{dV}{V} \quad \text{or} \quad \ln \frac{P_2}{P_1} = \ln \left(\frac{V_1}{V_2}\right)^{\gamma}$$

$$PV^{\gamma} = \text{constant}$$

$$TV^{\gamma-1} = \text{constant}$$

$$\frac{T}{P^{(\gamma-1)/\gamma}} = \text{constant}$$

$$\gamma \equiv \frac{c_p}{c_v}$$

$$(c_v + R)\frac{dT}{T} = R\frac{dP}{P}$$

Recalling that $c_p - c_v = R$, we obtain

$$c_p \ln \frac{T_2}{T_1} = R \ln \frac{P_2}{P_1} \tag{8-16}$$

These equations are the most convenient for some purposes, but for other applications it is more helpful to rearrange the equations into forms involving exponents and the ratio of heat capacities, $c_p/c_v = \gamma$. Examples of equivalent expressions are given in Table 8.2.

8.5 Bernoulli's Theorem

The Bernoulli effect appears when fluids move past stationary objects or objects move through fluids. In simplest form it is described as a decrease in pressure within a moving fluid. Bernoulli's equation is usually written, for a fluid of density ρ at height h moving with speed u, as

$$P + \frac{1}{2}\rho u^2 + \rho g h = \text{constant} \tag{8-17}$$

The sum of the pressure, the kinetic energy density, and the potential energy density is constant along a streamline (Fig. 8.8).

Bernoulli's equation is particularly fascinating in its physical effects. It also provides an excellent example of how the various principles of thermodynamics considered in this and previous chapters combine to give measurable effects.

It is tempting to interpret the equation as a sum of energy terms. Each has the dimensions of energy. Multiplying through by (a constant) volume yields

$$PV + \frac{1}{2}mu^2 + mgh = \text{constant}$$

However, PV is not equal to energy, and in particular, $\int V dP \neq \Delta E$.

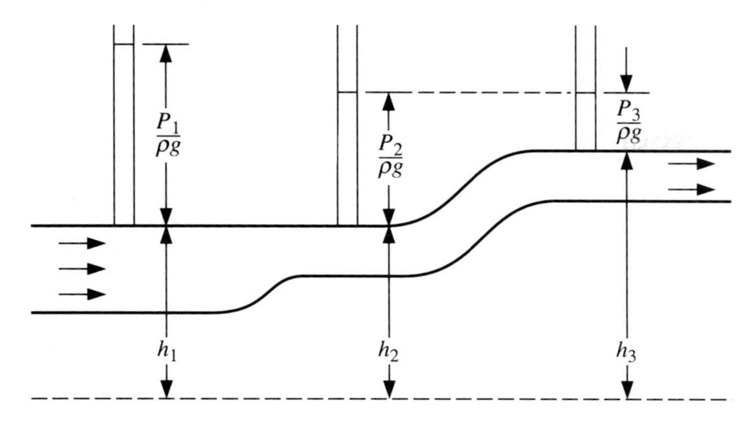

FIGURE 8.8 The pressure in a moving fluid decreases as the speed increases and as the height of the fluid increases.

If no energy is transferred to the fluid, and the fluid changes speed, there must be a conversion of internal energy into energy of flow. In a moving ideal gas flow, the velocity distribution is no longer symmetric. Letting $\mathbf{v}_o$ represent velocity in an ideal gas when $\mathbf{u} = 0$ and $\mathbf{v}_R$ the velocity relative to the center of mass moving with velocity $\mathbf{u}$ along the z axis, we find that the energy and temperature depend on

$$v_{xo}^2 + v_{yo}^2 + v_{zo}^2 \rightarrow v_{xR}^2 + v_{yR}^2 + (\mathbf{v}_{zR} + \mathbf{u})^2$$

The last term expands to

$$v_{zR}^2 + \mathbf{u}^2 + 2\mathbf{u} \cdot \Sigma \mathbf{v}_{zR}$$

and because the $\mathbf{v}_{zR}$ are symmetrically distributed about the motion of the center of mass, the sum vanishes. Hence we find the simple relationship,

$$v_o^2 \rightarrow v_R^2 + u^2$$

If the flow is isoergic, $v_o^2 = v_R^2 + u^2$. The temperature is determined by v_o^2 or by v_R^2. Hence if the flow is isoergic, $T_i = mv_o^2/3k > mv_R^2/3k = T_f$. However, gas flow is seldom isoergic.

By comparing motion of a surface past stationary fluid with motion of the fluid past a stationary object (for which the measured properties must be identical), we can see that the pressure gauge should move with the fluid, or have the pressure-sensitive surface parallel to the flow. A thermometer should move with the fluid to measure internal temperature.

There is a strong similarity of streamline flow between regions of different pressure to the Joule–Thomson experiment. One difference is that streamline flow is reversible, when viscosity may be neglected, whereas the Joule–Thomson experiment is strongly irreversible.

Consider a mass of fluid between plane A and plane B (Fig. 8.9) moving from a region of pressure P_1 to a region of pressure P_2. The flow is assumed to be

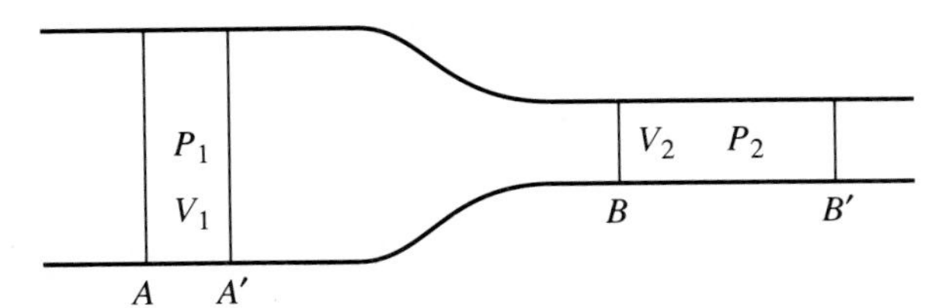

FIGURE 8.9 Work is done on the fluid element at P_1 by the trailing fluid, and the fluid element at pressure P_2 does work on the fluid in front of it.

inviscid (i.e., viscosity equal to zero). The pressure difference may be associated with a change in speed or a change in height, or both.

As a volume element of fluid V_1 moves across the plane A, the work done on the element of fluid by the fluid behind is $P_1 V_1$. The same quantity of fluid crosses the plane B, doing work on the fluid in front of it equal to $P_2 V_2$. The net work done on the fluid element, by fluid on either end, is

$$W_f = P_1 V_1 - P_2 V_2 = -\Delta(PV)$$

From this point, the problems of change of height and change of speed diverge. The principles underlying these are substantially different. We therefore analyze the two cases separately.

Change of Height

Consider an adiabatic flow with change in height. The gravitational field exerts a force on the element of fluid, transferring energy to or from the fluid as work.

$$W_g = m\mathbf{g} \cdot \Delta\mathbf{h} = -mg\,\Delta h$$

There is no thermal energy transfer, so the energy change is

$$\Delta E = Q + W_f + W_g = 0 - \Delta(PV) - mg\,\Delta h$$

and the enthalpy change is

$$\Delta H = \Delta E + \Delta(PV) = -mg\,\Delta h$$

If there is no change in speed, the energy changes are changes in internal energy only.

$$\Delta E = \Delta E_{\text{int}} \qquad \text{and} \qquad \Delta H = \Delta H_{\text{int}}$$

In a typical problem, the fluid is incompressible and the speed of flow is constant through a tube of constant cross section. Then

$$\Delta P = -\rho g\,\Delta h \qquad \text{and} \qquad \Delta(PV) = V\,\Delta P = -V\rho g\,\Delta h$$

The net work on the fluid is

$$W = W_f + W_g = -\Delta(PV) - V\rho g\,\Delta h = V\rho g\,\Delta h - V\rho g\,\Delta h = 0$$

The energy is constant along the flow.

$$\left\{\begin{array}{l}\text{adiabatic streamline}\\ \text{flow at constant speed}\\ \text{with change of height}\end{array}\right\} \qquad \begin{array}{l}\Delta E = \Delta E_{\text{int}} = 0\\ \Delta H = \Delta H_{\text{int}} = -V\rho g\,\Delta h\end{array} \qquad (8\text{-}18)$$

Adiabatic streamline flow between two heights, at constant speed, is isoergic.

Work done by the gravitational field is equal and opposite to the work done by the fluid at each end of the flow.

Change of Speed

The more interesting applications of Bernoulli's equation involve a change in speed. If the fluid is incompressible, the *equation of continuity* relates the speed of flow to the cross-sectional area of the flow. Fluid is neither lost nor gained, so

$$\{\text{incompressible}\} \qquad u_1A_1 = u_2A_2 \qquad (8\text{-}19)$$

This constraint does not apply for compressible fluids, but we can expect the speed to increase as the cross-sectional area decreases, even for an ideal gas.

It follows from the derivation above for change of height that if there is no fluid pressure at one end of the flow (e.g., the familiar problem of water escaping from a hole in the side of a can), the gravitational field does work on the fluid, which produces an increase of speed, as for other freely falling bodies. Because the fluid is in free fall (to the approximation of zero viscosity), the fluid above exerts no pressure on it.

$$\left\{\begin{matrix}\text{fluid in}\\ \text{free fall}\end{matrix}\right\} \qquad \begin{aligned} \Delta E = W = W_g = -V\rho g\Delta h = \frac{1}{2}\rho V\Delta(v^2) = \Delta(\frac{1}{2}mv^2) \\ \Delta H = \Delta E + \Delta(PV) = 0 \end{aligned} \qquad (8\text{-}20)$$

If the flow is horizontal, the only work done on the fluid element is by the fluid at either end.

$$\Delta E = Q + W_f = 0 - \Delta(PV)$$

In general, $\Delta(PV) \neq 0$, and therefore $\Delta E \neq 0$. However, as in the Joule–Thomson experiment,

$$\begin{aligned} \Delta H &= \Delta E + \Delta(PV) \\ &= -\Delta(PV) + \Delta(PV) \\ &= 0 \end{aligned} \qquad (8\text{-}21)$$

Adiabatic horizontal streamline flow is isenthalpic.

It is necessary to distinguish carefully between total energy and internal energy and between total enthalpy and internal enthalpy. The energy of flow is $\frac{1}{2}Mu^2$, where M may be taken as the molar mass, giving energy per mole. Then

$$\Delta E = Q + W = -\Delta(PV) = \Delta E_{\text{int}} + \frac{1}{2}M\Delta(u^2)$$

and

$$\Delta H = \Delta H_{\text{int}} + \frac{1}{2}M\Delta(u^2) = \Delta E + \Delta(PV) = 0$$

Therefore,

$$\Delta H_{\text{int}} = -\frac{1}{2}M\Delta(u^2) \qquad (8\text{-}22)$$

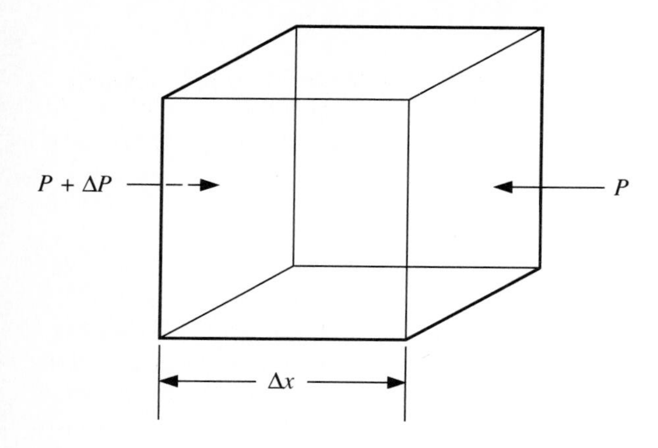

FIGURE 8.10 A higher pressure on the left of the element of fluid causes the element to accelerate to the right.

and

$$\Delta E_{\text{int}} = -\Delta(PV) - \frac{1}{2}M\Delta(u^2)$$

We see, therefore, a second difference between the Joule–Thomson experiment and streamline flow. In the former, $\Delta H = \Delta H_{\text{int}} = 0$, whereas in the latter, $\Delta H = 0 \neq \Delta H_{\text{int}}$. Hence we cannot use the Joule–Thomson coefficient, for example, to find $\Delta T/\Delta P$ in Bernoulli flow.

Bernoulli's equation is obtained most easily from Euler's equation, which is an expression of Newton's second law for fluid flow. An element of fluid with volume $A\,\Delta x$ (Fig. 8.10), acted on by a pressure difference ΔP in the direction of the flow, experiences a force

$$f = [P - (P + \Delta P)]\,A = -A\,\Delta P$$

The element will be accelerated so that in the time Δt required for the element to move its own length, Δx,

$$f = -A\,\Delta P = ma = (\rho A\,\Delta x)\,\frac{\Delta u}{\Delta t} = \rho A\,\frac{\Delta x}{\Delta t}\,\Delta u = \rho A u\,\Delta u$$

or in differential form,

$$dP + \rho u\,du = 0 \tag{8-23}$$

This is Euler's equation, in one dimension, for motion along a streamline.

For the special case of an ideal gas, with molecules of mass m,

$$\rho = m\frac{N}{V} = m\frac{P}{kT} \tag{8-24}$$

Three limiting cases are of particular interest: constant density (= constant volume), isothermal ideal gas, and adiabatic ideal gas. We retain the assumption of horizontal flow for the present.

Isochoric

If the fluid is incompressible, density is constant and Euler's equation integrates to give

$$\{\text{incompressible}\} \qquad P + \frac{1}{2}\rho u^2 = \text{constant} \tag{8-25}$$

Then, multiplying through by the constant volume,

$$V\,\Delta P = \Delta(PV) = -\frac{1}{2}V\rho\,\Delta(u^2) = -\frac{1}{2}M\,\Delta(u^2)$$

Then

$$\{\text{Incompressible}\} \qquad \begin{aligned} \Delta H_{\text{int}} &= \frac{1}{2}M\,\Delta(u^2) = \Delta(PV) \\ \Delta E_{\text{int}} &= -\Delta(PV) - \frac{1}{2}M\Delta(u^2) = 0 \end{aligned} \tag{8-26}$$

For a liquid, with only very small compressibility and coefficient of thermal expansion, there is a small temperature change as a consequence of the change of pressure. The internal energy change is the sum of the change due to change in temperature and the change directly due to change in pressure. These terms are (to a very good approximation) equal and opposite, giving no net change of internal energy.

For example, water at room temperature and atmospheric pressure has a cubic coefficient of compressibility of $\beta = 2.2 \times 10^{-4}$/kelvin and an isothermal compressibility of $\kappa = 4.57 \times 10^{-10}$/Pa. The rate of change of (internal) energy with temperature is $(\partial E/\partial T)_P = C_p - V\beta P$ and the change with pressure is $(\partial E/\partial T)_T = V(\kappa P - \beta T)$. A temperature drop of 1.57×10^{-8} K/Pa = 1.57 mK/atm gives the total change in energy for a change in pressure of $\Delta E = 75.3\ \text{J/K} \cdot 1.57 \times 10^{-8}\text{K/Pa} \cdot \Delta P - 1.179 \times 10^{-6}\ \text{J/Pa} \cdot \Delta P \approx 0$.

Isothermal

For an isothermal ideal gas, $\rho = cP$; density is proportional to pressure. Euler's equation becomes

$$dP + Pcu\,du = 0$$

or

$$\frac{dP}{P} = -cu\,du$$

which integrates to give

$$\ln\frac{P}{P_o} = -\frac{1}{2}cu^2$$

The density is $\rho = mP/kT$, so

$$c \equiv \frac{\rho}{P} = \frac{m}{kT}$$

Therefore, for isothermal flow of an ideal gas,

$$P = P_o e^{-mu^2/2kT} \tag{8-27}$$

Such flow is difficult to establish, in practice.

Adiabatic

Because thermal energy transfer tends to be slow, most gas flows are at least approximately adiabatic. For an ideal gas, the energy changes are quite different than calculated for isochoric flow. For example, if the fluid is a monatomic ideal gas, $E_{\text{int}} = \frac{3}{2}NkT$, so

$$\Delta E = -\Delta(PV) = -Nk\,\Delta T = -\frac{2}{3}\,\Delta E_{\text{int}}$$

Combining these equations gives

$$-\frac{5}{3}\,\Delta E_{\text{int}} = \frac{1}{2}Nm\,\Delta(u^2)$$

or

$$\Delta E_{\text{int}} = -\frac{3}{10}Nm\,\Delta(u^2) \tag{8-28}$$

and therefore

$$\left\{\begin{matrix}\text{adiabatic horizontal}\\ \text{streamline flow}\\ \text{of monatomic ideal gas}\end{matrix}\right\} \qquad \begin{aligned} \Delta T &= -\frac{1}{5}\frac{m}{k}\,\Delta(u^2) \\ \Delta E &= -\Delta(PV) = \frac{1}{5}Nm\,\Delta(u^2) \\ \Delta H &= 0 \end{aligned} \tag{8-29}$$

More generally, if the internal energy of a gas is

$$E_{\text{int}} = \frac{\mathcal{S}}{2}NkT$$

the temperature change in horizontal streamline flow is

$$\Delta T = -\frac{1}{\mathcal{S}+2}\frac{m}{k}\,\Delta(u^2)$$

and the energy change is

$$\Delta E = \frac{1}{\mathcal{S}+2}Nm\,\Delta(u^2)$$

and

$$\Delta H = 0$$

If there is also a change in height, the gravitational field does work on the gas.

$$W = W_f + W_g = -\Delta(PV) - Nmg\,\Delta h = \Delta E = \Delta E_{\text{int}} + \frac{1}{2}Nm\,\Delta(u^2)$$

Then

$$\left\{\begin{array}{l}\text{adiabatic}\\ \text{streamline}\\ \text{flow}\end{array}\right\} \quad \begin{aligned} \Delta T &= -\frac{1}{\mathcal{S}+2}\frac{1}{k}[mg\,\Delta h + \frac{1}{2}m\,\Delta(u^2)] \\ \Delta E &= \frac{1}{\mathcal{S}+2}Nm\,\Delta(u^2) - Nmg\,\Delta h \\ \Delta H &= -Nmg\,\Delta h \end{aligned} \tag{8-30}$$

We can calculate the properties of adiabatic flow from the equation

$$PV^\gamma = \text{constant} \tag{8-31}$$

and therefore

$$P^{1/\gamma} \sim \frac{1}{V}$$

Letting $\rho = CP^{1/\gamma}$ and using $\rho_o = mP_o/kT_o$, we find that

$$C = \frac{\rho_o}{P_o^{1/\gamma}} = \frac{(m/kT_o)P_o}{P_o^{1/\gamma}} = \frac{m}{kT_o}P_o^{(1-1/\gamma)}$$

Expansion of the gas and acceleration of the gas are separable problems. (Compare expansion of a spring in a gravitational field.) Inserting $\rho = CP^{1/\gamma}$ into Euler's equation gives

$$dP = -CP^{1/\gamma}u\,du$$

and

$$P^{-1/\gamma}\,dP = -Cu\,du$$

This integrates to

$$\frac{P^{1-1/\gamma}}{1-1/\gamma} - \frac{P_o^{1-1/\gamma}}{1-1/\gamma} = -\frac{m}{kT_o}P_o^{1-1/\gamma}\frac{u^2}{2}$$

or

$$P = P_o\left(1 - \frac{\gamma-1}{\gamma}\frac{\frac{1}{2}mu^2}{kT_o}\right)^{\gamma/(\gamma-1)} \tag{8-32}$$

Limiting Values

The effect of change in height adds a term, mgh, to the kinetic energy, $\frac{1}{2}mu^2$. Define r such that

$$r \equiv \frac{\frac{1}{2}mu^2 + mgh - (\frac{1}{2}mu_o^2 + mgh_o)}{kT_o} \tag{8-33}$$

Then the three equations above may be written more compactly, for any ideal gas, as

$$\begin{aligned}
&\{\text{isochoric}\} & P &= P_o(1 - r) \\
&\{\text{isothermal}\} & P &= P_o e^{-r} \\
&\{\text{adiabatic}\} & P &= P_o\left(1 - \frac{\gamma - 1}{\gamma}r\right)^{\gamma/(\gamma-1)}
\end{aligned} \tag{8-34}$$

Although these three equations look very different, they converge for small values of r (i.e., for changes in energy that are small compared to kT).

The exponential in the second equation may be approximated using

$$e^x = 1 + x + \frac{x^2}{2} + \cdots$$

and therefore

$$\{\text{isothermal}\} \qquad P = P_o(1 - r + \cdots)$$

Similarly, the right-hand side of the adiabatic equation may be expanded using

$$(1 + x)^n = 1 + nx + \frac{1}{2}n(n - 1)x^2 + \cdots$$

The binomial power becomes

$$\left(1 - \frac{\gamma - 1}{\gamma}r\right)^{\gamma/(\gamma-1)} = 1 - \frac{\gamma}{\gamma - 1}\frac{\gamma - 1}{\gamma}r + \cdots$$

and therefore

$$\{\text{adiabatic}\} \qquad P = P_o(1 - r + \cdots)$$

which agrees with the isochoric and isothermal solutions. The temperature in an adiabatic flow is

$$\{\text{adiabatic}\} \qquad T = T_o\left(1 - \frac{\gamma - 1}{\gamma}r\right)$$

and the density is

$$\{\text{adiabatic}\} \qquad \rho = \rho_o\left(1 - \frac{1}{\gamma}r + \cdots\right)$$

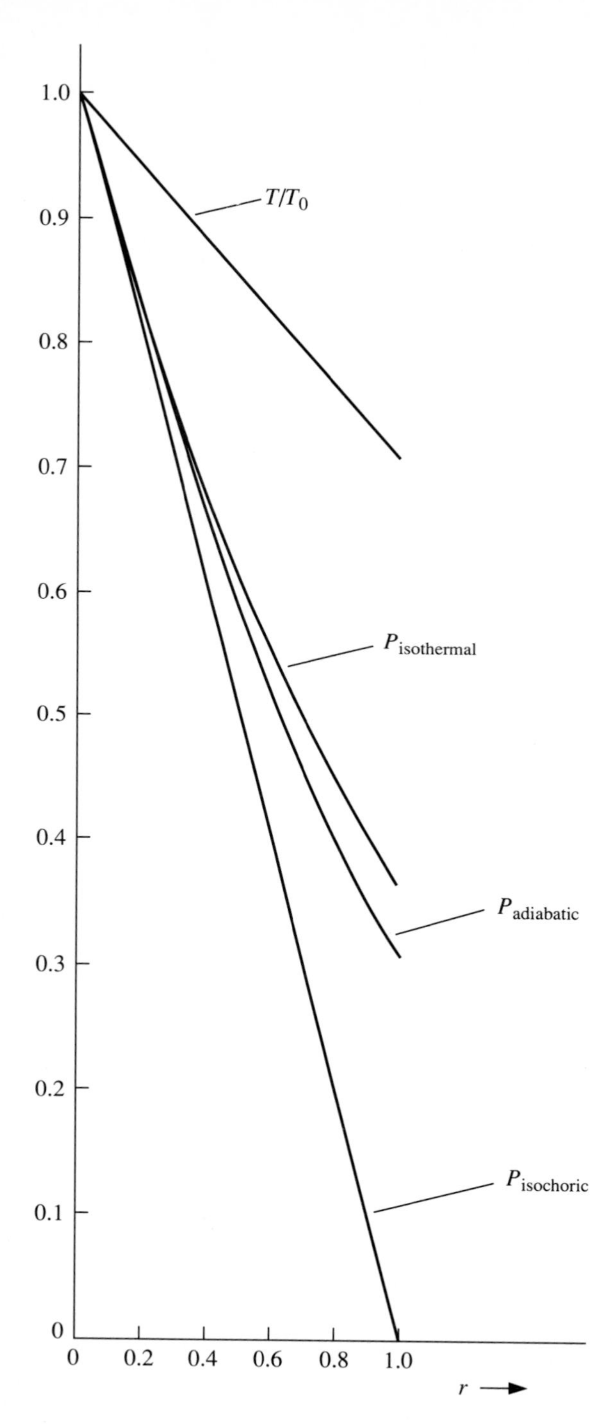

FIGURE 8.11 Plot of pressure and temperature of an ideal gas versus flow rate parameter, r, for constant density, constant temperature, and adiabatic flows. $r = \Delta(\text{K.E.} + \text{P.E.})/kT$.

A plot of P/P_0 versus r, for isochoric, isothermal, and adiabatic flows, is shown in Fig. 8.11, calculated for $\gamma = \frac{7}{5}$, which is appropriate for dry air. Note that $r = \frac{3}{2}$ would represent a flow equal to the rms speed of the gas molecules, but $r = 0.5$ corresponds to a flow equal to the rms value of v_z, the component of molecular speed along the direction of the flow. For comparison, $r = 1$ would represent an altitude of 8000 m (0°C).

Much of the confusion over the meaning of Bernoulli's equation arises from the question: Why should the speed of a fluid affect its pressure? The derivation from Euler's equation shows that this is the wrong question. An externally imposed pressure difference causes the change in speed or height. The flow is assumed inviscid. Therefore, it is reversible, and the same pressure increase is required to stop the flow as was required to initiate it. This higher stopping pressure is called the *stagnation pressure* or *ram pressure*.

The imposed pressure difference establishes a flow by pushing obstructions out of the way. The flowing gas is therefore not only at a lower pressure than the pump pressure, which is equal to the stagnation pressure; it is also lower than the ambient pressure for typical flow rates. That is why a ball is attracted into a streaming gas, against the pull of gravity, and why roofs are lifted off of houses by strong winds passing over the houses.

Bernoulli's equation is derived for points lying along a single streamline. In practice, it is typically applied to different streamlines. Because *each* streamline follows the same equation, Bernoulli's equation may be applied to *any* pair of points within the flow provided that those points are connectable, by their respective streamlines, to a region in which the pressure is the same for all the streamlines.

For example, the wind blowing over a house may cause a sufficiently great pressure difference to lift the roof. The air inside the house is not along the same streamline as air flowing over the roof. However, the air inside is at atmospheric pressure and the air flowing over the roof was accelerated, from atmospheric pressure, by a relatively small, steady pressure gradient. The reference pressure differences are small compared to the pressure difference associated with the flow, so Bernoulli's equation provides a very good approximation. Similarly, the air above a wing is not along the same streamline as the air below the wing, but the pressure of each region may be calculated, from its speed, relative to the same ambient pressure at some distance in front of the wing, so Bernoulli's equation provides exact pressure differences, to the approximation of inviscid, streamline flow.

Some additional complications appear in discussions of airflow over wings.[5] Air spills from the wing surface over the end, producing vortices. The effect of these is well known to pilots of small planes; venturing into a lingering vortex from a passing large plane on the runway can cause the small plane to flip over. Accurate calculations of wing lift must take into consideration turbulence and these vortex motions arising from edge effects, as well as the varying speed of air above and below the wing.

Because the speeds above and below the wing are different, it is convenient to substitute an average speed, then introduce an added "rotational" motion of air about the wing, such that the sum of the average motion and the rotational component gives the actual speed above and below the wing. This rotation is a mathematical artifact that contributes nothing additional to the forces on the wing beyond that calculated with the true speeds.

It is also necessary to point out that Bernoulli's equation is simply a convenient means of applying Newton's laws to fluid flows. The air passing a wing is deflected downward as the wing is pushed up. Without this effect there would be no speed difference and no Bernoulli effect. The lift can be calculated with or without Bernoulli's equation, but because it is the same effect being calculated, one should not try to separate the Bernoulli effect from Newton's second law. Otherwise, adding these supposedly independent effects would give twice the actual lift.

[5]See, for example, B. W. McCormick, *Aerodynamics, Aeronautics, and Flight Mechanics*, (New York: John Wiley & Sons, Inc., 1979).

SUMMARY

van der Waals' equation: $\left(P + \frac{n^2a}{V^2}\right)(V - nb) = nRT.$

or

$$\left(P' + \frac{3}{V'^2}\right)(3V' - 1) = 8T'$$

in terms of $P' = P/P_{c'}$ $V' = V/V_{c'}$ $T' = T/T_{c'}$ which illustrates the *law of corresponding states*.

The *Boyle point* is the temperature for which $\left(\frac{\partial(PV)}{\partial P}\right)_T = 0$ for low P.

Joule (or Gay–Lussac) *experiment* shows that $\left(\frac{\partial E}{\partial V}\right)_T = 0$ (for an ideal gas), and therefore also $\left(\frac{\partial E}{\partial P}\right)_T = \left(\frac{\partial H}{\partial P}\right)_T = \left(\frac{\partial H}{\partial V}\right)_T = 0.$

The *Joule–Thomson* coefficient is $\mu_{JT} = (\partial T/\partial P)_H$.

The pressure dependence of enthalpy of a real gas is given by

$\Delta H = -\int \mu_{JT} C_p \, dP$

$\mu_{JT} < 0$ at high T; $\mu_{JT} > 0$ at low T.

Adiabatic reversible expansion or compression of an ideal gas:

$$nC_v \, dT = dE = w = -P \, dV$$

$$c_v \ln \frac{T_2}{T_1} = R \ln \frac{V_1}{V_2} \qquad \text{and} \qquad c_p \ln \frac{T_2}{T_1} = R \ln \frac{P_2}{P_1}$$

$PV^\gamma = \text{constant}$ and $TV^{\gamma-1} = \text{constant}.$

Bernoulli's equation: $P + \frac{1}{2}\rho u^2 + \rho g h = $ constant along a streamline for nonviscous flow.

DERIVATIONS AND PROOFS

8.1 From the condition that $\partial P/\partial V = 0$ and $\partial^2 P/\partial V^2 = 0$, derive the following expressions for a van der Waals gas.

a. $b = \frac{1}{3}V_c$ **b.** $a = \frac{9R_cTV_c}{8}$

8.2 From the condition that $(V - V_c)^3 = 0$, derive the following expressions.

a. $a = 3V_c^2P_c$ **b.** $V_c = \frac{3RT_c}{8P_c}$

c. $a = \frac{27R^2T_c^2}{64P_c}$ **d.** $b = \frac{RT_c}{8P_c}$

8.3. Show that substitution of the equations given in derivations 1 and 2 into

$$\left(P'P_c + \frac{a}{V'^2V_c^2}\right)(V'V_c - b) = RT'T_c$$

leads to

$$\left(P' + \frac{3}{V'^2}\right)(3V' - 1) = 8T'$$

8.4. Starting with van der Waals' equation,
a. Find $PV = PV(V, T, a, b)$
b. Find $\left(\frac{\partial(PV)}{\partial P}\right)_T = \left(\frac{\partial(PV)}{\partial V}\right)_T \left(\frac{\partial V}{\partial P}\right)_T = f(a, b, V, T)\left(\frac{\partial V}{\partial P}\right)_T$
c. Show that, if PV is independent of P (a minimum, or at the Boyle point, with $P \approx 0$), then

$$RT = \frac{a}{b}\left(\frac{V - b}{V}\right)^2$$

d. Show from this that the Boyle point (which may be evaluated at low pressure) is

$$T_B = \frac{a}{bR}$$

8.5. Show that van der Waals' equation may be written

$$PV = RT - \frac{a}{V} + bP + \frac{ab}{V^2}$$

From this, show that
a. For low pressures,

$$\frac{PV}{RT} = 1 + \frac{P}{RT}\left(b - \frac{a}{RT}\right)$$

b. At the Boyle point,

$$\frac{PV}{RT} = \text{constant}$$

8.6. Show that, for low pressures, if $T > T_B$, then

$$\left(\frac{\partial(PV)}{\partial P}\right)_T > 0$$

8.7. Starting with the result of Derivation 8.5a,
a. Find $(\partial V/\partial T)_P$ in terms of a, P, and T.
b. From this result, show that

$$T\left(\frac{\partial V}{\partial T}\right)_P - V = \frac{2a}{RT} - b$$

8.8. Given the relationship

$$\left(\frac{\partial H}{\partial P}\right)_T = V - T\left(\frac{\partial V}{\partial T}\right)_P$$

and the equation

$$\left(\frac{\partial H}{\partial P}\right)_T = -C_p \mu_{JT}$$

show that

a. $\mu_{JT} = 0$ if $PV = RT$.

b. For a van der Waals gas, at low to moderate pressures,

$$\mu_{JT} = \frac{2a/RT - b}{C_p}$$

c. Therefore,

$$T_I = 2\frac{a}{Rb} = 2T_B$$

8.9. Derive the alternative equations for adiabatic reversible expansions and compressions given in Table 8.2.

The following problems are simplified by use of the Bridgman relations.

8.10. Streamline flow is isentropic (Chapter 9), so the change in temperature is given by $(\partial T/\partial P)_S = V\beta T/C_p$. Therefore, for an incompressible fluid undergoing adiabatic streamline flow ($dH_{int} = VdP$), the change in temperature is

$$dT = \frac{T\beta V}{C_p}\, dP$$

Show that

a. The change in energy with change in temperature is

$$\left(\frac{\partial E}{\partial T}\right)_P = C_p - PV\beta$$

b. The change in energy with change in pressure is

$$\left(\frac{\partial E}{\partial P}\right)_T = V(P\kappa - T\beta)$$

c. Therefore, the change in energy for an adiabatic change in pressure is

$$dE_{int} = \left[(C_p - PV\beta)\frac{T\beta}{C_p} - (T\beta - P\kappa)\right] V\, dP$$

8.11. Show that if $PV = nRT$, then

a. $\beta = 1/T, \kappa = 1/P$, and $(\partial E/\partial V)_T = 0$.

b. $\mu_{JT} = -(V/C_p)(1 - T\beta)$.

c. Show that if $\beta < 1/T$, the fluid warms on expansion. Show, from molecular considerations of the potential wells, that this is the behavior expected for liquids.

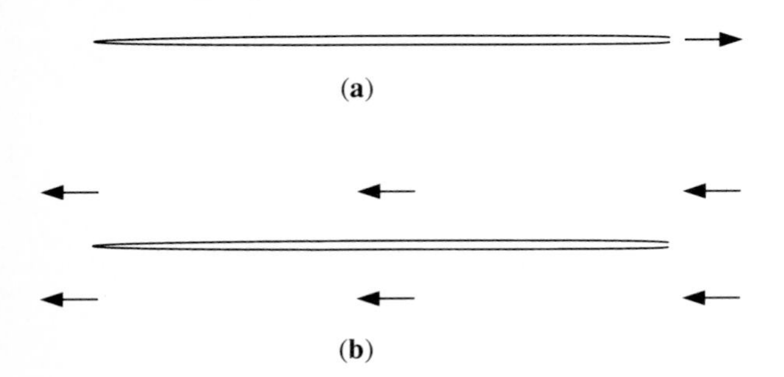

FIGURE 8.12 (a) The surfaces move through air. (b) The air moves across the surfaces.

8.12. Show that for an "incompressible" fluid (e.g., a liquid), the change of volume with pressure at constant S (entropy) is $dV = [(V\beta^2T/C_p)-\kappa]V\,dP$. This gives the order of magnitude of the percentage error in calculating changes in PV for a fluid such as water by assuming the liquid to be incompressible in adiabatic, reversible flow.

PROBLEMS

8.1. Find the final temperature if 32 g of SO_2 at 27°C is expanded adiabatically and reversibly from 1 atm to 0.25 atm. [c_p = 41 J/mol · K; c_v = 32 J/mol · K.]

8.2. Find the final temperature if 2.0 mol of N_2 is compressed adiabatically and reversibly from 1.0 atm and 25°C to 5.0 atm.

8.3. Find ΔH for an isothermal reversible expansion of 1 mol of oxygen from 25°C and 2.0 atm to $\frac{1}{2}$ atm.

8.4. Find the pressure exerted by 32 g of methane, CH_4, in a 2.0-L steel bomb at 127°C assuming that the gas obeys van der Waals' equation.

8.5. Find the volume occupied by 32 g of methane, CH_4, at a pressure of 2.0 atm at 127°C assuming that the gas obeys van der Waals' equation.

8.6. Air flows over the top surface of a wing at Mach 0.90 (i.e., at 90% of the speed of sound, which may be taken as 325 m/s, with an air density of 1.3 kg/m^3) and along the lower surface at Mach 0.70.

a. What is the pressure differential on the wing?

b. How much wing surface (approximately) must the plane have if the plane has a mass of 50.0 metric tons?

8.7. Wind of 75 mi/h is rated as being of hurricane strength. What is the lifting force of such a wind on a roof 10 m by 15 m?

8.8. Show how you could use measured concentrations at two points in a centrifuge tube to determine the density of particles in suspension. Be explicit as to what other measured values would be required in the calculation.

8.9. Consider airflow across a thin sandwich of two metal sheets sealed at the rear but with a small leak along the front edge (Fig. 8.12). Explain why the metal surfaces should, or should not, tend to bulge

a. When the sandwich moves through still air.

b. When air flows across the stationary sandwich.

8.10. D'Alembert showed (ca. 1743) that to the extent that viscosity can be neglected, the pressure is identical on the front edge and the trailing edge of any object in a streamline flow.[6] Explain why the pressure at each of these

[6]There is no "form drag" (due to unbalanced pressure distribution) on an object. See McCormick, op. cit., pp. 52–54.

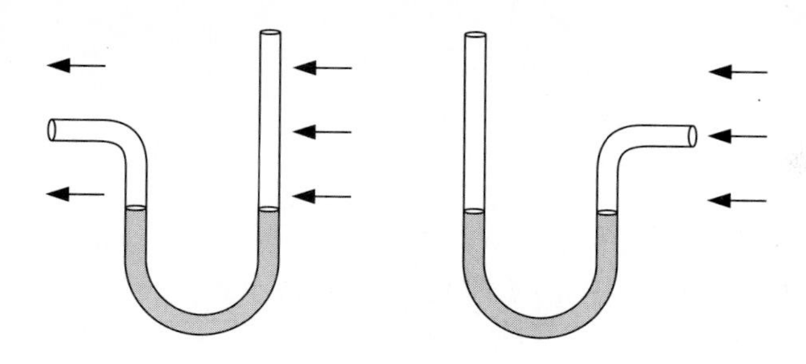

FIGURE 8.13 How will the readings of the two Pitot tubes compare in air?

points must be greater than the pressure in the flowing fluid, on the basis of the second law. (Carefully identify the directions of forces, the agent exerting each force, and what each force is exerted on.)

8.11. Predict (qualitatively) the effect of reversing the direction of a Pitot tube in air (Fig. 8.13). How will the readings of the tubes compare?

PART II

Second Law of Thermodynamics: The Irreversibility of Time

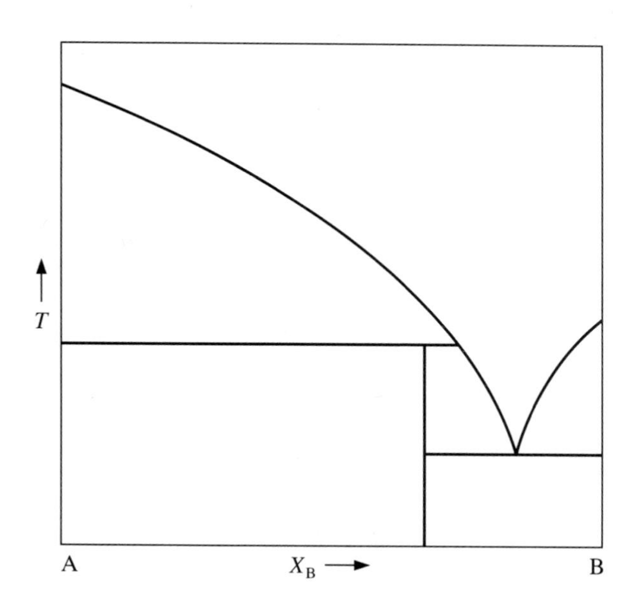

CHAPTER 9

The Direction of Change

The first law of thermodynamics, and its applications, may be described as accounting or bookkeeping. The total amount of energy is constant. Questions to be answered basically deal with whether energy is transferred between system and surroundings, and if so, how much and how.

The first law stipulates that a process cannot occur *unless* energy is conserved. If a process is permissible, under the first law, it is permitted to go in either direction. The first law provides no other information on whether the process is possible, or likely to occur. For example, there is an enormous amount of thermal energy stored in the water of the ocean. Can that thermal energy be extracted to drive ships across the ocean? It would seem attractive to extract thermal energy from outside air to heat our homes in winter. What limitations would we face in either of these attempts? The second law of thermodynamics provides information on the *direction of changes* in system and surroundings.

The second law has been stated in many different ways. Thomson (Lord Kelvin) gave one of the first clear statements:

> *A transformation whose* only *final result is that a system, of uniform temperature throughout, does work on the surroundings and decreases thermal energy within the system, is impossible.*

Clausius expressed the concept somewhat differently.

> *A transformation whose* only *final result is to transfer thermal energy from a body at a given temperature to a body at a higher temperature, is impossible.*

or equivalently, avoiding the implicit definition of "higher" temperature,

> *If thermal energy flows by conduction from a body A to another body B, then a transformation whose* only *final result is to transfer thermal energy from B to A is impossible.*

More abstract statements have also been proposed, such as Carathéodory's postulational approach expressed in terms of accessible states of a system and the mathematically phrased statement:

> *There exists a function, θ, that serves as an "integrating factor" for the thermal energy transfer in reversible processes, such that q_{rev}/θ has the properties of an exact differential; hence the integral of this ratio around any closed path is zero; $\oint q_{rev}/\theta = 0$.*

In the last statement, θ can be shown to have the properties of temperature.

Although any of these statements can serve as a starting point to derive any of the others, we will follow a more intuitive and general approach to reach an equally valid expression of the second law.

9.1 Minimum Work and Spilled Energy

To a first approximation, all forms of energy are equivalent and interconvertible. Examined more carefully, however, we discover that mechanical energy, or energy transferred as work, is more readily converted to another form of mechanical energy than is thermal energy. It is advantageous, therefore, to minimize the expenditure (i.e., transfer) of energy as work. Such energy is more valuable to us.

When the initial and final states of a system have been specified, it makes no difference to the system whether energy is added as Q or as W. Hence we prefer to minimize work done (or maximize the work extracted). However, in an *isothermal* process, it may be necessary to perform some work on the system to get from the initial state to the chosen final state. To move a box from the floor to a shelf, some energy must be transferred to the system as work if we define the system as the box plus the associated gravitational field.

The minimum work required is mgh, the work to increase the gravitational potential energy. No matter how clever or skilled you are, you cannot move the box to the shelf without performing this minimum amount of work on the system. However, it is not difficult to perform the same task with a *greater* expenditure of energy as work. Drag the box across the floor to the shelf, or drag it around the room in a large circle 10 times or 20 times, then lift it above the shelf and drop it onto the shelf.

The net work done on the system[1] is greater when we are clumsy. What happens to the excess energy? The box will have been warmed by friction, but it was stipulated that the process was to be isothermal. The process is not finished, therefore, until the box returns to its original temperature, which requires thermal energy transfer *from* the box *to* the surroundings: $Q < 0$.

By the "best" path,

$$\Delta E = mgh = Q + W$$

$$W = mgh$$

$$Q = 0$$

[1]Some of the work apparently done on the box was actually done on the floor or shelf, which (like us) is part of the surroundings. Only energy transferred from the surroundings to the system is counted as work done. Energy moved from one part of the surroundings to another part of the surroundings is not counted as work.

For *any* other path between the same initial state and final state,

$$\Delta E = mgh = Q + W$$
$$W > mgh$$
$$Q < 0$$

What happens, now, if the box is taken off the shelf and put on the floor? If the box is transferred to a carrier on a rope that passes over a pulley, lowering the box to the floor permits some other load to be raised to the shelf. The second box (and its associated gravitational field) is part of the surroundings, so the system (box 1) does work on the surroundings (box 2). In the ideal limit, as the box is lowered,

$$\Delta E = -mgh = Q + W$$
$$W = -mgh$$
$$Q = 0$$

But if the box is shoved onto a ramp and it slides down to the floor and then cools,

$$\Delta E = -mgh = Q + W$$
$$W > -mgh$$
$$Q < 0$$

For example, if the box is dropped, W will be (approximately) zero, which is greater than $-mgh$. If the box is lifted from the shelf, then dropped, W will be greater than zero. All possible paths may be divided into reversible and irreversible paths, as shown in Table 9.1.

When an ideal gas is compressed or expanded very slowly, the work done on the system approaches (from above) the limiting value

$$W = -\int P\,dV = -nRT \ln \frac{V_2}{V_1}$$

Table 9.1 Q, W, and ΔE for Raising and Lowering Box

	Reversible	Irreversible
Step 1: Raise box	$\Delta E = mgh$ $W = mgh$ $Q = 0$	$\Delta E = mgh$ $W > mgh$ $Q < 0$
Step 2: Lower box	$\Delta E = -mgh$ $W = -mgh$ $Q = 0$	$\Delta E = -mgh$ $W > -mgh$ $Q < 0$
Overall	$\Delta E = 0$ $W = 0$ $Q = 0$	$\Delta E = 0$ $W > 0$ $Q < 0$

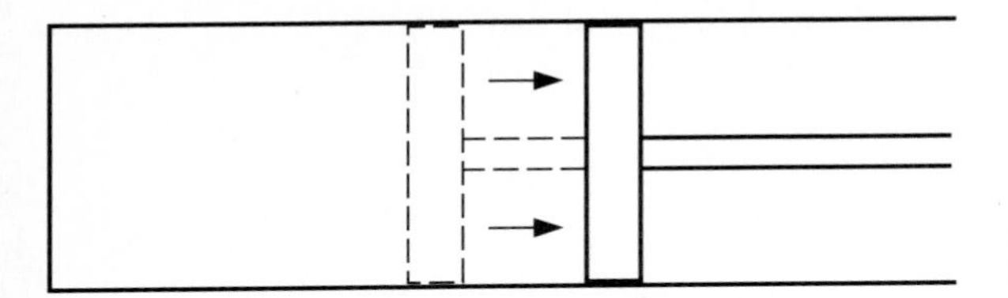

FIGURE 9.1 The piston moves very slowly, maintaining the gas in a series of equilibrium states. The process is reversible.

If a piston is given a large acceleration, and therefore a high speed, the effective pressure exerted by the gas on the piston is smaller, in an expansion, or larger, in a compression, than the equilibrium pressure (Fig. 9.1). In either case, it is necessary to do more work on the gas than in a reversible process; work done on the environment is less.

The initial and final states of the system are fixed when the process to be considered is defined. The states of the surroundings, on the other hand, are not specified and may differ markedly depending on how the process has been carried out.

To conduct a process between specified initial and final states of the system, with minimum requirement of work done on the system, or with maximum work done on the surroundings, accelerations or jerks should be avoided.[2] To avoid shocks, the process should be carried out slowly and reversibly.

Because the energy change for a process depends only on the initial and final states of the system, not on the path,

$$\Delta E = Q + W = Q_{\text{rev}} + W_{\text{rev}}$$

That is, $Q + W$ is the same for any path, including the reversible path. If W_{rev} is the minimum possible work, it follows that $Q_{\text{rev}} = \Delta E - W_{\text{rev}}$ is the maximum thermal energy transfer in the isothermal process.

There can be only one minimum value of W_{rev} and one maximum value of Q_{rev}, for any given pair of initial and final states. Therefore, W_{rev} and Q_{rev} are determined by the initial and final states; W_{rev} and Q_{rev} resemble changes in state functions.

Define two state functions, which for the moment can be labeled as A and B, such that

$$E = A + B$$

and still considering isothermal processes,

$$\{\text{constant } T\} \qquad \Delta E = \Delta A + \Delta B \tag{9-1}$$

That is, let

$$\{\text{constant } T\} \qquad \Delta A = W_{\text{rev}} \quad \text{and} \quad \Delta B = Q_{\text{rev}}$$

The three state functions E, A, and B are closely related, although clearly distinct. The minimum work done on the system in any isothermal process is $\Delta A = W_{\text{rev}}$. The maximum thermal energy transfer to the system in the isothermal process is $\Delta B = Q_{\text{rev}}$. The total energy transfer to the system is $\Delta E = \Delta A + \Delta B = W_{\text{rev}} + Q_{\text{rev}}$.

[2]This principle was once known as *Carnot's theorem*. It was proposed by Lazare Carnot, a distinguished mathematician and engineer, member of the Directory of France following the revolution, and a general and Minister of War under Napoleon.

In a cyclic process, energy transferred as work may quantitatively appear as mechanical energy (although it need not). Energy transferred as thermal energy can, at best, appear only partially as mechanical energy in a cyclic process. Energy converted into thermal energy may therefore be considered to be "spilled" energy. The energy that is "spilled" is still there, but like spilled milk, it is less useful.

Thermal energy is closely related to temperature, so it is helpful to introduce temperature explicitly into the definition of spilled energy. Replace the temporary symbol A with the symbol F, and replace the temporary symbol B with the product TS, where T is the temperature and S is the *spilled energy* (per kelvin).

Then

{constant T} $$\Delta A \equiv \Delta F = W_{\text{rev}} \leq W \qquad (9\text{-}2)$$

and

{constant T} $$\Delta B = T\Delta S = Q_{\text{rev}} \geq Q \qquad (9\text{-}3)$$

If a process is carried out reversibly, some minimum amount of work must be done on the system to change its volume, change its shape, or otherwise bring it to the specified final state. If a greater amount of work is done on the system than is required by this reversible limit, the excess work done by the surroundings must reappear in the surroundings as thermal energy, for we have been careful to specify both initial and final states of the system as fixed. It is this increased thermal energy in the surroundings, for a process that is cyclic for the system, that represents the "spilled energy" of the cyclic process.

9.2 Entropy

We extend the concept to noncyclic processes. For the change in state of the system, a measure of the spilled energy (per kelvin) in any isothermal process is

$$\Delta S = \frac{Q_{\text{rev}}}{T}$$

or, more generally,

{any process} $$dS = \frac{q_{\text{rev}}}{T} \qquad (9\text{-}4)$$

whether the actual process is reversible or not.

Expression in differential form allows consideration of processes in which temperature changes.

The state function, F, defined such that $\Delta F = W_{\text{rev}}$ in an isothermal process, is a very important function known as the *Helmholtz free energy*. It is considered in some detail in Chapter 11.

The spilled energy (per kelvin) function, S, is also very important. Clausius gave this function a Greek name indicating a "transformation" or "turning"; the word he chose is *entropy*.[3]

[3] Be particularly careful not to confuse *entropy*, S, with the function *enthalpy*, H.

Energy is never defined, except by specific examples of how it changes, or the value in one state relative to the value in another state. For example, $\Delta E = \Delta(\frac{1}{2}mv^2)$, or $\Delta E = \Delta(mgh)$. Similarly, there is not a unique definition of entropy (i.e., an absolute value of S) within the confines of classical thermodynamics. (We return to this question later, when we discuss the third law.) It follows that an absolute value of S has no utility in classical thermodynamics. If a value of S permitted the prediction of any measurable quantity, the measurement of the quantity would allow a calculation of S.

What is important is how entropy changes. The change in entropy is

$$dS = \frac{q_{rev}}{T}$$

Note that the defining equation does *not* assume that the process in question follows a reversible path. It says that for *any* path between states 1 and 2, the entropy change is calculated by finding a reversible path between the two states and finding $\int q_{rev}/T$ for that path. The integral gives ΔS for the arbitrary path. Because entropy is a state function, change in entropy of the system is $\Delta S = S_2 - S_1$, independent of the path.

Consider the very familiar process of melting a gram of ice at 0°C. The entropy change for the ice (the system) is

$$\Delta S = \frac{Q_{rev}}{T} = \frac{333.6 \text{ J/g}}{273.15 \text{ K}} = 1.221 \frac{\text{J}}{\text{g} \cdot \text{K}}$$

For the process to be reversible, the surroundings must be at the same temperature, and because $Q_{syst} = -Q_{surr}$, the entropy change of the surroundings is

$$\Delta S_{surr} = \frac{Q_{rev,surr}}{T_{surr}} = \frac{-333.6 \text{ J/g}}{273.15 \text{ K}} = -1.221 \frac{\text{J}}{\text{g} \cdot \text{K}}$$

It therefore appears that $\Delta S_{syst} + \Delta S_{surr} = 0$, and we might guess that entropy, also, is conserved. But there is a problem. How long would it take to melt a gram of ice by bringing it into contact with a thermal reservoir at 0°C? Have you tried melting an ice cube by packing it in ice? It would (at least) take an infinitely long time.

To get the ice to melt, it is necessary that the surroundings be somewhat warmer than the system. If

$$T_{surr} > T_{syst} \quad \text{and} \quad Q_{syst} = -Q_{surr}$$

then, considering only magnitudes (i.e., neglecting signs),

$$\{\text{absolute values}\} \qquad \Delta S_{syst} = \frac{Q_{syst}}{T_{syst}} > \frac{Q_{surr}}{T_{surr}} = \Delta S_{surr}$$

and (including signs)

$$\Delta S_{syst} + \Delta S_{surr} > 0 \tag{9-5}$$

The entropy change for system plus surroundings is greater than zero.

In variations of the process, such as trying to refreeze the water to ice, the temperature difference is always such that the higher temperature, and therefore smaller entropy change, is associated with the negative value of Q. That is, thermal energy always flows from the higher temperature to the lower temperature.

It follows that it must always be true that

$$\Delta S_{\text{system+surroundings}} > 0$$

with the recognition that in the limit of a fully reversible process (that may take forever), the total entropy change approaches the limiting value of zero. This statement is known as the *second law of thermodynamics*.

It follows, as a special case, that the entropy change of an isolated system is greater than, or equal to, zero.

The second law is as important as the first law of thermodynamics, conservation of energy. Like the conservation laws, the second law cannot be proved (although we will show later that it is really to be expected). It is followed in every experiment designed to test it, and having found no exceptions, we conclude that it will very likely be true in all future experiments.

The second law may be stated in three essentially equivalent forms:[4]

$$\Delta S_{\text{system}} + \Delta S_{\text{surroundings}} > 0$$

$$\Delta S_{(\text{system+surroundings})} > 0$$

The entropy of the universe is increasing, tending to a maximum.

The last form is due to Clausius.

There is an important difference between the first law and the second law that should be kept in mind. As will be shown later, the second law is a statistical law. Temperature has no real meaning for single particles; it is defined only when there are enough particles that an average value will remain constant through many, many internal changes and through interactions with the surroundings. Similarly, entropy is meaningful only as a statistical description. The law that the entropy of a system plus surroundings can only increase or remain constant is true statistically, for systems made up of many, many particles. The concept of entropy, as defined here, is valid for macroscopic samples of material. It has no meaning in single-particle interactions.[5]

An important relationship between entropy and other thermodynamic state functions is easily derived. In a reversible process on any system with states defined by two variables, work done on the system is

$$w_{\text{rev}} = -P\,dV$$

and

$$q_{\text{rev}} = T\,dS$$

[4]As mentioned previously, there are other forms as well, such as "Thermal energy always flows from a warmer body to a cooler body" and statements about efficiencies of heat engines. From any one of the statements, the others may be derived.

[5]Although the definition of $\Delta S = Q_{\text{rev}}/T$ cannot apply to single particles, the underlying principle of statistical probability proportional to the density of states is applicable and often important.

From

$$dE = q_{\text{rev}} + w_{\text{rev}}$$

it follows that

$$dE = T\,dS - P\,dV \tag{9-6}$$

Each of the quantities appearing in the last equation is a state function of the system, so the changes are independent of path. Therefore, because the equation is valid for one path (reversible), it must be valid for all paths. (If work is done other than $P\,dV$ work, there are additional terms.)

9.3 Equilibrium and the Direction of Change

The first law is an equality. Only processes for which energy is constant are allowed by the first law, and if this condition is satisfied, the first law will allow the process to go in either direction.

The second law is an inequality. It allows any process to go in one direction or the other, but usually not in both directions. The second law provides a test of whether a system is in equilibrium or, if not, in which direction the process will go. Of course, it makes no sense to undertake calculations, with the second law, for a process that is disallowed by the first law.

For an isolated system, energy and volume are constant. Break the system into two arbitrary parts (Fig. 9.2), with energies E_1 and E_2 and volumes V_1 and V_2. Then for the total system,

$$dE = dE_1 + dE_2 = 0 \qquad \text{and} \qquad dV = dV_1 + dV_2 = 0$$

From $E = E(S, V)$, or

$$dE = T\,dS - P\,dV$$

we obtain, by rearrangement, $S = S(E, V)$.

$$dS = \frac{1}{T}dE + \frac{P}{T}\,dV$$

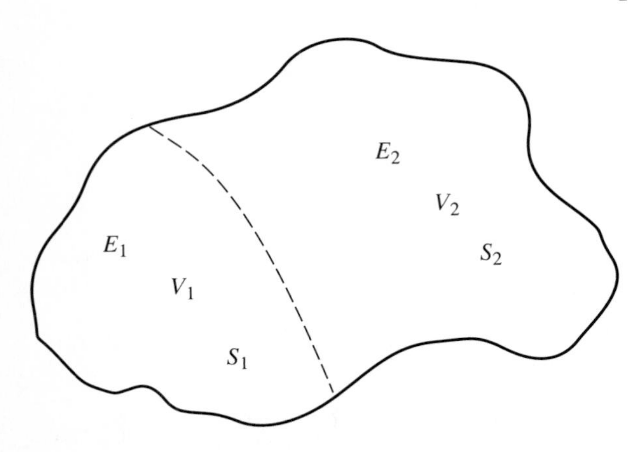

FIGURE 9.2 An isolated system is divided into two parts, of energies E_1 and E_2, volumes V_1 and V_2, and entropies S_1 and S_2.

Holding volume constant yields

$$\left(\frac{\partial S}{\partial E}\right)_V = \frac{1}{T} \tag{9-7}$$

and holding energy constant gives

$$\left(\frac{\partial S}{\partial V}\right)_E = \frac{P}{T} \tag{9-8}$$

The changes in entropy of the parts of the isolated system are

$$dS_1 = \left(\frac{\partial S_1}{\partial E_1}\right)_V dE_1 + \left(\frac{\partial S_1}{\partial V_1}\right)_E dV_1 = \frac{1}{T_1} dE_1 + \frac{P_1}{T_1} dV_1$$

and, because $dE_1 = -dE_2$ and $dV_1 = -dV_2$,

$$dS_2 = \left(\frac{\partial S_2}{\partial E_2}\right)_V dE_2 + \left(\frac{\partial S_2}{\partial V_2}\right)_E dV_2 = -\frac{1}{T_2} dE_1 - \frac{P_2}{T_2} dV_1$$

The total entropy change of the system is

$$dS = dS_1 + dS_2 = \left(\frac{1}{T_1} - \frac{1}{T_2}\right) dE_1 + \left(\frac{P_1}{T_1} - \frac{P_2}{T_2}\right) dV_1 \geq 0 \tag{9-9}$$

Consider first a process in which $dV_1 = 0$. Then the condition for equilibrium ($dS = 0$) for arbitrary transfer of energy, dE_1, is that $T_1 = T_2$. The two parts of the system, or any two adjacent systems, must be at the same temperature to be in equilibrium.

If the two parts of an isolated system are at the same temperature, then a further condition for equilibrium, with respect to volume changes, is that $P_1/T_1 = P_2/T_2$, and therefore $P_1 = P_2$. The two parts of the system, or any two systems in contact, must be at the same pressure to be in equilibrium.

If the two parts of an isolated system are not in thermal equilibrium (the temperatures are different), then

$$\left(\frac{1}{T_1} - \frac{1}{T_2}\right) dE_1 > 0 \tag{9-10}$$

Let T_1 be the higher temperature; $T_1 > T_2$. It follows that $dE_1 < 0$. That is, energy must flow, as thermal energy transfer, from the system with the higher temperature to the system with the lower temperature.

Similarly, if the temperatures are the same but the pressures are different,

$$\left(\frac{P_1}{T_1} - \frac{P_2}{T_2}\right) dV_1 > 0 \tag{9-11}$$

If $P_1 > P_2$, it follows that $dV_1 > 0$. The system at the higher pressure expands, compressing the system at the lower pressure.

None of the conclusions given above are surprising. Nevertheless, they are important. First, they show the relationship between these well-known conditions and the second law of thermodynamics. Second, when situations fall outside our realm of experience, it is important to have a reliable method of analysis.

For example, when experiments are conducted for which it may be convenient to assign negative values to T or P, the derivations above provide helpful guides in lieu of direct experience. Very similar mathematical analyses can be applied to other problems, such as systems where changes in chemical composition are possible by transport of material across a boundary. The second law predicts the direction of such transport.

The statement that the temperature must be the same in all parts of a system at equilibrium (or that there exists a state function that may be called temperature for which this condition must hold) is sometimes called the *zeroth law of thermodynamics*. If the existence of the temperature function is accepted, the zeroth law is a consequence of the second law, as shown above.

9.4 Calculation of Entropy Changes

The entropy change for the melting of ice was calculated above. The result may be generalized to any change of phase under equilibrium conditions. By definition, the entropy change is the thermal energy absorbed by the system, under equilibrium conditions, divided by the temperature of the system.

$$\{\text{constant } T\} \qquad \Delta S = \frac{Q_{\text{rev}}}{T}$$

A reversible change of phase is not only at constant temperature but also at constant pressure, so $Q_{\text{rev}} = \Delta H$. (No work other than $P\,dV$ work is likely in a phase change.) Entropy change is related to the enthalpy change. For a phase change, under reversible conditions,

$$\left\{\begin{matrix}\text{reversible constant}\\ T \text{ and } P\end{matrix}\right\} \qquad \Delta S = \frac{\Delta H}{T} \tag{9-12}$$

EXAMPLES 9.1

A. Find the change of entropy for a mole of water as it evaporates at its boiling point.
B. Cyclohexane, C_6H_{12}, has a heat of vaporization of 360 J/g at its normal boiling point of 80.7°C. Find the entropy change per mole for the vaporization of cyclohexane. □

Most liquids have an entropy change for vaporization at the normal boiling point of about 22 cal/mol · K or 92 J/mol · K. This is called the *Trouton constant*. Exceptions are found for liquids such as water, which has a higher value because of the strong hydrogen bonding that holds water molecules together in the liquid phase, and molecules such as hydrogen and helium, which have exceptionally weak intermolecular attractions and boil at very low temperatures.

If an ideal gas expands reversibly, at a constant temperature, $\Delta E = 0$ (because E depends only on temperature for an ideal gas), so $Q_{\text{rev}} = -W_{\text{rev}} = nRT \ln(V_2/V_1)$. The entropy change is

$$\Delta S = \frac{Q_{\text{rev}}}{T} = nR \ln \frac{V_2}{V_1} \tag{9-13}$$

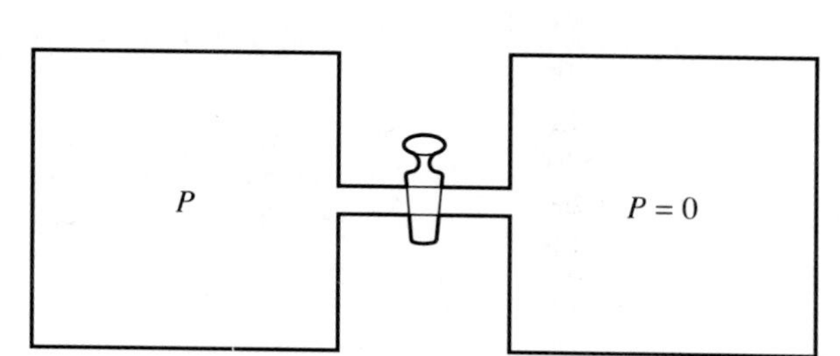

FIGURE 9.3 When the stopcock is opened, the gas at pressure P in the left container escapes, to fill the identical evacuated container on the right.

Because the process is reversible, the entropy change of the surroundings is equal and opposite.

$$Q_{\text{syst}} = -Q_{\text{surr}} \qquad \text{and} \qquad T_{\text{syst}} = T_{\text{surr}}$$

so

{reversible} $$\Delta S_{\text{surr}} = -nR \ln \frac{V_2}{V_1}$$

(Note the absence of T in these expressions for ΔS.) As expected in the limit of a reversible process, $\Delta S_{\text{syst}} + \Delta S_{\text{surr}} = 0$.

By contrast, consider the expansion of a gas from a container, where it is stored at some pressure, P, into an evacuated container (Fig. 9.3). The process is highly irreversible. The gas does no work on the surroundings during the expansion, and there is no thermal energy transfer. If the gas is ideal, there is no change in energy of the gas and no change in temperature. How can ΔS be calculated?

Recall that entropy is a state function, so the change in entropy depends only on the initial and final states, not on the path. The initial and final states of the system are the same as for the reversible, isothermal expansion of Fig. 9.1. Therefore, ΔS is calculated for the reversible path, and the entropy change is the same for any other path to reach the same endpoint. The entropy change for the adiabatic, isothermal, highly irreversible expansion of the ideal gas into an evacuated container in Fig. 9.3 is

$$\Delta S = nR \ln \frac{V_2}{V_1}$$

But this time the entropy change of the surroundings is different. Opening the stopcock to allow the gas to expand into the vacuum had no effect on the surroundings.

$$\Delta S_{\text{surr}} = 0$$

and therefore

$$\Delta S_{\text{syst}} + \Delta S_{\text{surr}} = nR \ln \frac{V_2}{V_1}$$

which is greater than zero, as expected.

The gas may be returned to its original state by compressing it isothermally (Fig. 9.4). This process will almost necessarily be reversible (because piston speeds are usually much less than the speed of sound). Then, for the compression,

---

ANSWERS 9.1

A. The heat of vaporization of water is 540 cal/g = 2260 J/g. The entropy change on vaporization is $\Delta H/T = 540$ cal/(373 g · K) = 1.45 cal/g · K or 2260 J/(373 g · K) = 6.06 J/g · K. Multiplying by the molar mass (18 g/mol) gives $\Delta S = 26$ cal/mol · K or 110 J/mol · K. This is higher than typical.

B. The molar mass of cyclohexane (84 g/mol) times the heat of vaporization per gram, divided by the boiling point (354 K), gives an entropy of vaporization of $84 \times 360/354 = 85$ J/mol · K. This is a more typical entropy change for vaporization at the normal boiling point.

$$\Delta S = nR \ln \frac{V_1}{V_2}$$

In this reversible process, the surroundings are intimately involved. The entropy change of the surroundings is equal and opposite to the entropy change of the system.

$$\{\text{reversible}\} \qquad \Delta S_{\text{surr}} = nR \ln \frac{V_2}{V_1}$$

and therefore

$$\Delta S_{\text{syst}} + \Delta S_{\text{surr}} = 0$$

as expected for the reversible path.

For the entire cyclic process of expansion and compression, the entropy change of the gas is zero, because the gas has returned to its original state.

$$\Delta S_{\text{syst}} = nR \ln \frac{V_2}{V_1} - nR \ln \frac{V_2}{V_1} = 0$$

But the surroundings have not returned to the original state. The entropy change of the surroundings is only the entropy change of the second step, because $\Delta S_{\text{surr}} = 0$ for the initial step. Hence for the complete cycle,

$$\Delta S_{\text{syst}} + \Delta S_{\text{surr}} = 0 + nR \ln \frac{V_2}{V_1} = nR \ln \frac{V_2}{V_1} > 0$$

Again, this is the expected result. For any complete cycle (of the system), the entropy of the system plus the surroundings increases.

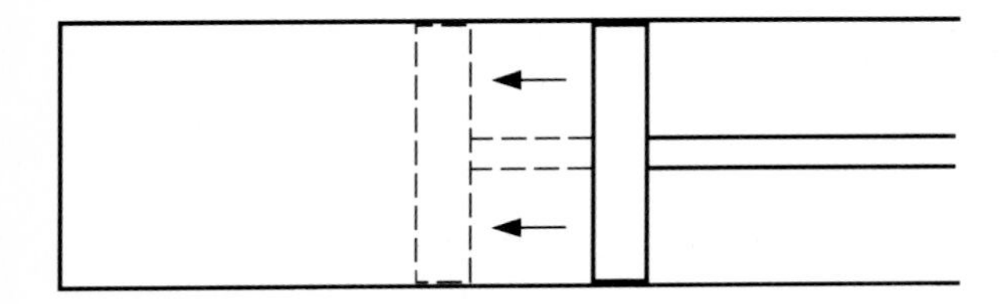

FIGURE 9.4 For sufficiently slow motion of the piston the compression is isothermal as well as reversible.

9.5 The Statistical Interpretation of Entropy

It is possible to define entropy and explore its properties without reference to molecular properties. Entropy is far more readily understandable, however, from its statistical interpretation. The formal mathematics of statistical mechanics is postponed to Part III, but some simple statistical models and concepts are appropriate here.

Consider an empty pair of identical containers (Fig. 9.3), connected by a large, open stopcock. Introduce a single molecule of nitrogen into one of the containers. Then at subsequent times (not too close together), the molecule would be equally likely to be found in the left container or the right container. There is no reason why one should be preferred over the other. Similarly, if N molecules are introduced, there is equal probability for any particular molecule to be found in the left container or in the right, so on the average, $N/2$ will be in one and $N/2$ in the other.

Start with N molecules in the left container and no molecules in the right container. When the stopcock is opened, gas molecules rush into the right container. Why? There is no difference in energy driving the molecules from one side to the other.

We know, from intuition based on experience, that when the stopcock is open, any particular molecule is equally likely to be found in either container. At any given instant approximately half the molecules will be in each container. There is no energy difference driving the molecules to this distribution. It is simply far more probable, when N is a large number, that $N/2$ molecules will be in each container than that all the molecules, or even a significant majority, will be in either container.

The state with equal numbers of molecules in each container has a higher entropy than the states with all the molecules, or a significant majority of the molecules, in either one of the containers.

> *Entropy is a measure of the* disorder, *or* randomness, *in a system. It is also directly related to the* probability *of a state. The most probable states, for a given set of external conditions, are the states of highest entropy.*

Classical mechanics treats individual elastic collisions as reversible. Imagine, then, a set of balls placed in the central channel at one end of a pinball board (Fig. 9.5). When the board is tipped, the balls undergo a series of collisions, deviating left or right as they make their way to the opposite end of the board. If edge effects are negligible, the result will be a bell-shaped distribution between channels at the far end. Now, if the board is tipped back, will the balls return to the central channel?

Intuition, based on experience, again tells us that the distribution resulting from the return trip will be more random, or uniform, than the starting (or intermediate) distribution. New random events are not likely to undo the effects of earlier random events.

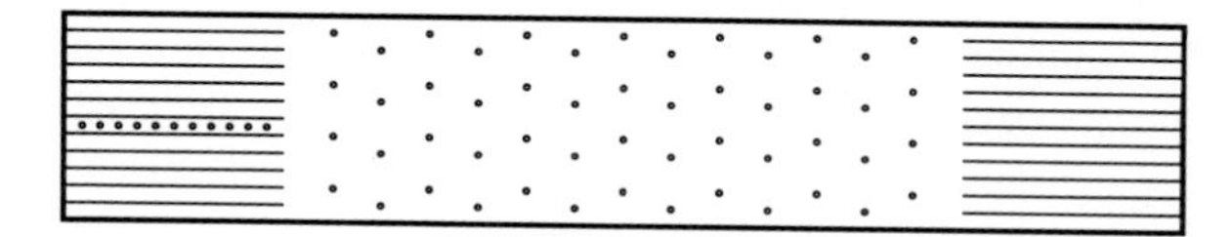

FIGURE 9.5 Balls are deviated left or right each time they strike a pin. A bell-shaped distribution curve is produced for the balls coming from any one channel.

Apparent asymmetry in the action of the pinball board is the asymmetry we introduced by our initial preparation of the sample. If we had randomly (and hence uniformly) distributed the balls between channels at the beginning, each tipping of the board would have the same effect—the balls would be randomly distributed into a uniform distribution.

A similar experiment could be carried out by preparing a uniform beam of molecules, all traveling with the same velocity. Quite quickly, because the velocities could not be made identical, there would be some collisions within the beam, causing greater disparities between velocities that would lead, in turn, to more collisions. Soon the uniform beam would acquire the characteristics of a Maxwell–Boltzmann velocity distribution. The entropy of the beam increases as the velocity distribution becomes more random.

Entropy differences can sometimes be predicted quite accurately from very simple models, without detailed calculations. More often, the sign of the entropy change is readily apparent. Three criteria are helpful.

1. Which state is more probable?
   This test would be sufficient to predict that the gas equally distributed between containers has a higher entropy than the same gas confined to either one of the containers.
2. Which state is more random, or disordered?
   Sometimes randomness, or disorder, is easy to recognize.
3. If the system is isolated and left to itself, which way could it go, between two recognizable end states?

If the isolated system could go from state A to state B, but could not, spontaneously, go from state B to state A, then state B must have higher entropy (by the second law). Although these three criteria are not independent, sometimes one gives more insight than another.

EXAMPLES 9.2

For each of the following pairs of states, indicate which would have the larger value of entropy.

A. Quartz (crystalline silica) or silica glass (“fused quartz”).

B. 100 g of lead at 100°C plus 100 g of lead at 0°C, or 200 g of lead at 50°C.

□

When entropy is expressed as a measure of the probability of states, the second law becomes a statement that without external interference, the universe, or any part of the universe, will probably go to a state of greater probability, which is essentially a tautology. In this sense the second law is a necessary property of the universe. The physical content of the second law may then be interpreted as the statement that the increase in probability, or disorder, of the system is a prescribed, monotonic mathematical function of ΔS, and $\Delta S = Q_{\text{rev}}/T$.

The quantitative relationship between entropy change and randomness or probability is suggested by the example of the expanding ideal gas. As the temperature is constant, there is no change in randomness of velocities or momenta. The only change is in the spatial distribution. Volume is a direct measure of the number of spatial states, or locations, available for each of the N molecules.

As shown above, after expansion from volume V_1 to volume V_2, the entropy is increased by $\Delta S = nR\ln(V_2/V_1)$, so if $V_2 = 2V_1$, $\Delta S = nR\ln 2 = Nk\ln 2$. The entropy increases in proportion to the logarithm of the number of states accessible to the system.

9.6 Entropy of Mixing

Mixing of dissimilar materials is a common example of a process that goes in one direction. If alcohol and water, or nitrogen and oxygen, are mixed, they will not spontaneously separate, without substantial changes in conditions (e.g., fractional distillation).

Consider the mixing of ideal gases. The particles do not interact (except by elastic collisions), so there are no complicating factors.

Two identical containers each hold n moles of an ideal gas, gas A in one container and gas B in the other (Fig. 9.6a). For example, there might be 3 mol of N_2 in the left container and 3 mol of O_2 in the right container, each at 1 atm. What is the entropy change when the divider between the containers is removed and the gases are allowed to mix? The process, for either gas, is exactly like that examined above.

For the nitrogen,

$$\Delta S = 3R\ln\frac{V_2}{V_1} = 3R\ln 2$$

and for the oxygen,

$$\Delta S = 3R\ln\frac{V_2}{V_1} = 3R\ln 2$$

The total entropy change for A plus B is

$$\Delta S_{\text{A+B}} = n_{\text{A}}R\ln 2 + n_{\text{B}}R\ln 2 = nR\ln 2$$

with $n \equiv n_{\text{A}} + n_{\text{B}}$.

What would the entropy change be if we had nitrogen in the left container and nitrogen in the right container? Each gas can diffuse into the other (Fig. 9.6b), but the final state is indistinguishable from the initial state, so there is no entropy change. When gas A is identical to gas B,

$$\Delta S_{\text{A=B}} = 0$$

The entropy was calculated in the first case as a sum of two expansions, which just happen to occur in the same total volume. The second case, where A and B are identical gases, suggests that there may be more involved, for here

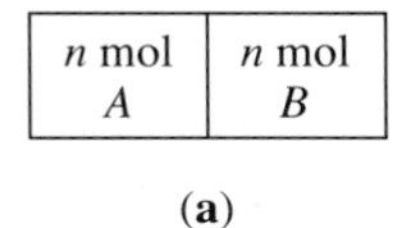

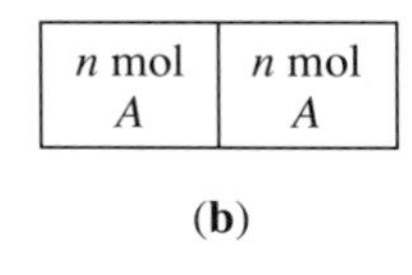

FIGURE 9.6 (a) Equal numbers of moles of two distinct gases, A and B, are allowed to mix in a fixed total volume ($V_{\text{A}} + V_{\text{B}}$). (b) The two gases are identical.

ANSWERS 9.2

A. The crystal is a more ordered arrangement. The amorphous glass has higher entropy.
B. Hot lead plus cold lead will interact to give lead at an intermediate temperature, but lead at the intermediate temperature will not spontaneously separate into hot lead and cold lead. The 200 g of lead at 50°C has more entropy.

---

the two gases expanded, yet the entropy change was zero. The overall pressure remains constant in each container for both experiments.

In the first experiment, the entropy increased because the dissimilar gases, A and B, underwent a mixing process. The entropy change, $\Delta S = nR\ln 2$, is an example of *entropy of mixing*.

The examples above are special cases, because the number of moles of each gas was the same. Suppose that gas A was initially under twice as great a pressure as gas B, at the same temperature and volumes (Fig. 9.7a); $n_A = 2n_B$. For example, there might be 6 mol of nitrogen in the container on the left and 3 mol of oxygen in the container on the right. The entropy changes, when the divider is removed, are

$$\Delta S_A = n_A R\ln 2 \qquad \text{and} \qquad \Delta S_B = n_B R\ln 2$$

so

$$\{A \neq B\} \qquad \Delta S_{A+B} = (n_A + n_B)R\ln 2 = 9R\ln 2 = 6.238R$$

This time, however, the entropy change would not be zero if the gases were identical (Fig. 9.7b). If A = B (e.g., 6 mol of nitrogen on the left and 3 mol of nitrogen on the right), the final state is the same as if A expanded to occupy two-thirds of the total volume and B is compressed to one-third of the total volume.

$$\Delta S_A = 6R\ln\frac{\frac{2}{3}}{\frac{1}{2}} = 6R\ln\frac{4}{3}$$

$$\Delta S_B = 3R\ln\frac{\frac{1}{3}}{\frac{1}{2}} = 3R\ln\frac{2}{3}$$

and

$$\{A = B\} \qquad \Delta S_{A+B} = \Delta S_A + \Delta S_B = 6R\ln\frac{4}{3} + 3R\ln\frac{2}{3} = 0.510R$$

Clearly, the mixing of different gases produces a greater increase in entropy than does the pressure equilibration.

Let n_1 be the number of moles in the left container and n_2 the number of moles in the right container. The entropy increase in Fig. 9.7a is then

$$\{\text{gas } 1 \neq \text{ gas } 2\} \qquad \Delta S = (n_1 + n_2)R\ln 2l$$

FIGURE 9.7 (a) $2n$ moles of gas A mix with n moles of gas B. (b) $2n$ moles of A mixes with n moles of A (i.e., identical gases).

The entropy increase in Fig. 9.7b is the sum for gases 1 and 2.

$$\Delta S_1 = n_1 R \ln \frac{n_1/(n_1+n_2)}{\frac{1}{2}} = n_1 R \ln 2 + n_1 R \ln \frac{n_1}{n_1+n_2}$$

$$\Delta S_2 = n_2 R \ln \frac{n_2/(n_1+n_2)}{\frac{1}{2}} = n_2 R \ln 2 + n_2 R \ln \frac{n_2}{n_1+n_2}$$

$$\{\text{gas 1} = \text{gas 2}\} \quad \Delta S = (n_1+n_2) R \ln 2 + \sum_i n_i R \ln \frac{n_i}{n_1+n_2} \tag{9-14}$$

The argument of the last logarithm is the *mole fraction*, X_i, and the n_i inside the summation may be replaced by nX_i, where $n \equiv \Sigma n_i$.

The difference between the two problems of Fig. 9.7a and b represents the effect of mixing indistinguishable gases, which is the entropy of mixing.

$$\begin{aligned} \Delta S_{\text{mixing}} &= nR\ln 2 - (nR\ln 2 + nR\sum X_i \ln X_i) \\ &= -nR\sum X_i \ln X_i > 0 \end{aligned} \tag{9-15}$$

This form is appropriate to any ideal solution.

EXAMPLES 9.3

A. Find ΔS if 5.0 mol of ideal gas A in volume V is connected to 5.0 mol of ideal gas B in another container of volume V.
B. Find ΔS if 10.0 mol of ideal gas A in volume V is connected to 5.0 mol of ideal gas B in another container of volume V.
C. Find ΔS if gas A (10.0 mol) and gas B (5.0 mol) are the same kind of gas. □

It is tempting to consider a situation in which two gases are almost identical. Then a very slight change in one gas relative to the other would result in a sudden change in entropy according to the calculations above. The difficulty with such an argument is that two collections of molecules are either identical or different. There is no small change in external conditions that can convert molecules that are different into molecules that are the same. Although we will not, at this point, pursue all the implications of molecules being indistinguishable, it is necessary to recognize that there really are significant entropy differences associated with any change in identifiability of one set as compared with another. Such differences appear in many guises.

9.7 Entropy and Temperature Changes

The calculations of entropy changes above have assumed a constant temperature for the system. When the temperature of the system changes, entropy changes may be found very easily *if* you remember a simple trick.

Start with the definition of entropy change.

$$dS = \frac{q_{\text{rev}}}{T}$$

ANSWERS 9.3

A. $\Delta S = n_A R \ln 2 + n_B R \ln 2 = 10R \ln 2 = 57.9$ J/K.
B. $\Delta S = n_A R \ln 2 + n_B R \ln 2 = 15R \ln 2 = 86.9$ J/K.
C. $\Delta S = 10R \ln \frac{\frac{10}{15}}{\frac{1}{2}} + 5R \ln \frac{\frac{5}{15}}{\frac{1}{2}}$

$$= 15R \ln 2 + 15\left(\frac{10}{15} R \ln \frac{10}{15} + \frac{5}{15} R \ln \frac{5}{15}\right)$$

$$= 10.397R + 15R(-0.2703 - 0.3662) = 0.849R = 7.063 \text{ J/K}.$$

The problem is to evaluate the integral over a temperature interval.

$$\Delta S = \int dS = \int \frac{q_{rev}}{T}$$

The right side looks discouraging, because T is changing. The key is to recall that heat capacity is $C = q/dT$, so

$$q_{rev} = C\,dT$$

(It is not yet necessary to decide whether pressure or volume or something else should be held constant as T is changed. The heat capacity, C, may be left unspecified for now.)

The equation is now

$$\Delta S = \int dS = \int \frac{q_{rev}}{T} = \int C \frac{dT}{T} \qquad (9\text{-}16)$$

If the change of heat capacity with temperature is known, the equation can be integrated. In practice, heat capacity is typically very nearly constant over moderate temperature intervals, so

$$\Delta S = C \int \frac{dT}{T} = C \ln \frac{T_2}{T_1} \qquad (9\text{-}17)$$

EXAMPLES 9.4

A. Find the entropy change when 10 g of water is warmed from 25°C to 50°C.
B. Find the entropy change when 1 mol of nitrogen gas is warmed from 25°C to 50°C at constant volume. □

Transfer of thermal energy between a system at T_1 and a second system at T_2, where $T_1 > T_2$, is clearly irreversible. The entropy change of each system may be found easily, however, through a simple thought experiment.

Imagine that a thermal insulator, or poor thermal conductor, is added between the two bodies and allowed to reach a steady state; that is, no further change with time occurs in the connecting insulator. The temperature of the insulator is then fixed at each point, varying from T_1, where it contacts the warmer system, to T_2, where it contacts the cooler system (Fig. 9.8).

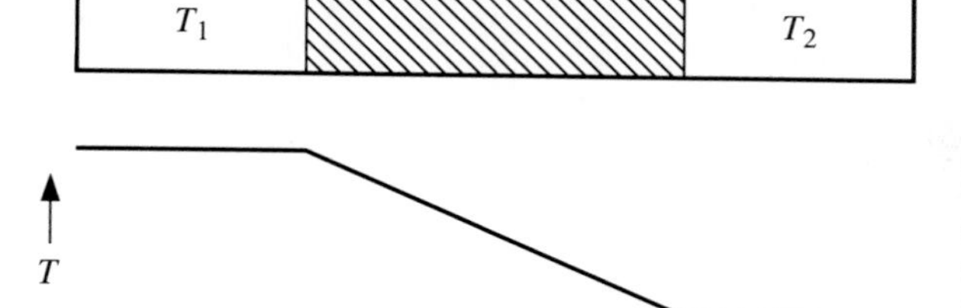

FIGURE 9.8 Thermal energy flows from the reservoir at T_1 through the insulator to the reservoir at T_2, reaching a steady state within the insulator.

Across any plane of the insulator (perpendicular to the heat flow) the thermal energy transfer is q/T on one side and $-q/T$ on the other. At each plane, the thermal energy transfer is reversible, so $\Delta S = 0$ for the insulator. The state of the insulator does not change with time.

At the interface of the insulator and the warmer system, the warmer system experiences an entropy change of

$$dS_1 = -\frac{q}{T_1} \qquad (q > 0)$$

and at the interface of the insulator with the cooler system, the change in entropy of the cooler system is

$$dS_2 = \frac{+q}{T} \qquad (q > 0)$$

The insulator has $dS = 0$, so the total change in entropy is

$$dS = dS_1 + dS_2 = \frac{-q}{T_1} + \frac{q}{T_2}$$

$$= q\left(\frac{1}{T_2} - \frac{1}{T_1}\right) > 0 \qquad (9\text{-}18)$$

The entropy change depends only on the amount of thermal energy transferred between the warmer system and the insulator and between the insulator and the cooler system.

The simple result obtained from the thought experiment leads to an important generalization. The entropy change of the warmer body must depend only on what happens to the warmer body, and the entropy change of the cooler body, or of the surroundings, must depend only on what happens to the cooler body or surroundings. This insight permits solution of many problems in which the state of the surroundings might be in question during the process occurring within the system.

For example, when the sun radiates 1 J of energy to the earth, the entropy change of the sun (assuming a temperature of 6000 K) is

$$\Delta S_{\text{sun}} = -\frac{1}{6000}\frac{\text{J}}{\text{K}}$$

and the entropy change of the earth (assuming a temperature of 300 K) is

---

ANSWERS 9.4

A. The heat capacity of 10 g of water is 10.0 cal/K or 41.8 J/K and $\Delta S = C \ln(323/298) = C \times 8.06 \times 10^{-2} = 0.806$ cal/K or 3.37 J/K.
B. For a diatomic gas near room temperature, c_V is $\frac{5}{2}R = 21$ J/mol · K, so $\Delta S = C \ln(323/298) = 1.7$ J/mol · K.

$$\Delta S_{\text{earth}} = \frac{1}{300} \frac{\text{J}}{\text{K}}$$

The total increase of the universe, for the process, is the sum of these two terms.[6]

9.8 Elasticity and Entropy

Elastic materials achieve their resilience in quite different ways. We may contrast solids, such as glass or steel, with ideal gases and with common polymeric elastomers, such as rubber. The properties of gases follow from kinetic theory and thermodynamics (Chapters 6 to 8). Polymeric materials are best understood through statistical mechanics (Section 20.8).

A perfectly elastic sphere, dropped from a height h onto a perfectly elastic surface, will rebound to the height h. Real materials are not perfectly elastic. The real sphere, striking a surface with speed v_1, rebounds with speed $v_2 = \epsilon v_1$, where ϵ is the *coefficient of restitution,* which is always less than 1. The fraction of the kinetic energy lost is $1 - \epsilon^2$. Glass on glass gives a large value of ϵ; steel on steel is slightly less; other materials are typically significantly lower.

Elasticity can also be measured by stretching a wire. Within the elastic limit, $f = -\,kx$ for an extension x in length. It is not so easy to measure typical small deviations from elasticity in this experiment. Springs are generally more susceptible to theoretical and experimental analysis. In a coiled spring, the internal strain is primarily a torsion rather than a stretch; this torsion allows the spring to uncoil slightly, and therefore to change the length of the coil.

When an ideal spring is stretched adiabatically, work is done on the spring. The energy of the spring is increased. That increased energy appears as potential energy of distortion. There is no change in temperature. The process is reversible and isentropic.

If the spring returns to its equilibrium length reversibly, acting against a variable external force, the spring does work on the surroundings, reducing the internal potential energy with no change in temperature. The contraction is reversible and isentropic.

If the spring is allowed to contract without external restraint, the process is adiabatic but not reversible. Internal potential energy is transformed into thermal energy. Temperature and entropy increase.

In the solid coiled spring, work is done by, or against, the force associated with the potential energy of crystal lattice distortion as the wire is subjected

[6]Even this simple example has some hidden complexities. The underlying assumption is that the radiant energy is transferred as thermal energy. This is appropriate for any radiator, or absorber, in equilibrium with a radiation field. Absorbers of solar radiation are not in equilibrium with the radiation field, so photovoltaic processes are possible. The energy received is then transferred to the earth, in part, as work rather than as thermal energy transfer. When the photovoltaic energy is then degraded, at 300 K, to thermal energy, the end state for the earth is as assumed above, so the sum of the entropy changes for the earth is equal to that calculated above.

to changing torsion. We may neglect work against the atmosphere; the volume change is very small.

The spring may be contrasted with expansion or compression of an ideal gas. There is no internal potential energy in an ideal gas. When the ideal gas expands adiabatically and reversibly, thermal energy is decreased as the gas does work on the surroundings. Temperature decreases. The loss of entropy associated with decreased thermal motion is compensated by an entropy increase arising from increased volume.

Adiabatic compression of the gas will necessarily be reversible to a good approximation. Spatial entropy is reconverted to thermal entropy, as volume decreases and temperature rises.

In the ideal gas, work is done by, or against, the force, area times pressure, associated with the momentum of the gas molecules, influenced by the gas density. As will be shown in Section 20.8, reversible adiabatic stretching of a polymeric elastomer redistributes entropy among internal degrees of freedom: the temperature increases, increasing thermal entropy, while spatial entropy decreases as the polymer becomes more ordered, or "crystalline."

SUMMARY

If a transformation is permissible under the first law, the second law tells which direction it can go, without external motivation.

Entropy is defined by its change: $dS = q_{\text{rev}}/T$.

For any change (defined by its end points): q_{rev} is a maximum, $q_{\text{rev}} \geq q$ and w_{rev} is a minimum, $w_{\text{rev}} \leq w$.

$dS_{\text{system+surroundings}} \geq 0$.

$dE = T\,dS - P\,dV$.

$$\left(\frac{\partial E}{\partial S}\right)_V = T; \qquad \left(\frac{\partial S}{\partial V}\right)_E = \frac{P}{T}.$$

Therefore, $dS = \left(\frac{1}{T_1} - \frac{1}{T_2}\right) dE_1 \geq 0$ and $dS = \left(\frac{P_1}{T_1} - \frac{P_2}{T_2}\right) dV_1 \geq 0$.

The *Trouton constant* is ΔS at the normal boiling point. For "normal" liquids, the Trouton constant is approximately 92 J/mol · K.

$$\Delta S_{\text{mixing}} = -nR \sum X_i \ln X_i > 0$$

For a (continuous) change of temperature,

$$\Delta S = \int \frac{q_{\text{rev}}}{T} = \int C \frac{dT}{T} = C \ln \frac{T_2}{T_1}$$

For transfer of thermal energy between two (fixed) temperatures,

$$dS = q\left(\frac{1}{T_1} - \frac{1}{T_2}\right)$$

QUESTIONS

9.1. Compression of a gas is reversible provided that the speed of the piston is much less than the speed of sound in the gas. How would friction (of the piston) change this, if at all? (**Note:** Be sure you have clearly defined the system.)

9.2. Consider two telescoping cylinders, one on the left containing ideal gas A and closed at the right end with a membrane that is permeable to gas B, but not to A; the other cylinder contains ideal gas B and is closed at the left end with a membrane that is permeable to gas A but not to B. (The first membrane is to the right of the second membrane.) What entropy change, if any, occurs when the two cylinders are collapsed, so that the initial state of pure gas A in volume V and pure gas B in a separate volume V is changed to A + B in the same volume V?

DERIVATIONS AND PROOFS

9.1. Justify each step in the following derivation, starting from $dE = TdS - PdV$.

$$\begin{aligned} TdS &= dH - VdP \\ &= \left(\frac{\partial H}{\partial T}\right)_P dT + \left[\left(\frac{\partial H}{\partial P}\right)_T - V\right] dP \\ &= T\left(\frac{\partial S}{\partial T}\right)_P dT + T\left(\frac{\partial S}{\partial P}\right)_T dP \end{aligned}$$

$$\left(\frac{\partial S}{\partial T}\right)_P = \frac{1}{T}\left(\frac{\partial H}{\partial T}\right)_P$$

$$\left(\frac{\partial S}{\partial P}\right)_T = \frac{1}{T}\left[\left(\frac{\partial H}{\partial P}\right)_T - V\right]$$

$$\frac{1}{T}\frac{\partial^2 H}{\partial P \partial T} = \frac{1}{T}\frac{\partial^2 H}{\partial T \partial S} - \frac{1}{T}\left(\frac{\partial V}{\partial T}\right)_P - \frac{1}{T^2}\left[\left(\frac{\partial H}{\partial P}\right)_T - V\right]$$

$$\left(\frac{\partial H}{\partial P}\right)_T = V - T\left(\frac{\partial V}{\partial T}\right)_P$$

9.2. Prove that

$$\left(\frac{\partial E}{\partial V}\right)_T + P = T\left(\frac{\partial P}{\partial T}\right)_V$$

(The derivation is similar to Derivation 9.1.)

9.3. Show, from the prior two derivations, that

$$TdS = C_v dT + T\left(\frac{\partial P}{\partial T}\right)_V dV$$

9.4. A condition for equilibrium is that no adiabatic process is possible for which $dS > 0$. Justify this statement. Is it also necessary to assume $P_{\text{int}} = P_{\text{ext}}$ and/or to assume $dV = 0$? Would it be sufficient to assume $w = -PdV$? Explain. (What is the P appearing here?)

PROBLEMS

9.1. Define an adiabatic process, in symbols and in words.

9.2. **a.** State the first law of thermodynamics.
b. State the second law of thermodynamics.
c. What is the value of ΔE for a complete cycle? Explain.
d. What is the value of ΔS for a complete cycle? Explain.

9.3. Find Q, W, and ΔS for 2 mol of N_2 when it is compressed from 2 atm to 4 atm at 25°C.

9.4. 50 g of steam at 120°C is added to 1 kg of water initially at 20°C. (The system is steam plus water.) Find
a. The final temperature.
b. ΔS.

9.5. 0.1 kg of ice at −10°C is added to 1 kg of water at 30°C in a suitably insulated container. (The system is ice plus water.) Find
a. ΔH.
b. ΔS.
c. The final temperature.

9.6. For the process of subliming 5g of ice at −10°C, warming the vapor to 200°C, and compressing it isothermally to 4 atm, find
a. ΔS.
b. ΔH.
c. ΔE.

9.7. 3.0 mol of He is compressed adiabatically and reversibly from 0.05 atm to 0.20 atm. The initial temperature was 27°C. Find

a. Q. **b.** W.
c. ΔE. **d.** ΔH.
e. ΔS. **f.** Final T.

9.8. Consider a fluid in (unstable) equilibrium at the intermediate volume predicted by van der Waals' equation below the critical point. If a small clump of gas undergoes a momentary excursion to a slightly higher pressure, does the second law predict that the clump will expand or contract? Is this in agreement with the behavior predicted in discussion of the instability (Section 8.1)?

9.9. Materials in column B and materials in column A below can be interconverted. Which has the higher value of S? Indicate B (before) or A (after) or S (the same).

	B	A
a.	1 mol of NaCl +1 kg of H_2O	1058.5 g of 1 m NaCl solution
b.	500 g of ice (0°C) + 100 g of steam (100°C)	600 g of H_2O at 40°C
c.	CO_2 at 25°C, 1 atm	dry ice at 1 atm
d.	ordinary iron	magnetized iron

9.10. The entropy of a gas is typically substantially greater than of liquids and solids. It has been suggested, therefore, that a reasonable estimate of ΔS of reaction can often be obtained by looking at the change in number of moles of gas in the reaction, using the Trouton constant (about 92 J/mol·K) as a rough estimate of the entropy of the gas. Similarly, combination reactions tend to decrease entropy; dissociations increase entropy. Predict the sign of $\Delta S_{\text{reaction}}$ for the following reactions.
a. $Ag_2O(s) \rightarrow 2Ag(s) + \frac{1}{2}\, O_2(g)$
b. $2C(s) + O_2(g) \rightarrow 2CO(g)$
c. $2H_2(g) + O_2(g) \rightarrow 2H_2O(l)$
d. $n\ C_2H_4(g) \rightarrow -(C_2H_4)_n-$ (polyethylene)

CHAPTER **10**

Heat Engines and Heat Pumps

The industrial revolution of the eighteenth and nineteenth centuries quite suddenly changed the quality of life in a number of ways. Railroads and steamships altered the means and speeds for transporting goods and people. Factory production of goods increased available quantities and lowered prices. Increased leisure time, improved communication, and an economy that could support research and education fostered the rapid development of the totally new sciences of chemistry and physics, including studies of electricity and magnetism that laid the foundation for the twentieth century emphasis on electrical controls, communications, and computations.

The British Isles were the birthplace of the industrial revolution, which was in large part an invention born of necessity. Importation of "china" from the Far East, followed by domestic production of hard ceramics, led to replacement of wooden dishes with cleanable ware, which improved public health and caused an increase in population.[1] The subsequent shortage of fire wood and the clearing of forests, as well as the use of coal in smelting iron, encouraged a turn to coal. As surface coal was consumed, the miners dug deeper, striking water that had to be pumped from the mines. It was this immediate problem—how to power pumps to lift water from coal mines—that led to the invention of crude steam engines that could consume a small fraction of the coal and thereby make more coal accessible.[2]

In France, Sadi Carnot, son of Lazare Carnot who was a leader in the Republic, found his status as an officer in the French army strongly influenced by the waxing and waning fortunes of the Republic. In 1820, soon after the restoration of the monarchy, Sadi retired from the French army on half pay, at age 24, and undertook to develop a theory of heat engines.

The French recognized that Newcomen, Watt, and their compatriots had moved Great Britain far ahead of the continent in industrialization. That technological lead provided the economic strength that made the British Empire dom-

[1]C. T. Currelly, *I Brought the Ages Home,* (Toronto: Royal Ontario Museum, 1976).

[2]E. Mendoza, Introduction to: S. Carnot, *Reflections on the Motive Power of Fire* (New York: Dover Publications, Inc., 1960).

inant through the nineteenth century. By the end of that century the entire world had been jolted by the burst of production, trade, and new political thought attributable primarily to English-speaking entrepreneurs.

The scene had been set for revolutionary changes in life style. Even though commerce and communication had increased significantly during the millennium from the Dark Ages through the Renaissance, there had been remarkably little direct effect on the common people, who worked farms close to subsistence level. Development of the horse collar had permitted substitution of horses and mules for oxen, and an improved plow blade design increased plowing efficiency, but otherwise, life continued in much the same pattern for centuries.

By the beginning of the twentieth century, however, automobiles, telephones, electric lights, and surplus income began their penetration to farms and small towns, rapidly changing life at all levels in western civilization. The new technology led soon afterward to even more dramatic changes in Asia and Africa.

It was said, during the early years, that the steam engine did more for thermodynamics than thermodynamics did for the steam engine. More recently, however, thermodynamic principles have been critically important in the design and development of new engines of higher efficiency and improved environmental acceptability. The theory developed by Sadi Carnot, corrected and recast in modern terminology, provides an easy introduction to the important principles underlying all engines that involve conversions between random, or thermal, energy and directed motion, or work.

10.1 Carnot Cycle

Assume that thermal reservoirs are available at some fixed, high temperature, T_H (boiler temperature), and a lower temperature, T_C (cooling water). An engine (the system, Fig. 10.1) takes in thermal energy at T_H, does work on the surroundings, and rejects excess thermal energy into the cooling water at T_C. All steps are reversible.

The system undergoes a cycle, consisting of alternating isothermal and adiabatic steps.

1. Isothermal: at T_H.
2. Adiabatic: Temperature falling from T_H to T_C.
3. Isothermal: At T_C.
4. Adiabatic: Temperature rising from T_C to T_H.

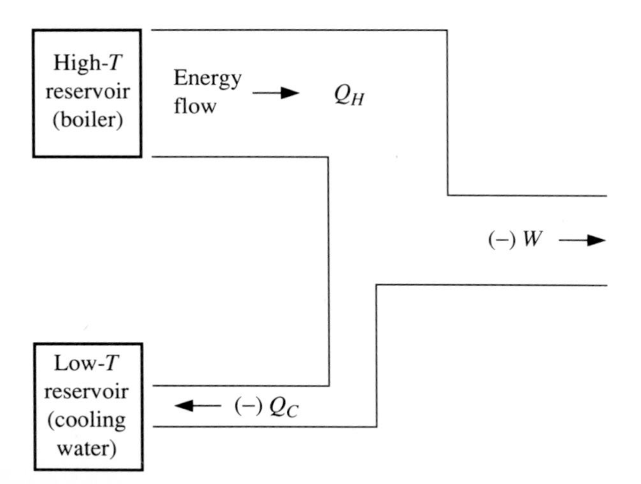

FIGURE 10.1 Thermal energy input, Q_H, at T_H, is output as work and as thermal energy transfer at T_C. For the engine cycle, $\Delta E = 0$.

Step 1. $\Delta S \equiv \Delta S_H = \dfrac{Q_{\text{rev}}}{T} = \dfrac{Q_H}{T_H}.$

Step 2. $\Delta S \equiv \Delta S_a = \displaystyle\int \frac{q_{\text{rev}}}{T} = \int \frac{0}{T} = 0.$

Step 3. $\Delta S \equiv \Delta S_C = \dfrac{Q_{\text{rev}}}{T} = \dfrac{Q_C}{T_C}.$

Step 4. $\Delta S \equiv \Delta S_{a'} = \displaystyle\int \frac{q_{\text{rev}}}{T} = \int \frac{0}{T} = 0.$

For the complete cycle, $\Delta S = 0$ and $\Delta E = 0$. Therefore,

$$Q_H + Q_C + W = 0$$

$$\frac{Q_H}{T_H} + \frac{Q_C}{T_C} = 0 \tag{10-1}$$

Substitution of $-T_C Q_H / T_H$ for Q_C in the first equation gives

$$-W = Q_H + Q_C = Q_H - \frac{T_C}{T_H} Q_H = \frac{T_H - T_C}{T_H} Q_H$$

and therefore

$$\frac{-W}{T_H - T_C} = \frac{Q_H}{T_H} = -\frac{Q_C}{T_C}$$

Let $\Delta S_H \equiv \zeta$ (zeta). Then the equations are conveniently written

$$\begin{aligned} Q_H &= \zeta T_H \\ -Q_C &= \zeta T_C \\ -W &= \zeta(T_H - T_C) \equiv \zeta\, \Delta T \end{aligned} \tag{10-2}$$

The *input* to the engine is Q_H, the thermal energy from the boiler. The *output* from the engine is $-W$, the (net) work done by the engine on the surroundings. The *efficiency,* ε, is

$$\varepsilon = \frac{\textit{output}}{\textit{input}} = \frac{-W}{Q_H} = \frac{\zeta(T_H - T_C)}{\zeta T_H} = \frac{\Delta T}{T_H} = 1 - \frac{T_C}{T_H} \tag{10-3}$$

For example, Carnot analyzed an engine operating with an ideal gas. Then:

Step 1. For an isothermal, reversible expansion of an ideal gas at T_H, $\Delta E_H = 0$ and therefore

$$Q_H = -W_H = nRT_H \ln \frac{V_2}{V_1} > 0 \qquad (V_2 > V_1)$$

$$\Delta S_H = \frac{Q_{\text{rev}}}{T} = \frac{Q_H}{T_H} = nR \ln \frac{V_2}{V_1} > 0$$

Step 2. In an adiabatic, reversible expansion,

$$dE_a = w_a$$

$$n\text{c}_v \, dT = -P\,dV = -nRT \frac{dV}{V}$$

$$\text{c}_v \frac{dT}{T} = -R \frac{dV}{V}$$

$$\text{c}_v \ln \frac{T_C}{T_H} = -R \ln \frac{V_3}{V_2} \qquad (T_C < T_H;\ V_3 > V_2)$$

$$\Delta S_a = \int dS = \int \frac{q_{\text{rev}}}{T} = \int \frac{0}{T} = 0$$

$$\Delta E_a = n\text{c}_v(T_C - T_H) = W_a$$

Step 3. Isothermal: $\Delta E_C = 0$ and therefore

$$Q_C = -W_C = nRT_C \ln \frac{V_4}{V_3} < 0 \qquad (V_4 < V_3)$$

$$\Delta S_C = \frac{Q_{\text{rev}}}{T} = \frac{Q_C}{T_C} = nR \ln \frac{V_4}{V_3} < 0$$

Step 4. Adiabatic:

$$\text{c}_v \ln \frac{T_H}{T_C} = -R \ln \frac{V_1}{V_4} \qquad (T_H > T_C; V_1 < V_4)$$

$$\Delta E_{a'} = n\text{c}_v(T_H - T_C) = W_{a'}$$

$$\Delta S_{a'} = 0$$

Because $\oint dE = 0$, it follows that $\Delta E_a + \Delta E_{a'} = 0$ and therefore $W_a + W_{a'} = 0$. Comparing equations from the two adiabatic steps gives

$$-R \ln \frac{V_3}{V_2} = \text{c}_v \ln \frac{T_C}{T_H} = -\text{c}_v \ln \frac{T_H}{T_C} = +R \ln \frac{V_1}{V_4}$$

so

$$\frac{V_2}{V_3} = \frac{V_1}{V_4}$$

and

$$\ln \frac{V_2}{V_1} = -\ln \frac{V_4}{V_3}$$

Then, letting $\zeta \equiv nR \ln \dfrac{V_2}{V_1} = \Delta S_{H'}$, we obtain

$$Q_c = -W_c = nRT_c \ln \frac{V_4}{V_3} = -nRT_c \ln \frac{V_2}{V_1} = -\zeta T_c$$

$$Q_H = -W_H = nRT_H \ln \frac{V_2}{V_1} = \zeta T_H$$

$$Q_a = Q_{a'} = 0 \qquad W_a = -W_{a'}$$

$$W = \Sigma W_i = W_H + W_C = -\zeta(T_H - T_C) < 0$$

EXAMPLES 10.1

A. Find the efficiency of a Carnot engine, as described above, operating between 25 and 100°C.
B. Will the efficiency increase or decrease if T_H is raised to 150°C? □

10.2 Carnot's Theorem

For a reversible heat engine, operating on a Carnot cycle between T_H and T_C with an ideal gas as the working fluid, the efficiency depends only on the temperatures, T_H and T_C, but the efficiencies are substantially lower than would be desired. At the same time, the conditions imposed on the engine are severe. Real engines are not restricted to reversible steps and certainly are not limited to ideal gases as the working fluid. It is therefore worth investigating the question of how much better an engine can perform if these restrictions are relaxed.

Assume that an improved engine has been developed described by the primed variables Q'_H, Q'_C, and W'. The new engine will also be operated between T_H and T_C for comparison with the idealized engine. To facilitate the comparison, the original, reversible engine will be operated in reverse mode. The two engines are connected, and interact, as shown in Fig. 10.2.

The reversible engine provides exactly the amount of thermal energy, at T_H, required to operate the more efficient engine. That is, the absolute values of Q_H and Q'_H are made equal, so $Q'_H = -Q_H$. This can be accomplished by adjusting the size of the piston, length of stroke, and/or the amount of gas in the cylinder.

The object of the design of the primed engine was to obtain a higher efficiency, so it will be assumed that

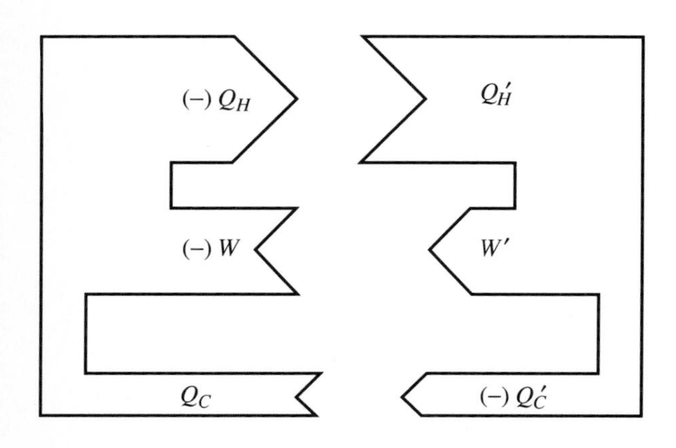

FIGURE 10.2 The reversible heat engine, driven by the work input W, provides thermal energy input, $-Q_H = Q'_H$, to the "improved" engine, which provides the output $-W'$, assumed to be greater in magnitude than W.

{absolute values} $$\varepsilon' = \frac{W'}{Q'_H} > \frac{W}{Q_H} = \varepsilon$$

Then, because $Q_H = -Q'_H$, it follows that $|W'| > |W|$. That is, the primed engine provides more output, $-W'$, than is required, W, to operate the reversible engine.

Clearly, this is a delightful situation. When the improved engine is once set in motion it drives the reversible engine, which in turn provides all the thermal energy at T_H needed to drive the new engine, However, the new engine also has excess output, in the form of work, or directed energy, that may be diverted to driving a generator to provide electricity, or powering an automobile, or whatever task is at hand. The result is too good to be true. The combination of machines in Fig. 10.2 represents a perpetual motion machine that violates the second law of thermodynamics.

The only assumption made was that $\varepsilon' > \varepsilon$. The consequences are unreasonable, so we conclude that the assumption is untenable. The efficiency of the "new, improved" engine cannot be greater than the efficiency of the reversible engine. The efficiency of the second engine must be less than, or at best equal to, the efficiency of the reversible engine initially analyzed.

{absolute values} $$\varepsilon' = \frac{W'}{Q_H} \leq \frac{W}{Q_H} = \varepsilon$$

This conclusion, which was reached by Carnot about 1824, is known as *Carnot's theorem:*

The maximum possible efficiency of any heat engine operating between two fixed temperatures, T_H and T_C, is equal to the efficiency of a reversible engine operating between those temperatures:

$$\varepsilon = \frac{-W}{Q_H} \leq \frac{T_H - T_C}{T_H} \qquad (10\text{-}4)$$

EXAMPLES 10.2

A. Suppose that a gasoline engine operates between a maximum temperature of 1000°C and a coolant temperature of 80°C. What is the maximum possible efficiency of the engine, by Carnot's theorem?

B. Federal regulations restrict many power plants to a maximum operating temperature of 600°C. If the cooling water is 30°C, what is the maximum possible efficiency of the power plant? □

10.3 Thermodynamic Temperature

An important consequence of the equations derived for the Carnot cycle is that the ratio of temperatures is equal to a ratio of thermal energy transfers.

$$\frac{T_H}{T_C} = \frac{Q_H}{-Q_C} \qquad (10\text{-}5)$$

ANSWERS 10.1

A. $\varepsilon = \Delta T/T = (373 - 298)/373 = 0.2011 = 20\%$
B. $\varepsilon = (423 - 298)/423 = 0.296 = 30\%$. Raising T_H has a larger percentage effect on the numerator than on the denominator, as is shown by the alternative form, $\varepsilon = 1 - T_C/T_H$.

ANSWERS 10.2

A. $\varepsilon = \Delta T/T_H = 920/1273 = 72\%$.
B. $\varepsilon = 570/873 = 65\%$.

Lord Kelvin pointed out that this provides an operational thermodynamic definition of the absolute temperature scale. Following his suggestion, the modern temperature scale, now known as the Kelvin scale, is based on absolute zero and only one other fixed point, which has been chosen as the triple point of water (see Section 13.2), defined to be 273.16 K. This choice makes the size of the degree, called a kelvin, equal to 1 degree on the Celsius scale.

There are many different ways in which temperature can be measured, as we saw in Chapter 2. The Kelvin scale provides a "thermodynamic temperature," T. A unique conversion between this scale and any other arbitrary temperature scale can be found, employing changes in state functions of an arbitrary system and the amount of thermal energy transferred to or from a system at a constant temperature. This assures us that when we write an equation involving the temperature, T, the meaning is not subject to uncertainties concerning the particular scale that we, or someone else, may choose for measuring and expressing that temperature.

Assume that we have established an arbitrary temperature scale, $\tau = \tau(T)$, that increases as T increases but need not be a linear function of T. For some reversible thermal energy transfer, $q = TdS$, divide by dP (at constant T) to obtain

$$\left(\frac{q}{dP}\right)_T = T\left(\frac{\partial S}{\partial P}\right)_T$$

It can be shown[3] that

$$\left(\frac{\partial S}{\partial P}\right)_T = -\left(\frac{\partial V}{\partial T}\right)_P$$

Also,

$$\left(\frac{\partial V}{\partial T}\right)_P = \left(\frac{\partial V}{\partial \tau}\right)_P \frac{d\tau}{dT}$$

Furthermore, because τ and T are uniquely related, a derivative at constant τ is

[3]From the Bridgman relations, $(\partial S)_T = -(\partial T)_S = \beta V$, $(\partial T)_P = -(\partial P)_T = 1$, and $(\partial V)_P = -(\partial P)_V = \beta V$. Therefore, $(\partial S/\partial P)_T = \beta V/(-1)$ and $(\partial V/\partial T)_P = \beta V/1$. Alternatively, it is shown in Chapter 11 that the Gibbs free energy function has the derivatives $(\partial G/\partial T)_P = -S$ and $(\partial G/\partial P)_T = V$. Taking second derivatives yield

$$\frac{\partial}{\partial P}\frac{\partial G}{\partial T} = -\frac{\partial S}{\partial P} = \frac{\partial}{\partial T}\frac{\partial G}{\partial P} = \frac{\partial V}{\partial T}.$$

the same as a derivative at constant T. Therefore,

$$\left(\frac{q}{dP}\right)_\tau = -T\left(\frac{\partial V}{\partial \tau}\right)_P \frac{d\tau}{dT}$$

so

$$\frac{d \ln T}{d\tau} = \frac{-(\partial V/\partial \tau)_P}{(q/dP)_\tau} \tag{10-6}$$

In this last equation, q is the amount of thermal energy transferred to the system from the surroundings at some fixed value of τ; dP is the resultant change in the equilibrium pressure of the system; $(\partial V/\partial \tau)_P$ is the change in volume of the system with change in τ, measured at constant pressure; and $d\tau$ is an arbitrary change in temperature on the τ scale.

The equation uniquely defines the relationship between the thermodynamic temperature, T, and the completely arbitrary scale, τ, in terms of readily measurable changes in an arbitrary system. For example, the system need not be an ideal gas or any other special substance, and we need not know T to define the T scale.

Whether the thermodynamic scale is defined by this equation or by the ratio equation given previously, T is defined subject to an arbitrary multiplicative constant, or $\ln T$ is subject to an arbitrary additive constant. In other words, T is fully defined except for the scale, or the size of the degree. The choice of the value (273.16 K) for the triple point of water then completes the definition.

10.4 Reversed Heat Engines

The proof of Carnot's theorem given above tacitly assumed that it is possible, at least in principle, to build a heat engine that can run backwards, taking in energy as work and transferring thermal energy to the surroundings at a higher temperature (T_H) than it absorbs thermal energy from the surroundings (T_C). Such an engine can extract thermal energy from outside and put thermal energy into a room that is warmer than outside.

EXAMPLE 10.3

If 1 kJ of electric energy is purchased from the power company and passed through a resistance coil heater, it releases 1 kJ of thermal energy to the room. Therefore, the resistance coil has essentially 100% efficiency, or coefficient of performance, for this task. If the temperature outside is 0°C and inside it is 25°C, what is the efficiency, or coefficient of performance, $-Q_H/W$, for putting thermal energy into the room (i.e., how much thermal energy at 25°C is put into the room for the expenditure of 1 kJ of electric energy)? □

It appears at first glance that something is wrong with the proposed scheme. Calculations show that more energy is put into the high-temperature reservoir than is fed to the machine as work. The difference, however, is obtained as thermal energy from the low-temperature reservoir. Considering only absolute values for the moment, an amount of energy,

ANSWER 10.3

K = output/input $= -Q_H/W = \zeta T_H/(\zeta\,\Delta T) = 298/25 = 11.82$, so if $W = 1$ kJ, $-Q_H = 11.82$ kJ. The efficiency, or coefficient of performance, is 11.82 = 1182%.

Q_C, is taken in at T_C; an amount of energy, W, is added as work; and the sum of these, Q_H, is expelled as thermal energy at T_H.

Such heat engines, running backward, are called *refrigerators, air conditioners,* or *heat pumps*. The greater the temperature difference, $T_H - T_C$, the more work must be supplied to pump the thermal energy from T_C to T_H. If the temperature interval is small, very large quantities of thermal energy can be pumped to the higher level with very little expenditure of energy (as work).

EXAMPLES 10.4

A. For 1 kJ of electric energy put in, how much thermal energy (Q_C) can be extracted from a deep-freeze unit at −10°C if the effective compressor temperature is 30°C?

B. How much thermal energy is put into the room in example A? □

10.5 Other Heat Engine Cycles

The earliest practical adaptation of steam power was Savery's *steam pump*, developed in 1698. It was highly inefficient but could pump water through a height of several meters. Steam from a boiler was admitted to a large chamber, with valves (Fig. 10.3).

Spraying the chamber with water produced a partial vacuum. The pressure differential lifted water into the chamber. The subsequent charge of steam to the chamber expelled the water, rewarmed the chamber, and filled the chamber with steam again.

In 1705 Newcomen introduced his *atmospheric engine* (Fig. 10.4), which shared fundamental design concepts with Savery's engine. The chamber was replaced by a cylinder and piston, so when the steam-filled cylinder was cooled, by spraying water inside, the piston was driven down by atmospheric pressure. The opposite end of the rocker arm holding the piston lifted water and a counter-

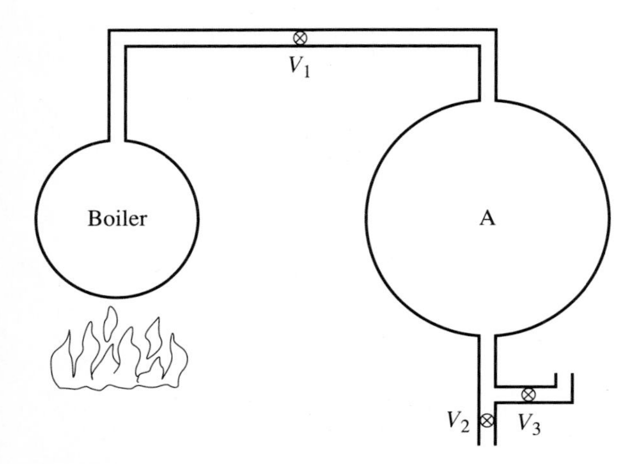

FIGURE 10.3 Savery's steam pump.

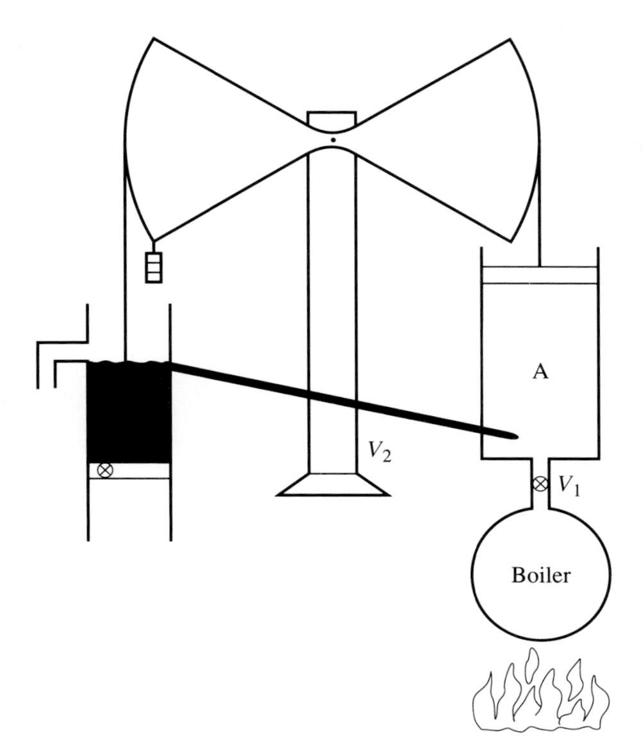

FIGURE 10.4 Newcomen engine.

weight. Automation of the valves was soon introduced and the engine was widely employed through most of the eighteenth century.

While repairing Newcomen engines, James Watt recognized the inefficiency of cooling the cylinder. Between 1763 and 1782 he introduced major improvements in design that increased both the efficiency and the range of applicability of steam engines. A separate condenser was added (Fig. 10.5). Steam was exhausted from the insulated, hot cylinder to the cooler condenser. The engine was made double acting; steam was admitted first to one side of the piston, then the other, driving the piston back and forth. Also, the reciprocating motion of the piston shaft was converted to rotary motion, opening the way to applications in industry and transportation, with drive belts and drive wheels.

Stirling Engine

The *Stirling engine,* patented in 1816, operates with a noncondensable gas (e.g., air), maintaining the principle introduced by Watt of keeping hot and cold portions of the engine distinct. It relies on a linkage equivalent to a cam shaft to move pistons in two cylinders out of phase, with a *regenerator* section between the two cylinders to capture, store, and return thermal energy as gas moves through it. The steps, in idealized form, are (Figs. 10.6 and 10.7):

1. The gas expands isothermally in cylinder B, absorbing $Q_H = nRT_H \ln(V_H/V_C) > 0$ at T_H from the burning fuel.
2. The gas is transferred isochorically (at constant V) from cylinder B, at T_H, to cylinder A, at T_C, passing through the regenerator and giving up $Q_R < 0$.
3. Cool gas is compressed isothermally in cylinder A, giving off $Q_C = nRT_C \ln(V_C/V_H) < 0$ at T_C.
4. The cool gas is transferred isochorically to cylinder B, passing through the regenerator, which warms it to T_H, supplying $Q_{R'} = -Q_R > 0$.

ANSWERS 10.4

A. $Q_C/W = T_C/\Delta T = 263/40 = 6.575$. $Q_C = 6.6$ kJ.
B. $Q_H/W = T_H/\Delta T = 303/40 = 7.575$. $Q_H = 7.6$ kJ.

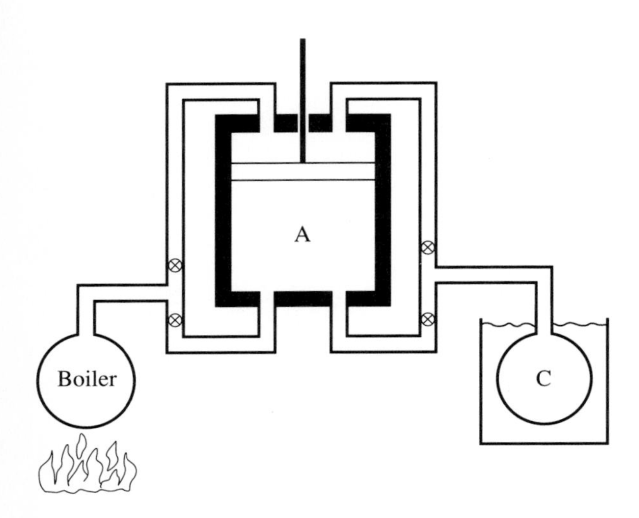

FIGURE 10.5 Watt's double-acting steam engine.

The net work done by the engine (gas) is $-W = Q_H + Q_C$ and the theoretical efficiency is

$$\varepsilon = \frac{-W}{Q_H} = \frac{\Delta T}{T_H} \tag{10-7}$$

as for a Carnot cycle. Renewed interest in the Stirling engine has developed in

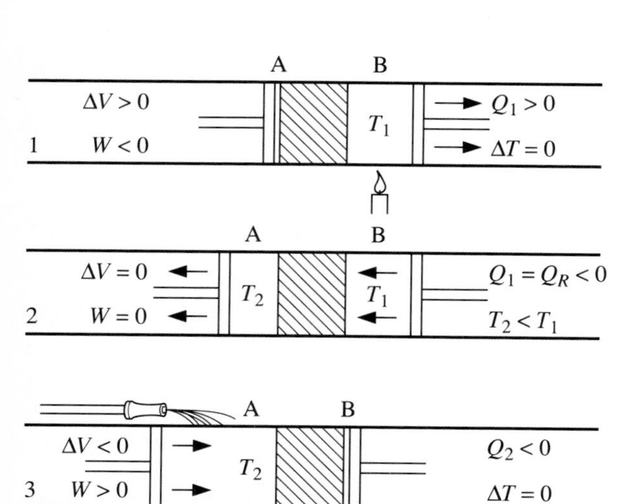

FIGURE 10.6 Stirling engine (schematic). The four steps of the cycle are (1) an isothermal expansion, driven by thermal energy input at the upper temperature; (2) an isochoric transfer that transfers thermal energy to the regenerator; (3) an isothermal compression, with cooling, at the lower temperature; and (4) an isochoric transfer, with thermal energy transfer to the gas from the regenerator.

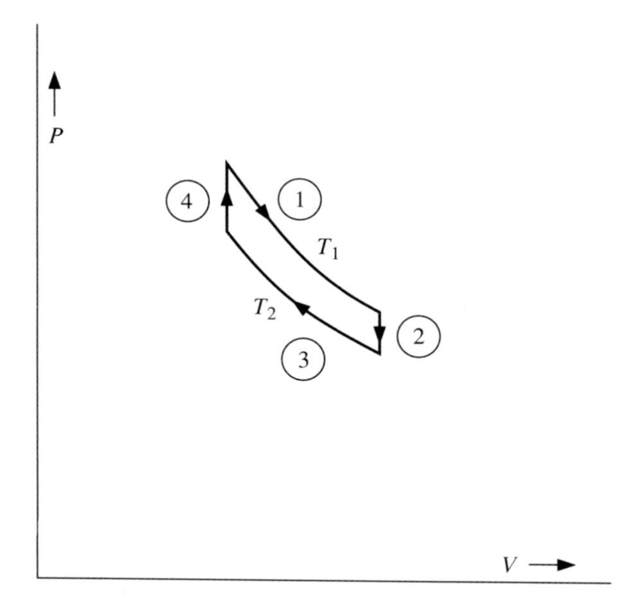

FIGURE 10.7 *PV* diagram for a Stirling engine. Steps 1 and 3 are isothermal; 2 and 4 are isochoric.

recent decades because external combustion offers advantages for controlling air pollution. Difficulties arise in the gearing required.

Rankine Cycle

Real steam engines do not operate with an ideal gas as the working fluid. An approximation to the operation of an engine with a condensable fluid is the *Rankine cycle*. The *PV* diagram is given in Fig. 10.8. The liquid-vapor equilibrium line and isotherms (dashed) are shown for reference.

A simplified representation of a Rankine engine, operating as a heat engine, consists of four primary steps (Fig. 10.9):

1. A pump compresses the liquid and feeds it to the boiler.

$$\Delta V \approx 0 \qquad W \approx 0 \qquad Q \approx 0 \qquad \Delta T \approx 0 \qquad \Delta P > 0$$

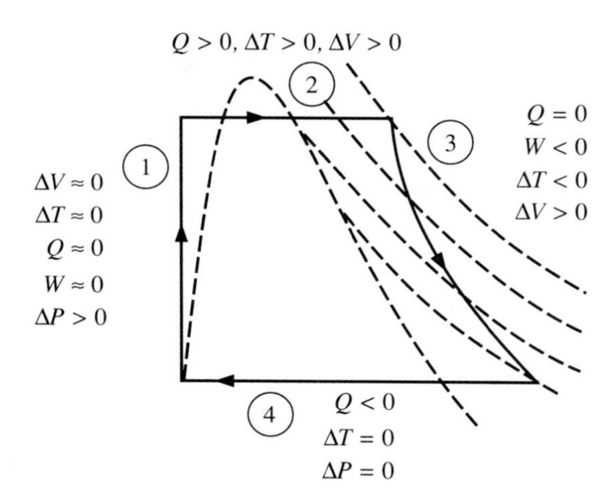

FIGURE 10.8 *PV* diagram for the Rankine cycle (simplified). Step 3 is the adiabatic power stroke.

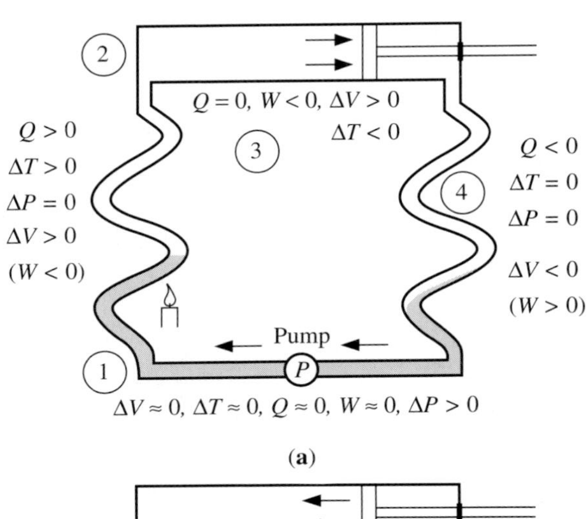

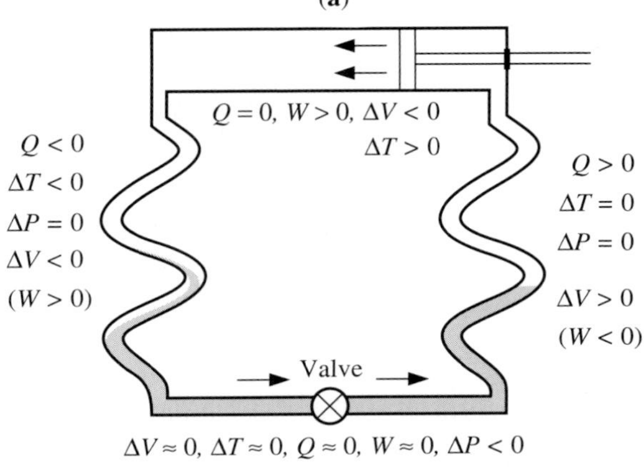

FIGURE 10.9 Simplified representation of the Rankine cycle as a heat engine and as a heat pump or refrigerator.

2. The liquid is heated to the boiling point, vaporized, and superheated, isobarically.

$$\Delta P = 0 \qquad T_C \rightarrow T_H \qquad \Delta T > 0 \qquad Q_1 > 0 \qquad W = -P\,\Delta V < 0$$

3. The superheated vapor expands adiabatically against a piston, cooling to the condensation point. This is the power stroke.

$$\Delta P < 0 \qquad T_H \rightarrow T_C \qquad \Delta T < 0 \qquad W < 0 \qquad Q = 0$$

4. The cooled vapor condenses to liquid, giving off thermal energy.

$$\Delta T = 0 \qquad Q_2 < 0 \qquad W = -P\,\Delta V > 0$$

The work done in steps 2 and 4 is approximately equal and opposite, so the work is that of the power stroke of step 3. For the full cycle,

$$Q = Q_1 + Q_2 = -W$$

The efficiency is

$$\varepsilon = \frac{-W}{Q_1} = \frac{Q_1 + Q_2}{Q_1} = 1 - \frac{-Q_2}{Q_1} \tag{10-8}$$

However, Q_1 is taken in over a range of temperatures, $T_C \rightarrow T_H$, so the efficiency is substantially lower than for an ideal Carnot cycle.

Calculation of theoretical efficiency for the Rankine cycle requires specific information about the heat of vaporization of the working fluid and the behavior of the vapor near the condensation point, where it is nonideal. In practice, values are taken from *steam tables* (or equivalent tables for refrigerants), which give volume, enthalpy, and entropy values as a function of temperature, pressure, and phase.

Otto Cycle

By treating the exothermic chemical reaction of fuel with air as thermal energy input, a gasoline engine may be treated as if it operated with an ideal gas, or air. The *air standard Otto cycle* alternates adiabatic and isochoric steps. These are preceded and followed by intake and exhaust, for which the number of moles of gas is not constant. However, the intake and exhaust steps of the cycle may be ignored; there is no energy transfer and the two steps effectively cancel each other in other respects.

The initial steps are compression followed by combustion:

1. Adiabatic: $V_1 \rightarrow V_2, T_C \rightarrow T'$ $\quad (V_2 < V_1;\ T' > T_C)$.
2. Isochoric: $T' \rightarrow T_H, Q_H > 0$ $\quad (T_H > T')$.

The pressure increases as thermal energy is transferred from a series of external reservoirs at temperatures ranging from T' to T_H, in simulation of the combustion process inside a gasoline engine. In a gasoline engine, ignition occurs shortly before "top dead center," so the combustion process is approximately isochoric as the piston passes through the end of its compression stroke. T_H is the maximum temperature.

The power stroke is rapid and therefore approximately adiabatic. The fourth step, at the end of the piston stroke, would be approximately isochoric, although in practice the exhaust and intake steps occur at this time.

3. Adiabatic: $V_2 \rightarrow V_1, T_H \rightarrow T''$ $\quad (V_1 > V_2;\ T'' < T_H)$. This is the power stroke.
4. Isochoric: $T'' \rightarrow T_C,\ Q_2 < 0$ $\quad (T_C < T'')$.

The pressure-volume relationships of the Otto cycle are represented graphically in Fig. 10.10.

As in the Carnot cycle, $Q = 0$ for the two adiabatic steps, so

$$\Delta E = 0 = Q_1 + Q_2 + W$$

The input is Q_1 (step 2) and the output is W (sum for steps 1 and 3). The efficiency is

$$\varepsilon = \frac{-W}{Q_1} = \frac{Q_1 + Q_2}{Q_1}$$

but again Q_1 is thermal energy taken in over a range of temperatures ($T' \rightarrow T_H$), so the efficiency must be less than for an ideal Carnot cycle.

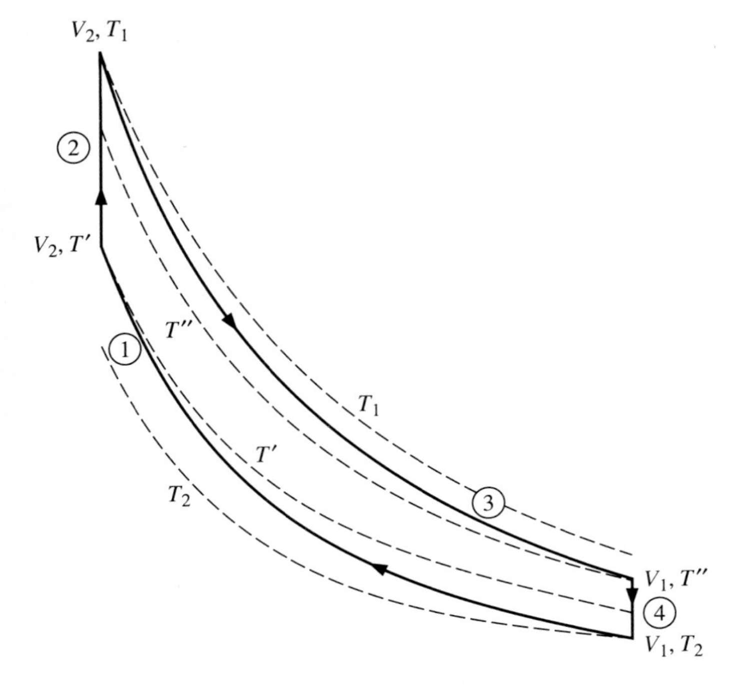

FIGURE 10.10 Otto cycle PV diagram. Steps 1 and 3 are adiabatic, crossing between isotherms (dashed).

For step 2,

$$Q_1 = \Delta E \text{ (step 2)} = nc_v(T_H - T')$$

and for step 4,

$$Q_2 = \Delta E \text{ (step 4)} = nc_v(T_C - T'')$$

The efficiency is

$$\varepsilon = \frac{Q_1 + Q_2}{Q_1} = 1 + \frac{Q_2}{Q_1} = 1 - \frac{T'' - T_C}{T_H - T'} \tag{10-9}$$

and the temperatures are related by the equations for adiabatic, reversible expansions of an ideal gas,

$$c_v \ln \frac{T_C}{T'} = R \ln \frac{V_2}{V_1} \qquad \text{and} \qquad c_v \ln \frac{T_H}{T''} = R \ln \frac{V_1}{V_2}$$

Therefore, $T_H T_C = T'T''$, where T_H and T_C, the maximum and minimum temperatures, are usually assumed to be fixed, but T' and T'', the intermediate temperatures, are adjustable and can control the efficiency (see Section 10.6).

Efficiency of an Otto cycle engine is usually expressed in terms of the *compression ratio*, $r = V_1/V_2$, which is typically of the order of magnitude of 10. In terms of r,

$$T'' - T_C = T_H\left(\frac{1}{r}\right)^{R/C_v} - T'\left(\frac{1}{r}\right)^{R/C_v}$$

so

$$\frac{T'' - T_C}{T_H - T'} = \left(\frac{1}{r}\right)^{R/c_v}$$

Substituting $c_p - c_v = R$ and $\gamma = c_p/c_v$, $R/c_v = \gamma - 1$ and the efficiency is

$$\varepsilon = 1 - \left(\frac{1}{r}\right)^{\gamma - 1}$$

In terms of the four temperatures,

$$\varepsilon_{\text{Otto}} = \frac{T_H - T'' - (T' - T_C)}{T_H - T'} < \frac{T_H - T_C}{T_H} = \varepsilon_{\text{Carnot}} \tag{10-10}$$

EXAMPLE 10.5

Assuming that the working gas has the properties of air ($\gamma = \frac{7}{5}$) and that the compression ratio, r, is 8, what is the theoretical efficiency of this Otto cycle? □

Diesel Cycle

Although the engine patented in 1892 by Rudolf Diesel has many similarities to the standard gasoline engine, there are distinct differences that help give it an advantage for heavy-duty service. First, there are no spark plugs, so spark plug fouling is not a problem.

Second, because only air is present during compression, the compression ratio may be increased without fear of preignition. This permits higher efficiency, although it requires more massive engine blocks. Fuel is injected at the appropriate time into the air that has been heated by compression, causing fuel ignition. The rate of fuel injection controls rate of burning, so a lower grade of fuel (approximately equivalent to kerosene) can be burned.

However, the goal sought by Diesel in his original design was not achieved in practice. His intent was to inject the fuel at the point of maximum compression and obtain a power stroke of constant temperature and constant pressure, thus avoiding rising temperatures and achieving Carnot efficiency by having the thermal energy input, from the reaction, at the high temperature.

The *Diesel cycle* resembles a mix of Rankine and Otto cycles (Fig. 10.11). Ignoring intake and exhaust steps again, the steps are:

1. Adiabatic: $V_1 \to V_2, T_C \to T'$ $(V_2 < V_1$ and $T' > T_C)$.
2. Isobaric: $V_2 \to V_3, T' \to T_H$ $(V_3 > V_2$ and $T_H > T')$, $Q_1 > 0$.
3. Adiabatic: $V_3 \to V_H, T_H \to T''$ $(V_1 > V_3$ and $T'' < T_H)$.
4. Isochoric: $T'' \to T_C$ $(T_C < T''$; $\Delta V = 0)$, $Q_2 < 0$.

For the full cycle,

$$\Delta E = 0 = Q_1 + Q_2 + W$$

The energy transfer terms are

$$Q_1 = \Delta H \text{ (step 2)} = nc_p(T_H - T')$$

ANSWER 10.5

$\varepsilon = 1 - (1/r)^{\gamma - 1} = 1 - (\frac{1}{8})^{7/5-1} = 1 - 8^{-0.4} = 0.5647 = 56\%$. The actual efficiency would be expected to be significantly lower.

---

and

$$Q_2 = \Delta E \text{ (step 4)} = n c_v (T_C - T'')$$

The efficiency is

$$\varepsilon = \frac{Q_1 + Q_2}{Q_1} = \frac{c_p(T_H - T') - c_v(T'' - T_C)}{c_p(T_H - T')}$$

$$= 1 - \frac{1}{\gamma}\frac{T'' - T_C}{T_H - T'} \tag{10-11}$$

This is less easily evaluated than for the Otto cycle.

Define a *compression ratio*, $r_C = V_1/V_2$, and an *expansion ratio*, $r_E = V_1/V_3$, or a *cutoff ratio* (before fuel injection is cut off), $r_2 = V_3/V_2 = r_C/r_E$. From the adiabatic steps,

$$c_v \ln \frac{T_C}{T'} = R \ln \frac{V_2}{V_1} = -R \ln r_C$$

and

$$c_v \ln \frac{T''}{T_H} = R \ln \frac{V_3}{V_1} = -R \ln r_E$$

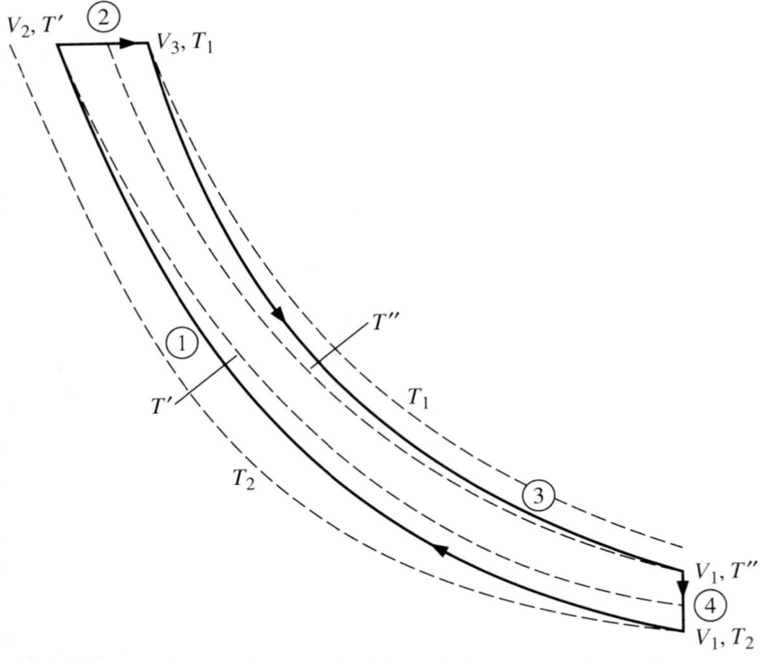

FIGURE 10.11 Diesel cycle. The adiabats (steps 1 and 3) cross between isotherms.

Then, with some manipulation,

$$\varepsilon = 1 - \frac{1}{\gamma}\frac{(1/r_E)^\gamma - (1/r_C)^\gamma}{1/r_E - 1/r_C} = 1 - \frac{1}{\gamma}\frac{r_2^\gamma - 1}{r_2 - 1}\frac{1}{r_C^{\gamma-1}} \qquad (10\text{-}12)$$

EXAMPLE 10.6

Assume that the working gas has the properties of air ($\gamma = \frac{7}{5}$), that the compression ratio, r_C, is $\frac{15}{1}$, and that the cutoff ratio is $r_2 = r_C/r_E = 3$. What is the predicted efficiency of the Diesel engine? □

Diesel engines have also been made that operate on a two-stroke cycle. Air is compressed in the first stroke, fuel is injected and the gases expand in the second stroke. At the end of the stroke, with the gases approximately at atmospheric pressure, the valves are opened and exhaust gases are blown out with a fan, admitting fresh air for the next cycle at the same time.

Servel Electrolux Refrigerator

An interesting variation on heat pumps is commonly known as the *Servel system*. The driving action is the input of thermal energy at a high temperature. Through any of several options, thermal absorption and ejection processes are accompanied by transport by convection. Two inherent advantages of the system, therefore, are the absence of motors and compressors and the operation on thermal energy input (e.g., burning natural gas).

Thermal energy sources, such as natural gas and other fuels, are less expensive than electricity, per joule, because electric power is produced[4] through a Carnot-limited process with corresponding inefficiency. The coefficient of performance of the Servel process for refrigeration is

$$K_R = \frac{Q_C}{-Q_H} \qquad (10\text{-}13)$$

For comparison, a Carnot refrigerator has a coefficient of performance of

$$K_{R_C} = \frac{Q_C}{-(Q_H + Q_C)} = \frac{Q_C}{W} = \frac{-Q_H}{W}\frac{Q_C}{-Q_H} > \frac{Q_C}{-Q_H} = K_{R_S}$$

In principle, the smaller coefficient of performance of the Servel refrigerator is exactly compensated by the advantage of not having to convert the fuel source to electric energy through a Carnot cycle.

In practice, there are many factors to be considered. Differences in temperatures of operation affect the net cost of fuel. There are differences in cost of equipment. The Servel system is noiseless. It may be operated from bottled gas or other heat source. However, the lifetime is often limited by problems of corrosion and oxidation.

[4]There is an alternative. Chemical reactions run in electrochemical cells, called *fuel cells*, producing electric power with efficiencies that can approach 100%. They are not limited by Carnot's theorem, because the energy is not degraded to thermal energy along the way. The technology of fuel cells has been well developed for special applications, such as space travel, but not adequately for large-scale electric power production.

ANSWER 10.6

$1 - \frac{1}{1.4}\,\frac{3^{1.4}-1}{3-1}\,\frac{1}{15^{0.4}} = 56\%$. In practice the advantage of the Diesel engine lies more in durability and (at times) less-expensive fuel than in higher theoretical efficiency.

10.6 Power and Efficiency

Carnot's argument is elegant. As before, let $\Delta S_H \equiv \zeta$ (*zeta*), which is positive for a heat engine or negative for a heat pump. Regardless of the working fluid (if any), or the details of operation of the cycle, if thermal energy is transferred reversibly at T_H and at T_C,

$$Q_H = \zeta T_H$$
$$-Q_C = \zeta T_C$$

Therefore, if the intervening steps ($T_H \to T_C$ and $T_C \to T_H$) are adiabatic,

$$-W = \sum Q_i = Q_H + Q_C = \zeta(T_H - T_C)$$

so

$$\frac{-W}{Q_H} = \frac{\Delta T}{T_H}$$

and

$$\frac{W}{Q_C} = \frac{\Delta T}{T_C}$$

Furthermore, this represents the maximum possible efficiency ($-W/Q_H$) for any cyclic heat engine operating between the same two temperatures, T_H and T_C.

Real heat engines and heat pumps do not achieve the performance levels calculated from Carnot's theorem.[5] Some of the discrepancy can be attributed to friction and extraneous heat leakage, but for many engines the friction is small and insulation is good. The major limitation is inherent in the cycle, as defined.

Consider a basic reversible heat engine operating between thermal reservoirs at T_H and T_C. To achieve Carnot efficiency, the engine must take in thermal energy at T_H. But if the working fluid of the engine is at T_H, no thermal energy will flow from the reservoir. Similarly, if the engine cools its working fluid, by adiabatic expansion, to T_C, no thermal energy will flow to the reservoir at T_C. The *theoretical efficiency* of the engine is $\Delta T/T_H$, but a single cycle would require an infinite amount of time, so the *power* output is zero.

To make the engine work, the highest and lowest temperatures of the working fluid must differ from the reservoir temperatures. Let the working fluid vary from $T_H - \delta T$ to $T_C + \delta T$. The efficiency is then

[5]On the other hand, most heat engines are not cyclic, so it becomes more difficult to compare actual performance with a theoretical model. For example, if "waste heat" warms the incoming air and fuel, less energy must be supplied in the compression step and a real gain may be achieved in performance.

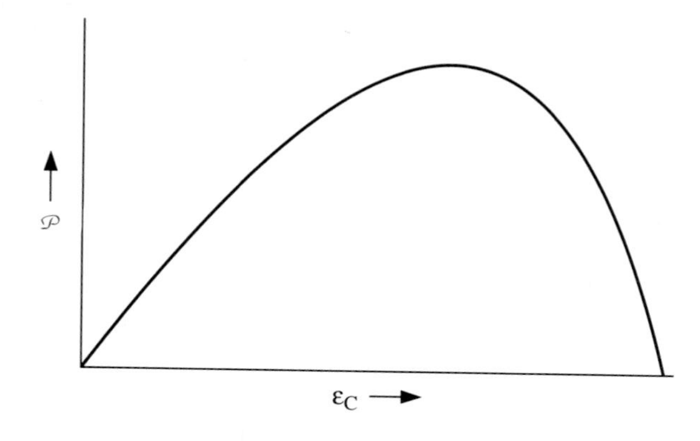

FIGURE 10.12 Power output versus efficiency for a typical heat engine. Calculated for $\varepsilon_C = 0.64$; $\varepsilon_{CA} = 0.40$.

$$\varepsilon = \frac{(T_H - \delta T) - (T_C + \delta T)}{T_H - \delta T} = \frac{\Delta T - 2\delta T}{T_H - \delta T} < \frac{\Delta T}{T_H} \tag{10-14}$$

The thermal energy flow into the engine is at the rate

$$\mathcal{P}_{\text{in}} = \frac{Q_H}{\Delta t} = \kappa \frac{\delta T}{\Delta L}$$

where ΔL is an effective distance the energy must be transferred by conduction and κ is an effective, or average, thermal conductivity over this path.

As the temperature difference, δT, increases, the power input per cycle increases, as shown in Fig. 10.12. However, the efficiency is decreasing, from the maximum value given by Carnot's theorem to zero when $T_H - \delta T = T_C + \delta T$. The power output is zero when $\delta T = 0$, it goes through a maximum, and it becomes zero again with $\delta T = \frac{1}{2}(T_H - T_C)$.

Similarly, for a heat pump, the thermal power in and out depends on δT, the difference between working fluid and reservoir temperatures at T_H and T_C. The coefficient of performance is

$$\frac{Q_H}{-W} = \frac{T_H + \delta T}{\Delta T + 2\delta T}$$

or (10-15)

$$\frac{Q_C}{W} = \frac{T_C - \delta T}{\Delta T + 2\delta T}.$$

It has been shown[6] that for a wide variety of conditions of operation of heat engines, the efficiency under conditions of maximum power output is given, at least approximately, by

$$\varepsilon = \varepsilon_{\text{CA}} = \frac{-W}{Q_1} = \frac{\sqrt{T_H} - \sqrt{T_C}}{\sqrt{T_H}} \tag{10-16}$$

conveniently designated as the *Curzon–Ahlborn* efficiency.

For the Otto cycle, maximum efficiency occurs for a compression ratio $r \equiv V_1/V_2 = (T_H/T_C)^{1/(\gamma-1)}$, but this gives zero power. Maximum power out-

[6]See F. L. Curzon and B. Ahlborn, *Am. J. Phys.* **43,** 22 (1975); M. J. Ondrechen, B. Andresen, M. Mozurkewich, and R. S. Berry, *Am. J. Phys.* **49,** 681 (1981); H. S. Leff, *Am. J. Phys.* **55,** 602 (1987), and references given there.

put occurs for $r = (T_H/T_C)^{1/2(\gamma-1)}$, and for this ratio, the efficiency is equal to the Curzon–Ahlborn efficiency.

Optimum power output is achieved in the Diesel cycle when

$$T' = T'' = (T_H^\gamma T_C)^{1/(\gamma+1)}$$

In this case, the efficiency depends on γ, and therefore on the properties of the working fluid. However, if $\gamma = 1.4$, as for air, the efficiency is almost indistinguishable from ε_{CA}.

10.7 The Third Law of Thermodynamics

The third law has comparatively little practical application, but is theoretically significant. It provides an experimental confirmation of our interpretation of entropy and provides a convenient benchmark for compilations of thermodynamic data.

One form of the third law of thermodynamics is

It is impossible by any procedure, no matter how idealized, to reduce any system to the absolute zero in a finite number of operations.

A proof of the impossibility of attaining absolute zero has often been given based on the second law of thermodynamics. The best possible coefficient of performance of a refrigerator is that of a Carnot cycle, from which the amount of thermal energy extracted per cycle from a reservoir at T_2 is

$$Q_2 = W\frac{T_2}{T_1 - T_2}$$

The ratio $T_2/(T_1 - T_2) \to 0$ as $T_2 \to 0$, so the amount of thermal energy transferred, for any given work input, goes to zero as the lower temperature approaches absolute zero.

As simple as the proof above appears, it is irrelevant to the third law. The concern is *not* with the amount of thermal energy transferred but with the temperature change produced by the process. Replacing q_2 by $C_v\,dT$, we have

$$dT = w\frac{T_2}{C_v}\frac{1}{T_1 - T_2} \tag{10-17}$$

The heat capacity is proportional to T^3 at low temperatures (Section 20.5), so the right-hand side does not vanish. If a Carnot refrigerator could be made to operate with maximum coefficient of performance when the lower temperature approaches zero kelvin, it should be possible to reach absolute zero, according to the second law.

Consideration of the processes by which cooling can occur, including transfer of thermal energy from the system to a Carnot engine, leads to the conclusion that two mechanisms exist. A system may be cooled by thermal energy transfer to a lower-temperature thermal reservoir, or a system may be cooled without thermal energy transfer (i.e., by an adiabatic process). The former mechanism is not possible if the system to be cooled is approaching absolute zero because then no lower-temperature thermal reservoir is available. The latter mechanism, which is the cooling mechanism in the Carnot cycle, is the only kind of process available for approaching 0 K. A reversible, adiabatic process is isentropic.

Let A and B be any two states (phases or substances) that can be interconverted. Then the entropy of A at any temperature T_1 is

$$S_A(T_1) = S_A^\circ + \int_0^{T_1} C_v(A) \frac{dT}{T}$$

and the entropy of B, at T_2, is

$$S_B(T_2) = S_B^\circ + \int_0^{T_2} C_v(B) \frac{dT}{T}$$

where S_A° and S_B° represent entropy of A and B at absolute zero. Select T_1 and T_2, and label states A and B, such that

$$S_A(T_1) = S_B(T_2) \qquad \text{and} \qquad T_1 > T_2$$

Then if rates are suitable, we may hope to carry out a reversible, adiabatic (hence isentropic) process that will take us from state A, at T_1, to state B, at the lower temperature, T_2.

$$A(T_1) \rightarrow B(T_2)$$

$$S_A(T_1) = S_A^\circ + \int_0^{T_1} C_v(A) \frac{dT}{T} = S_B^\circ + \int_0^{T_2} C_v(B) \frac{dT}{T} = S_B(T_2) \tag{10-18}$$

and

$$S_B^\circ - S_A^\circ = \int_0^{T_1} C_v(A) \frac{dT}{T} - \int_0^{T_2} C_v(B) \frac{dT}{T}$$

If $T_2 = 0$, the second integral vanishes, so

$$\Delta S^\circ = S_B^\circ - S_A^\circ = \int_0^{T_1} C_v(A) \frac{dT}{T} \tag{10-19}$$

If $T_1 > 0$, such a process should work and it should be possible to reach 0 K. Experimentally, however, it is found that $C_v(A)$ vanishes faster than $1/T$ and, correspondingly, that the entropy difference at 0 K vanishes; $S_B^\circ - S_A^\circ = 0$ (including any entropy of reaction).

Entropy of reaction cannot be measured directly at zero kelvin. Not only is it not possible to reach 0 K (although temperatures of millikelvin and below can be reached), but more important, chemical reactions cannot be studied at such low temperatures. The rates of chemical reactions depend exponentially on temperature, so few, if any, reactions could proceed at temperatures even close to 0 K.

However, entropy of reaction at 0 K can be measured indirectly by measuring the entropy change for reactants warmed from zero to a higher temperature, such as room temperature, measuring the entropy of reaction at room temperature (e.g., from the temperature dependence of the electrochemical potential), and measuring the entropy change for products cooled from room temperature to 0 K.

$$\Delta S_0 = \int_0^T \frac{C_p(\text{reactants})}{T}\, dT + \Delta S_T + \int_T^0 \frac{C_p(\text{products})}{T}\, dT \tag{10-20}$$

Because C_p goes to zero faster than T goes to zero, evaluation of the integrals does not depend on reaching 0 K.

Nernst observed that all calculated entropies of reaction seemed to approach zero as the lower temperature limit approached zero. His generalization, called *Nernst's heat theorem,* may be stated as:

> *The entropy of reaction of all chemical reactions approaches zero as the temperature approaches zero kelvin.*

Nernst's law is typically extended to:

> *Every substance in its lowest energy state, at absolute zero, has the same entropy,*

or, more explicitly,

> *All pure, perfect crystals have zero entropy at* 0 K.

The last form is purely arbitrary. Because each substance has, ideally, the same value of entropy and there is no thermodynamic measurement that can give this absolute value, it is permissible to assign any value whatsoever; $S = 0$ is a convenient choice. (The choice is consistent with the statistical definition of entropy but not with the way in which such values are determined experimentally.)

Nernst's heat theorem, and exceptions to the theorem as stated, are to be expected on the basis of statistical mechanics. Exceptions are found for substances that do not occupy the lowest-accessible energy state as they are cooled. For example, the lowest-energy state for a crystal of carbon monoxide, CO, has all the CO molecules aligned parallel. Crystals in which some, or many, of the CO molecules are aligned backward, as OC, relative to the rest of the crystal have only very slightly higher energy and therefore are well populated when the gas is cooled and frozen. There is a substantial energy barrier to rotation of CO molecules within the crystal, so as the temperature is lowered, misaligned molecules do not reorient.

The energy difference between the "perfect" CO crystal and a crystal in which the CO molecules are randomly aligned is very small. In principle, any arbitrarily small ΔE is possible for such misalignments in crystals, depending on the nature of the molecules. The entropy difference, on the other hand, is not negligible. For the CO example, the entropy difference arising from this misalignment is

$$S_0 = R \ln 2 \tag{10-21}$$

which is 5.76 J/mol · K. Entropy difference of misalignment appears in several substances that are pseudosymmetric and affects the value of ΔS_0 of reaction.

If it is impossible to reach 0 K in any finite number of steps, Nernst's heat theorem (with allowance for degeneracy) must be valid. If Nernst's heat theorem (with allowance for degeneracy) is valid, it is impossible to reach 0 K in any finite number of steps.

Because of the validity of Nernst's heat theorem, cooling any system to absolute zero by an adiabatic process is doomed to failure. The exceptions are of no help, for they represent only higher entropy states. Thus Nernst's heat theorem may be taken as an alternative statement of the third law of thermodynamics.

The third law can take on several different forms, as we have seen. It may be helpful to summarize the components of this generalization, including aspects presented above, without attempting to rank these components in order of importance. For completeness, reference is included to the Gibbs free energy function, G, defined in Chapter 11.

1. It is an experimental observation that $\partial \Delta H/\partial T$ and $\partial \Delta S/\partial T$ remain finite as $T \to 0$ (and therefore also $\partial \Delta G/\partial T$ is finite, where $G \equiv H - TS$).
2. At $T = 0, \Delta G \equiv \Delta H - T\Delta S = \Delta H$.
3. $\partial \Delta H/\partial T$ and $\partial \Delta G/\partial T \to 0$ as $T \to 0$, but $\partial \Delta H/\partial T$ and $\partial \Delta G/\partial T$ have opposite signs for $T \approx 0$. $\partial \Delta H/\partial T > 0$ for $T \approx 0$.
4. $\Delta S_{\text{reaction}} = 0$ at $T = 0$.
5. $\Delta S = 0$ for *all processes* at $T = 0$, including change of pressure. Because $(\partial S/\partial P)_T = -(\partial V/\partial T)_P$ it follows that the thermal coefficient of expansion, $\beta = -(1/V)(\partial V/\partial T)_P \to 0$ for $T \to 0$.
6. $S \to S_0$, a finite value, as $T \to 0$.
7. C_p and $C_v \to 0$ as $T \to 0$; otherwise, $\Delta S = \int \; C dT/T \to \infty$. This is independently confirmed by experiment and theory.
8. Planck proposed that $\partial S/\partial n_i = 0$ at $T = 0$. This is equivalent to the assumption that $S = 0$ at $T = 0$.
9. It is not possible to reach $T = 0$.

SUMMARY

Carnot cycle: Alternating isothermal and adiabatic steps:
At high temperature, $T_H : Q_H = \zeta T_H \qquad (\zeta \equiv \Delta S_H)$.
At low temperature, $T_C : -Q_C = \zeta T_C$.
Total work per cycle: $-W = \zeta(T_H - T_C) = \zeta \, \Delta T$.

The efficiency, $\varepsilon = -\dfrac{W}{Q_H} = \dfrac{\Delta T}{T}$, is the same for all Carnot engines operating between the same temperatures, T_H and T_C, and is the maximum possible for these temperatures.

Thermodynamic temperature is defined by $T_H/T_C = -Q_H/Q_C$.

The Stirling engine, alternating isothermal and isochoric steps, has the same theoretical efficiency as the Carnot engine.

Other cycles transfer thermal energy over a range of temperatures and have lower theoretical efficiencies than the Carnot cycle. In practice, most engines are noncyclic and efficiencies are limited by other considerations. For example, Curzon and Ahlborn have shown that a practical expression for heat engine efficiencies is

$$\varepsilon_{CA} = \frac{\sqrt{T_H} - \sqrt{T_C}}{\sqrt{T_H}}$$

The *third law of thermodynamics* may be expressed as the impossibility of reaching absolute zero by any process, or that $\Delta S \to 0$ as $T \to 0$ for any physical or chemical process. The latter is often approximated as $S \to 0$ as $T \to 0$.

PROBLEMS

10.1. Find the efficiency of a Carnot heat engine operating between 0 and 100°C.

10.2. A reversible Carnot heat pump is used to refrigerate a "deep-freeze" unit. The temperature in the vicinity of the compressor is 30°C. The temperature inside the unit is −10°C.

a. Find the amount of electrical energy required to remove 1000 J of thermal energy from inside the freezer.

b. How much thermal energy will be added to the room in this process?

c. Demonstrate a simple check on the consistency of your answers to parts a and b. Which thermodynamic principle is applicable?

10.3. A reversible Carnot heat engine, or heat pump, is available for general use, operating between reservoirs at 24 and 5°C.

a. Find the efficiency of the device if it acts as a heat engine.

b. To do 100 J of work on the surroundings, how many joules of thermal energy must be removed from the reservoir at 24°C?

c. To do 100 J of work on the surroundings, how many joules of thermal energy must be added to the reservoir at 5°C?

d. If the engine is operated as a heat pump, how many joules of thermal energy can be added to the reservoir at 24°C if 100 J of work is done on the engine?

e. How many joules of thermal energy can be removed from the reservoir at 5°C if 100 J of work is done on the engine?

10.4. Answer the following true (T), false (F), or may be true or false (M).

Two reversible Carnot heat engines, designated A and B, operate between T_H and T_C. Step H (Q_H and W_H) takes place at T_H; step C (Q_C and W_C) takes place at T_C. T_H is greater than T_C. Represent Q and W for the adiabatic steps ($T_H \rightarrow T_C$ and $T_C \rightarrow T_H$) by Q_{HC} and W_{HC} and by Q_{CH} and W_{CH}.

a. $Q_H(\text{A}) = Q_H(\text{B})$.

b. $W(\text{A}) = W(\text{B})$ (total work per cycle).

c. $-W(\text{A}) = Q(\text{A})$.

d. $Q_H(\text{A})$ depends on whether the gas is real or ideal.

e. $Q_H(\text{A}) = -Q_C(\text{A})$.

f. Step H is an expansion if the devices act as heat pumps.

g. $Q_{HC}(\text{A}) = -Q_{CH}(\text{A})$.

h. $W_{HC}(\text{A})$ depends on what gas is used, even if the gas is known to be ideal.

10.5. Plot H versus S for an ideal gas undergoing a Carnot cycle. Label each segment according to the scheme: step 1 is at T_H, and so on, with $T_H > T_C$.

10.6. Make plots (as in problem 10.5) of

a. E versus T. **b.** S versus T.

c. P versus V. **d.** H versus P.

10.7. Find the Curzon–Ahlborn efficiency of a heat engine operating between 0 and 100°C. Compare with the Carnot efficiency.

10.8. Find the Curzon–Ahlborn efficiency of a heat engine operating between 30 and 600°C. Compare with the Carnot efficiency.

10.9. Find the Curzon–Ahlborn efficiency of an automobile engine operating between 1000 and 100°C.

CHAPTER **11**

Free Energy

The second law provides a powerful theoretical test for equilibrium conditions and for the direction of change. If the entropy change of the system plus the surroundings is zero for any process, the system is in equilibrium with respect to its surroundings for that process. If the entropy change of system plus surroundings would be negative for a process, that process will not occur no matter how long we wait. Only those processes can occur for which the entropy change of the system plus surroundings would be positive.

Unfortunately, the second law is often inconvenient to apply directly because it requires analysis of changes in the surroundings as well as the system. This is reminiscent of the problem encountered with the first law. It is often inconvenient to calculate energy change for the system because measured thermal energy transfer, Q, has to be corrected for work done against the atmosphere. The problem with energy was solved by defining a new function, $H = E + PV$. The change in enthalpy, ΔH, is equal to Q for all processes at constant pressure, provided that $w = -P\,dV$.

The same technique is helpful in applications of the second law. A new state function is defined, called the *free energy,* that automatically corrects for changes in the surroundings for certain common conditions, so calculations need be done only on the system. However, we are sometimes interested primarily in constant-volume conditions (for which E is a convenient function) and sometimes in constant-pressure conditions (for which H is more convenient), so two different functions are defined, each bearing the label of free energy.

11.1 Helmholtz Free Energy

In searching for the entropy function, a function F was defined by the equation $E = F + TS$, or

$$F = E - TS \qquad (11\text{-}1)$$

Then

$$dF = dE - T\,dS - S\,dT$$

The change in E does not depend on the path,

$$dE = q + w = q_{\text{rev}} + w_{\text{rev}}$$

so

$$dF = q_{\text{rev}} + w_{\text{rev}} - T\,dS - S\,dT$$

But $q_{\text{rev}} = T\,dS$, so if temperature is constant,

{constant T} $$dF = w_{\text{rev}} \qquad (11\text{-}2)$$

as originally intended.

The most common form of work is $P\,dV$ work; it includes work done on the system by the atmosphere and work of expansion and compression in steam engines, gasoline engines, and so on. Other forms of work, however, including electrochemical work, work against electric or magnetic fields, work against surface tension, and so on, may also be important. It is convenient, therefore, to divide W, the work done on the system, into two parts,

$$w = -P\,dV + w' \qquad (11\text{-}3)$$

or

$$W = -\int P\,dV + W'$$

where $W' = \int w'$ is the total non$-P\,dV$ work. (W' is sometimes called *useful work* and sometimes *electrical work,* but neither is adequately descriptive of the distinction between W and W'.)

If the path is reversible,[1]

{reversible} $$w_{\text{rev}} = -P\,dV + w'_{\text{rev}}$$

If volume, as well as temperature, is constant, $P\,dV = 0$ and

{T,V} $$dF = w'_{\text{rev}} \qquad (11\text{-}4)$$

At constant temperature and volume, the change in F is equal to the amount of reversible, non$-P\,dV$ work (i.e., the minimum non$-P\,dV$ work) that must be done on the system to get from the initial state to the final state.

Criteria for Equilibrium and Spontaneous Processes

From $dF = dE - d(TS)$, setting $dE = q + w$,

[1]Note that constant pressure does *not* ensure that $-P\,dV$ will give the reversible work of expansion. See the footnote at the end of Section 11.3.

$$dF = q + w - T dS - S dT$$

Replace $T\ dS$ with q_{rev}. If $w = -P\ dV$, we have

$\{w' = 0\}$ $$dF = q - q_{rev} - P\,dV - S\,dT$$

and thus at constant volume and constant temperature,

$\{T, V; w' = 0\}$ $$dF = q - q_{rev}$$

But q_{rev} is the maximum, so $q - q_{rev}$ is negative or, in the limit that the process is reversible, equal to zero.

$\{T, V; w' = 0\}$ $$dF \leq 0 \qquad (11\text{-}5)$$

At constant temperature and volume, with no non$-P\,dV$ work done on the system, the free energy, F, of the system must decrease for any process, except that in the limit of a reversible process, F may be constant.

$\{T, V; w' = 0; \text{reversible}\}$ $$dF = 0 \qquad (11\text{-}6)$$

The function F is called the *work function* or the *Helmholtz free energy.* It has often been represented by the symbol A. The Helmholtz free energy, F, provides a direct test of whether a system is in equilibrium and whether a process can or cannot occur given sufficient time.

11.2 Gibbs Free Energy

Although the Helmholtz free-energy function, F, achieves much of what is required, it shares the difficulty, with E, of requiring constant volume for its convenient application. That difficulty is removed in much the same manner as for the energy. Define a new function, G, called the *Gibbs free energy,*

$$G \equiv F + PV$$

$$G \equiv H - TS \qquad (11\text{-}7)$$

or

$$G \equiv E + PV - TS$$

Replacing dE with $q + w$, we find that the total differential is

$$dG = q + w + P\,dV + V\,dP - T\,dS - S\,dT$$

At constant temperature, letting $T\,dS = q_{rev}$ and $w = -P\,dV + w'$, we obtain

$\{T\}$ $$dG = q - q_{rev} + w' + V\,dP$$

Therefore, if the process is reversible, at constant temperature and at constant pressure,

{T, P; reversible} $$dG = w'_{rev} \quad (11\text{-}8)$$

The work done on an electrochemical cell, for example, at constant temperature and pressure under reversible conditions, is equal to ΔG.

Criteria for Equilibrium and Spontaneous Processes

From the definition,

$$dG = dE + P\,dV + V\,dP - S\,dT - T\,dS$$

Replacing dE with $q + w$, we have

$$dG = q + w + P\,dV + V\,dP - S\,dT - T\,dS$$

so at constant temperature and pressure,

{T, P} $$dG = q + w + P\,dV - T\,dS$$

If $w' = 0$, then $w = -P\,dV$ and

{T, P; w′ = 0} $$dG = q - q_{rev}$$

and, again, because q_{rev} is a maximum,

{T, P; w′ = 0} $$dG \leq 0 \quad (11\text{-}9)$$

At constant temperature and pressure, with no non−$P\,dV$ work done on the system, the free energy, G, of the system must decrease for **any** *process, except that, in the limit of a* reversible *process, G may be constant.*

{T, P; w′ = 0; reversible} $$dG = 0 \quad (11\text{-}10)$$

The function G is called the Gibbs free energy. It has, until recently, most often been represented by the symbol F in the United States.

The reader must be aware that the symbol F may represent either F or G, depending on the author and the date of writing.

There is a complete parallel between G and F. The Gibbs free energy, G, plays the same role at constant temperature and pressure as the Helmholtz free energy, F, at constant temperature and volume. Each provides a test of whether a process is or is not possible, without calculating changes for the surroundings. And each is constant, or equal to non−$P\,dV$ work, if and only if the process occurs at equilibrium (and constant temperature and pressure or volume).

EXAMPLES 11.1

A. Calculate ΔG and ΔF for the process of vaporizing 1 mol of water at 1 atm and 100°C.

B. Calculate ΔG and ΔF for the process of melting 18 g of ice at 1 atm and 0°C. □

In the derivations above it was assumed that the state is completely determined by two functions. In principle these could be any two of the state functions T, P, V, E, S, H, and so on, although some are better for differentiating some sets of states than are others. Often chemical composition or external variables, such as applied fields, may be significant. Such additional conditions are easily added to the basic equations already found, as shown in Section 11.6.

11.3 Applications of Free-Energy Functions

From the definition of G,

$$G \equiv E + PV - TS$$

and its differential,

$$dG = dE + P\ dV + V\ dP - T\ dS - S\ dT$$

and letting $dE = q_{\text{rev}} + w_{\text{rev}}$, it follows that[2] if $w_{\text{rev}} = -P\,dV$, then

$$dG = V\ dP - S\ dT$$

From this, two very important equations are obtained:

$$\left(\frac{\partial G}{\partial P}\right)_T = V \tag{11-11}$$

and

$$\left(\frac{\partial G}{\partial T}\right)_P = -S \tag{11-12}$$

Note that P goes with V, and T goes with S, as is required for dimensional consistency.

The corresponding equations for F are

$$dF = q_{\text{rev}} + w_{\text{rev}} - T\ dS - S\ dT$$

and $q_{\text{rev}} = T\ dS$, so if $w_{\text{rev}} = -P\,dV$,

$$dF = -P\ dV - S\ dT$$

and therefore

$$\left(\frac{\partial F}{\partial V}\right)_T = -P \tag{11-13}$$

and

$$\left(\frac{\partial F}{\partial T}\right)_V = -S \tag{11-14}$$

[2]An alternative formulation of the restriction is given at the end of this section.

ANSWERS 11.1

A. $\Delta G = \Delta H - T\Delta S$. P is constant, so $\Delta H = Q$ and $\Delta S = Q_{\text{rev}}/T$, so $\Delta G = Q - T(Q_{\text{rev}}/T) = Q - Q_{\text{rev}} = 0$, as expected for a reversible process. $\Delta F = \Delta E - T\Delta S = Q + W - T(Q_{\text{rev}}/T) = W$. The work is estimated easily. $W = -\int P dV = -P\Delta V = (PV)_{\text{initial}} - (PV)_{\text{final}} = -(PV)_{\text{final}}$ because the volume of liquid is negligible compared to the volume of vapor. $\Delta F = -(PV)_{\text{final}} = -nRT = -1 \times R \times 373\text{ K} = -3.1\text{ kJ}$. ΔF is not zero; the process does not occur at constant volume.

B. $\Delta G = \Delta H - T\Delta S = Q - Q_{\text{rev}} = 0$ for the reversible process at constant pressure ($Q = \Delta H$). $\Delta F = \Delta E - T\Delta S = Q + W - Q_{\text{rev}} = W = -\int P dV = -P\Delta V$. The volume of water is about 0.1 cm^3/g less than the volume of ice. Per mole, ΔV is therefore about $1.8\text{ cm}^3 = 1.8 \times 10^{-6}\text{ m}^3$, so $P\Delta V$ is about 1×10^5 Pa $\times 1.8 \times 10^{-6}\text{ m}^3 = 0.18$ J; therefore $\Delta F = -0.18$ J.

The first equation of each pair has an obvious interpretation. The free energy (Gibbs or Helmholtz) increases when the system is compressed and decreases when the system expands, at constant temperature. Specifically, G increases with pressure, at constant temperature, because V is always positive, and similarly F decreases with V, at constant temperature, because P is always positive.

EXAMPLES 11.2

A. Find ΔG and ΔF for the isothermal expansion of 3 mol of an ideal gas from a pressure of 4 atm to a pressure of 1 atm at 27°C.

B. Find ΔG and ΔF for the isothermal expansion of 3 mol of an ideal gas from a volume of 1 m^3 to a volume of 4 m^3 at 27°C.

C. Find ΔG and ΔF for 10 g of water at 25°C if the pressure is increased from 1 atm to 10 atm. □

The second equation of each pair shows that the change in free energy with temperature, at constant pressure or volume, is equal to minus the entropy. Entropy was defined by means of the equations

$$dS = \frac{q_{\text{rev}}}{T} \quad \text{and} \quad S = \frac{E - F}{T}$$

These tell how entropy changes, but do not give the absolute value; absolute values are not known for E or F. Without an absolute value for S, the change in F or G with temperature cannot be found.[3]

It is, however, possible to find values of ΔS and from those values to determine how ΔG and ΔF change with temperature. For example, consider a chemical reaction such as production of gaseous SO_3,

$$SO_2 + \frac{1}{2}O_2 \rightarrow SO_3$$

[3]The third law (Section 10.7) provides a logical path for assigning a value to S, but the values assigned do not lead to experimentally verifiable new results. Also, certain components of S and of F or G can be evaluated, as will be shown in Part III.

The free-energy change is the final value minus the initial value, or

$$\Delta G_{\text{reaction}} = G_{SO_3} - G_{1/2\ O_2} - G_{SO_2}$$

The individual terms on the right are unknown, but $\Delta G_{\text{reaction}}$ is an observable quantity. (For example, it may be obtained from the work term, W' = electric charge × voltage, in an electrochemical cell reaction.)

The change in $\Delta G_{\text{reaction}}$ with temperature is

$$\left(\frac{\partial(\Delta G_{\text{react}})}{\partial T}\right)_P = \left(\frac{\partial G_{SO_3}}{\partial T}\right)_P - \left(\frac{\partial G_{(1/2)O_2}}{\partial T}\right)_P - \left(\frac{\partial G_{SO_2}}{\partial T}\right)_P$$
$$= -(S_{SO_3} - S_{(1/2)O_2} - S_{SO_2})$$
$$= -\Delta S_{\text{reaction}}$$

Similar arguments can be made for any other process, leading to the general result

$$\left(\frac{\partial \Delta G}{\partial T}\right)_P = -\Delta S \tag{11-15}$$

Although the change in free energy with temperature of a single substance cannot be known, the change with temperature of a *change* in free energy can be measured. This is all that we need or can make use of.

In the same way, we obtain

$$\left(\frac{\partial \Delta G}{\partial P}\right)_T = \Delta V \tag{11-16}$$

Corresponding equations for the Helmholtz free energy are

$$\left(\frac{\partial \Delta F}{\partial T}\right)_V = -\Delta S \tag{11-17}$$

and

$$\left(\frac{\partial \Delta F}{\partial V}\right)_T = -\Delta P \tag{11-18}$$

It is impractical to tabulate ΔG values for all possible chemical reactions. The problem is solved, as for enthalpy changes, by tabulating $\Delta G^\circ_{\text{formation}}$ values from reactions that form a compound from the elements. For any element (in its standard state), $\Delta G^\circ_f = 0$.

In the reaction considered above,

$$\Delta G^\circ_{\text{formation}(SO_2)} = G^\circ_{SO_2} - G^\circ_S - G^\circ_{O_2}$$

As before, there is no way of determining absolute values for the free-energy terms on the right. However, each of these can be found relative to the appropriate

---

ANSWERS 11.2

A. $\Delta G = \int V dP = \int nRT dP/P = nRT \ln(P_2/P_1) = 3 \times 300R \times \ln\frac{1}{4} = -10.4$ kJ. $\Delta F = \int -P dV = -\int nRT dV/V = -nRT \ln 4 = -10.4$ kJ. [**Note**: For an ideal gas at constant T, $\Delta(PV) = 0$, so $\Delta G = \Delta F$.]
B. $\Delta G = nRT \ln(P_2/P_1) = -nRT \ln(V_2/V_1) = -3 \times 300R \ln 4 = -10.4$ kJ. $\Delta F = \Delta G$.
C. $\Delta G = \int V dP = V\Delta P$ (because $V \approx$ constant for the liquid). $\Delta G = 10 \times 10^{-6} \times (9 \times 1 \times 10^5$ Pa$) = 9$ J. $\Delta F = \int P dV \approx 0$. [$\kappa = -(1/V)(\partial V/\partial P)_T$, so $dV = -\kappa V dP$ and $\int P dV = -\int \kappa V P dP = -\kappa V \Delta(P^2/2)$ with $\kappa = 4.57 \times 10^{-10}$, so $\Delta F \approx 1.8$ mJ.]

---

elements. Although $G^\circ_{SO_2}$, G°_S , and $G^\circ_{O_2}$ are unknown, these quantities drop out.

$$\Delta G^\circ_{\text{formation (SO}_2)} = (G^\circ_{SO_2} - G^\circ_S - G^\circ_{O_2}) - (G^\circ_S - G^\circ_S) - (G^\circ_{O_2} - G^\circ_{O_2})$$
$$= \Delta G^\circ_{\text{formation (SO}_2)} - \Delta G^\circ_{\text{formation (S)}} - \Delta G^\circ_{\text{formation (O}_2)}$$

where

$$\Delta G^\circ_{\text{formation (S)}} \equiv G^\circ_S - G^\circ_S \equiv 0$$
$$\Delta G^\circ_{\text{formation(O}_2)} \equiv G^\circ_{O_2} - G^\circ_{O_2} \equiv 0$$

The free energies of formation are observable quantities. They have been tabulated, for common compounds, permitting the free-energy change to be calculated for any chemical reaction involving the same compounds.

EXAMPLE 11.3

The free energies of formation of SO_3 and SO_2 are

$$\Delta G^\circ_{\text{formation (SO}_3)} = -370 \text{ kJ/mol}$$
$$\Delta G^\circ_{\text{formation (SO}_2)} = -300 \text{ kJ/mol}$$

Find ΔG° for the reaction

$$SO_2 + \frac{1}{2}O_2 \rightarrow SO_3$$

□

Free-energy values depend on temperature and pressure. Corrections can be calculated for changes in pressure by means of the equation $(\partial G/\partial P)_T = V$ or $(\partial \Delta G/\partial P)_T = \Delta V$. Corrections can be found for changes in temperature for ΔG, by means of the equation $(\partial \Delta G/\partial T)_P = -\Delta S$.

The derivations above assumed that $w_{rev} = -P dV$. Yet the equations involve only state functions, and changes in state functions are independent of path. An alternative to the condition $w_{rev} = -P dV$ is to require that dG must be a change between equilibrium states described by the state functions V, P, S, and T, any two of which fully specify the state. Then the transition from the initial to the final state can, in principle, be carried out along the path[4] for which $w = w_{\text{rev}} = -P\,dV$. The change, ΔG, cannot depend on path, so the result applies for all paths between the specified end states. If there are additional variables, such as concentrations, the extent of a chemical reaction, or a change in applied field, these variables are held constant in evaluating the partial derivatives.

For many practical problems, PV is constant. Then $\Delta G = \Delta F$ and the equations may be expressed equally well in terms of Gibbs or Helmholtz free energy. Certain other problems are expressed more simply in terms of the Helmholtz free energy, as shown in Part III. The dependence of radiation energy density on temperature is one such example.

Radiation in equilibrium with surrounding matter at a temperature T exerts a pressure on the walls. This pressure is given by classical electromagnetic theory, or by quantum statistical mechanics (Section 20.9), as

$$P = \frac{1}{3}\frac{E}{V} \qquad \text{or} \qquad PV = \frac{1}{3}E$$

Take the derivative of F/T with respect to temperature:

$$\frac{\partial(F/T)}{\partial T} = \frac{1}{T}\left(\frac{\partial F}{\partial T}\right)_V - \frac{F}{T^2} = -\frac{TS + F}{T^2} = -\frac{E}{T^2}$$

Both E and F are extensive properties. Replacing[5] $(\partial F/\partial V)_T$ by F/V and recalling that $(\partial F/\partial V)_T = -P$, we find

$$PV = \frac{1}{3}E = -F$$

Again taking the derivative of F/T with respect to T, this time replacing F/T by $-\frac{1}{3}E/T$, gives

$$-\frac{E}{T^2} = \frac{\partial(F/T)}{\partial T} = -\frac{1}{3}\left(\frac{1}{T}\frac{\partial E}{\partial T} - \frac{E}{T^2}\right)$$

which gives

$$d\ln E = 4\,d\ln T$$

or

[4]The distinction between Q and Q_{rev}, and especially between W and W_{rev}, is important but often subtle. Consider, for example, the conversion of supercooled water, at $-20°C$, to ice at -20°C. Because P is constant (at 1 atm), the work done on the ice is $-P\,\Delta V \approx -0.18$ J/mol. To find $Q(= \Delta H)$, apply Kirchhoff's law, warming the water to 0°C, freezing it, and cooling the ice; this gives $Q = \Delta H \approx 5.3$ kJ/mol. Although this Q is evaluated for a reversible path, it is *not* Q_{rev} for the isothermal process. To find Q_{rev}, calculate ΔS for the same path as ΔH; $\Delta S = C_p(\text{liq})\ln(273/253) - \Delta H/273 + C_p(\text{ice})\ln(253/273) \approx -19$ J/mol·K. Then $Q_{\text{rev}} = T\,\Delta S \approx -4.86$ kJ/mol. The difference is $Q - Q_{\text{rev}} \approx -415$ J/mol. Therefore, recognizing that $Q + W = Q_{\text{rev}} + W_{\text{rev}}$, $W_{\text{rev}} - W \approx W_{\text{rev}} \approx -415$ J/mol, compared to $W = 0.18$ J/mol. This large value of W_{rev}, compared to W, is consistent with our understanding of the ice-water system. To freeze the water *reversibly* at $-20°C$, it would be necessary to increase the pressure to a very large value to make the liquid stable at -20°C, then slowly lower the pressure to atmospheric. $W_{\text{rev}} = -\int P\,dV$. The pressure required is on the order of kilobars; hence $|\,W_{\text{rev}}\,| \gg |\,W\,|$.

[5]See Problem 11.9.

ANSWER 11.3

$\Delta G^\circ_{\text{formation}} = 0$ for any element, so subtraction yields $\Delta G^\circ_{SO_2+(1/2)O_2 \rightarrow SO_3} = -370 - (-300) = -70$ kJ/mol. The negative value for the reaction shows that the process can go in the direction indicated, and therefore probably will, at least if given adequate time and/or an appropriate catalyst.

$$E \sim T^4 \tag{11-19}$$

which is the Stefan–Boltzmann equation. It was obtained experimentally by Stefan in 1879 and derived five years later by Boltzmann.

11.4 Transformations and Thermodynamic Potentials

The four state functions with units of energy—E, H, G, and F—are often called *thermodynamic potential functions*. They are related as shown by the geometric mnemonic diagram of Fig. 11.1. For a system of fixed composition, fixing two state functions determines the state and all other state functions. Therefore, each potential can be written as a function of any two state functions. For example, we found it convenient to express the energy as $E = E(T, V)$ and enthalpy as $H = H(T, P)$. However, each of the potentials is a "natural" function of a particular pair of variables.

$$\begin{aligned} E &= E(S, V) \\ H &= H(S, P) \\ F &= F(T, V) \\ G &= G(T, P) \end{aligned} \tag{11-20}$$

Each has similar properties for describing equilibrium states and spontaneous processes. We expect a wagon at rest to roll downhill, but not uphill, because it moves such that $\Delta E < 0$, just as $\Delta G < 0$ is often the appropriate criterion for spontaneous chemical reactions or phase changes.

The relationship among the potential functions and their respective independent variables is readily shown through a mathematical transformation called a *Legendre transform*. Let $f(x, y)$ be any function of two variables. Then

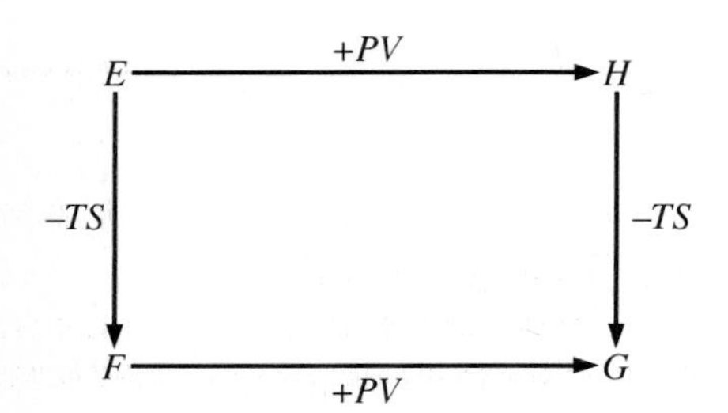

FIGURE 11.1 The thermodynamic potentials are linked by the differences PV and TS, as shown.

$$df = \left(\frac{\partial f}{\partial x}\right)_y dx + \left(\frac{\partial f}{\partial y}\right)_x dy$$

Represent the partial derivatives by u and z.

$$\left(\frac{\partial f}{\partial x}\right)_y = u \qquad \left(\frac{\partial f}{\partial y}\right)_x = z$$

Then

$$df = u\, dx + z\, dy \tag{11-21}$$

Differentials of the products ux and zy are

$$d(ux) = u\, dx + x\, du$$
$$d(zy) = z\, dy + y\, dz$$

Therefore,

$$\begin{aligned} d(f - ux) &= df - d(ux) \\ &= u\, dx + z\, dy - u\, dx - x\, du \\ &= z\, dy - x\, du \end{aligned} \tag{11-22}$$

and

$$\begin{aligned} d(f - zy) &= df - d(zy) \\ &= u\, dx + z\, dy - z\, dy - y\, dz \\ &= u\, dx - y\, dz \end{aligned} \tag{11-23}$$

The original function, $f(x, y)$, is replaced by the function $f_a = f - ux = f_a(y, u)$, and by the function $f_b = f - zy = f_b(x, z)$.

Applying the Legendre transform to thermodynamic potentials yields

$$dE = TdS - PdV$$
$$d(E + PV) = TdS - PdV + PdV + VdP = TdS + VdP$$

so if we define $H \equiv E + PV$,

$$dH = TdS + VdP$$

Similarly,

$$dF = d(E - TS) = TdS - PdV - TdS - SdT = -SdT - PdV$$
$$dG = d(H - TS) = TdS + VdP - TdS - SdT = VdP - SdT$$

The relationships are summarized below.

$$
\begin{aligned}
E &= E(S, V) & dE &= T\,dS - P\,dV \\
H &= H(S, P) & dH &= T\,dS + V\,dP \\
F &= F(T, V) & dF &= -P\,dV - S\,dT \\
G &= G(T, P) & dG &= V\,dP - S\,dT
\end{aligned}
\tag{11-24}
$$

$$
\begin{aligned}
\left(\frac{\partial E}{\partial S}\right)_V &= T & \left(\frac{\partial E}{\partial V}\right)_S &= -P \\
\left(\frac{\partial H}{\partial S}\right)_P &= T & \left(\frac{\partial H}{\partial P}\right)_S &= V \\
\left(\frac{\partial F}{\partial V}\right)_T &= -P & \left(\frac{\partial F}{\partial T}\right)_V &= -S \\
\left(\frac{\partial G}{\partial P}\right)_T &= V & \left(\frac{\partial G}{\partial T}\right)_P &= -S
\end{aligned}
\tag{11-25}
$$

The conditions for equilibrium are

$$
\begin{aligned}
&\{\text{constant } S,\ V\} & dE &= 0 \\
&\{\text{constant } S,\ P\} & dH &= 0 \\
&\{\text{constant } T,\ V\} & dF &= 0 \\
&\{\text{constant } T,\ P\} & dG &= 0
\end{aligned}
\tag{11-26}
$$

The four corresponding conditions for any process to occur are

$$
\begin{aligned}
&\{\text{constant } S,\ V\} & dE &< 0 \\
&\{\text{constant } S,\ P\} & dH &< 0 \\
&\{\text{constant } T,\ V\} & dF &< 0 \\
&\{\text{constant } T,\ P\} & dG &< 0
\end{aligned}
\tag{11-27}
$$

11.5 Criteria for Spontaneous Processes

There is an apparent paradox built into the equations and inequalities above. Consider the common example, from mechanics, of a cart that will roll downhill but not uphill. Represent the state of the cart by two variables, S and V, so that $E = E(S, V)$, then require constant S and constant V as the process conditions. Neither the entropy nor volume of the cart changes. It follows that $E = E(S, V)$ is also constant; $\Delta E \not< 0$ when the cart rolls downhill; yet we recognize that the process is spontaneous.

For a complete description we should set $E = E(S, V; h, v)$, recognizing that the energy of the system (defined as the cart plus its associated gravitational field) depends not only on its internal variables (S and V) but also on the height, h, of the cart and on its speed, v. Then as the cart rolls downhill, $\Delta h < 0, \Delta v > 0$, and $\Delta E = 0, \Delta S = 0, \Delta V = 0$. The cart can roll back up a similar hill. If there is friction, $\Delta T > 0$ and $\Delta S > 0$. Then, to reach a final state for which $\Delta S = 0$ (for the process), the cart must lose thermal energy to its surroundings: $\Delta S < 0$ and $\Delta E < 0$. The overall process (rolling downhill and losing friction-generated thermal energy to the surroundings) is a spontaneous process for which $\Delta E < 0$ at constant S and constant V.

Although mechanics problems are not typically analyzed in this way, the results are consistent with the expectations of mechanics. If the entropy and volume of a system are unchanged, the system can change energy only by interacting with the environment. It is critical that all fields not part of the system, but interacting with the system, be considered. If work is done on the system by a gravitational field or an electromagnetic field, the equation becomes

$$\{\text{constant } S,\ V\} \qquad dE = w'_{\text{rev}}$$

For example, a ball falling in a gravitational field gains $\Delta E = \frac{1}{2}mv^2$ when work, $w' = -mg\,\Delta h$, is done by the field on the ball, for which S and V are constant. For an adiabatic process, the change in energy of the system, at equilibrium, is equal to the work done on the system by the surroundings (which may be positive or negative). Under the restriction that $\Delta S = 0$ and $\Delta V = 0$, the falling ball acts as a particle and $\Delta E = W'$, which is the work-energy theorem.

For a spontaneous process without change in speed,

$$\{\text{constant } S,\ V\} \qquad dE = w'_{\text{rev}} < w' \tag{11-28}$$

For example, if a box falls off the shelf, $W'_{\text{rev}} = 0$, but $W' = mgh$. The box is warmed by the fall, then gives off thermal energy in returning to its original temperature. The energy change of the system is less than the work done on the system (other than $P\,dV$ work, which is zero at constant volume).

The enthalpy criterion for equilibrium is very similar. There must be no change in the internal state: $\Delta S = 0$; and the only work done is against constant pressure (typically, atmospheric pressure).

Changes in free-energy functions involve a compromise, or a conflict, between two very different effects. On the one hand, a system will tend to move toward a state of lower energy, or enthalpy, as discussed above. On the other hand, a system will tend to move toward greater randomness, or disorder, or probability, as measured by an increase in entropy. At low temperature, the lowering of energy (or enthalpy) is more important; at high temperature more energy is available, so the push toward higher entropy is dominant.

At constant temperature,

$$\Delta F = \Delta E - T\,\Delta S$$

$$\Delta G = \Delta H - T\,\Delta S$$

If T is small, ΔE or ΔH dominates; if T is large, ΔS dominates. Processes tend to be spontaneous (ΔF or ΔG negative) when energy or enthalpy decreases and/or when entropy increases.

11.6 Additional Variables in Thermodynamic Potentials

In each problem there is some *process variable* that describes the change in state. Often this is in addition to the usual thermodynamic state functions. It may be a variable describing the phase (how much ice and how much liquid water are present in the system) or the extent of a chemical reaction or some other external variable. The equations and inequalities of Section 11.5 assumed that only two variables were required to specify the state of the system, but earlier discussion of work (Section 3.2) showed that a variety of additional variables may appear. When that happens, additional terms appear in the differentials of the potential functions.

We considered above the example of $E = E(S, V, h, v)$ for a system consisting of a cart and its associated gravitational field. Two additional examples are changes in chemical concentrations and the progress of an electrochemical reaction. The method of handling these is representative of how other variables can also be included.

If the numbers of moles of the components of a solution are included as variables, $E = E(S, V, n_i)$, the differential of the energy becomes

$$dE = \left(\frac{\partial E}{\partial S}\right)_{V,n_i} dS + \left(\frac{\partial E}{\partial V}\right)_{S,n_i} dV + \sum_i \left(\frac{\partial E}{\partial n_i}\right)_{S,V} dn_i$$

Define $\mu_i = (\partial E/\partial n_i)_{S,V}$. Then the equation becomes

$$dE = T\,dS - P\,dV + \sum_i \mu_i\,dn_i \tag{11-29}$$

After Legendre transforms,

$$dH = T\,dS + V\,dP + \sum_i \mu_i\,dn_i$$

$$dF = -S\,dT - P\,dV + \sum_i \mu_i\,dn_i$$

$$dG = V\,dP - S\,dT + \sum_i \mu_i\,dn_i$$

It follows that

$$\mu_i = \left(\frac{\partial E}{\partial n_i}\right)_{S,V} = \left(\frac{\partial H}{\partial n_i}\right)_{S,P} = \left(\frac{\partial F}{\partial n_i}\right)_{T,V} = \left(\frac{\partial G}{\partial n_i}\right)_{T,P} = \bar{G}_i \tag{11-30}$$

The function μ_i, which was introduced earlier without an adequate definition, is identical to the partial molar free energy, $\bar{G}_i$. It is also equal to rates of change of energy, enthalpy, and Helmholtz free energy with respect to the number of moles of the *i*th component, but with different variables held constant.

It is often convenient to change the units of μ, expressing it per particle. Then

$$\mu_i \equiv \left(\frac{\partial G}{\partial N_i}\right)_{T,P} \quad \text{and} \quad G \equiv \mu N \tag{11-31}$$

Partial molar quantities are defined as rates of change of a property of the system (i.e., solution) with respect to the number of moles of a component, under conditions of constant temperature and pressure. Therefore, the chemical potential is equal to the partial molar (Gibbs) free energy, but it is *not* equal to partial molar energy, enthalpy, or Helmholtz free energy.

For some purposes it is convenient to regard the product $(-)PV$, which appears in the definitions of some of the thermodynamic potential functions, as simply another potential. (It has, in fact, been called *the thermodynamic potential.*) Noting that it is $-P$ that appears naturally in the Legendre transforms, we choose the negative sign for the product and introduce the symbol Ω.

$$\Omega \equiv -PV = E - TS - \mu N \tag{11-32}$$

Then, for example, $\partial\Omega/\partial\mu = -N$. The function Ω will be particularly helpful in Part III.

Work done in electrical circuits is equal to the product of charge, $\mathcal{Q}$, and the electric potential, or voltage, $\mathcal{E}$. The charge, in turn, is a product of the number of moles of electrons times the charge on an electron. One mole of electric charge is called a *faraday,* with symbol $\mathcal{F}$. For n moles of charge, the charge transferred is $\mathcal{Q} = n\mathcal{F}$ and the work done *by* the reacting system is $W = n\mathcal{F}\mathcal{E}$, or the work done *on* the reacting system is $W = -n\mathcal{F}\mathcal{E}$. Such work is included in W'.

The differential of the energy, when an electrochemical reaction is included, becomes

$$dE = TdS - PdV - \mathcal{F}\mathcal{E}\,dn$$

The minus sign shows that when $\mathcal{E}$ is positive, the cell is doing work on the surroundings ("producing electricity"), so negative work is done on the system (the electrochemical cell). For finite changes, under the usual restrictions,

$$\begin{aligned}
&\{\text{constant } S,\ V\} \qquad &\Delta E &= -n\mathcal{F}\mathcal{E} \\
&\{\text{constant } S,\ P\} \qquad &\Delta H &= -n\mathcal{F}\mathcal{E} \\
&\{\text{constant } T,\ V\} \qquad &\Delta F &= -n\mathcal{F}\mathcal{E} \\
&\{\text{constant } T,\ P\} \qquad &\Delta G &= -n\mathcal{F}\mathcal{E}
\end{aligned} \tag{11-33}$$

The number n is the *number of equivalents* or the number of moles of electrons transferred if the reaction proceeds as written.

11.7 Maxwell Relations

We begin with the definitions,

$$H = E + PV$$
$$F = E - TS$$
$$G = H - TS$$

to which is added the first-law equation,

$$dE = q + w + \sum_i \mu_i\,dn_i$$

and the definition of entropy change,

$$dS = \frac{q_{\text{rev}}}{T}$$

We substitute for w a generalized expression (Section 3.3),

$$w = -P\,dV + \sum_i X_i\,dx_i$$

From these equations we find, by Legendre transforms, a series of equations, such as

$$\begin{aligned}
dE &= T\,dS - P\,dV + \sum_i \mu_i\,dn_i + \sum_i X_i\,dx_i \\
d(E - \sum_i \mu_i n_i) &= T\,dS - P\,dV - \sum_i n_i\,d\mu_i + \sum_i X_i\,dx_i \\
dH = d\,(E + PV) &= T\,dS + V\,dP + \sum_i \mu_i\,dn_i + \sum_i X_i\,dx_i \\
dG = d(H - TS) &= V\,dP - S\,dT + \sum_i \mu_i\,dn_i + \sum_i X_i\,dx_i
\end{aligned} \tag{11-34}$$

and so on. Each of these is an exact differential, so partial derivatives of the coefficients may be equated (Section 4.4).

From the first equation we obtain

$$\left(\frac{\partial T}{\partial V}\right)_{S,n_i} = -\left(\frac{\partial P}{\partial S}\right)_{V,n_i} \quad \text{and} \quad \left(\frac{\partial T}{\partial n_i}\right)_{S,V,n_j} = \left(\frac{\partial \mu_i}{\partial S}\right)_{n_i,V}$$

and

$$-\left(\frac{\partial P}{\partial n_i}\right)_{S,n_j,V} = \left(\frac{\partial \mu_i}{\partial V}\right)_{n_i,S}$$

(Constant n_i in these equations implies that all mole numbers are constant; constant n_j implies that all mole numbers are constant except n_i.) Similarly, from the second equation,

$$\left(\frac{\partial T}{\partial V}\right)_{S,\mu_i} = -\left(\frac{\partial P}{\partial S}\right)_{V,\mu_i} \quad \text{and} \quad \left(\frac{\partial T}{\partial \mu_i}\right)_{S,V,\mu_j} = -\left(\frac{\partial n_i}{\partial S}\right)_{\mu_i,V}$$

and so on.

Clearly, such mathematical manipulations generate a large number of equalities; there are three equations per exact differential, plus any others that may arise in special instances from the generalized work terms, $\sum_i X_i\,dx_i$. These equalities of derivatives are called the *Maxwell relations*. They often aid in expressing a problem in terms of known quantities.

Some of the most helpful Maxwell relations can be recalled quickly through use of a mnemonic device introduced by Born and since adapted to other variables (Fig. 11.2). Arrange the thermodynamic potentials, $E = E(S, T), F = F(T, V), G = G(T, P)$, and $H = H(P, S)$, around the sides of a square alphabetically. Then add the independent variables V,T,P, and S at the corners, with arrows (again alphabetically) from S to T and from P to V. Then the direction of the arrow determines the sign. For example, the arrows are from S to T and from P to V, corresponding to the additive terms $+T\,dS$ and $-P\,dV$.

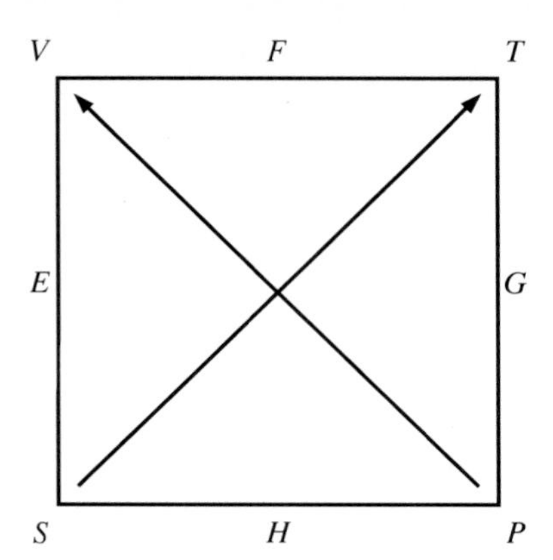

FIGURE 11.2 Born's mnemonic.

Now, recognizing the geometric pattern

$$\begin{matrix} V & \\ S & P \end{matrix} \quad \text{and} \quad \begin{matrix} & T \\ S & P \end{matrix}$$

we can write

$$\left(\frac{\partial V}{\partial S}\right)_P = \left(\frac{\partial T}{\partial P}\right)_S$$

Similarly, by rotating the square, we find that

$$\left(\frac{\partial S}{\partial V}\right)_T = \left(\frac{\partial P}{\partial T}\right)_V \qquad \left(\frac{\partial P}{\partial S}\right)_V = -\left(\frac{\partial T}{\partial V}\right)_S \qquad \text{and} \qquad \left(\frac{\partial S}{\partial P}\right)_T = -\left(\frac{\partial V}{\partial T}\right)_P$$

inserting the minus signs because S and V, or P and T, are at opposite ends of their respective arrows.

If we are interested in μ and N, which hold positions mathematically equivalent to T and S, respectively, we can modify the diagram as shown in Fig. 11.3.

From this figure we read such relationships as

$$\left(\frac{\partial V}{\partial N}\right)_P = \left(\frac{\partial \mu}{\partial P}\right)_N \qquad \text{and} \qquad \left(\frac{\partial \mu}{\partial V}\right)_N = -\left(\frac{\partial P}{\partial N}\right)_V$$

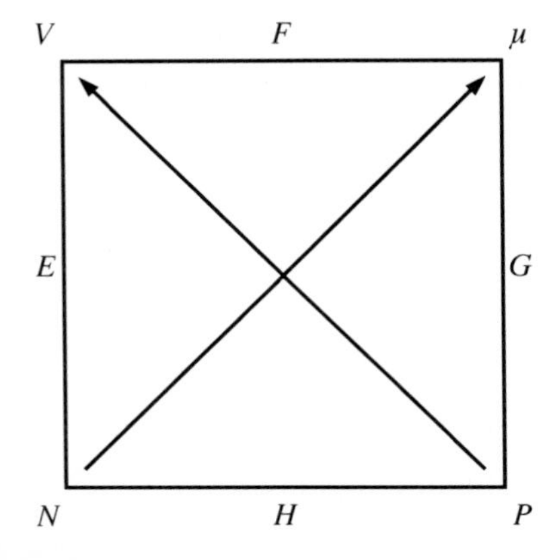

FIGURE 11.3 Modified Born mnemonic diagram, assuming that $H = H(N, P)$, $G = G(\mu, P)$, and so on.

and so on. The possibilities are extended further with standard mathematical equalities:

$$\left(\frac{\partial x}{\partial y}\right)_z = \left(\frac{\partial y}{\partial x}\right)_z^{-1} \tag{11-35}$$

and

$$\left(\frac{\partial x}{\partial y}\right)_z = \frac{(\partial x/\partial w)_z}{(\partial y/\partial w)_z} \quad \text{and} \quad \left(\frac{\partial x}{\partial y}\right)_z = -\frac{(\partial z/\partial y)_x}{(\partial z/\partial x)_y} \tag{11-36}$$

A more extensive discussion of the Maxwell relations and their applications is given by Callen (ref. 6, Chapter 7).

The Bridgman relations provide an alternative method for systems of constant composition.

SUMMARY

Definitions: $F \equiv E - TS$ and $G \equiv H - TS$.
At constant temperature, $dF = w_{\text{rev}}$

$\{T, V\}$ $dF = w'_{\text{rev}}$ and $\{T, P\}$ $dG = w'_{\text{rev}}$

$\{T, V;\ w' = 0\}$ $dF \leq 0$ and $\{T, P;\ w' = 0\}$ $dG \leq 0$

$\{S, V;\ w' = 0\}$ $dE \leq 0$ and $\{S, P;\ w' = 0\}$ $dH \leq 0$

Especially important: $\left(\frac{\partial G}{\partial P}\right)_T = V$ and $\left(\frac{\partial G}{\partial T}\right)_P = -S$.

Also, $\left(\frac{\partial F}{\partial V}\right)_T = -P$ and $\left(\frac{\partial F}{\partial T}\right)_V = -S$

and $\left(\frac{\partial \Delta G}{\partial P}\right)_T = \Delta V$ and $\left(\frac{\partial \Delta G}{\partial T}\right)_P = -\Delta S$.

The free energy provides (usually) the most convenient test of equilibrium and the direction in which a process can go spontaneously.

Define ΔG_{form} as the free energy change to form any substance from its chemical elements (at constant temperature and pressure). Then

$$\Delta G_{\text{reaction}} = \sum_{\text{products}} \Delta G_{\text{form}} - \sum_{\text{reactants}} \Delta G_{\text{form}}$$

for any reaction. Standard states are defined for all substances.

The thermodynamic "potentials," E, H, F, and G are related by *Legendre transforms*; each has its own natural variables.

The *chemical potential* is $\mu = \left(\frac{\partial G}{\partial n_i}\right)_{T,P,n_j} = \bar{G}_i$

but also $\mu = \left(\frac{\partial E}{\partial n_i}\right)_{S,V,n_j} = \left(\frac{\partial H}{\partial n_i}\right)_{S,P,n_j} = \left(\frac{\partial F}{\partial n_i}\right)_{T,V,n_j}$

Other variables may appear in the equations; for example,

$$dE = T\,dS - P\,dV + \sum_i \mu_i\,dn_i - \mathcal{F}\mathcal{E}\,dn + \cdots.$$

The *Maxwell relations* provide additional helpful equations.

PROBLEMS

11.1. 3.0 mol of He is compressed at 25°C from 0.50 atm to 3.50 atm. Find
a. Q. **b.** W.
c. ΔE. **d.** ΔH.
e. ΔS. **f.** ΔG.

11.2. 0.25 mol of N_2 at 27°C expands into a vacuum from an initial volume of 5.0 L to a final volume of 15.0 L. Find
a. Q. **b.** W.
c. ΔE. **d.** ΔH.
e. ΔS. **f.** ΔG.

11.3. Calcite and aragonite are two distinct crystalline forms of calcium carbonate. Typically, calcite is formed by inorganic deposition and aragonite by biological mechanisms. Each has the formula $CaCO_3$, but calcite has a density of 2.711 g/cm^3, compared to 2.93 g/cm^3 for aragonite. For the transition of aragonite to calcite at room temperature and 1 atm, $\Delta G = -795$ J/mol. What applied pressure would make aragonite the stable phase at 25°C?

11.4. A vessel containing 5.0 mol of an ideal gas, A, is connected to an identical vessel containing 5.0 mol of another ideal gas, B, and the two are allowed to reach equilibrium. For the process described, at 27°C, find
a. ΔG (gas A). **b.** ΔG (gas B). **c.** ΔG (A + B).

11.5. A vessel containing 5.0 mol of an ideal gas, A, is connected to an identical vessel containing 10.0 mol of another ideal gas, B, and the two are allowed to reach equilibrium. For the process described, at 27°C, find
a. ΔG (gas A). **b.** ΔG (gas B). **c.** ΔG (A + B).

11.6. Solve Problem 11.5 if gas A and gas B are the same (e.g., both He).

11.7. The heat of vaporization of benzene, C_6H_6, is 33.9 kJ/mol. 100 cm^3 liquid benzene is in equilibrium with benzene vapor at 98°C and a pressure P. Under these conditions, what is ΔG for vaporizing 1.0 g of benzene?

11.8. Find ΔG for the freezing of 1.0 g of water at -20°C.

11.9. Starting with the generalized form of the first law equation, show that if $E \sim V$ and $F \sim V$ (which should be expected if intensive properties such as P and T are constant),then

$$\left(\frac{\partial E}{\partial V}\right)_T = \frac{E}{V} \quad \text{and} \quad \left(\frac{\partial F}{\partial V}\right)_T = \frac{F}{V}$$

and
a. $F = -PV$
b. $\mu = 0$

11.10. An alternative derivation of the Stefan–Boltzmann equation begins with

$$dS = \left(\frac{\partial S}{\partial T}\right)_V dT + \left(\frac{\partial S}{\partial V}\right)_T dV$$

and

$$dE = TdS - PdV = \left(\frac{\partial E}{\partial T}\right)_V dT + \left(\frac{\partial E}{\partial V}\right)_T dV$$

to obtain

$$dS = \frac{1}{T}\left[\left(\frac{\partial E}{\partial T}\right)_V dT + \left(\frac{\partial E}{\partial V}\right)_T dV\right] + \frac{P}{T} dV$$

Collecting coefficients and setting

$$\frac{\partial^2 S}{\partial V \partial T} = \frac{\partial^2 S}{\partial T \partial V}$$

followed by the substitution $P = \frac{1}{3}\frac{E}{V}$ gives an equation in E/V from which the Stefan–Boltzmann equation is obtained by integration.

Complete the derivation, using the assumptions of Problem 11.9.

CHAPTER 12

Physical Equilibria

Ideal gases give simple equations, but most substances are solid or liquid or in solid or liquid solution. The importance of thermodynamics lies in its applicability to such real substances. By linking the properties of real substances to the thermodynamic potentials (E, H, F, and G) and other thermodynamic variables (T, P, S, and V), the power of thermodynamics is brought to bear on chemical and physical equilibrium processes of all materials. This chapter is concerned with physical equilibrium, involving change of phase.

12.1 Dalton's Law and Vapor Pressure

Early in the nineteenth century, John Dalton recognized that when gases are mixed, each gas typically acts independently, as if it alone were present at a pressure called the *partial pressure,* smaller than the total pressure of the gaseous mixture. The partial pressure of a gas is equal to the pressure it would have if it were in the container, at the same temperature, by itself. That is, in the approximation of ideal gases, the partial pressure, p_i, is equal to

$$p_i = \frac{n_i RT}{V} \tag{12-1}$$

The partial pressure, p_i, of a gas may also be written as the *mole fraction* of the gas times the total gas pressure, P.

$$p_i = X_i P = \frac{n_i}{n_1 + n_2 + n_3 + \cdots} P \tag{12-2}$$

The sum of the mole fractions is equal to 1 and the sum of the partial pressures of all gases present is equal to the total pressure of the gas.

$$P = \sum_i p_i \quad (12\text{-}3)$$

Any solid or liquid loses some atoms or molecules from the surface to the vapor phase. The rate of loss may be great or may be negligibly small, depending primarily on the temperature and on how tightly the molecules are bound in the condensed phase.

For example, in liquid water the molecules are moving with an average speed of about 600 m/s, but because of collisions between the molecules, it takes hours for a particular molecule to find its way from the bottom or center of a glass of water to the surface. At the surface, unbalanced forces act on the molecules to pull them back toward the liquid. Nevertheless, about 10^{22} molecules escape per second from each square centimeter of surface of water at room temperature. Water, or even ice, disappears from a container unless water molecules are returning to the surface at a comparable rate.

The number of molecules striking a surface may be calculated from the effusion equation (Section 7.3). It is $\frac{1}{4}(N/V)\bar{v}$ per unit area and time. The density of molecules of a vapor, N/V, is determined by the partial pressure of the vapor and by the temperature. Equilibrium is established when the partial pressure of the vapor is just sufficient that the number of molecules returning to the surface is equal to the number escaping from the surface. That partial pressure of vapor is called the *vapor pressure* of the liquid or solid for the given temperature.

The vapor pressure of a liquid or solid at any chosen temperature is the pressure of pure vapor that would exist in equilibrium with the liquid or solid at that temperature.

The vapor pressure may be thought of as the tendency for molecules to escape from the condensed phase. Note, in particular, that a solid or liquid has a vapor pressure whether or not any vapor is present! If a solid or liquid is confined by a piston that exerts a pressure greater than the vapor pressure, there will be no vapor present, but the substance still has a vapor pressure.

The vapor pressure is not equal to the pressure of vapor unless the vapor is pure and equilibrium exists between the vapor and the condensed phase.

If the vapor is impure, the total pressure of vapor is greater than the vapor pressure of any of the components. There is only a slight dependence of vapor pressure on total applied pressure, as discussed in Section 12.4.

Although other gases do not (to the ideal gas approximation) appreciably affect the equilibrium pressure of any gas in contact with its condensed phase, the extraneous gases do affect the rate at which equilibrium is achieved. Water molecules escaping from a surface strike air molecules and bounce back to the surface, an effect that is balanced by water molecules that are kept from striking the surface of the liquid by air molecules that obstruct the path. If a container is evacuated of air, liquid water in the container evaporates rapidly, cooling the liquid and probably causing it to freeze.[1]

Approximately 7×10^{-20} J, or less, is required for an individual water molecule to escape from the surface of the liquid. The average kinetic energy

[1]The rate of loss will exceed 10^{22} molecules, or 3×10^{-4} kg per second because the liquid under reduced pressure will boil. Vapor then forms within the liquid as well as at the upper surface.

of molecules in the liquid, however, is only about 5×10^{-21} J at room temperature, so most molecules reaching the surface do not escape. Only the most energetic molecules can break away. These carry with them more than their fair share of the energy in the liquid, so evaporation cools the condensed phase.

A familiar example of evaporation and a partial pressure is the presence of water vapor in air at ambient temperature.

> Absolute humidity *is the partial pressure of water in the air, which is less than, or at most equal to, the vapor pressure of water at the same temperature.*

> Relative humidity *is the ratio of the absolute humidity to the vapor pressure of water.*

The common description of relative humidity as a ratio of the amount of water in the air to the amount "the air can hold" misrepresents the effect of air. Ambient air pressure is not a consideration. Air molecules have negligible effect on the partial pressure of water and on the vapor pressure of water.

Vapor pressure is a measure of the tendency of molecules to escape from the condensed phase; it can be called the *escaping tendency.* As long as vapors are ideal, the vapor pressure is a satisfactory measure of the escaping tendency, readily connectable to the thermodynamic potentials and to the pressure. In practice, however, many vapors deviate significantly from ideal behavior, as shown by the fact that liquids and solids exist. It is therefore necessary to develop a better measure of escaping tendency, applicable for nonideal vapors. That development is postponed until Section 12.3.

12.2 Equilibrium Between Two Pure Phases

If two phases, such as liquid water and water vapor, are in equilibrium at constant temperature and pressure, and a small amount of water evaporates, most thermodynamic properties of the system are changed, including E, H, F, V, S, and other derived functions, such as c_p and c_v. However, the Gibbs free energy is constant.

At equilibrium between two pure phases there is no change in Gibbs free energy for passage of a small quantity of material from one phase to the other; the molar free energy is the same in the two phases, A and B.

$$\{T, P\} \qquad d\text{G} = \text{G}_\text{B} - \text{G}_\text{A} = 0$$

A change in temperature or pressure disturbs the equilibrium. For example, as the temperature increases, more liquid evaporates (per second per unit of area), but an increase in pressure of pure vapor can restore the equilibrium, under new conditions. The molar free energies must be the same for the two phases before the change *and* the same for the two phases after the change, hence the change in free energy must be the same for each phase (Fig. 12.1).

$$d\text{G}_\text{A} = d\text{G}_\text{B}$$

For each phase (whether solid, liquid, or gas) the change in free energy is a function of the changes in temperature and pressure.

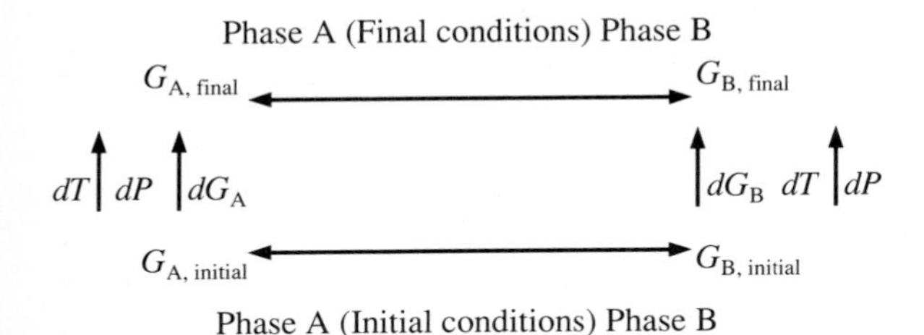

FIGURE 12.1 Initial equilibrium between phases A and B, at T_1 and P_1, is replaced by equilibrium between A and B at T_2 and P_2.

$$d\,\mathrm{G_A} = \mathrm{v_A}\,dP - \mathrm{s_A}\,dT$$

$$d\,\mathrm{G_B} = \mathrm{v_B}\,dP - \mathrm{s_B}\,dT$$

Equating the two changes in free energy, $d\mathrm{G}_A$ and $d\mathrm{G}_B$, gives

$$\mathrm{v_A}\,dP - \mathrm{s_A}\,dT = \mathrm{v_B}\,dP - \mathrm{s_B}\,dT$$

At equilibrium, the temperature and pressure are the same throughout the entire system (both phases), as was shown in Section 9.3.

Rearranging the equation gives

$$\frac{dP}{dT} = \frac{\mathrm{s_B} - \mathrm{s_A}}{\mathrm{v_B} - \mathrm{v_A}} = \frac{\Delta S}{\Delta V} \qquad (12\text{-}4)$$

Because the system is at equilibrium, at constant temperature and pressure,

$$\Delta G = \Delta H - \Delta(TS) = \Delta H - T\,\Delta S = 0$$

so

$$\Delta S = S_\mathrm{B} - S_\mathrm{A} = \frac{H_\mathrm{B} - H_A}{T} = \frac{\Delta H}{T}$$

This gives the *Clapeyron equation* in its usual form,

$$\frac{dP}{dT} = \frac{\Delta H}{T\Delta V} \qquad (12\text{-}5)$$

The ratio of change in pressure to change in temperature for two pure phases in equilibrium is equal to the ratio of the enthalpy change to volume change for the phase transition, divided by the temperature.

EXAMPLES 12.1

A. The Clapeyron equation is an expression for the *change* in pressure with change in temperature. How is that consistent with the assumption that $\Delta G = 0$ because temperature and pressure are constant?

B. The normal boiling point of toluene is 110.6°C. Assuming a Trouton constant, Δs, of 22 cal/mol · K = 92 J/mol · K and neglecting the volume of liquid in comparison to vapor, estimate the change in pressure required to raise the boiling point of toluene 1°C.

C. The density of solid benzene is 1.012 g/cm^3 and the density of the liquid at the freezing point (5.5°C) is 0.879 g/cm^3. The heat of fusion is 127 J/g.

a. If the pressure is doubled (from 1 atm to 2 atm), will the freezing point of benzene increase or decrease?
b. How much will the freezing point change? □

As suggested by the toluene example above, it is often an adequate approximation to treat the vapor as an ideal gas, even at or below its normal boiling point, and to neglect the volume of the liquid, typically on the order of 0.1% of the volume of the vapor. With these approximations,

$$\Delta \text{v} = \frac{RT}{P}$$

and

$$\frac{dP}{P} = d \ln P = \frac{\Delta \text{H}}{RT^2}\, dT \tag{12-6}$$

This equation is usually called the *Clausius–Clapeyron equation.*

For finite change in temperature and pressure, the Clausius–Clapeyron equation can be integrated, assuming ΔH to be constant, to give

$$\ln \frac{P_2}{P_1} = \ln \frac{p_2}{p_1} = \frac{\Delta \text{H}(T_2 - T_1)}{RT_1T_2} = \frac{\Delta \text{H}\, \Delta T}{RT_1T_2} \tag{12-7}$$

The pressures, P_2 and P_1, are the system pressures on pure phases in equilibrium. For impure vapor, these are replaced by the partial pressures, p_2 and p_1, at the two temperatures.

The accuracy of the integral form of the Clausius–Clapeyron equation may be improved somewhat by replacing ΔH at either T_1 or T_2 by an average value of ΔH. For example, with Kirchhoff's law, ΔH may be found at the average temperature, $(T_2 - T_1)/2$. However, there are still fundamental difficulties with this equation that limit its applicability for exact work and that limit assurance in this equation as a basis for derivation of other relationships for condensed phases. The approximations encountered in the derivation of the equation are removed by replacing the partial pressures by a more general thermodynamic measure of escaping tendency.

12.3 Definition of Fugacity as Escaping Tendency

There are two quite different ways of looking at an equilibrium across a phase boundary. In the preceding section, emphasis has been placed on the change in thermodynamic variables, especially G, T, and P. An alternative approach is to look at the dynamic equilibrium, comparing the rate at which molecules move from A to B with the rate at which they move from B to A. To be consistent with thermodynamic arguments, however, it is not an absolute rate that is required, but only the relative rate, which may be evaluated at equilibrium, where the rates become equal and the time dependence irrelevant.

The escaping tendency from any phase is influenced primarily by the temperature and by the binding energy of molecules within the phase or, for gases, by the temperature and density. To bring the full power of thermodynamics to bear on such problems, the escaping tendency must be related to the thermodynamic

---

ANSWERS 12.1

A. The Clapeyron equation gives a slope, dP/dT, at a point. ΔG, ΔH, and ΔS are measured at that point for equilibrium at constant temperature and pressure.

B. Because ΔT is small, the equation may be written $\Delta P = (\Delta H / T\,\Delta V)\,\Delta T$, or $\Delta P = \dfrac{92\ \text{J/mol}\cdot\text{K} \times 1\text{K}}{8.314 \times 383.75/(1 \times 10^5\text{Pa})\ \text{J/mol}} = 2.88 \times 10^3\ \text{Pa} = 21.9$ torr.

C. a. The freezing point rises. For solid $\rightarrow$ liquid, $\Delta H > 0$, $\Delta V > 0$, and $T > 0$, so $dP/dT > 0$.

b. $\Delta T = \dfrac{1 \times 10^5\ \text{Pa} \times 278.65\ \text{K} \times 0.1495 \times 10^{-6}\ \text{m}^3/\text{g}}{127\ \text{J/g}} = 3.32 \times 10^{-2}\ \text{K}$

---

functions without assumptions about ideal-gas behavior or other approximations. We define such an escaping tendency and label it *fugacity*, from the same Latin root that gives us "fugitive" (i.e., to escape). Fugacity is given the symbol f.

Fugacity is defined by two equations, one that tells how it changes and one that gives an initial value so that absolute values of the fugacity can be measured or calculated. The first equation, giving the change in fugacity in terms of changes in the Gibbs free energy between states 1 and 2, is

$$\Delta G = RT \ln \frac{f_2}{f_1} \tag{12-8}$$

or equivalently, for the ith substance in any state,

$$G_i = RT \ln f_i + B(T) \tag{12-9}$$

where $B(T)$ is an unknown function of temperature. Changes in Gibbs free energy, at constant temperature, are well defined, so if f_1 is known, these equations enable us to find f_2.

The second equation "pins" one end of the fugacity scale. Set fugacity equal to vapor pressure in the limit of very low pressure, where the vapor is ideal.

$$\lim_{P\to 0} \frac{f}{p} = 1 = \frac{f_0}{p_0} \tag{12-10}$$

A value of p_0 can always be found, sufficiently low that $f_0 = p_0$.

If the substance behaves as an ideal gas,

{Ideal gas} $$\Delta G = RT \ln \frac{p}{p_0} = RT \ln \frac{f}{f_0} \tag{12-11}$$

and therefore,

{Ideal gas} $$f = p \tag{12-12}$$

Fugacity is equal to partial pressure at all values of the total pressure or partial pressures for an ideal gas.

For real substances that are not ideal gases,

$\{p_0 \approx 0,\ \text{so}\ f_0 = p_0\}$ $$\Delta G = RT \ln \frac{f}{f_0} = RT \ln \frac{f}{p_0}$$

and from $(\partial G/\partial P)_T = V$, $dG = V\,dP$ at constant temperature, so

$$\{p_0 \approx 0, \text{ so } f_0 = p_0\} \qquad \Delta\text{G} = RT \ln \frac{f}{f_0} = \int_{P_o}^{P} V\,dP \qquad (12\text{-}13)$$

Therefore, f is known as a function of P if V is known as a function of P.

In some respects the fugacity may be considered to be an idealized vapor pressure, but caution is required. For example,

$$\{\text{nonideal gas}\} \qquad f\text{v} \neq RT$$

A valid relationship may be found quite easily, however. For pressures that are not too great,

$$\{p \text{ small}\} \qquad \frac{f}{p} = \frac{p}{p_I} \qquad (12\text{-}14)$$

Note that f differs from p, the actual (partial) pressure, in the opposite direction that the ideal pressure, $p_I = RT/\text{v}$, differs from p.

To evaluate fugacity of a real gas, define an *ideal volume,* $\text{v}_I \equiv V_I/n \equiv RT/P$ for a pure substance; this is the volume the gas would occupy, per mole, at the temperature T and pressure P, if it were ideal. Define a volume discrepancy,

$$\alpha \equiv \text{v}_I - \text{v} = \frac{RT}{P} - \text{v}$$

Then

$$d\text{G} = RT\,d\ln f = \text{v}\,dP \equiv \text{v}_I\,dP - (\text{v}_I - \text{v})\,dP = \text{v}_I\,dP - \alpha\,dP$$

and

$$d\ln f = \frac{1}{RT}\left(\frac{RT}{P}\,dP\right) - \frac{1}{RT}\alpha\,dP = d\ln p - \frac{1}{RT}\alpha\,dP$$

Therefore,

$$\ln\frac{f}{f_o} = \ln\frac{p}{p_o} - \frac{1}{RT}\int_{P_o}^{P}\alpha\,dP$$

From $f_o = p_o = P_o$, we find that $\ln(f/f_o) - \ln(p/p_o) = \ln(f/p)$ and

$$\ln\frac{f}{p} = -\frac{1}{RT}\int_{P_o}^{P}\alpha\,dP \qquad (12\text{-}15)$$

The volume discrepancy, α, is found experimentally as a function of pressure from a very low pressure, P_o, where the gas is ideal, to the pressure in question. Then the integral of α over pressure gives $\ln(f/p)$, from which f may be found for any P.

Most gases are found to have constant α at low pressures, as would be expected. For example, a pure van der Waals gas, at sufficiently low pressure that a/V^2 may be neglected in comparison to P, has a volume

$$\mathrm{v} = \frac{RT}{P} + b$$

and $\alpha = \mathrm{v}_l - \mathrm{v} = -b$, which is constant with pressure at these low pressures. Because the gas is pure, $p = P$, and because P_o is very low, $\Delta P = P - P_o = P$.

When α is independent of pressure,

$$\ln \frac{f}{p} = -\frac{\alpha P}{RT} = -\frac{P}{RT}\left(\frac{RT}{P} - \mathrm{v}\right) = -1 + \frac{\mathrm{v}P}{RT} = \frac{p}{p_I} - 1$$

For low pressures, f does not differ greatly from p, so

$$\ln \frac{f}{p} = \ln \frac{p + (f - p)}{p} = \ln\left(1 + \frac{f - p}{p}\right) = \frac{f - p}{p} = \frac{f}{p} - 1$$

Equating the two expressions for $\ln(f/p)$ gives the equation above:

$\{p \text{ small}\}$
$$\frac{f}{p} = \frac{p}{p_I}$$

12.4 Fugacity, Equilibrium, and the Exact Clausius–Clapeyron Equation

The change in Gibbs free energy in going from phase A to phase B (e.g., from liquid phase to vapor phase) is

$$\Delta\mathrm{G} = RT \ln \frac{f_\mathrm{B}}{f_\mathrm{A}} \tag{12-16}$$

At equilibrium, $\Delta G = 0$ and therefore $f_\mathrm{B} = f_\mathrm{A}$.

If the phases are not pure, the Gibbs free energy must be replaced by the effective value, the partial molar free energy, $\bar{\mathrm{G}}_i$. For the ith substance in phases A and B,

$$\Delta\bar{\mathrm{G}}_i = \bar{\mathrm{G}}_{i,\mathrm{B}} - \bar{\mathrm{G}}_{i,\mathrm{A}}$$

and the condition for equilibrium with respect to the component is that $\Delta\bar{\mathrm{G}}_i = 0$, or

$$f_{i,\mathrm{A}} = f_{i,\mathrm{B}} \tag{12-17}$$

For each component in equilibrium, the fugacity must be the same in every phase.

The fugacity provides a rigorous criterion for equilibrium, whether a phase is an ideal gas, nonideal gas, liquid, or solid, and whether pure or impure. To the approximation that the vapor is ideal, fugacity is equal to vapor pressure. Then the condition for equilibrium is that the vapor pressure of any component in each phase in equilibrium must be the same.

EXAMPLES 12.2

A. The vapor pressure of limestone (density 2.7 g/cm^3 for calcite) is very small at 25°C, so it should be a good approximation to set fugacity equal to vapor pressure.

a. If the pressure on limestone is increased by 1000 atm, will the vapor pressure of limestone increase or decrease?

b. By what percentage will the vapor pressure change?

B. The vapor pressure of ice (density 0.917 g/cm^3) at low pressure is 4.58 torr at 0°C. What is the vapor pressure of ice under 1 atm total pressure? □

When the Clausius–Clapeyron equation is derived without the usual assumption of ideal gas behavior for the vapor in equilibrium with the condensed phase, two changes are apparent: the total pressure, or partial pressure of the substance, is replaced by the fugacity, f, and the *heat of vaporization*, $\Delta\text{H} = \text{H}_v - \text{H}_c$, is replaced by the change in enthalpy in going from the condensed phase to the "escaped" state (i.e., to the gas at very low pressure, p^*). As shown in Section 8.3, the enthalpy difference of the vapor, $\text{H}_v^* - \text{H}_v$, in going from an initial pressure P to a final pressure $P^* = 0$, is

$$\text{H}_v^* - \text{H}_v = -\int_P^0 \mu_{\text{JT}}\text{C}_p \, dP \tag{12-18}$$

With these changes, the Clausius–Clapeyron equation changes from

$$\frac{d \ln p}{dT} = \frac{\Delta\text{H}}{RT^2}$$

to

$$\left(\frac{\partial \ln f}{\partial T}\right)_P = \frac{\text{H}_v^* - \text{H}_c}{RT^2} \tag{12-19}$$

The rate at which the log of the fugacity changes with temperature when the pressure applied to the condensed phase is maintained constant is equal to the molar enthalpy, H_v^*, at zero pressure or at a pressure sufficiently low that the vapor behaves as an ideal gas, minus the molar enthalpy, H_c, of the condensed phase, divided by RT^2.

To the approximation that $\Delta\text{H} \equiv \text{H}_v^* - \text{H}_c$ is constant over $\Delta T = T_2 - T_1$, the exact Clausius–Clapeyron equation integrates to

$$\ln \frac{f_2}{f_1} = \frac{\Delta\text{H}\,\Delta T}{RT_1T_2} \tag{12-20}$$

To derive this more exact Clausius–Clapeyron equation, let P^* be chosen as any pressure sufficiently low that the pure vapor will behave as an ideal gas at this pressure. The fugacity, f, and free energy, G_v, of the vapor in equilibrium with the condensed phase (at the vapor pressure of the condensed phase) are also the fugacity and free energy of the condensed phase; $\text{G}_c = \text{G}_v$. These differ from the free energy at P^* by

$$\text{G}_c - \text{G}_v^* = \Delta\text{G} = RT \ln \frac{f}{f^*}$$

ANSWERS 12.2

A. $RT \ln \dfrac{f_2}{f_1} = \Delta\mathrm{G} = \mathrm{v}\,\Delta P$, and R, T, v, and ΔP are positive, so $f_2 > f_1$; the vapor pressure increases.

$$\ln \frac{f_2}{f_1} = \left(\frac{1}{2.7} \times 10^{-6} \frac{m^3}{\mathrm{g}} \times 100 \frac{\mathrm{g}}{\mathrm{mol}}\right) \times \frac{10^3 \times 1 \times 10^5 \ \mathrm{Pa}}{8.314 \times 298 \ \mathrm{J/mol}} = 1.49$$

so $f_2/f_1 = 4.46$, or a 346% increase in vapor pressure (for a thousandfold increase in applied pressure).

B. $RT \ln \frac{p_2}{p_1} = \left(\dfrac{10^{-6} \ \mathrm{m}^3}{0.917 \ \mathrm{g}} \times \dfrac{18 \ \mathrm{g}}{\mathrm{mol}}\right) \times 1 \times 10^5 \ \mathrm{Pa}$; $\ln p_2/p_1 = 8.64 \times 10^{-4}$; $p_2/p_1 = 1.000864$; $p_2 = 4.58(4)$ torr. It is because the vapor pressure increases so slowly with pressure that we can usually consider it to be independent of pressure.

---

or

$$\ln f = \frac{\mathrm{G}_c - \mathrm{G}_v^*}{RT} + \ln f^*$$

Take the temperature derivative of $\ln f$, holding constant the applied pressure.

$$\{P\} \qquad \frac{d \ln f}{dT} = \frac{1}{RT}\left[\left(\frac{\partial \mathrm{G}_c}{\partial T}\right)_P - \left(\frac{\partial \mathrm{G}_V^*}{\partial T}\right)_P\right] - \frac{\mathrm{G}_c - \mathrm{G}_v^*}{RT^2} + \frac{d \ln f^*}{dT}$$

The last term is zero, because $f^* = P^*$ was chosen as a fixed pressure. Make the substitution (for G_v^* and for G_c),

$$\left(\frac{\partial \mathrm{G}}{\partial T}\right)_P = -\mathrm{S} = \frac{\mathrm{G} - \mathrm{H}}{T}$$

which gives

$$\{P\} \qquad \frac{d \ln f}{dT} = \frac{1}{RT}\left(\frac{\mathrm{G}_c - \mathrm{H}_c}{T} - \frac{\mathrm{G}_v^* - \mathrm{H}_v^*}{T}\right) - \frac{\mathrm{G}_c - \mathrm{G}_v^*}{RT^2}$$

and therefore

$$\left(\frac{\partial \ln f}{\partial T}\right)_P = \frac{\mathrm{H}_v^* - \mathrm{H}_c}{RT^2}$$

This is the exact Clausius–Clapeyron equation, applicable to substances that are not ideal gases.

The exact form of the Clausius–Clapeyron equation provides two advantages. First, it gives a precise prescription for calculating changes in fugacity (or vapor pressure, to the extent $f_2/f_1 = p_2/p_1$) with temperature. It includes the deviation of the gas from ideal behavior by replacing vapor pressure with fugacity and by including a correction for the change of enthalpy of the vapor with pressure. The usual heat of vaporization, ΔH_{vap}, is replaced by

$$\Delta H_{\text{vap}} \rightarrow H_v^* - H_c = \Delta H_{\text{vap}} + \int_P^{P^*} -\mu_{JT} C_p \, dP \qquad (12\text{-}21)$$

The second advantage of the more exact equation is that it provides a solid starting point for later derivations involving condensed phases where assumptions about ideal gas behavior would be totally inappropriate, such as in equilibria between two or more condensed phases.

12.5 Colligative Properties and Dilute Solutions

Avogadro's law and the ideal gas law apply to all gases at low pressures, because gas particles are far enough apart that the interactions between them are usually negligible. Molecules bounce off each other and the walls, but the time duration of the interactions is negligible.

By contrast, liquid molecules are never outside the range of interaction of other molecules. Properties of the liquid, such as boiling point, freezing point, and viscosity, depend almost entirely on the nature and strength of the interactions between molecules rather than on properties of the isolated molecules.

It is particularly surprising, therefore, that a certain class of properties of liquids, called *colligative* properties, depend solely on the *number* of "foreign" molecules, without regard to size, shape, mass, polarity, or other specific properties of either the solute molecules or the solvent molecules. Just as all gases follow the ideal gas law at sufficiently low pressures, all solvent-solute systems follow the colligative laws for sufficiently low concentrations.

Included in the colligative or "collective" properties are changes of freezing point and boiling point of the solvent, vapor pressure of solvent (relative to pure solvent) and of solute, distribution of a solute between two immiscible solvents, and osmotic pressure. The equations for colligative properties will be derived, starting with the empirical generalization known as the *dilute solution postulate*.

Equations that will be derived (not in this order) include:

1. Raoult's law: $p_1 = p_1^\circ X_1$

The vapor pressure of a solvent is proportional to the mole fraction of solvent.

2. Henry's law: $p_2 = k c_2$

The vapor pressure of a solute is proportional to the concentration of solute.

3. Nernst's distribution law: $\dfrac{c_{2,\text{A}}}{c_{2,\text{B}}} = K_D$

The ratio of concentrations of a solute between two immiscible solvents, A and B, is a constant, independent of concentrations.

4. Osmotic pressure: $\Delta P = \dfrac{RT}{\overline{\text{V}}_1} X_2$

The additional pressure, on a solution, required to establish equilibrium with pure solvent across a semipermeable membrane, is proportional to the mole fraction of solute in the solution.

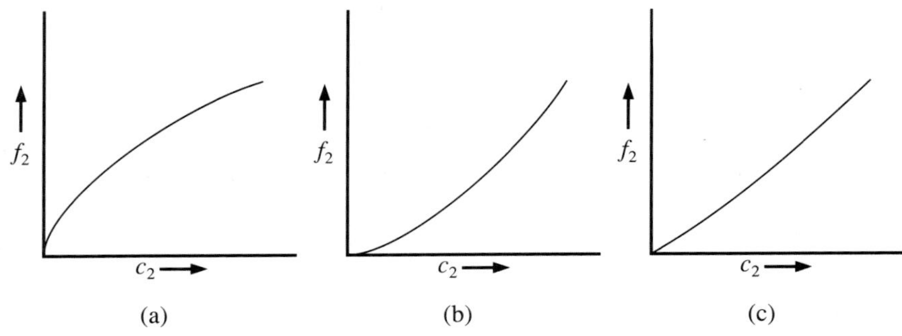

FIGURE 12.2 Possible behaviors of fugacity, f, versus concentration, c, as the concentration approaches zero. Only finite, nonzero slopes are found in real dilute solutions.

5. Change in boiling point or freezing point: $\Delta T = \dfrac{RT^2}{\Delta H} X_2$

The elevation or depression of a freezing point or boiling point is proportional to the mole fraction of solute.

Dilute-Solution Postulate

If fugacity, or vapor pressure, of a solute is plotted against concentration, the fugacity must go to zero when the concentration goes to zero. It is less certain how the fugacity approaches zero. The slope, df/dc, might be zero; the slope might be infinite; or the slope could have a finite, nonzero value. The experimental evidence is quite clear (Fig. 12.2).

For any solute, the slope of fugacity against concentration has a finite, nonzero slope at the origin.

$$\{c_2 \to 0\} \qquad \frac{df_2}{dc_2} = k \qquad (0 < k < \infty) \qquad (12\text{-}22)$$

Apparent exceptions to the dilute-solution postulate arise only when the solute changes. For example, the fugacity of acetic acid versus concentration would approach zero with a slope of zero, because the acetic acid dissociates in dilute solution into hydrogen ions and acetate ions ($AcO^- \equiv CH_3{-}COO{-}^-$).

$$AcOH \to H^+ + AcO^-$$

However, if the fugacities of the components actually present near zero concentration, H^+ and AcO^-, are plotted against concentration, the dilute-solution postulate is satisfied.

Henry's Law

It is a small step from the dilute-solution postulate to Henry's law. The derivative, evaluated near $c_2 = 0$, may be written

$$\{c_2 \approx 0\} \qquad \frac{df_2}{dc_2} = \frac{f_2 - 0}{c_2 - 0} = k$$

Therefore,

$$f_2 = kc_2 \tag{12-23}$$

which is Henry's law in its most general form.

The derivation of Henry's law assumes that $c_2 \approx 0$. Therefore, f_2 is also very small and the fugacity, f_2, can be replaced by the partial pressure, p_2, to obtain Henry's law in the more common form,

$$p_2 = kc_2 \tag{12-24}$$

Henry's law must be valid in the limit of zero concentration. In practice, it is usually a good approximation for moderately large concentrations, especially if the constant, k, is adjusted for a best fit.

EXAMPLES 12.3

A. The Henry's law constant for oxygen in water at 20°C is 225 atm, for c_2 in g/100 g. What is the concentration of oxygen in water in equilibrium with air (20% oxygen)?

B. Carbon tetrachloride has a vapor pressure of 109 torr at 25°C and is practically insoluble in water. Is the Henry's law constant of CCl_4 in water large or small?

□

Nernst's Distribution Law

If a solute is present in two immiscible phases, A and B, Henry's law for each solution is

$$f_{2,A} = k_A c_{2,A}$$

$$f_{2,B} = k_B c_{2,B}$$

At equilibrium,

$$f_{2,A} = f_{2,B}$$

and therefore

$$\frac{c_{2,A}}{c_{2,B}} = \frac{k_B}{k_A} = K_D \tag{12-25}$$

where K_D is a constant because it is a ratio of two constants. K_D is called the *distribution coefficient*.

EXAMPLE 12.4

The distribution coefficient, K_D, for the salt mercuric bromide, $HgBr_2$, between water and benzene is 0.90; the salt is slightly more soluble in benzene than in water. If 50 cm^3 of an aqueous solution that is 0.010 M in $HgBr_2$ is to be extracted with 150 cm^3 of benzene:

a. How much of the salt is left in the water if the extraction is performed in one step (all 150 cm^3 of benzene at once)?

ANSWERS 12.3

A. $p_2 = kc_2,\ c_2 = p_2/k = 0.2$ atm/(225 atm) $= 8.88 \times 10^{-4}$(g/100g).
B. $k = p_2/c_2 = 109$ torr/($c_2 \approx 0$); k is very large.

b. How much of the salt is left in the water if the extraction is performed in three steps, each with 50 cm^3 of benzene? □

Raoult's Law

The first three equations have dealt with the solute. Raoult's law gives the fugacity, or vapor pressure, of the solvent as a function of concentration. The derivation begins with the free energy of the solution in terms of the partial molar free energies, $\bar{G}_i$, of the components, with mole numbers n_i. For convenience, three components will be assumed.

The partial molar free energy of the ith component is

$$\mu_i \equiv \bar{G}_i \equiv \left(\frac{\partial G}{\partial n_i}\right)_{T,P,n_j}$$

The free energy of the solution is the sum of the free energies of the components, or

$$G = n_1\bar{G}_1 + n_2\bar{G}_2 + n_3\bar{G}_3 = \mu_1 n_1 + \mu_2 n_2 + \mu_3 n_3$$

and therefore the total differential of G is

$$dG = \mu_1\, dn_1 + n_1\, d\mu_1 + \mu_2\, dn_2 + n_2\, d\mu_2 + \mu_3\, dn_3 + n_3\, d\mu_3$$

As shown in Section 11.6, the total differential of $G(T, P, n_i)$ may also be written

$$dG = V\,dP - S\,dT + \sum_i \left(\frac{\partial G}{\partial n_i}\right)_{T,P,n_j} dn_i = V\,dP - S\,dT + \Sigma \mu_i\, dn_i$$

Subtracting the last equation from the preceding equation gives

$$0 = n_1\, d\mu_1 + n_2\, d\mu_2 + n_3\, d\mu_3 - V\,dP + S\,dT \qquad (12\text{-}26)$$

which is known as the *Gibbs–Duhem equation*. It relates changes in the chemical potentials, μ_i, to changes in pressure and temperature of the solution, at a given point on the concentration curve.

At constant temperature and pressure the Gibbs–Duhem equation gives

$$\{T, P\} \qquad n_1\, d\mu_1 + n_2\, d\mu_2 + n_3\, d\mu_3 = 0$$

for small changes in the mole numbers, n_i.

Substitution of

$$\{T\} \qquad d\mu_i \equiv d\bar{G}_i \equiv RT\, d\ln f_i$$

gives, after dividing through by RT,

$\{T, P\}$ $$n_1\, d \ln f_1 + n_2\, d \ln f_2 + n_3\, d \ln f_3 = 0$$

Dividing by $n_1 + n_2 + n_3$ gives the respective mole fractions, X_i.

$\{T, P\}$ $$X_1\, d \ln f_1 + X_2\, d \ln f_2 + X_3 d \ln f_3 = 0 \tag{12-27}$$

The assumption of a dilute solution is introduced by requiring that X_2 and X_3 be very small, and therefore that

$$\frac{df_2}{dX_2} = \frac{f_2}{X_2} \quad \text{so} \quad dX_2 = X_2 \frac{df_2}{f_2} = X_2\, d \ln f_2$$

and similarly,

$$X_3\, d \ln f_3 = dX_3$$

Therefore,

$\{T, P\}$ $$X_1\, d \ln f_1 + dX_2 + dX_3 = 0$$

Let f_1° be the fugacity of the pure solvent and assume X_1 to be constant, as it is always nearly equal to 1 for the pure solvent *and* the dilute solution. Then integrating the equation from pure solvent ($X_1 = 1$, $X_2 = X_3 = 0$) to the very dilute solution gives

$\{T, P\}$ $$X_1 \ln \frac{f_1}{f_1^\circ} + X_2 + X_3 = 0$$

The equation may be further simplified with some small-value approximations. Write

$$\ln \frac{f_1}{f_1^\circ} = \ln \frac{f_1^\circ - (f_1^\circ - f_1)}{f_1^\circ} = -\frac{f_1^\circ - f_1}{f_1^\circ} = \frac{f_1}{f_1^\circ} - 1$$

Therefore,

$$X_1 \frac{f_1}{f_1^\circ} - X_1 + X_2 + X_3 = 0$$

or

$$f_1 = f_1^\circ \left(1 - \frac{X_2}{X_1} - \frac{X_3}{X_1} \right) = f_1^\circ (1 - X_2 - X_3)$$

for $X_1 \approx 1$. The sum of the mole fractions is 1, so $(1 - X_2 - X_3)$ is equal to X_1. Therefore,

$$f_1 = f_1^\circ X_1 \tag{12-28}$$

which is the exact form of Raoult's law.

Because the solution is assumed to be dilute, $X_1 \approx 1$ and therefore $f_1 \approx f_1^\circ$, so it must be a very good approximation that

$$\frac{f_1}{f_1^\circ} = \frac{p_1}{p_1^\circ}$$

ANSWERS 12.4

Number of moles of salt $= n = 50/1000\ L \times 0.010\ M = 5 \times 10^{-4}$ mol. If $n_w =$ number of moles of salt left in the water at equilibrium, then $\dfrac{n_w/50}{(n - n_w)/V_B} = 0.9.$

a. $\dfrac{n_w/50}{(5 \times 10^{-4} - n_w)/150} = 0.9$ and $n_w = 1.15 \times 10^{-4}$ mol $= 24\%$.

b. $V_B = 50$, so $n_w/(n - n_w) = 0.9$ and $n_w = (n \times 0.9)/1.9$.

First: $n_w = \dfrac{5 \times 10^{-4} \times 0.9}{1.9} = 2.37 \times 10^{-4}$ mol

Second: $n_w = \dfrac{2.37 \times 10^{-4} \times 0.9}{1.9} = 1.12 \times 10^{-4}$ mol

Third: $n_w = \dfrac{1.12 \times 10^{-4} \times 0.9}{1.9} = 5.3 \times 10^{-5}$ mol $= 11\%$

Successive extractions with small volumes is more effective than a single extraction, in this example leaving only 11% of the original amount in contrast to 23% with a single extraction.

With this substitution,

$$p_1 = p_1^\circ X_1 \qquad (12\text{-}29)$$

which is the more conventional form of Raoult's law.

Henry's law and Raoult's law are similar but have significant differences. Henry's law, $p_2 = k c_2$, may be expressed in any concentration units for the solute. The constant, k, depends on the solvent, solute, temperature, and concentration units and cannot be predicted.

The constant appearing in Raoult's law is p_1°, the vapor pressure of pure solvent, independent of what the solute(s) may be. The concentration, however, *must* be expressed as a mole fraction.

EXAMPLES 12.5

A. If 1 g of alcohol (C_2H_5OH; molar mass = 46.07 g) is added to 100 g of water, at 100°C, what is the vapor pressure of the water?
B. If 1 g of NaCl (molar mass = 58.4 g) is added to 100 g of water, at 100°C, what is the vapor pressure of the water? □

The remarkable aspect of Raoult's law is the total lack of dependence on the nature of the solute. This may be illustrated with a specific example of toluene, $C_6H_5{-}CH_3 \equiv \phi{-}CH_3$, as solvent and styrene, $\phi{-}CH{=}CH_2$, as solute. For a given amount of toluene solvent, the decrease in vapor pressure of toluene is proportional to the number of styrene molecules in the solution. It is tempting to interpret this in terms of a certain fraction of the solution volume, or of the surface area of the solution, being occupied by molecules other than toluene, but that model is untenable.

Allow some or all of the styrene molecules to polymerize with each other, to give polystyrene,

$$\left(\begin{array}{cc} H & H \\ | & | \\ -C- & C- \\ | & | \\ \phi & H \end{array} \right)_n$$

Now there are exactly the same number of $\phi - \overset{|}{C}H - \overset{|}{C}H_2$ "foreign" units in the toluene, but they are grouped into a smaller number of independent molecules, and the effect on the vapor pressure of the toluene is substantially reduced, being proportional to the number of molecules of solute, not to the bulk of the solute.

Osmotic Pressure

Many plant and animal membranes and certain synthetic organic and inorganic membranes allow some molecules or ions to pass through but prevent others from passing, either because of relative size or electric charge. Such membranes are called *semipermeable*.

Consider a dilute solution on one side of a membrane that is permeable to the solvent but not to the solute, with pure solvent on the other side of the membrane. As solvent diffuses into the solution, the escaping tendency of solvent in the solution increases, approaching the value for pure solvent. However, no matter how dilute the solution becomes, the escaping tendency of solvent must be greater in the pure solvent than in solution.

Equilibrium may be established between pure solvent and solution across a membrane, bringing the fugacity of solvent in the solution up to the value for pure solvent, if pressure is applied on the solution only. The pressure required, to raise the fugacity from $f_1^\circ X_1$ to f_1°, is given by

$$\Delta G = RT \ln \frac{f_1^\circ}{f_1^\circ X_1} = \int \bar{v}_1 \, dP \qquad (12\text{-}30)$$

The volume, $\bar{v}_1$, is the effective volume per mole of the solvent, which is the partial molar volume of solvent. In differential form the preceding equation becomes

$$-RT d \ln X_1 = \bar{v}_1 \, dP$$

If there is a single solute, then $X_1 = 1 - X_2$ and

$$-RT d \ln X_1 = RT dX_2 = \bar{v}_1 \, dP$$

Integrating this equation from pure solvent to solution with solute concentration X_2 gives

$$\Delta P = \frac{RT}{\bar{v}_1} X_2 \qquad (12\text{-}31)$$

which is the equation for osmotic pressure.

EXAMPLES 12.6

A. What is the osmotic pressure of a water solution containing 1 g of NaCl per 100 g of water at 25°C?

B. If the dilute solution approximations are valid, what mole fraction of solute is required in tree sap to achieve an osmotic pressure sufficient to lift water 30 m up the tree? □

Osmosis will be recognized as a particular form of diffusion in which it is the solvent rather than the solute that is diffusing to where the solvent has

ANSWERS 12.5

A. $X_2 = (1/46.07)/(100/18) = 3.9 \times 10^{-3}$, $X_1 = 0.996$, so $p_1 = 0.996$ atm.
B. NaCl is dissociated into Na^+ and Cl^-, so $X_2 = [2 \times (1/58.4)/(100/18)] = 6.16 \times 10^{-3}$, and $X_1 = 0.994$, $p_2 = 0.994$ atm.

ANSWERS 12.6

A. For Na^+ and Cl^-:

$$X_2 = \frac{2/58.4}{100/18} = 6.16 \times 10^{-3}$$

and

$$\Delta P = \frac{8.314 \times 298}{18 \times 10^{-6}} \times 6.16 \times 10^{-3} = 8.489 \times 10^5 \text{ Pa} = 8.5 \text{ atm}$$

B. $P = \rho g h$; an adequate approximation is 10 m of water per atm; $P = 3$ atm $= 3 \times 10^5$ Pa.

$$X_2 = \frac{3 \times 10^5 \text{ Pa} \times 18 \times 10^{-6} \text{ m}^3}{8.314 \text{ J} \times 298 \text{ K}} = 2.18 \times 10^{-3}$$

a lower concentration. Osmosis contributes to the transport of water into plants and the transport of water inside living systems. It cannot account for many of the "active transport" processes, by which sodium, potassium, and other ions are moved across living membrane systems from lower to higher concentrations, with the expenditure of energy supplied by chemical reactions.

The equation derived for osmotic pressure gives the pressure difference required to establish equilibrium across the membrane when a concentration difference is present. If the applied pressure on the impure phase exceeds the osmotic pressure, the fugacity of solvent in the solution will exceed that in pure solvent. Solvent then moves from the solution to the pure phase. The process, called *reverse osmosis,* is being applied to desalination, providing potable water from saline water. Energy must be supplied in the form of the work required to pressurize the solution and maintain the pressure differential as solvent moves across the membrane.

Largely because of the requirements for dimensional consistency, the osmotic pressure equation superficially resembles the ideal gas law. This may be a helpful mnemonic. Representing the mole fraction by N_2, we have

$$\Delta P \overline{v}_1 = N_2 RT$$

Change of Boiling Point or Freezing Point

If two pure phases are in equilibrium, the equilibrium is upset when solute is added to one of the phases. Equilibrium may be restored by changes of concentration, by applying pressure to the impure solution, or by changing the temperature. A change of temperature affects both phases but not by exactly the same amount, so equilibrium can be reestablished.

We start with f_1°, the fugacity of pure solvent, equal to f_1, the fugacity of solvent in the second (initially also pure) phase. Then change the temperature of each phase and add solute to the second phase, requiring that the fugacity of the solvent remain the same across the phase boundary at all times. Thus

$$d \ln f_1^\circ = d \ln f_1$$

$$d \ln f_1^\circ = \left(\frac{\partial(\ln f_1^\circ)}{\partial T}\right)_{X_2} dT \tag{12-32}$$

$$d \ln f_1 = \left(\frac{\partial(\ln f_1)}{\partial T}\right)_{X_2} dT + \left(\frac{\partial(\ln f_1)}{\partial X_2}\right)_T dX_2$$

From $X_1 d \ln f_1 + dX_2 + dX_3 = 0$, upon omitting the third component, the last term becomes

$$\left(\frac{\partial \ln f_1}{\partial X_2}\right)_T dX_2 = -\frac{dX_2}{X_1} = -dX_2$$

because $X_1 \approx 1$. The first derivative on each side is replaced by means of the Clausius–Clapeyron equation.

$$\frac{d \ln f_1}{dT} = \frac{\mathrm{H}_1^* - \mathrm{H}_1}{RT^2}$$

Therefore, the condition for equilibrium is

$$\frac{\mathrm{H}_1^* - \mathrm{H}_1^\circ}{RT^2} dT = \frac{\mathrm{H}_1^* - \mathrm{H}_1}{RT^2} dT - dX_2$$

or

$$dX_2 = \frac{\mathrm{H}_1^\circ - \mathrm{H}_1}{RT^2} dT \tag{12-33}$$

The numerator is the difference in enthalpy between the impure phase and the pure phase, which is (apart from sign) the heat of vaporization, the heat of fusion, or the heat of sublimation. The sign is in the direction of impure → pure.

EXAMPLES 12.7

A. What is the sign of ΔT when salt is added to an ice-water mixture?
B. What is the sign of ΔT when salt is added to a water liquid–vapor equilibrium?
C. What is the sign of ΔT if CCl_4 is added to boiling water? □

SUMMARY

The *partial pressure* of a gas (treated as ideal) is $p_i = n_i RT/V$.
The sum of the partial pressures is the total pressure of a gas mixture.
The *vapor pressure*, of a liquid or solid, is the partial pressure of ideal gas that could exist in equilibrium with it.
For an ideal gas, $\Delta G = \int V dP = nRT \ln(P_2/P_1)$. If the vapor is ideal, $\Delta G = nRT \ln(p_2/p_1)$, where p_1 and p_2 are vapor pressures.

ANSWERS 12.7

A. Ice is the pure phase, so $H_1^\circ - H_1 = H_{ice} - H_{liq} < 0$ and therefore $\Delta T < 0$.
B. Vapor is the pure phase, so $H_1^\circ - H_1 > 0$ and $\Delta T > 0$.
C. Liquid water is the pure phase, so $H_1^\circ - H_1 < 0$ and $\Delta T < 0$.

The theory is extended by defining *fugacity* (= escaping tendency) such that $f = p$ when $p \approx 0$ (and therefore vapor is ideal). Then $\Delta G = nRT \ln(f_2/f_1) = \int V dP$, in general.

For most purposes, fugacity can be assumed equal to the vapor pressure, or the partial pressure at equilibrium.

Clapeyron equation: $\dfrac{dP}{dT} = \dfrac{\Delta H}{T\,\Delta V}$ for equilibrium between phases.

Clausius–Clapeyron equation:

$$d \ln P = \frac{\Delta H\, dT}{RT^2} \quad \text{or} \quad \ln \frac{p_2}{p_1} = \frac{\Delta H\, \Delta T}{RT_1T_2}$$

or, in more exact form,

$$\frac{\partial \ln f}{\partial T} = \frac{(H_v^* - H_c)\, dT}{RT^2}$$

where H_v^* is the molar enthalpy of the vapor at low pressure, where $f = p$.

The fugacity of any substance must be the same in every phase in equilibrium.

Gibbs–Duhem equation: $\sum n_i\, d\mu_i = V dP - S dT$.

Dilute solution hypothesis: As $c_2 \to 0$, $\partial f_2/\partial c_2 = k$ (k finite, $\neq 0$).

Dilute solution equations: Raoult's law: $f_1 = f_1^\circ X_1$

Henry's law: $f_2 = k c_2$

Nernst's distribution law: $\dfrac{c_{2,A}}{c_{2,B}} = K_D \left(= \dfrac{k_B}{k_A}\right)$

Osmotic pressure: $\Delta P = \dfrac{RT}{\overline{V}_1} X_2$

Change in boiling point or freezing point: $X_2 = \dfrac{(H^\circ - H)\,\Delta T}{RT^2}$

QUESTIONS

12.1. Boiling, with formation of bubbles, occurs when the vapor pressure is greater than the applied pressure. The vapor pressure of water at room temperature is about 25 torr; the vapor pressure of mercury is about 10^{-5} torr. Explain why water is observed to boil when the container is evacuated but there is no visible boiling of mercury no matter how good the vacuum.

12.2. Describe the process of respiration in terms of Dalton's law of partial pressures and Henry's law. Recall that oxygen dissolves in the blood in the lungs and is delivered to cells throughout the body. Carbon dioxide is formed throughout the body and is picked up by the blood and delivered to the lungs.

12.3. The malady known as diver's bends occurs when bubbles of nitrogen gas form in the bloodstream when a diver returns too rapidly to the surface from a deep dive in water. Explain the process in terms of Henry's law.

12.4. What would happen to the boiling point of motor oil if some water were added to the crankcase?

PROBLEMS

12.1. The vapor pressure of water in equilibrium with ice at (approximately) 0°C is 4.579 torr. Find the vapor pressure of ice at −5°C.

12.2. Find the vapor pressure of liquid water at −5°C.

12.3. Compare the answers to Problems 12.1 and 12.2 with the value obtained from ΔG, calculated independently.

12.4. The melting/freezing point of pure water, under its own vapor pressure (4.579 torr), is 273.16 K. The density of water at this temperature is 0.999 841 g/cm^3 and the density of ice may be taken as 0.917 g/cm^3. What is the melting point of pure water under 1 atm pressure?

12.5. The Henry's law constant for N_2 in water at 0°C is 4.09×10^7 for concentration as mole fraction and pressure in torr. For O_2 the value is 1.91×10^7.

a. What effect should saturating water with air, at 0°C and 1 atm, have on the freezing point of water? [An adequate approximation is 80% N_2 and 20% O_2 for air.]

b. What should be the difference between the freezing point of pure water under its own vapor pressure (4.579 torr at 0°C) and the freezing point of water saturated with air at atmospheric pressure?

12.6. The solubility of CCl_4 (molar mass = 154 g) in water at 27°C is about 0.9 g/L, and the vapor pressure of CCl_4 is about 100 torr. What would be the equilibrium vapor pressure of CCl_4 above a beaker containing CCl_4 covered with a layer of water?

12.7. A substance with a molar mass of 200 g is suspected of forming a dimer in water solution. 1.0 g is added to 1.0 L of water. Explain how, in principle, the extent of dimerization could be determined for the solution by means of

a. A measurement of vapor pressure of the water.

b. A measured change in freezing point of the solution.

c. A measured change in boiling point of the solution.

d. The osmotic pressure.

12.8. Calculate the magnitude of each effect in Problem 12.7. Which, if any, of the effects would be of appropriate magnitude to allow determination of the extent of dimerization?

12.9. Does the Henry's law constant for oxygen in water increase or decrease with increasing temperature? Explain.

12.10. The boiling points of the three xylene isomers (C_8H_{10}) are all approximately 140°C. Estimate the vapor pressure of xylene at 25°C.

12.11. The vapor pressure of solid nitrous oxide, N_2O, is represented over an appropriate temperature range by $\ln p_s = 21.703 - 2774/T$, and the vapor pressure of the liquid phase by $\ln p_l = 17.393 - 1988/T$ when pressures are in torr (760 torr = 1 atm = 1.01×10^5 Pa).

a. What is the triple point of nitrous oxide?

b. What is the heat of sublimation at the triple point?

c. What is the heat of vaporization at the triple point?

d. What is the heat of fusion at the triple point?

12.12. Seawater is typically about a 0.55-*M* solution of various salts (about 90% NaCl, 10% $MgCl_2$, plus sulfates and other halides). What is the minimum amount of work required to purify 1 mol (18 g) of seawater at 25°C?

CHAPTER 13

Phase Diagrams

The equations of physical equilibrium allow predictions of how vapor pressures, freezing points, and boiling points change with concentration for very dilute solutions. They do not tell us the actual boiling point, freezing point, or vapor pressure, or how the equilibrium temperatures vary at higher concentrations of solutes. Such information must be obtained from experiment. It is conveniently presented in graphic form, called a *phase diagram*.

13.1 Gibbs' Phase Rule

Elementary textbooks often talk about three (or four) "states" of matter, meaning solid, liquid, and gas (and sometimes plasma). Such "states" are more properly identified as *phases*.

> A phase *is a homogeneous sample of matter, or a collection of equivalent homogeneous samples.*

A gas, or any mixture of gases, is a single phase. A pure liquid or a liquid solution is a phase. A single crystal or a collection of equivalent crystals, such as granulated sugar or pure table salt, is a phase.

Under nonequilibrium conditions, or when other variables such as pressure differ within a phase (e.g., because of gravity), a phase may have continuously varying properties within its boundaries. Only homogeneous phases will be considered here.

At every phase boundary, the equilibrium conditions must be satisfied. For example, if liquid water is in equilibrium with its vapor, the water vapor has a pressure (or partial pressure) equal to the vapor pressure of the liquid. If alcohol is added to the water, the partial pressure of alcohol vapor is equal to the vapor pressure of the alcohol in the solution.

> *Let p represent the number of phases and c the number of (independent) components in the system.*

If there are p phases, there are $(p-1)+(p-2)+(p-3)+\cdots+(p-p)$ possible phase boundaries. These give $p-1$ *independent* conditions for equilibrium. Other boundaries only duplicate the conditions established by the first $p-1$ boundaries.

For example, four phases, a, b, c, and d, can have boundaries a/b, a/c, a/d, b/c, b/d, and c/d; $3+2+1=6$ boundaries. However, equilibrium between a and b and between b and c ensures equilibrium between a and c. Then equilibrium between c and d ensures equilibrium between a and d and between b and d. Hence a/b, b/c, and c/d equilibria suffice for these and for a/c, a/d, and b/d.

When more than one substance is present, the concentration of each substance can vary in each of the p phases. For c components, only $c-1$ concentrations must be specified. The remaining concentration is found by difference.

The total number of variables, then, is $c-1$ concentrations in each of p phases, plus "external" variables, usually chosen as temperature and pressure (although in special cases there may be electric, magnetic, or other fields in addition).

There are $p(c-1)+2$ *variables for* c *components in* p *phases.*

The conditions on the equilibrium are the $p-1$ phase boundary conditions, for each of the c components.

There are $c(p-1)$ *conditions for* c *components in* p *phases.*

Given v independent variables and e equations, a unique solution exists if, and only if, $v=e$. If $v>e$, an infinite number of solutions is possible. If $v<e$, there is, in general, no simultaneous solution for all the equations.

The number of degrees of freedom is $\mathscr{f}=v-e$,

equal to the number of additional conditions that can be imposed and still have a unique solution.

Gibbs pointed out that if c components are distributed in p phases in equilibrium, the number of degrees of freedom is

$$\mathscr{f}=v-e=p(c-1)+2-c(p-1)=c-p+2$$

The number of degrees of freedom *for* c *components in* p *phases is*

$$\mathscr{f}=c-p+2 \tag{13-1}$$

if there are two external variables, such as T *and* P.

This equation, called *Gibbs' phase rule,* provides a logical basis for the construction of phase diagrams. It must be modified if the number of external variables is less than two (e.g., if temperature and/or pressure has been fixed) or greater than two because of external fields.

EXAMPLE 13.1

Magnesium sulfate can exist as the anhydrous salt, $MgSO_4$, or as any of the following solid hydrates: $MgSO_4 \cdot H_2O$, $MgSO_4 \cdot 4H_2O$, $MgSO_4 \cdot 5H_2O$, $MgSO_4 \cdot$

$6H_2O$, and $MgSO_4 \cdot 7H_2O$. After careful preparation of samples of the anhydrous salt and of each of the hydrates, Stu Dent places a sample of each in an open vial in a desiccator, to store them for later measurements. To avoid dehydration, he adds a little water in a dish in the bottom of the desiccator. What will he find when he returns to the desiccator after some time? □

The question of how many components are in a system is often a trivial question, but sometimes may be difficult. For example, if ammonia, NH_3, and hydrogen chloride, HCl, gases are brought together, they react to form ammonium chloride, NH_4Cl, but there are still only two *independent* component variables, whether these are chosen as the equivalent amounts of NH_3 and HCl, or as NH_3 and NH_4Cl, or as HCl and NH_4Cl. By contrast, if pure NH_4Cl is introduced into a container, it produces NH_3 and HCl, but the number of components is one, because the amounts of NH_3 and HCl are not independently variable.

13.2 One-Component Phase Diagrams

The simplest type of phase equilibrium is a single, pure substance: $c = 1$. Then

$$\mathscr{f} = c - p + 2 = 3 - p \tag{13-2}$$

The maximum number of phases that can coexist is three. If the number of phases is less than three, there are one or two degrees of freedom, as shown in Table 13.1. A point represents zero degrees of freedom; a line (or lines) represents one degree of freedom, and an area represents two degrees of freedom.

A rough picture of the phase diagram for water can be constructed with no further information. Let temperature and pressure be the two external variables. At high temperature and low pressure, water is a gas, which is a single phase, represented by an area. Similarly, at high pressure and low temperature, water is a solid, also a single phase and therefore an area. Somewhere between must be a third area representing liquid (Fig. 13.1).[1]

The boundaries between the areas representing single phases are lines. A line represents one degree of freedom or, from Table 13.1, two phases. Along such lines, the two phases adjacent to the line are in equilibrium. The slopes of the lines can be calculated from the Clapeyron equation or the Clausius-Clapeyron equation. From the Clapeyron equation, inverted,

$$\frac{dT}{dP} = \frac{T\Delta V}{\Delta H} \tag{13-3}$$

Table 13.1 Degrees of freedom and representations when $\mathscr{f} = 3 - p$

Number of phases	1	2	3
$\mathscr{f}$ **(degrees of freedom)**	2	1	0
Representation	Area	Line(s)	Point(s)

[1]Such diagrams are often drawn with axes reversed: P vertical and T horizontal. The choice here is consistent with more of the two-component diagrams that follow. Additional solid phases, at high pressures, are omitted.

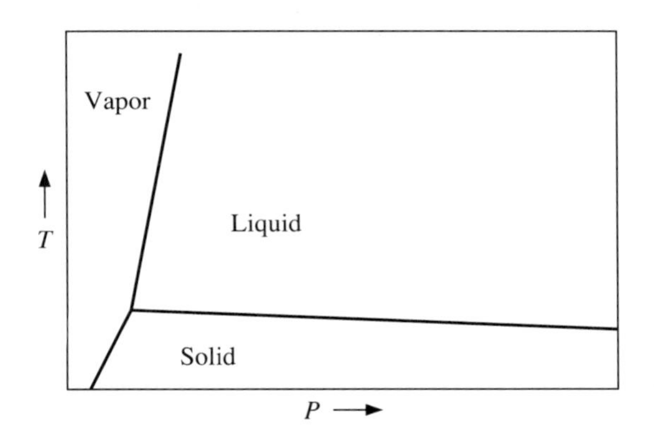

FIGURE 13.1 Phase diagram for water (not to scale).

it is clear that the slope is positive if ΔV and ΔH have the same sign; otherwise, the slope is negative.

For the transition from liquid to vapor, ΔV is large and positive and ΔH is positive, so the slope is positive (and large). The equilibrium line is (to a very good approximation) a plot of vapor pressure of the liquid as a function of temperature (independent of the total, or applied, pressure). Similarly, the line separating solid and vapor is a plot of vapor pressure of the solid, which is a function of temperature.

Ice is unusual in that the volume of the solid is greater than the volume of the liquid. For the transition from solid to liquid, ΔV is negative, ΔH is positive, and the slope is negative (but very small).

Where solid–liquid and liquid–solid lines meet the solid–vapor line, the three adjacent areas are in equilibrium. The point is called a *triple point*. Such points provide convenient primary or secondary standards for temperature scales, because they are fixed. The number of degrees of freedom is zero.

The liquid–vapor equilibrium line is of finite length. By raising the temperature, then changing pressure and lowering the temperature again, it is possible to make an "end run," moving from gas to liquid or back without undergoing a phase transition. The terminal point of the liquid-vapor line is the critical point. A fluid above its critical point is not distinguishable as a liquid, no matter how great the pressure.

13.3 Two-Component Phase Diagrams at Fixed Pressure: Solution–Vapor

The phase rule for two components is

$$\mathscr{f} = c - p + 2 = 4 - p \tag{13-4}$$

The number of degrees of freedom and their representations are shown in Table 13.2.

Table 13.2 Degrees of freedom and representations when $\mathscr{f} = 4 - p$

Number of phases	1	2	3	4
$\mathscr{f}$ **(degrees of freedom)**	3	2	1	0
Representation	Volume	Area(s)	Line(s)	Point(s)

ANSWER 13.1

The number of components is two, $MgSO_4$ and H_2O (neglecting air). Temperature is fixed by the laboratory environment. The number of degrees of freedom, $f = c - p + 1 = 3 - p$, can be positive or zero but cannot be negative, so the maximum number of phases that can coexist is $p = 3$. One phase is water vapor and one phase is liquid water, so only one of the six solids can survive the storage conditions. The others add or lose water to convert to the form that is stable under the set conditions. If no excess water had been added, two solids could coexist. If air is counted as a component, atmospheric pressure provides another condition, and the result is the same.

Although it is possible to make three-dimensional models of phase equilibria, and sometimes it is desirable to do so, phase diagrams are usually limited to two dimensions. This requires that the maximum number of degrees of freedom be reduced by one. If either temperature or pressure is fixed, two-component phase equilibria are again described by Table 13.1.

Benzene and toluene form solutions that are sufficiently close to ideal that Raoult's law applies at any concentration. At high temperatures both compounds are gases and a mixture is a single phase, represented by an area (Table 13.1). At a somewhat lower temperature there must be another area that corresponds to the single phase of liquid solutions.

EXAMPLES 13.2

The normal boiling point of benzene is 80.1°C and of toluene 110.6°C. Assuming a Trouton constant of 90 J/mol · K, find

A. The heats of vaporization of benzene and toluene.

B. The vapor pressure of pure benzene and pure toluene at 100°C.

C. The composition of a benzene–toluene solution that has a combined vapor pressure of 1 atm at 100°C. □

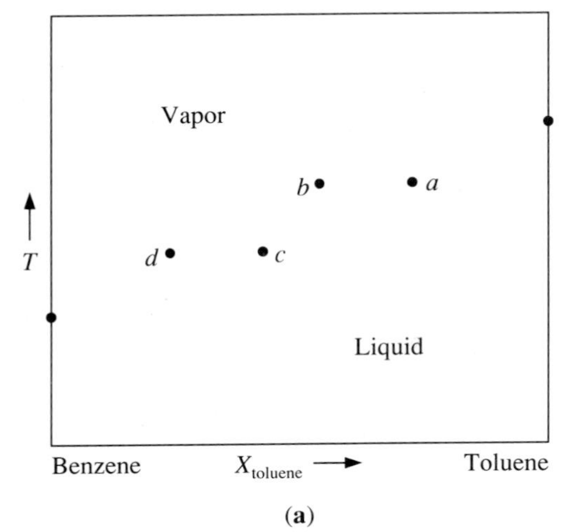

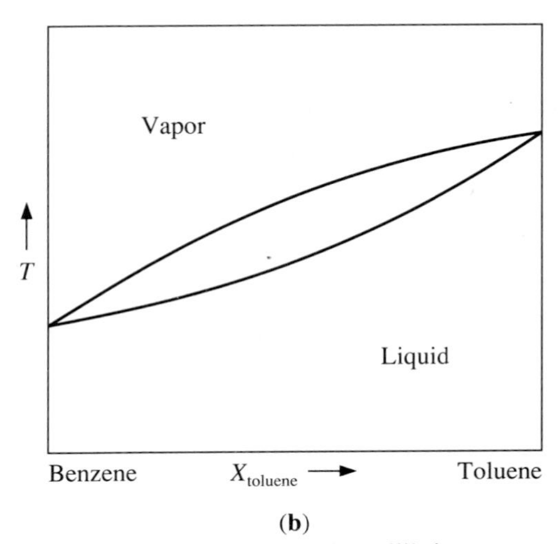

FIGURE 13.2 Benzene–toluene equilibrium (constant pressure): (a) Calculated equilibrium states for liquid and vapor; (b) phase diagram. The area between the curved lines is not part of the equilibrium phase diagram. There can be no phase of such temperature and composition.

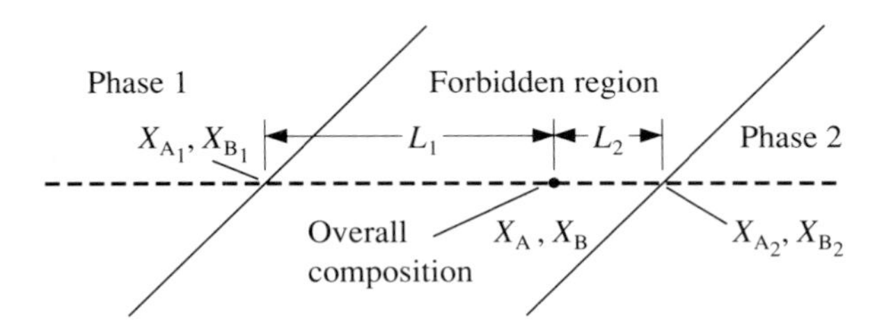

FIGURE 13.3 Lever-arm rule. The amount of each phase times the distance to the overall composition is the same for each phase in equilibrium [e.g., wt (phase 1) $\times L_1$ = wt (phase 2) $\times L_2$].

Two calculated equilibria for benzene and toluene are shown in Fig. 13.2a. One pair of points, at 100°C, is a liquid solution that is 25.3% benzene in equilibrium with vapor that is 45.0% benzene. The other pair of points is at 90°C, with liquid of 57.0% benzene and vapor of 76.6% benzene. The liquid values must lie along a liquid–vapor equilibrium line at the edge of the liquid phase. The vapor values also lie along a liquid–vapor equilibrium line, at the edge of the vapor-phase area. When these points are connected with the respective boiling points for pure benzene and toluene, the diagram appears as shown in Fig. 13.2b, where a portion of the figure is not accessible. It is appropriately called a *forbidden region*.

The vapor in equilibrium with a liquid, at any temperature, is richer than the liquid in the more volatile component. Because of this difference, there is a region between the two liquid–vapor equilibrium lines that corresponds neither to liquid nor to vapor. As may be seen from Table 13.1, there cannot be an area representing simultaneous existence of two (or more) phases. The area lying between the two liquid–vapor equilibrium lines is not really a part of the equilibrium phase diagram. There can be no phase with composition and temperature that falls within this forbidden region.[2]

A mixture that is 50% toluene and 50% benzene, brought to a temperature of 95°C, would fall in the forbidden region, so it is unstable. It separates into a liquid phase and a vapor phase. These two phases, separated by the forbidden region, then exist in equilibrium.

Phases in equilibrium lie on an isotherm, a horizontal line in the diagram, as shown in Fig. 13.3. The amount of material in each of the two phases in equilibrium can be calculated from the *lever-arm rule*.

If the overall composition (i.e., the composition of the total mixture) has a percentage concentration C of component A that falls within a forbidden region, the material divides into two phases: an amount m_1 of phase 1 of concentration C_1 (of A) and an amount m_2 of phase 2 of concentration C_2 (of A), such that

$$m_1C_1 + m_2C_2 = mC \equiv C(m_1 + m_2)$$

and therefore

$$(C - C_1)m_1 = (C_2 - C)m_2 \tag{13-5}$$

The amount of one phase, multiplied by the distance of the composition of that phase from the overall composition, is equal to the amount of the other phase, in equilibrium with it, multiplied by the distance of the composition of the second phase from the overall composition.

[2]The forbidden region between the lines is often called a "two-phase" region, but there is *no* phase present with composition and temperature falling in this region. The only material present falls within the liquid or vapor areas.

ANSWERS 13.2

A. The estimated heats of vaporization are $90 \times 353.25 = 31.8$ kJ/mol · K for benzene and $90 \times 383.75 = 34.5$ kJ/mol for toluene. (Actual values are 30.76 kJ/mol and 33.47 kJ/mol.

B. At 100°C, the vapor pressure, in atm, is given approximately by

$$\ln \frac{p^\circ}{1 \text{ atm}} = \frac{31{,}800 \text{ J/mol} \times 19.9 \text{ K}}{8.314 \text{ J/mol} \cdot \text{K} \times 353 \text{ K} \times 373 \text{ K}} = 0.578$$

so $p^\circ = 1.78$ atm. (Actual value $= 1.78$ atm.) For toluene,

$$\ln \frac{p^\circ}{1 \text{ atm}} = \frac{-34{,}500 \text{ J/mol} \times 10.6 \text{ K}}{8.314 \text{ J/mol} \cdot \text{K} \times 373 \text{ K} \times 384 \text{ K}} = -0.307$$

so $p^\circ = 0.736$ atm. (Actual value $= 0.732$ atm.)

C. $p_B = 1.78$ atm X_B; $p_T = 0.736$ atm X_T; and $p_B + p_T = 1$ atm. Also, $X_B + X_T = 1$, so $1.78X_B + 0.736(1 - X_B) = 1$. Solving for X_B gives 0.253 and for X_T, 0.747. A solution that is 74.7% toluene and 25.3% benzene will have a total vapor pressure of 1 atm. The vapor will be $0.747 \times 0.736 = 55.0\%$ toluene and $0.251 \times 1.80 = 45.0\%$ benzene.

Amounts are measured in whatever units have been chosen for the horizontal axis of the phase diagram.

EXAMPLE 13.3

If a solution of overall composition 60% toluene and 40% benzene ($X_T = 0.60, X_B = 0.40$) is warmed to $100°C$, what will be the composition of the phase(s) actually present? □

13.4 Two-Component Phase Diagrams at Constant Pressure: Immiscible

The form of the liquid–solid diagram for benzene and toluene can be predicted qualitatively from some simple calculations involving only properties of the individual compounds. The assumption underlying the calculations is that benzene and toluene are immiscible in the solid phase, which is expected because the molecules have different shapes.

EXAMPLES 13.4

The freezing point of benzene is 5.5°C and the freezing point of toluene is −95°C. The heats of fusion are 9837 and 6619 J/mol, respectively. Calculate the rate of depression of freezing point with mole fraction of impurity, dT/dX_2:

A. For benzene, with toluene as the impurity.

B. For toluene, with benzene as the impurity. □

The solid phase is the pure phase, so addition of toluene depresses the freezing point of benzene. Similarly, benzene depresses the freezing point of toluene. This information is depicted in Fig. 13.4a.

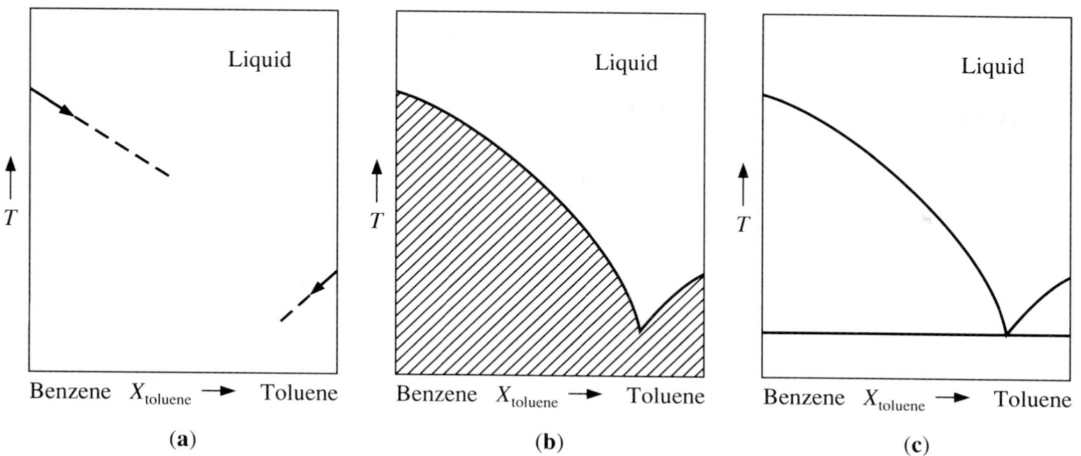

FIGURE 13.4 Solid–liquid phase diagram for immiscible solids. (a) Each substance depresses the freezing point of the other. (b) The freezing-point curves meet at a point called the eutectic. (c) The completed conventional diagram. Mole fraction toluene versus temperature.

For larger mole fractions of impurity, the decrease in freezing point is no longer adequately approximated by a linear function. The line along the edge of the liquid area gives the solubility of the impurity in the major component, which varies logarithmically with temperature.

EXAMPLE 13.5

If solutions are not dilute, the dilute-solution postulate cannot be applied, but Henry's law may still be a reasonable approximation. Equating the total differentials of the free energies in the two phases (one of which is pure) yields the equation

$$\left(\frac{\partial \ln f^\circ}{\partial T}\right)_c dT = \left(\frac{\partial \ln f}{\partial T}\right)_c dT + \left(\frac{\partial \ln f}{\partial c}\right)_T dc$$

where c may now be arbitrary concentration units. Show that substitution of the Clausius–Clapeyron equation and Henry's law permits integration to find the solubility as a function of temperature. □

Extrapolation of the solid–liquid equilibrium curves, called solubility curves, gives the form shown in Fig. 13.4b, with the forbidden region shown hatched. The low point, at which the solubility curves intersect, is called a *eutectic point*. It is customary to draw a horizontal line through the eutectic point. Although such a line has no significance to the phase diagram, it provides a convenient marker, especially in some diagrams where details may be difficult to discern.

Pure toluene and pure benzene solids are single phases and are therefore represented on the phase diagram by areas. However, the solubility of the other compound in the pure crystals is very small, so the width of the areas is generally negligible. It would seem that the areas should at least be represented by double vertical lines, or by a thicker line, but common practice is to draw these areas as single lines, relying on the context to show that they are areas of very small width. The complete solid–liquid phase diagram, in conventional form, is shown in Fig. 13.4c.

ANSWER 13.3

From the diagram, drawing a horizontal line at 100°C, or from the calculation of Example 13.2, the liquid phase must have $X_B = 0.253$ and $X_T = 0.747$ and the vapor phase $X_B = 0.450$ and $X_T = 0.550$. If n_l is the number of moles of liquid and n_v the number of moles of vapor, $0.253n_l + 0.450n_v = 0.40(n_l + n_v) =$ original amount of benzene and $0.747n_l + 0.550n_v = 0.60(n_l + n_v) =$ original amount of toluene. These simultaneous equations give $n_l/n_v = 0.34$. The solution partially evaporates; on a molar basis, 74.6% appears as vapor ($X_B = 0.450$, $X_T = 0.550$) and 25.4% remains as liquid, but of composition $X_B = 0.253$ and $X_T = 0.747$. There is no phase at 100°C with composition $X_B = 0.40$, $X_T = 0.60$.

ANSWERS 13.4

A. $\dfrac{dT}{dX_T} = \dfrac{RT^2}{-\Delta H_f(B)} = \dfrac{8.3(278.7)^2}{-9837} = -65.6$ K is the rate of change of melting point with mole fraction, for small mole fractions of toluene in benzene.
B. For toluene,

$$\frac{dT}{dX_B} = \frac{RT^2}{-\Delta H_f(T)} = \frac{8.3(179.1)^2}{-6619} = -40.3 \text{ K}$$

is the rate of decrease, with mole fraction of benzene.

ANSWER 13.5

The Clausius–Clapeyron equation gives

$$\left(\frac{\partial \ln f^\circ}{\partial T}\right)_c dT - \left(\frac{\partial \ln f}{\partial T}\right)_c dT = \frac{H^* - H^\circ}{RT^2} - \frac{H^* - H}{RT^2}$$

which simplifies to

$$\left(\frac{\partial \ln(f/f^\circ)}{\partial T}\right)_C dT = \frac{H^\circ - H}{RT^2}$$

Substituting this form into the original equation, and letting $f = kc$, gives the equation

$$\left(\frac{\partial \ln kc}{\partial c}\right)_T dc = \frac{H^\circ - H}{RT^2}$$

which integrates (assuming that k is constant with concentration) to the solubility equation:

$$\ln \frac{c_2}{c_1} = \frac{\Delta H \Delta T}{RT_1T_2}$$

where ΔH is the *heat of solution* and $\Delta T = T_2 - T_1$.

At the eutectic temperature, pure toluene, pure benzene, and the liquid eutectic mixture may exist in equilibrium. With three phases in equilibrium, there are no degrees of freedom remaining. The eutectic point is fixed (and usually little affected by changes in pressure).

Immiscible liquids have liquid–vapor phase diagrams like the solid–liquid equilibrium of immiscible solids just considered. For example, CCl_4 and water are immiscible as liquids. The vapors mix freely, so the liquid–vapor phase diagram is as shown in Fig. 13.5.

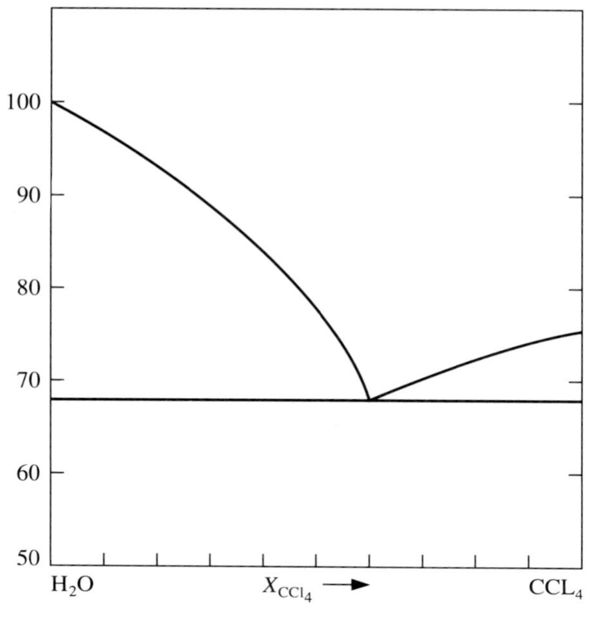

FIGURE 13.5 Liquid–vapor equilibrium for immiscible liquids.

EXAMPLES 13.6

Carbon tetrachloride boils at 76.8°C with a heat of vaporization of 29.9 kJ/mol. Find

A. The rate of depression of the boiling point of CCl_4 by addition of water.
B. The rate of depression of the boiling point of water by the addition of CCl_4.

□

Substances miscible both as solids and as liquids have solid–liquid phase diagrams like the liquid–vapor diagrams of miscible liquids. For example, silver and gold have sufficiently similar properties that they dissolve in each other in any proportion. The solid–liquid phase diagram is shown in Fig. 13.6.

When the lower-temperature phases are partially miscible, the very narrow area representing pure material becomes visibly broadened. Silver and copper show such a solid–liquid equilibrium (Fig. 13.7).

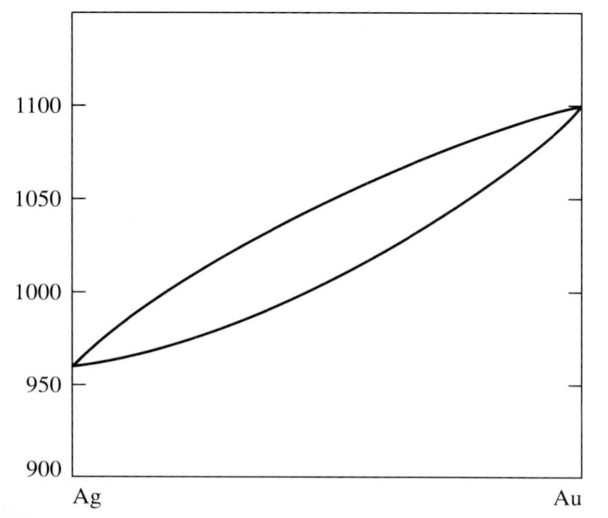

FIGURE 13.6 Solid–liquid equilibrium for miscible solids.

ANSWERS 13.6

For small concentrations of impurities:

A. $$\frac{dT}{dX_w} = \frac{8.3(350)^2}{-29,900} = -34.1 \text{ K.}$$

B. $$\frac{dT}{dX_{CCl_4}} = \frac{8.3(373)^2}{-40,644} = -28.5 \text{ K.}$$

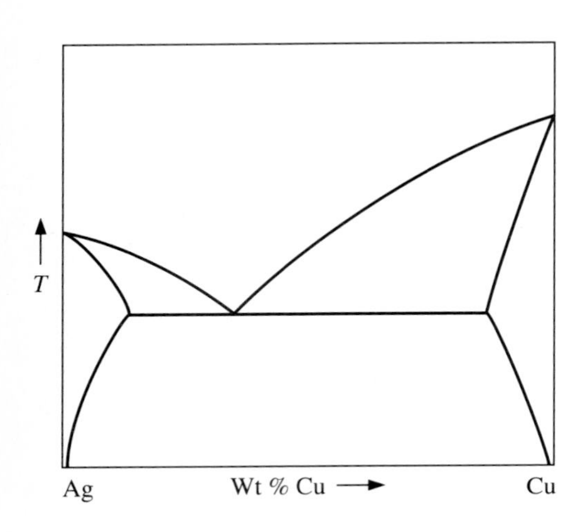

FIGURE 13.7 Solid–liquid equilibrium for partially miscible solids.

13.5 Two-Component Phase Diagrams at Fixed Temperature

Liquid–vapor equilibria viewed at fixed temperature require a reorientation, because the vertical axis represents increasing pressure. The vapor phase is at the bottom of the diagram and liquid is at the top. If the solutions are ideal, obeying Raoult's law, the vapor pressure of each component varies linearly with concentration, as shown in Fig. 13.8. The total pressure is the sum of the partial pressures and therefore is also linear with concentration.

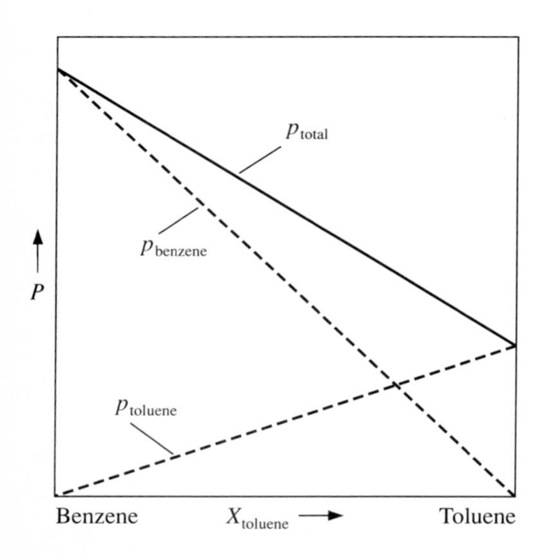

FIGURE 13.8 Partial pressures (dashed lines) and total pressures of benzene–toluene solutions in equilibrium with vapor at 100°C.

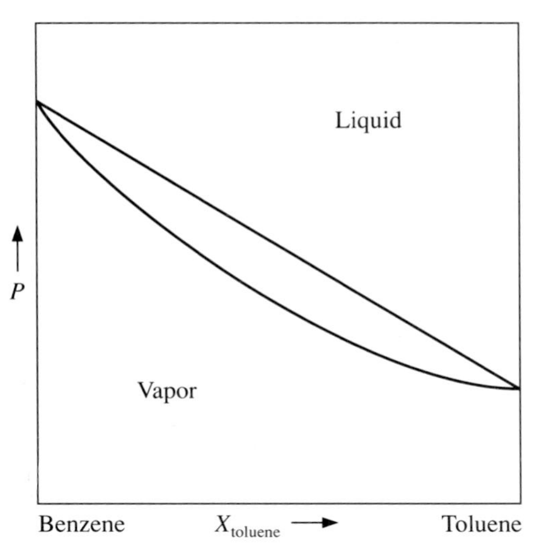

FIGURE 13.9 Liquid–vapor phase diagram at constant temperature for benzene–toluene (100°C).

The phase diagram for liquids forming an ideal solution appears as shown in Fig. 13.9. Because the upper liquid–vapor equilibrium line, at the edge of the liquid phase, is the sum of the partial pressures of the liquid components, it is a straight line.[3] The composition of the vapor is the ratio of the partial pressure of one component to the sum of the partial pressures, or

$$X_{1,v} = \frac{p_1}{p_1 + p_2} = \frac{p_1^\circ X_1}{p_1^\circ X_1 + p_2^\circ X_2} \tag{13-6}$$

which gives the curved line forming the edge of the vapor area. In this expression, the symbols 1 and 2 are interchangeable; in an ideal solution, either component can be solvent or solute.

If two liquids are partially miscible, and if each solution adheres to Raoult's law for the major constituent and Henry's law for the minor constituent, the form of the phase diagram is predictable from the solubilities, as shown in Fig. 13.10.

Consider component A as solvent, obeying Raoult's law. The fugacity of A follows the solid straight line from $X_A = 1, f_A = f_A^\circ$ toward $X_A = 0$, $f_A = 0$

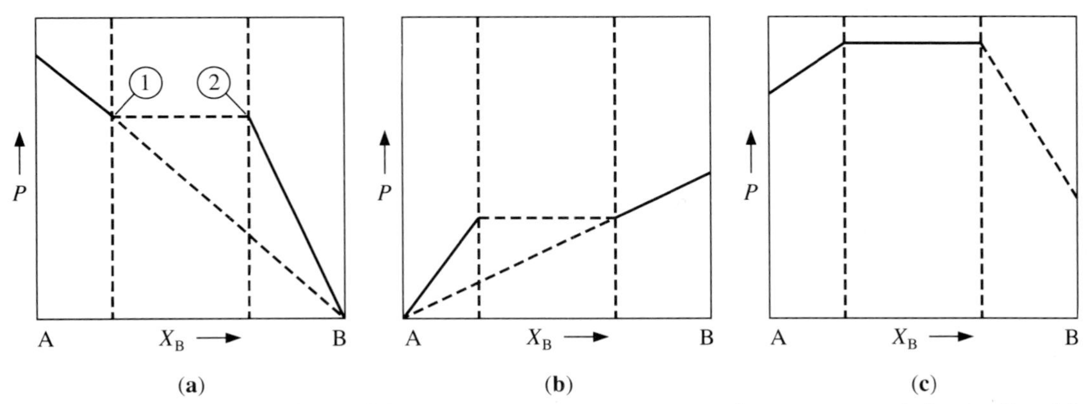

FIGURE 13.10 Partially miscible system at constant temperature for components following Raoult's law and Henry's law.

[3]The partial pressures in the vapor phase are $p_1 = p_1^\circ X_1$ and $p_2 = p_2^\circ X_2$, giving a total pressure of $p_1^\circ X_1 + p_2^\circ X_2 = p_1^\circ X_1 + p_2^\circ(1 - X_1) = p_2^\circ + (p_1^\circ - p_2^\circ)X_1$, showing that the pressure is linear with composition of the liquid. The vapor has the same pressure as the liquid but a different composition, so pressure is not linear with composition of the vapor.

(Fig. 13.10a). Point 1 is in equilibrium with point 2 and therefore is at the same pressure as point 2, where A is the solute (in B). From point 2, A as solute follows the solid straight line, obeying Henry's law, to $X_A = 0$, $f_A = 0$.

Component B follows similar behavior (Fig. 13.10b). The fugacity falls, along the solid line, to the saturated solution of A in B. The saturated solution of B in A is in isobaric equilibrium (horizontal line). Below this point the fugacity of B, as solute, falls linearly to zero. The total pressure (Fig. 13.10c) is the sum of the partial pressures, or fugacities, from Fig. 13.10a and b. In this special case, each Henry's law constant is proportional to the ratio of the solubilities; $k_{A/B} = f_A^\circ X_{B/A}/X_{A/B}$.

13.6 Variations in Two-Component Phase Diagrams

Although not all phase diagrams of two-component systems are as simple as the examples considered above, nearly all can be regarded as relatively simple combinations of the features that have been discussed. For example, two elements, or compounds, may react to form a compound of intermediate composition.

The most basic definition of a chemical compound is that it is a material of definite composition.

A vertical line in a phase diagram represents a constant composition and is interpreted as a chemical compound.

In Fig. 13.11, vertical lines appear at each edge, representing pure A and pure B. The vertical line between represents a compound of composition AB_2. The left and right portions of the phase diagram are simple eutectic phase equilibria, like those already considered (Figs. 13.4, 13.5, and 13.7).

A variation on the simple eutectic diagram is shown in Fig. 13.12. The vertical line near the center clearly represents a compound, but as the temperature is raised, the compound reaches a temperature at which it establishes equilibrium with pure A and with the melt. This is called an *incongruent* melting point of the compound; it "melts" to give a liquid and another solid.

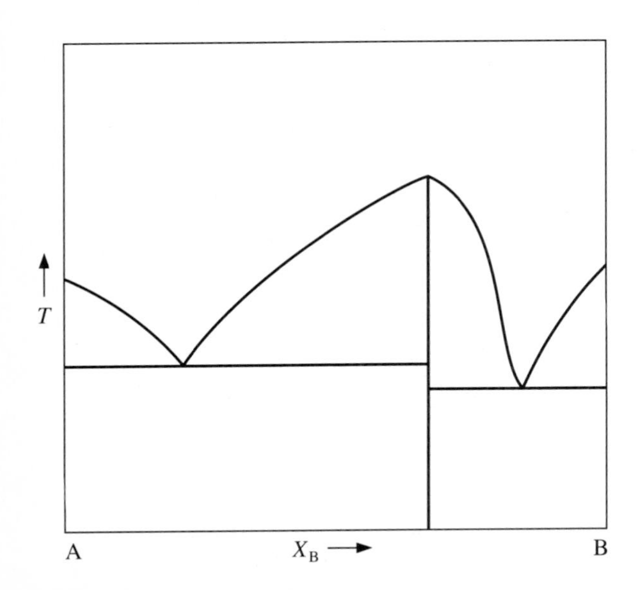

FIGURE 13.11 Compound formation with immiscible solids.

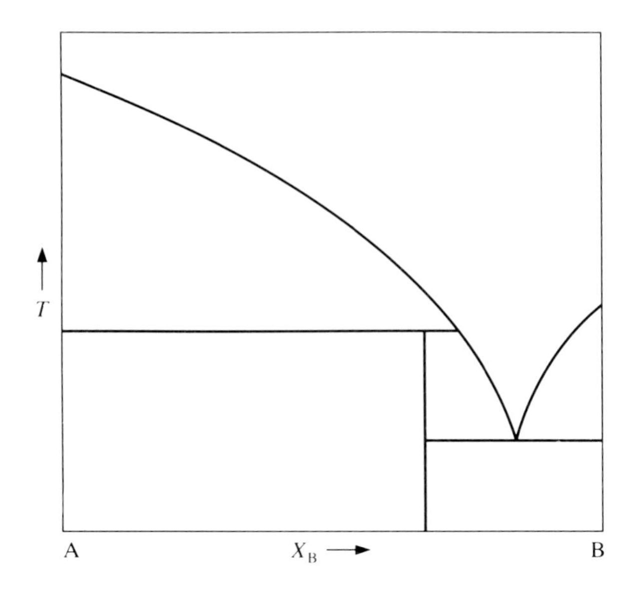

FIGURE 13.12 Incongruent melting point.

A somewhat different variation may appear when both high- and low-temperature phases are miscible. As shown in Fig. 13.13, there may be a boiling point (or melting point) maximum or minimum at some intermediate concentration. Measurements on liquid–vapor systems have shown that the composition at which the maximum or minimum occurs varies as the pressure changes, so these points do not represent definite compounds. The maximum- or minimum-boiling mixtures are called *azeotropes,* sometimes distinguished, respectively, as *positive* and *negative* azeotropes.

Although the azeotropic composition is not a compound, at any constant pressure the system behaves as if it were. The left and right portions are equivalent to the simple equilibrium of Fig. 13.6, for liquid–vapor equilibrium. The azeotropic composition boils at a constant temperature as if it were a pure compound.

Probably the best known example of an azeotrope is the 95% ethanol, 5% water mixture, which boils at 78.1°C, slightly below the boiling point of pure ethanol (78.4°C). To obtain absolute alcohol, a third component, traditionally benzene, is added. A benzene–alcohol mixture can be separated from the water by distillation, and the alcohol is then separated from the benzene by a second distillation.

Even the most complicated phase diagrams can usually be easily interpreted by means of two elementary principles. First, as pointed out above, horizontal lines are added as pointers but have no real significance to the phase equilibrium. They cross forbidden regions, so if they are removed it will be clear that areas above and below are both forbidden regions.

Second, forbidden areas separate areas that represent single phases. Therefore, a progression from left to right across a phase diagram at any temperature must alternately encounter single-phase areas and forbidden regions. With these two rules, and a moderate amount of experience to judge when single lines represent areas of narrow width, nearly any phase diagram can be read quickly and easily.

13.7 Thermal Analysis: Heating and Cooling Curves

Early information on phase equilibria came almost entirely from visual inspection of materials, with the naked eye or with an optical microscope, and from measurements of rates of heating and cooling. Newer techniques, such as x-ray

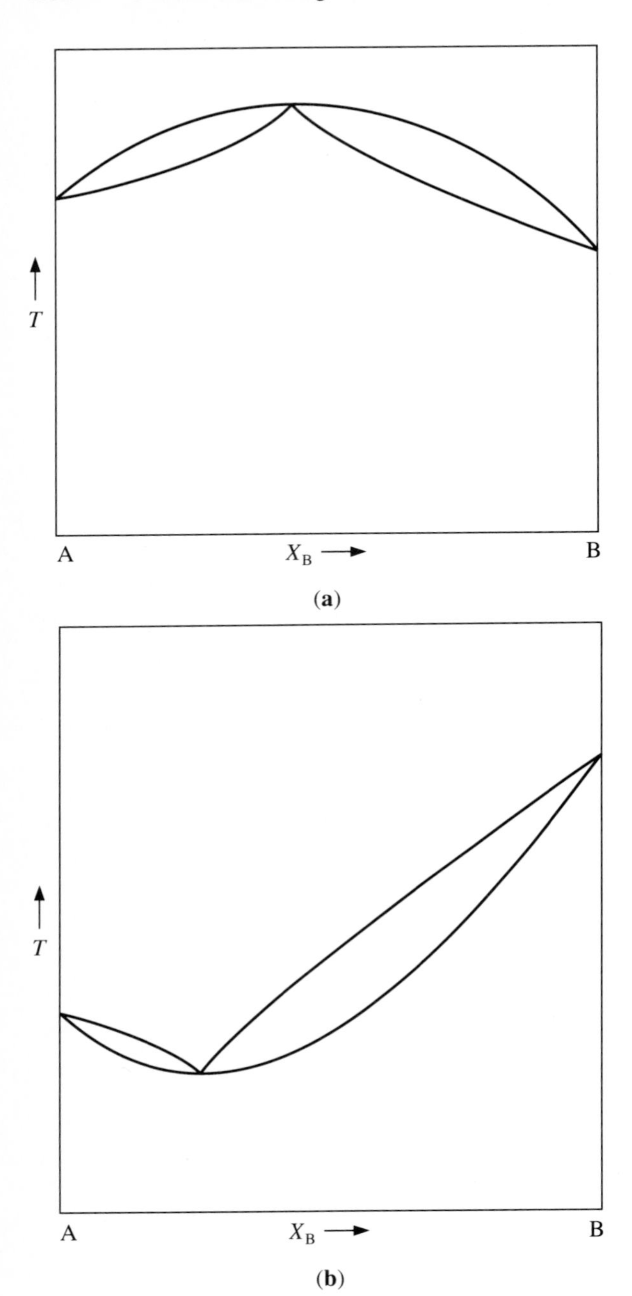

FIGURE 13.13 Maximum-boiling and minimum-boiling point azeotropes.

diffraction, vibrational spectroscopy, and magnetic resonance spectroscopy, have partially superseded the original methods, but thermal analysis is still an important experimental method.

A simple example is provided by the solid–liquid eutectic diagram in Fig. 13.14a. When melt of composition represented by point y is cooled, the temperature falls uniformly until the solid–liquid equilibrium curve is encountered at point a. Then solid B begins to freeze out (point b), releasing heat of fusion that slows down the cooling of the sample.

As more B is frozen out, the melt becomes richer in A, following the equilibrium curve as shown by the dashed line. When the eutectic temperature is

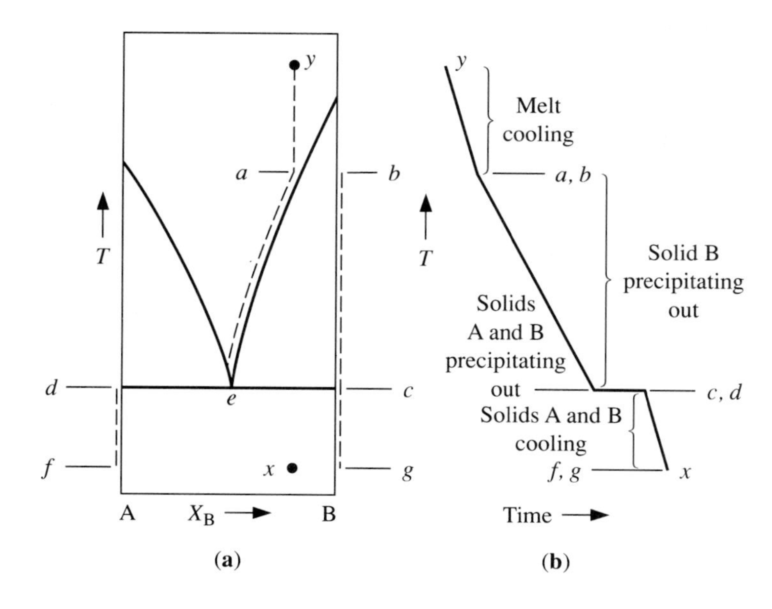

FIGURE 13.14 (a) As the melt (point y) cools, it precipitates out solid B (b), then at the eutectic (e) also precipitates out solid A. Final overall composition and temperature (points f and g) is represented by x. (b) The cooling curve shows breaks where the vertical line ($y \cdots x$) intersects lines of the diagram.

reached, solid A also begins to freeze out (point d), along with additional solid B (point c). The appearance of this mixture of two solids differs from a casual mixture of A and B, which led some early interpreters of phase equilibria to regard this mixture as if it were a compound. (In some very old phase diagrams, a solid or dotted vertical line is drawn below the eutectic point.)

As long as solid A, solid B, and the melt are all present ($p = 3$; points d, e, and c), there are no degrees of freedom, so the temperature remains fixed. Eventually, all the melt disappears, in favor of solids A and B. The temperature of the two solid phases then drops, with the compositions following the dashed lines toward points f and g. The final solid mixture has the overall composition of point x (which is the same as point y), but there is no phase present that has this composition. The equilibrium state is pure A and pure B, in amounts given by the lever-arm rule about point x.

If the solids are partially miscible (Fig. 13.15), the behavior is very similar, although the composition of the solids in equilibrium with the melt depends on temperature.

If cooling is carried out at what would otherwise be a reasonably steady rate, the actual rate of temperature decrease varies with time, as shown in Fig. 13.14b and 13.15b and d. Consideration of the causes of the breaks in the curve permits the following convenient generalizations:

> *Draw a vertical line down the phase diagram at the overall composition of the mixture present. At a temperature corresponding to an intersection of this vertical line with any line of the phase diagram, the cooling curve will show a break and a change in slope. If the line intersected is sloping, the new T versus t curve will be sloped; if the line intersected is horizontal, the new T versus t curve will be horizontal.*

The cooling curve of miscible systems, as shown in Fig. 13.16, behaves similarly to those already discussed. When the temperature reaches the equilibrium line, the lower temperature phase separates out. This depletes the high-temperature phase in that component, so the composition of the high-temperature phase shifts.

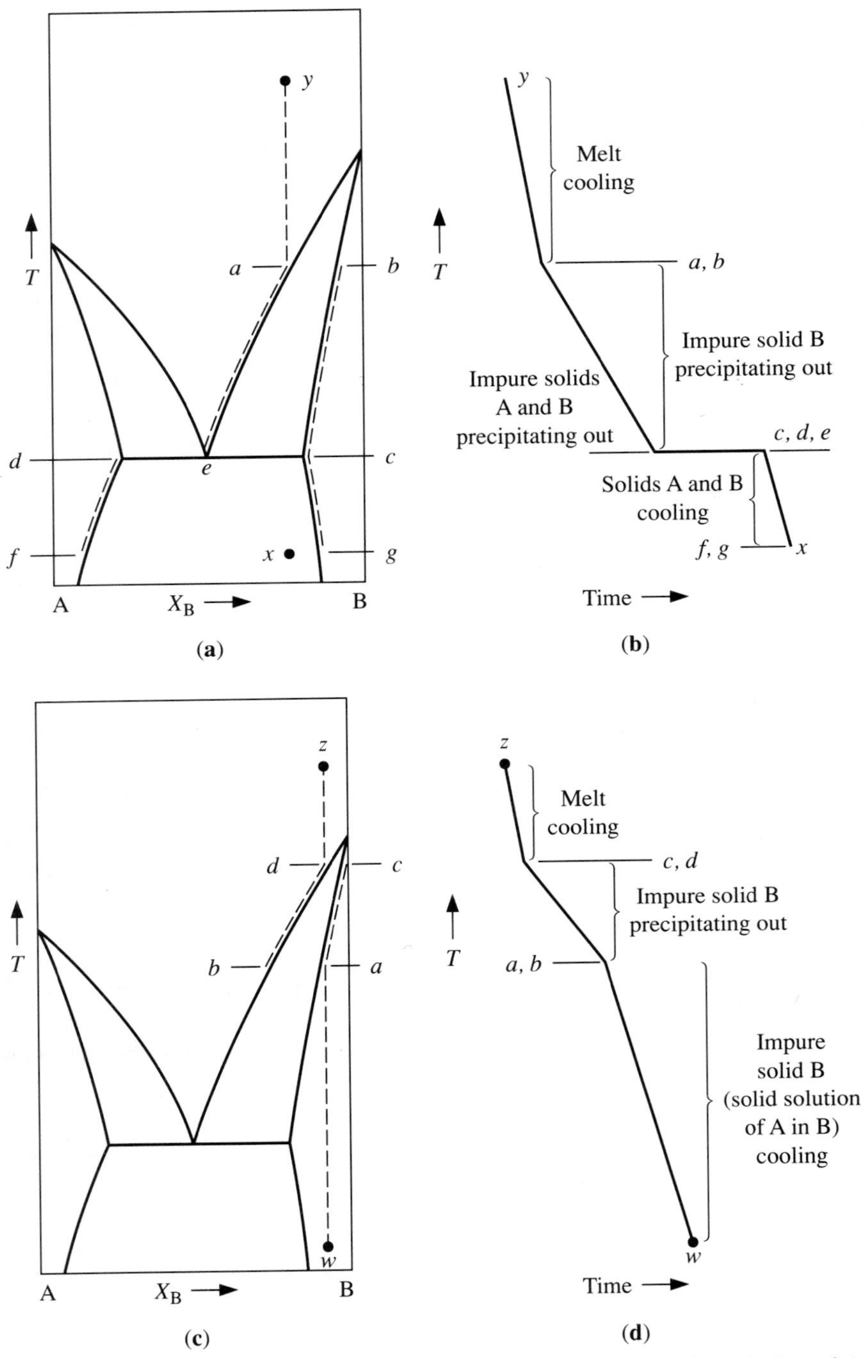

FIGURE 13.15 (a) When the melt (y) cools, impure B (a solid solution of A in B; point b) precipitates out; then at the eutectic (e) impure A also precipitates out. (b) The cooling curve for the system from y to $x(= f + g)$ shows breaks at points b and e. (c) If the initial composition is at z, the system cools to the solid solution of composition w. (d) The cooling curve for the "impure B" melt has no horizontal segment.

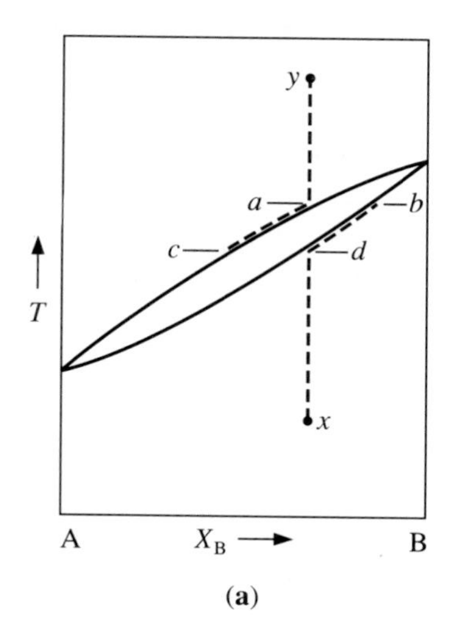

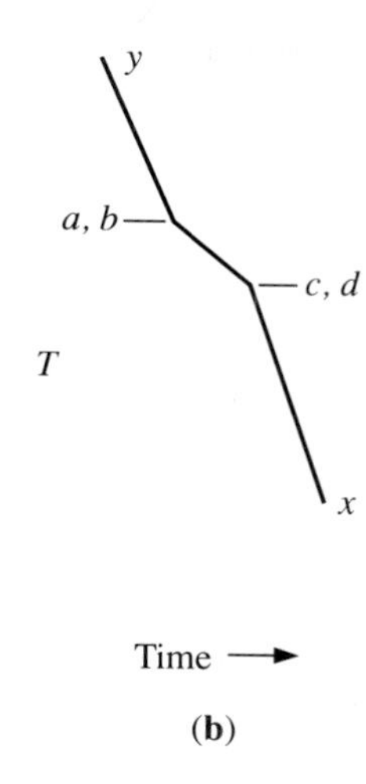

FIGURE 13.16 Miscible system: (a) Path followed as the melt cools; (b) temperature versus time cooling curve.

As shown by the lever-arm rule, the relative amounts of the two phases is continually changing as the compositions of the phases change, until eventually all of the high-temperature phase is gone; everything is in the single, low-temperature phase, which then cools with no further changes.

The phase diagram assumes equilibrium at all points. Some of the most important applications of phase diagrams, however, involve metastable states. For example, iron and carbon form a series of phases at different temperatures (Fig. 13.17). Austenite, or γ-ferrite, is a hard, brittle (metastable) steel. When it cools slowly it produces the eutectoid mixture of α-ferrite and cementite (Fe_3C), called pearlite, at about 725°C. Iron quenched rapidly from a high temperature consists primarily of austenite and is suitable for files and other cut-

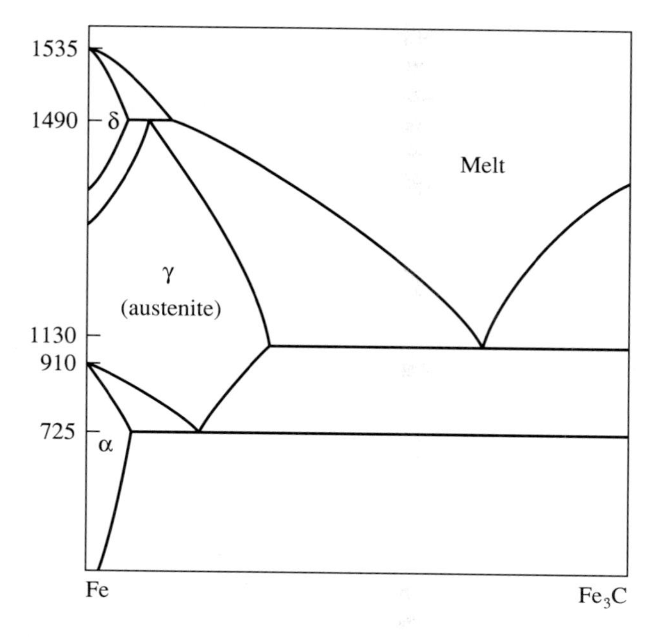

FIGURE 13.17 Iron–carbon phase diagram (not to scale). The α- and δ-ferrites are body-centered (plus carbon in solution); γ-ferrite (austenite) is face-centered. The minimum at 725°C, formed from a solid solution, is called a *eutectoid*. The maximum at 1490°C is called a *peritectic*. Temperatures are approximate.

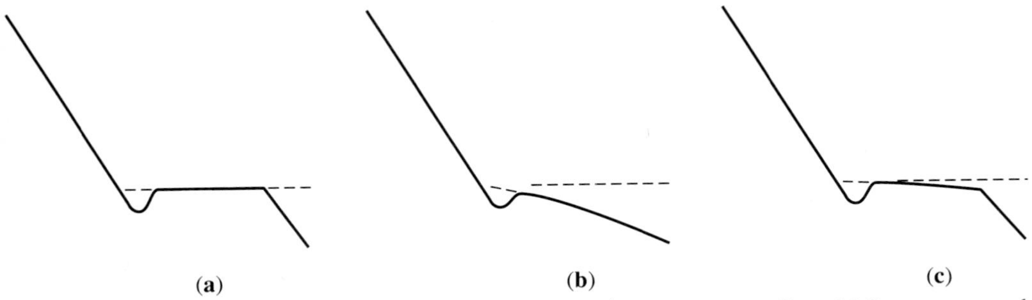

FIGURE 13.18 Cooling curves at the freezing point, showing supercooling: (a) Pure compound; (b) impure material; (c) nearly pure compound.

ting tools. When the steel is subsequently reheated slowly, some pearlite (eutectoid mixture, plus residual α-ferrite or cementite) forms and remains when the material is quenched, giving soft steel. The process is called *tempering* (softening).

A different type of metastability arises when a phase transition is initiated. Typically, substances in one phase are not in the correct geometrical relationship to form a new phase, and the transition may be relatively improbable on short time scales. Water vapor may be cooled well below its condensation point if there are no ice nuclei or foreign particles that initiate nucleation. Similarly, water that is boiled typically superheats, then as a bubble is nucleated and grows a "bump" occurs that may splatter water from the container.

For cooling curves, the lag in nucleation typically takes the form of supercooling, as shown in Fig. 13.18. The temperature of the liquid phase drops below the point at which solid should begin to precipitate. When the first solid does appear, the release of the heat of fusion raises the temperature to the appropriate melting point. The sample then continues to freeze, then cool, more or less in equilibrium, as the temperature falls.

Cooling curves provide a sensitive test of the purity of a compound that will freeze near room temperature. If the substance is pure, the composition follows the vertical line of pure compound and the freezing of the compound produces a horizontal line. If there is impurity, the composition strikes the liquid–solid equilibrium curve along a sloping portion and the cooling curve changes slope without becoming horizontal, as shown in Fig. 13.18b and c.

13.8 Purification Methods

Purification of a chemical involves separation of components of a mixture under conditions at or near the equilibrium conditions described by a phase diagram. Fractional distillation, fractional crystallization, and chromatography are common examples of the application of phase equilibria to purification.

Fractional Distillation

In a simple distillation, as in Fig. 13.19a, the first vapor to come off is richer in the lower-boiling component, but as more vapor comes off, the composition of the distillate moves steadily back toward the composition of the original liquid. If all the liquid is evaporated, the distillate has exactly the composition of the original liquid.

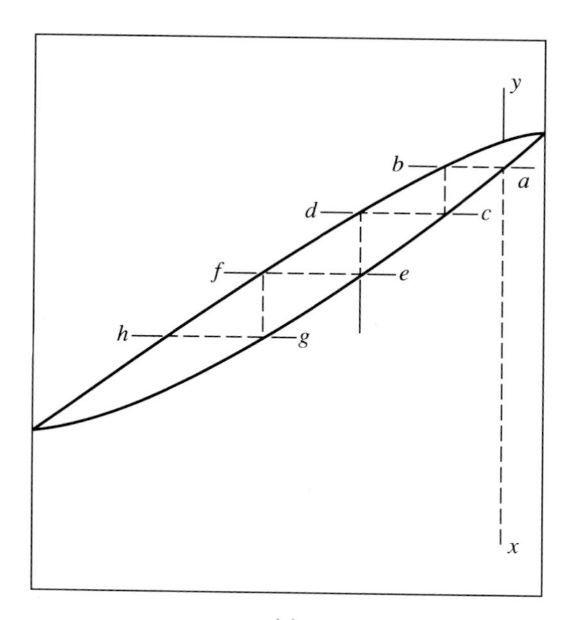

(a)

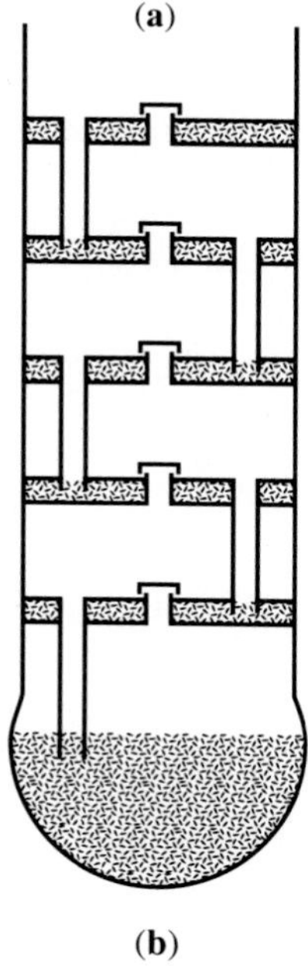

(b)

FIGURE 13.19 Fractional distillation. (a) Each liquid sample can only be in equilibrium with vapor along an isotherm (horizontal). (b) A bubble-cap column is a conceptually important design.

Separation of components by distillation requires a succession of steps, separated in space. The principle is best illustrated by a bubble-cap distillation column, shown in Fig. 13.19b. Suppose the material in the pot initially has the composition indicated by e, in Fig. 13.19a. The vapor above the pot will be of composition f. This vapor condenses on the first plate as liquid of the same composition, g. Vapor above the first plate is then of composition h. However, as the column approaches equilibrium, removal of the more volatile A changes the composition of the material in the pot toward composition c, with corresponding shifts of composition of liquid on the first plate to $d = e$ and on the second plate to $f = g$. With achievement of equilibrium and a sufficient number of plates, the vapor above the top plate is essentially pure A and the material left in the pot is essentially pure B.

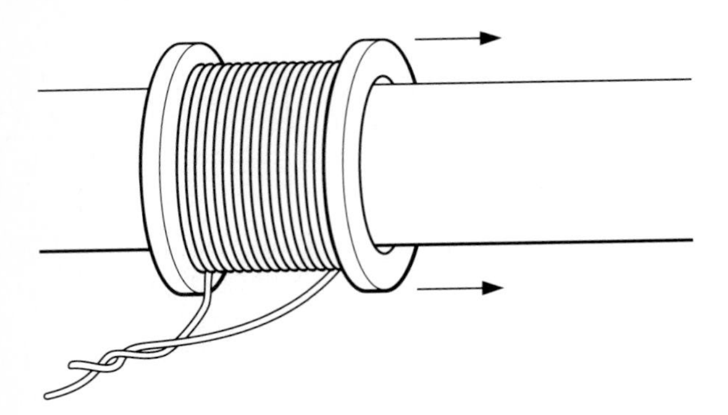

FIGURE 13.20 Zone refining. Impurities concentrate in the molten phase and are swept along toward the end.

To achieve complete separation would require an infinite number of plates of a perfect bubble-cap column. In practice, a reasonable number of stages (e.g., 100, or sometimes as few as 10 or even 5) gives a satisfactory separation of most mixtures.

The bubble-cap column is not the best design for most purposes. An empty tube roughly emulates a bubble-cap column of one plate. If the tube is filled with metal or glass pieces that encourage condensation and reevaporation it is possible to achieve the equivalent of a 100-plate bubble-cap column in a distillation column that is only on the order of a meter in length. Columns are rated by the *number of theoretical plates* and by the *height equivalent to a theoretical plate* (HETP).

The full efficiency of a distillation column can be realized only in very slow distillations. The removal of distillate at the top must not disturb the equilibrium throughout the column. Initial attainment of equilibrium in a column operating at 100 theoretical plates normally requires several hours.

Fractional Crystallization

A significant fraction of chemicals to be purified are liquids. Generally, these have undergone a prior purification step by fractional distillation, so another distillation is of marginal value. However, the probability is high that the impurities are less soluble in the solid phase than in the liquid, so fractional crystallization can be effective. Often solids are purified by crystallization from a solute, which is later removed by evaporation.

Many solids have negligible solubility for other substances, so the first material precipitated from solution is the pure major component and pure material continues to come out until the eutectic point is reached. This is in strong contrast with distillation equilibria, where many steps (i.e., theoretical plates) are required to achieve separation.

The principal practical difficulty in fractional crystallization is maintaining equilibrium between the liquid and solid phases. A precipitating solid often traps some of the liquid, crystallizing around pockets of liquid or adsorbing liquid on the surface of the crystals as they form. Careful design of the apparatus and slow crystallization can alleviate the problem.

A particularly effective method for purification of solids by fractional crystallization is called *zone refining*. The technique was developed for purification of germanium and silicon, where it was important to achieve levels of 99.99% and better. Zone refining opened the way to modern electronics by providing the ultrapure semiconductors required to understand the mechanisms of conduction, and for mass production of transistors and integrated chips.

Zone refining takes advantage of the very general principle that impurities are less soluble in a solid phase than in the liquid phase. A heating coil is placed around a cylinder of the "pure" semiconductor, melting a small cross section (Fig. 13.20). Impurities accumulate in this molten section. As the coil then moves along the cylinder, the molten section moves, sweeping impurities with it to the end of the cylinder. At the completion of each pass the heater is returned to the initial point for another pass until the required purity level is achieved.

Chromatography

Another class of equilibria is between a substance dissolved in a foreign liquid or adsorbed on the surface of a foreign solid and the substance in the vapor phase or in solution. Each such equilibrium has a distribution coefficient, which may or may not be constant with concentration and pressure. Establishment of equilibrium causes the mobile phase to become either more or less concentrated in certain components. By passing the mobile phase through a column or across a piece of paper, a very large number of equilibrium stages can be achieved, yielding a high purity of a selected component or a clear separation, by time of emergence, of two or more components of a mixture.

The technique was originally developed for separation of colored substances in a column packed with a solid, through which the solution passed. The result was a vertical display of the colors, leading to the name *chromatography.* Common examples today include columns of silica gel or alumina for separating polar and nonpolar materials and chromatography on strips of paper and on the surface of thin films of adsorbent. *Vapor-phase chromatography* vaporizes liquids or solids and carries the vapors in a carrier gas through a long, thin, coated tube. Chromatography is an important analytical tool that is also adaptable to production of large quantities of material.

13.9 Three-Component Phase Diagrams

When the number of components is increased to three, the phase rule becomes $\mathscr{f} = 3 - p + 2 = 5 - p$, so a single phase would require four dimensions. Even if pressure or temperature is fixed, three dimensions are needed. Two-dimensional representations of three-dimensional figures can be drawn, but they typically show only one surface. Alternatively, a third dimension may be represented in a plane figure by adding contour lines.

To obtain a phase diagram that can be fully represented in two dimensions, it is necessary to fix two variables. Often both temperature and pressure are fixed and special coordinates are chosen to depict the concentration variables. For example, the total amount of solvent may be fixed and moles of solute plotted along horizontal and vertical axes (Fig. 13.21). Pure solute would be at infinity along either axis, but the diagram is drawn to provide information only about the solutions, and especially the limits of solubility.

More often, triangular coordinates are selected, with each vertex representing a pure component (Fig. 13.22). By simple geometry it can be shown that the sum of the distances of a point from the three sides is the same for any point. (Draw three smaller triangles, with the three sides as bases and the arbitrary point at the apex of each. Because the sum of the areas of the three triangles is equal to the area of the original triangle, the sum of the three altitudes is equal to the altitude of the original.)

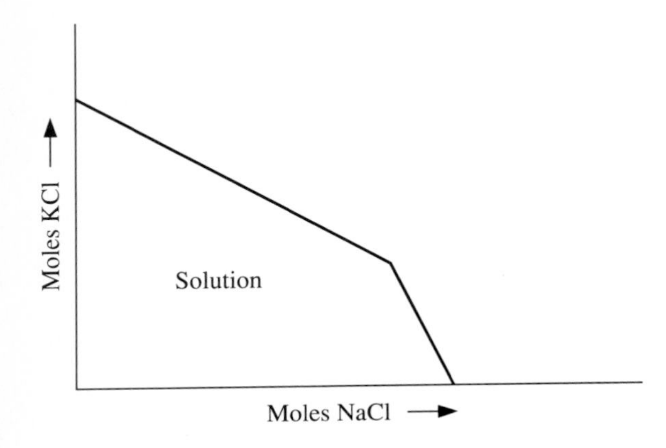

FIGURE 13.21 Three-component phase equilibrium with a constant amount of solvent.

The phase diagram of water, ether, and acetone is shown in Fig. 13.23a. Water and ether are nearly immiscible, but acetone is fully miscible with each. The range of possible solution concentrations falls in the area at the top of the triangle, extending around the forbidden region at the bottom. Water-rich solutions may coexist with ether-rich solutions, on opposite sides of the forbidden region, but the simple diagram does not tell which concentrations can coexist.

In two-component diagrams, it was sufficient to draw isothermal lines to connect the phases on each side of a forbidden region. The entire three-component diagram, as drawn, is an isothermal plane, so additional information must be provided in the form of *tie lines,* as shown in Fig. 13.23b. Like the horizontal lines drawn through eutectic points, these tie lines are indicator lines, not separating regions but showing connections between points lying in different phase regions. Tie lines are straight and do not cross. They are determined experimentally.

More complex three-component equilibria are exhibited by water, sodium chloride, and potassium chloride (Fig. 13.24) and by water, sodium sulfate, and sodium chloride (Fig. 13.25). In each case the only single-phase area is at the top (water-rich solutions) and the points representing the pure solid compounds.

Two types of forbidden areas appear. Those with tie lines are equilibria between two phases, a single pure solid and the water solutions of various compositions. The remaining clear areas, which are triangles with straight sides, are "three-phase" forbidden areas; if the composition falls in any of these regions the system will split into three separate phases, with compositions represented by the corners of the triangles.

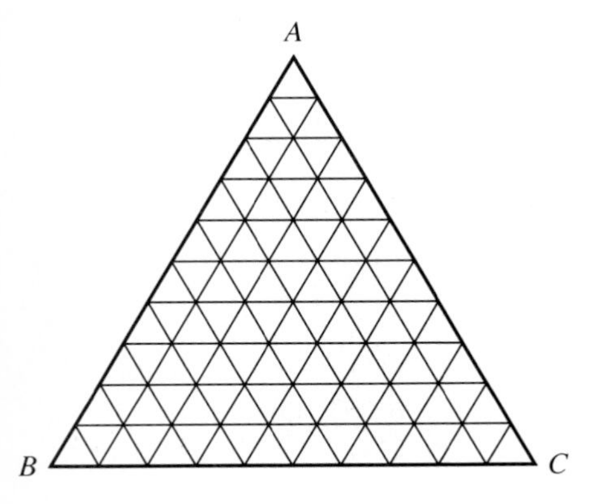

FIGURE 13.22 Triangular coordinate system for a three-component diagram.

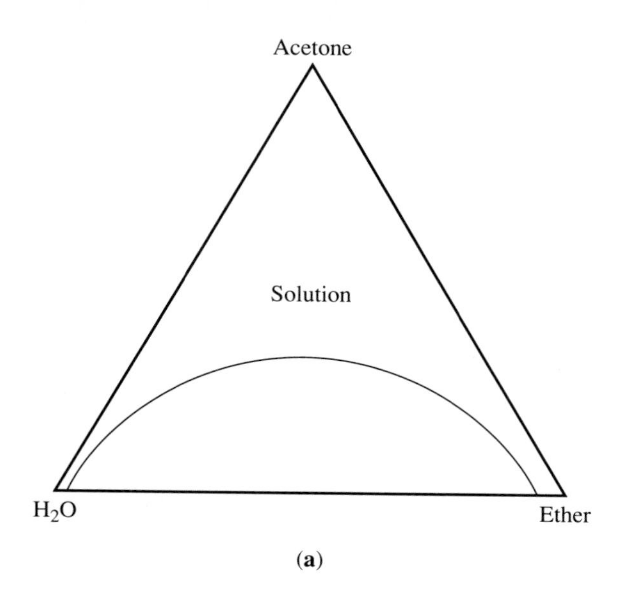

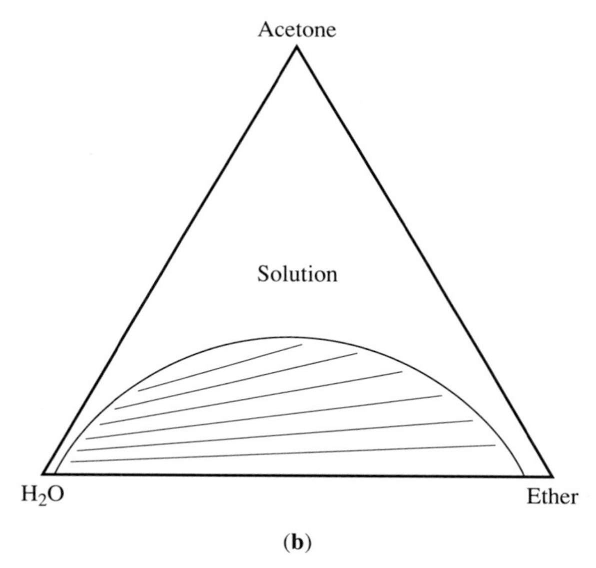

FIGURE 13.23 Three-component phase diagram. Solubility diagram for water, acetone, and diethyl ether at 30°C. (a) Solution area extends on both sides of a forbidden region of immiscibility. (b) Tie lines must be added to show compositions in equilibrium.

SUMMARY

A phase is a (locally) homogeneous sample of matter, or a collection of equivalent homogeneous samples.

If c components are distributed among p phases, there are $\mathcal{f}$ degrees of freedom: $\mathcal{f} = c - p + 2$ (allowing for variation in temperature and pressure).

If $c \geq 2$, it is necessary to fix one (or more) variables to produce a phase diagram in two dimensions.

Phase diagrams represent $\mathcal{f} = 2$ as an area; $\mathcal{f} = 1$ as one or more lines; $\mathcal{f} = 0$ as one or more points.

A chemical compound is a substance of fixed composition, so it appears as a vertical line (or an area enclosing an implicit vertical line).

Areas, representing single phases, are necessarily separated, horizontally, by forbidden regions.

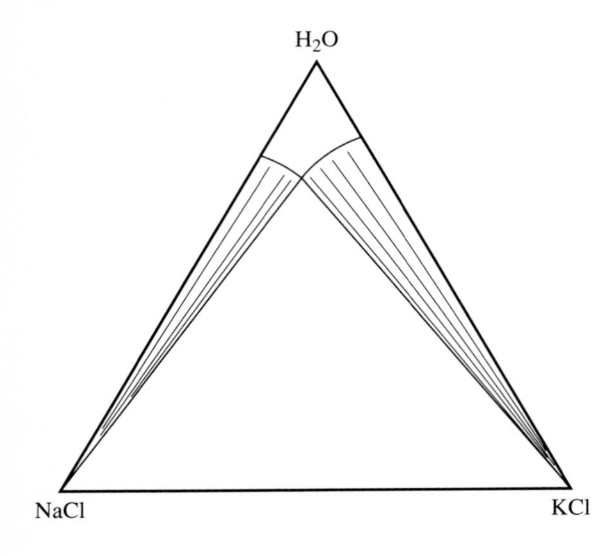

FIGURE 13.24 The area at the top is aqueous solution. Other regions are forbidden.

Lever-arm rule: If the overall composition, C, falls within a forbidden region, the system splits into amounts m_1 and m_2, of composition C_1 and C_2, such that $(C - C_1)m_1 = (C_2 - C)m_2$.

Typically, an impurity is less soluble in a solid than in a liquid, so the impurity depresses the freezing point. Two depressed freezing point curves meet to form a eutectic, or a minimum-freezing mixture of fixed composition.

If substances are miscible in both high- and low-temperature phases, they may still form an azeotrope, either maximum- or minimum-boiling.

Cooling curves show changes of slope where a vertical line, drawn through the (fixed) sample composition, intercepts a line on the phase diagram. The cooling curve shows a horizontal plateau if the line intersected is horizontal (e.g., a eutectic temperature).

Fractional distillation is equivalent to a set of successive separations, described as "plates."

Purification by crystallization may be accomplished in one step, but is often repeated. Zone refining is a form of fractional crystallization.

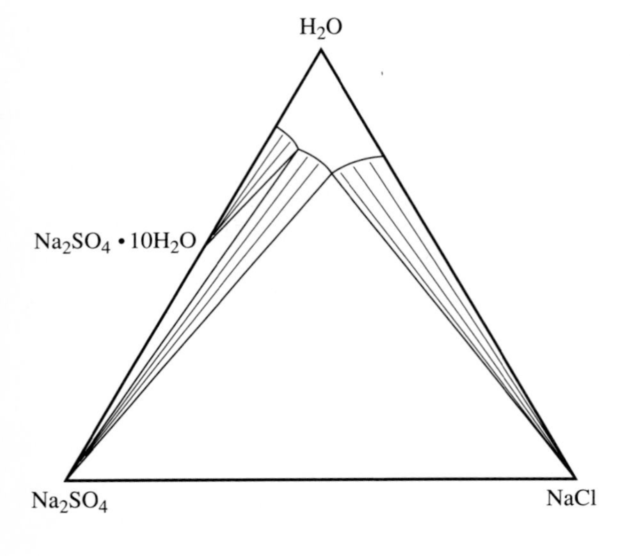

FIGURE 13.25 Sodium sulfate forms a compound.

Chromatography is establishment of successive adsorption–desorption equilibria along a column.

When there are three components, special coordinate systems are required, such as triangular coordinates, and diagrams are drawn for constant temperature and pressure, or as three-dimensional models. Often tie lines, experimentally determined, must be added to remove ambiguities.

PROBLEMS

13.1. Sodium sulfate forms anhydrous crystals as well as the familiar decahydrate ($Na_2SO_4 \cdot 10H_2O$) and a less familiar heptahydrate ($Na_2SO_4 \cdot 7H_2O$). How many of these can exist in equilibrium with an aqueous solution at an arbitrary temperature (e.g., 25°C) under 1 atm pressure?

13.2. The critical point of carbon dioxide occurs at 31°C and 73 atm. The triple point (solid–liquid–vapor) is at −57°C and 5.3 atm. At the triple point the solid is more dense than the liquid.

a. Sketch the phase diagram for carbon dioxide, labeling axes clearly.

b. Label areas, lines, and specific points, indicating what phases are present.

c. How does your phase diagram correlate with your experience concerning dry ice and CO_2 gas? Why is the temperature −80°C maintained by dry ice?

13.3. Bismuth and lead melt at 268°C and 322°C, respectively. They form no compounds but do form solid solutions with compositions ranging from 0 to 10 and 90 to 100 mol % lead. The eutectic temperature is 120°C and the eutectic mixture contains 60 mol % lead. Draw the phase diagram. Label axes clearly and label appropriate areas to indicate phases present.

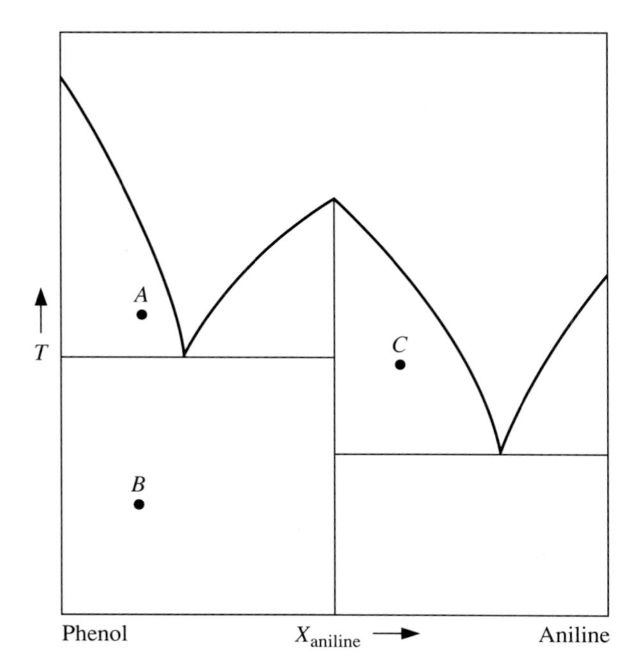

FIGURE 13.26 Solid–liquid phase diagram of phenol and aniline.

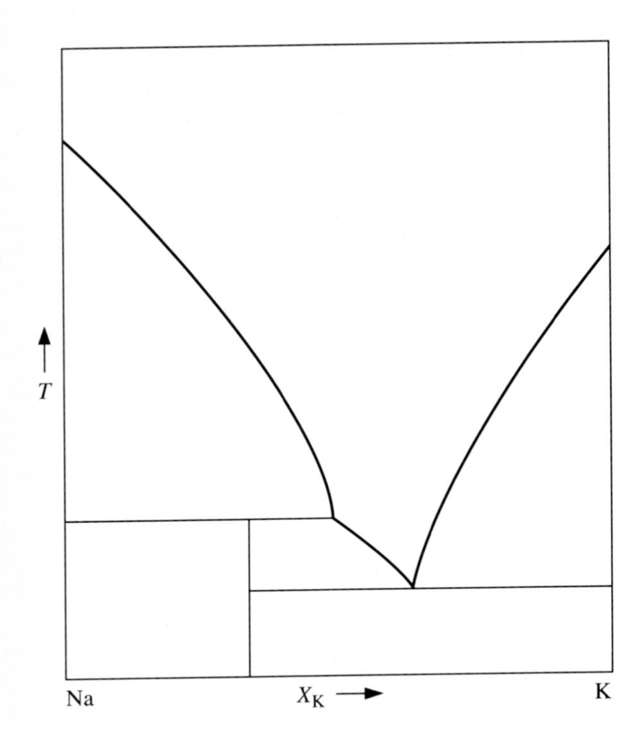

FIGURE 13.27 Solid–liquid phase diagram of sodium and potassium.

13.4. The solid–liquid phase diagram for phenol (P, C_6H_5OH) and aniline (A, $C_6H_5NH_2$) is shown in Fig. 13.26. If the overall composition and temperature corresponds to point
a. A **b.** B
c. C
describe what phase(s) will actually be present.

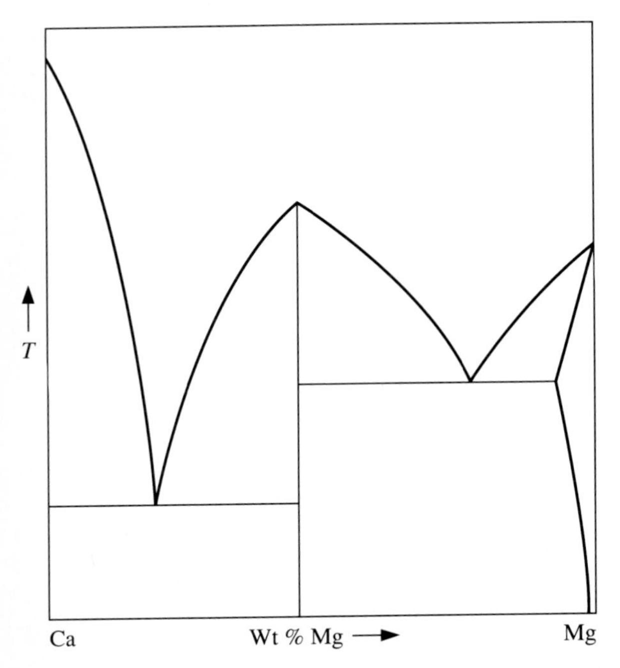

FIGURE 13.28 Solid–liquid phase diagram for magnesium and calcium.

13.5. The solid–liquid phase diagram for sodium and potassium is shown in Fig. 13.27. Interpret the diagram, indicating the phases present when the overall composition and temperature fall into each of the regions of the diagram.

13.6. The solid–liquid phase diagram for Mg and Ca is shown in Fig. 13.28. Identify each area of the diagram that represents a single phase, and indicate the nature of that phase.

13.7. The solid–liquid diagram for silver and mercury is shown in Fig. 13.29. Identify the phases shown in the diagram. The melting point of mercury is −38.87°C. Sketch the probable shape of the phase diagram in the vicinity of the melting point of mercury. (The compound Ag_2Hg_3, designated gamma-1, is present in most dental alloys, although the alloys typically include copper and tin.)

13.8. What criteria can distinguish a liquid from a gas? Discuss the applicability, or nonapplicability, of each criterion to a fluid above its critical point.

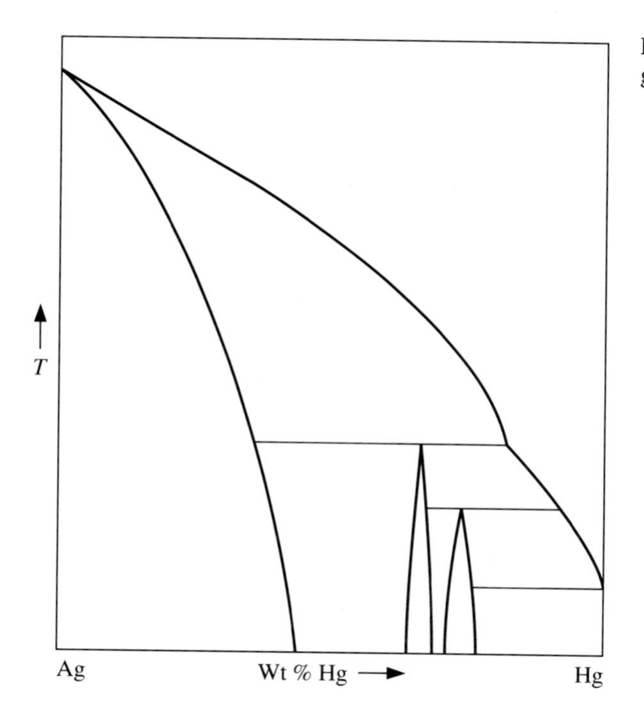

FIGURE 13.29 Solid–liquid phase diagram for silver and mercury.

CHAPTER 14

Chemical Equilibrium

Chemical reactions have obvious importance in chemistry, polymer science, and biophysics, but are also important in many areas of solid-state and materials science and in earth and planetary sciences. Recently, astrophysicists have found that outer space is surprisingly rich in organic compounds, undergoing formation and decomposition reactions.

Whether a chemical reaction will or will not occur depends on several considerations. Some reactions can be predicted to be spontaneous[1] because the products have much less energy than the reactants, or because there is only a small energy difference but the products have significantly greater entropy than the reactants. Yet many reactions predicted to be spontaneous do not actually occur; the *rate* of reaction may be small or zero.

For example, graphite is more stable than diamond, so the process

$$C_d \rightarrow C_g$$

would be expected to be spontaneous. However, the rate of reaction is close enough to zero that diamonds have been around for some billions of years and are likely to last for additional billions of years if left near 300 K. In contrast to this rate-limited reaction, the reaction of hydrogen and oxygen to form water is very rapid and, once started, will continue until the concentration of one, or both, of the reactants is close to zero. We can predict that the reverse reaction,

$$2H_2O \rightarrow 2H_2 + O_2$$

will not occur because it is energetically unfavorable.

[1]The term *spontaneous* must not be associated with *instantaneous*. A spontaneous process is simply one that can go in the direction under consideration, given sufficient time, and perhaps an appropriate catalyst. Under the same conditions, therefore, a spontaneous process *cannot* go in the reverse direction, as measured by macroscopic parameters, except as it is externally driven (e.g., by an electric current or by photons).

The Gibbs free energy function provides a guide to whether a reaction can occur, given sufficient time and, if necessary, an appropriate catalyst.[2] Thermodynamics makes no predictions concerning the rates of (spontaneous) reactions.

14.1 Free Energy of Reaction

It was shown earlier that a necessary and sufficient condition for a process to be spontaneous is that the Gibbs free energy change should be negative, under conditions of constant temperature and pressure and with no work, other than $P\,dV$ work, done on the system.

$$\begin{Bmatrix} \text{constant } T \text{ and } P \\ w' = 0 \end{Bmatrix} \qquad \Delta G < 0$$

Consider a completely general reaction represented by

$$a\mathrm{A} + b\mathrm{B} \rightarrow c\mathrm{C} + d\mathrm{D}$$

That is, a moles of substance A can react with b moles of substance B to yield c moles of substance C and d moles of substance D (or some equivalent set of values for other numbers of reactants and products).

The chemical equation makes no assumptions and provides no information about the quantities or concentrations of any of the reactants or products actually present.

The criterion for spontaneity may be written in terms of the partial molar free energies or chemical potentials.

$$\Delta G = \sum_{\text{product}} n_i \bar{G}_i - \sum_{\text{reactants}} n_j \bar{G}_j$$

$$= c\bar{G}_C + d\bar{G}_D - a\bar{G}_A - b\bar{G}_B < 0 \qquad (14\text{-}1)$$

The states of reactants and products are assumed to be constant before, during, and after the reaction.

A reaction that is spontaneous for one set of values of concentrations or pressures is nonspontaneous for other values. If the concentration of any substance is zero, the partial molar free energy of that substance goes to $-\infty$. Therefore, if the concentration of any product is zero, $\Delta G_{\text{react.}} = -\infty$ and the process is spontaneous. If the concentration of any reactant is zero, ΔG for the reaction goes to $+\infty$; the process then cannot be spontaneous.

The relationship is more quantitative when concentrations or pressures are replaced with fugacities. Recall that

$$\bar{G}_i = RT \ln f_i + B_i(T)$$

where $B_i(T)$ is an unknown function of temperature, for the ith substance, that will drop out of the later equations. The greater the pressure or concentration of

[2] A catalyst acts much like a lubricant. It will allow processes that should occur to go more rapidly but will not change the direction of a reaction.

any substance, the greater its fugacity and therefore the greater the partial molar free energy.

The value given in thermodynamic tables is ΔG_f°, the free energy of formation of the substance from the elements when each reactant and product is in its standard state.[3] As shown in Section 11.3, ΔG° of reaction is obtained by combining the ΔG_f° values.

$$\Delta G^\circ = c\,\Delta G_{f\mathrm{C}}^\circ + d\,\Delta G_{f\mathrm{D}}^\circ - a\,\Delta G_{f\mathrm{A}}^\circ - b\,\Delta G_{f\mathrm{B}}^\circ$$

More generally, because the standard state may be a solution, free energies are replaced by partial molar free energies of each compound and the free energy of reaction is written

$$\Delta G^\circ = c\bar{G}_\mathrm{C}^\circ + d\bar{G}_\mathrm{D}^\circ - a\bar{G}_\mathrm{A}^\circ - b\bar{G}_\mathrm{B}^\circ \qquad (14\text{-}2)$$

or for arbitrary concentrations or pressures,

$$\Delta G = c\bar{G}_\mathrm{C} + d\bar{G}_\mathrm{D} - a\bar{G}_\mathrm{A} - b\bar{G}_\mathrm{B} \qquad (14\text{-}3)$$

EXAMPLES 14.1

A. From Table 14.1, ΔG_f° of ozone, O_3, is 163 kJ. Is the reaction

$$\tfrac{3}{2}O_2 \rightarrow O_3$$

spontaneous or nonspontaneous if oxygen and ozone are each present in their standard states (partial pressures of 1 atm)? [Assume 25°C unless otherwise indicated in problems.]

B. Is the reaction

$$CO + \tfrac{1}{2}O_2 \rightarrow CO_2$$

a spontaneous reaction when each of the gases is present at 1 atm partial pressure? □

Subtracting ΔG° from ΔG gives

$$\Delta G - \Delta G^\circ = c(\bar{G}_\mathrm{C} - \bar{G}_\mathrm{C}^\circ) + d(\bar{G}_\mathrm{D} - \bar{G}_\mathrm{D}^\circ) - a(\bar{G}_\mathrm{A} - \bar{G}_\mathrm{A}^\circ) - b(\bar{G}_\mathrm{D} - \bar{G}_\mathrm{D}^\circ)$$

Replace $\bar{G}_i - \bar{G}_i^\circ$ with $RT\ln(f_i/f_i^\circ)$. Then

$$\Delta G - \Delta G^\circ = cRT\ln\frac{f_\mathrm{C}}{f_\mathrm{C}^\circ} + dRT\ln\frac{f_\mathrm{D}}{f_\mathrm{D}^\circ} - aRT\ln\frac{f_\mathrm{A}}{f_\mathrm{A}^\circ} - bRT\ln\frac{f_\mathrm{B}}{f_\mathrm{B}^\circ}$$

Now replace each ratio of fugacities with the symbol a, called the *activity*.

$$a_i \equiv \frac{f_i}{f_i^\circ} \qquad (14\text{-}4)$$

[3] The values of ΔG° depend on the choice of standard states, but standard states are seldom explicitly stated. Users of the tables are assumed to be familiar with the conventions for standard states discussed in the following section.

Table 14.1 Standard Free Energies of Formation[a]

Compound	State	ΔG_f°	Compound	State	ΔG_f°
AgBr	c	−95.94	HCl	g	−95.31
AgCl	c	−109.72		aq	−131.17
AgI	c	−66.316	HI	g	1.57
Ag_2O	c	−10.82		aq	−51.67
Al_2O_3	c	−1581.	H_2O	g	−228.60
Br_2	g	3.14		liq	−237.19
C	diamond	2.87	H_2S	g	−33.01
	g	−877.916	I_2	g	19.4
CCl_4	liq	−68.62	NH_3	g	−16.59
CF_4	g	−66.245		aq	−26.7
CO	g	−137.16	NH_4Cl	c	−203.9
CO_2	g	−394.40	N_2O	g	104.19
	aq	−386.2	NO	g	86.61
$CaCO_3$	calcite	−1128.76	NO_2	g	51.25
	aragonite	−1127.71	N_2O_4	g	97.74
$CaCl_2$	c	−750.2	NaCl	c	−384.0
	aq	−815.38		aq	−393.04
CaF_2	c	−1161.9	NaOH	c	−381.5
	aq	−1106.0		aq	−419.170
$Ca(NO_3)_2$	c	−741.99	O_3	g	163.4
	aq	−774.04	SiO_2	c	−805.0
CaO	c	−604.2	SO_2	g	−300.2
$Ca(OH)_2$	c	−909.2	SO_3	g	−370.4
	aq	−867.64	C_2H_2	g	209.0
FeO	c	−244.	C_2H_4	g	68.12
Fe_2O_3	c	−741.0	C_2H_6	g	−32.89
Fe_3O_4	c	−1014.2	C_2H_5OH	g	−168.6
HBr	g	−53.26		liq	−166.2

[a] Values are in kJ/mol, for 25°C and 1 atm, from Circular 500, National Bureau of Standards, "Selected Values of Chemical Thermodynamic Properties," or from JANAF Tables. Standard states for liquids (liq) and crystalline solids (c) are pure materials at 1 atm; for gases (g) the pure gas at 1 atm in the "ideal gas state" (extrapolated from low pressure). "aq" refers to the (hypothetical ideal) 1 *m* solution in water.

Then

$$\Delta G = \Delta G^\circ + RT(c \ln a_C + d \ln a_D - a \ln a_A - b \ln a_B) \quad (14\text{-}5)$$

or

$$\Delta G = \Delta G^\circ + RT \ln \frac{a_C^c a_D^d}{a_A^a a_B^b} \quad (14\text{-}6)$$

ANSWERS 14.1

A. $\Delta G_f^\circ = 163$ kJ is greater than zero, so the reaction to form ozone is not spontaneous when each of the gases is at 1 atm.
B. To obtain ΔG° for the reaction of CO with oxygen, the values of ΔG_f° of the two compounds are subtracted.

$$\begin{aligned}\Delta G^\circ &= \Delta G_f^\circ(CO_2) - \left[\Delta G_f^\circ CO) + \frac{1}{2}\Delta G_f^\circ(O_2)\right] \\ &= -394.40 - (-137.16 + 0) \\ &= -257.24 \text{ kJ}\end{aligned}$$

ΔG is negative and the reaction is spontaneous.

This equation relates the free-energy change of the reaction, ΔG, to the tabulated value, ΔG°. The logarithmic term,

$$\ln \mathscr{Q} \equiv \ln \frac{a_C^c a_D^d}{a_A^a a_B^b} \tag{14-7}$$

is a correction factor for the difference in concentrations or pressures, of reactants and products, from the standard state values assumed in obtaining ΔG°. It may take any value from $-\infty$ to $+\infty$.

EXAMPLES 14.2

A. What value of $\mathscr{Q}$ is necessary to make the reaction of oxygen at 1 atm, to form ozone at 1 atm, a spontaneous reaction?
B. From Table 14.1, the reaction of H_2 gas with solid I_2 gives $\Delta G_f^\circ(HI) = 1.57$ kJ/mol. What value of $\mathscr{Q}$ would make this reaction spontaneous? □

Often it is helpful to be able to characterize the progress of a process by a single variable, λ. We will show how a chemical reaction can be thus described.

Let a general chemical reaction be represented by

$$aA + bB \rightarrow cC + dD$$

The equation tells nothing about how many moles of any substance are actually present, if any. Assume at some instant that there are n_j moles of the jth substance present; n_A moles of A, n_B moles of B, and so on.

$$\begin{array}{cccc} aA & + bB & \rightarrow cC & + dD \\ n_A & n_B & n_C & n_D \end{array}$$

As the reaction proceeds,

$$dn_j = \nu_j \, d\lambda \tag{14-8}$$

where $\nu_A = -a$, $\nu_B = -b$, $\nu_C = c$, $\nu_D = d$, and λ is the progress variable.

Hence

$$d\lambda = -\frac{dn_A}{a} = -\frac{dn_B}{b} = \frac{dn_C}{c} = \frac{dn_D}{d} \tag{14-9}$$

$$dG = \sum \mu_j \, dn_j = \sum \mu_j \nu_j \, d\lambda \tag{14-10}$$

The quantity[4]

$$\mathcal{A} = -\left(\frac{\partial G}{\partial \lambda}\right)_{T,P} = -\sum \mu_j \nu_j \tag{14-11}$$

is called the *affinity* of the reaction. If

$$\begin{aligned} &\mathcal{A} > 0 \text{ the reaction is spontaneous forward.} \\ &\mathcal{A} = 0 \text{ the reaction is at equilibrium.} \\ &\mathcal{A} < 0 \text{ the reaction is spontaneous backwards.} \end{aligned} \tag{14-12}$$

14.2 Standard States and Activities

Recall that fugacity has units of pressure. We defined an *activity* that is a ratio of fugacities and therefore dimensionless.

The ratio of the fugacity of a substance to the standard state fugacity is the activity, a_i.

$$a_i \equiv \frac{f_i}{f_i^\circ} \tag{14-4}$$

The activity varies with the fugacity, so activity depends on pressure and/or concentration. It does *not* depend explicitly on temperature; f_i and f_i° refer to the same temperature. The standard reference state for fugacity, and hence for the activity, is selected for convenience, depending on the phase of the substance.

For any substance, with standard state at 1 atm, the effect of pressure on activity can be found from volume and pressure information.

$$\int_{P=1 \text{ atm}}^{P} V\,dP = \Delta G = RT \ln \frac{f}{f^\circ} = RT \ln a \tag{14-13}$$

Solids

The fugacity of a solid is nearly independent of pressure. The pure solid, under 1 atm pressure, is chosen as the standard state. The activity of the solid is $a_i = f_i/f_i^0 = 1$.

The activity of a solid is (usually) equal to 1.

The activity of a solid is lowered if impurities are added. The activity may be raised slightly by introducing strain into the solid or by increasing the pressure.

[4]Some authors reverse the sign of $\mathcal{A}$

ANSWERS 14.2

A. To make the reaction spontaneous, $\Delta G \leq 0$, so $RT \ln \mathcal{Q} \leq -163.4$ kJ. At 25°C, $RT = 2479$ J/mol, so $\ln \mathcal{Q} \leq -65.91$, or $\mathcal{Q} \leq 2.4 \times 10^{-29}$. The ozone must be at a very low pressure.
B. By similar reasoning, $RT \ln \mathcal{Q} \leq -1570$ J and therefore $\ln \mathcal{Q} \leq -0.633$, so $\mathcal{Q} \leq 0.531$. The pressure of HI need be only slightly less than the pressure of H_2 in the presence of solid I_2.

The activity of a solid may also be affected by particle size. For very small particles, surface effects are not negligible compared to bulk properties.

Gases

For an ideal gas, fugacity is equal to pressure. Taking the standard state as the pure gas at 1 atm, we obtain

$$a = \frac{f}{f^\circ} = \frac{P(\text{atm})}{1 \text{ atm}} = |P(\text{atm})| \tag{14-14}$$

The activity of an ideal gas is equal to the numerical value of the pressure, expressed in atmospheres.

For a nonideal gas it might seem convenient to choose the standard state such that $f^\circ = 1$ atm. However, standard states are normally selected at a pressure of 1 atm, which for a nonideal gas differs from a fugacity of 1 atm. The standard state of a real gas is therefore chosen as the (fictitious) ideal gas state for which $f^\circ = P = 1$ atm. A correction factor, γ, is introduced to compensate for the nonideality of the real gas.

$$a = \frac{f}{f^\circ} = \gamma P(\text{atm}) \tag{14-15}$$

The activity coefficient[5] is usually close to 1 atm^{-1}, especially at low pressures. It is numerically equal to the ratio of the fugacity of the real gas to the fugacity of the idealized gas.

EXAMPLES 14.3

A. Find ΔG for the reaction of oxygen (at 0.20 atm) to form ozone, when the ozone is present at 1 ppm (hence 10^{-6} atm). [Assume ideal gases, or $|\gamma_i| = 1$.]
B. Find ΔG for the reaction of hydrogen gas with solid iodine to form HI if the hydrogen is at a partial pressure of 1 atm and the HI is at a partial pressure of $\frac{1}{2}$ atm. □

[5]Activity is dimensionless, so the activity coefficient has units of reciprocal pressure (or concentration, for other phases). In practice, units are sometimes neglected and the activity of a real gas is written

$$a = \gamma \, |P(\text{atm})|$$

treating γ as a dimensionless factor.

Liquids

The standard state for a liquid is the pure liquid. Therefore,

$$f_1 = f_1^\circ X_1$$

for the solvent when the solution is dilute, with f_1° the fugacity of the pure liquid solvent and X_1 the mole fraction of the solvent. Over the range for which Raoult's law is valid,

$$a_1 = \frac{f_1}{f_1^\circ} = X_1 \qquad (14\text{-}16)$$

The activity of a liquid (solvent) is equal to its mole fraction.

Outside the range of validity of Raoult's law, an activity coefficient is required.

$$a_1 = \gamma_1 X_1 \qquad (14\text{-}17)$$

Solutes

The choice of standard state for a solute is not taken as the pure solute, because it is preferable to obtain the relationship

$$a_2 = \gamma_2 c_2 \qquad (14\text{-}18)$$

with the activity coefficient, γ, numerically equal to 1 for the range of concentrations for which Henry's law is valid.

The activity of a solute is equal to the concentration (usually in molarity) in dilute solutions.

EXAMPLES 14.4

A. 3.42 g of sucrose ($C_{12}H_{22}O_{11}$, molar mass 342 g) is added to 1 liter of water. What is the activity of the sucrose?
B. What is the activity of the water in the sucrose solution? [What approximations are required?] □

The standard state for a solute may now be found by working backward. Assuming Henry's law, which will at least be true in the limit of infinite dilution,

$$f_2 = k c_2$$

Then

$$a_2 = \frac{f_2}{f_2^\circ} = \frac{k c_2}{f_2^\circ} = \gamma_2 c_2$$

ANSWERS 14.3

A. $\mathcal{Q} = \dfrac{(P(O_3))}{[P(O_2)]^{3/2}} = \dfrac{10^{-6}}{(0.20)^{3/2}} = 1.1\times10^{-5}$; $RT \ln \mathcal{Q} = -28.3$ kJ. $\Delta G^\circ =$ 163.4 kJ, so $\Delta G = 163.4 - 28.3 = 135.1$ kJ. This is not spontaneous under these conditions.

B. $\mathcal{Q} = \dfrac{(P(\mathrm{HI}))^2}{P(H_2)a(I_2)} = \dfrac{(\frac{1}{2})^2}{1 \times 1} = \dfrac{1}{4}$; $RT \ln \mathcal{Q} = -3.44$ kJ; $\Delta G^\circ = 1.57$ kJ; $\Delta G = -1.87$ kJ

ANSWERS 14.4

A. 3.42 g = 0.010 mol, so the concentration is 0.010 M and the activity is 0.010, assuming Henry's law to be valid to this concentration.
B. Assume that the partial molar volume times mole number of the sugar is negligible; then there is 1 L of water. 1000 g/(18 g/mol) = 55.56 mol. The mole fraction of the sucrose is $X_2 = 0.01/55.56 = 1.8 \times 10^{-4}$. The mole fraction of the water is $1 - X_2$, which is 0.99(982) = 1.00, so the activity is $a = 1.00$.

When $c_2 \to 0, |\gamma_2| = 1$. It follows that $f_2^\circ = k$. The standard state of the solute is the state for which the fugacity of the solute is numerically equal to the Henry's law constant of the solute (for whatever concentration units have been chosen, but usually molarity).

Deviations from Henry's law may be quite large at high concentrations. For ionized solutes, the deviations are appreciable even for dilute solutions (1 M and well below). Although activity coefficients cannot be calculated or determined exactly for individual ions, approximate values may be calculated from the theory of Debye and Hückel, discussed below.

For studies involving changes of temperature, molality is a better choice than molarity. For some purposes, concentrations are better expressed as mole fractions.

14.3 Equilibrium Constants

For certain values of the activities, ΔG is zero and the reaction is in equilibrium.

{at equilibrium} $$\Delta G = 0 = \Delta G^\circ + RT \ln \frac{a_C^c a_D^d}{a_A^a a_B^b} \tag{14-19}$$

or

$$\Delta G^\circ = -RT \ln \left(\frac{a_C^c a_D^d}{a_A^a a_B^b} \right)_{\text{at equilibrium}} \tag{14-20}$$

The standard free energy change for any reaction is a constant value, typically available from a thermodynamic table. Therefore, the argument of the logarithm, $\mathcal{Q}$, must also be a constant for all sets of activities that satisfy an equilibrium condition.

Represent this constant value of $\mathcal{Q}$, for equilibrium conditions, by K_{eq}.

$$K_{eq} \equiv \left\{ \frac{a_C^c a_D^d}{a_A^a a_B^b} \right\}_{\text{at equil}} \tag{14-21}$$

Note that this does not say that the individual activities must have specified values at equilibrium. The only requirement is that the ratio

$$\mathcal{Q} = \frac{\prod\limits_{\substack{i \\ \text{products}}} a_i}{\prod\limits_{\substack{j \\ \text{reactants}}} a_j} \tag{14-22}$$

must be equal to the constant, K_{eq}, when the reaction is in equilibrium.

EXAMPLES 14.5

A. Find the equilibrium constant for the gaseous reaction

$$\tfrac{1}{2}N_2 + \tfrac{3}{2}H_2 \rightarrow NH_3$$

from $\Delta G°$, in Table 14.1, and write the expression for $\mathcal{Q}(a)$ in terms of gas pressures.
B. Using values of ΔG_f° from Table 14.1 for the reaction

$$2NO_2 \rightarrow N_2O_4$$

find $\Delta G°$ for the reaction, and from this find the equilibrium constant. Give the expression for $\mathcal{Q}(a)$ in terms of pressures.
C. There is an interesting paradox in the defining equation for the equilibrium constant,

$$\Delta G° = -RT \ln K_{eq}$$

The left-hand side refers to standard-state conditions for reactants and products; the right-hand side refers to equilibrium conditions for reactants and products. These are *not* the same. How can the two values be equal? □

The equilibrium constant depends on how the chemical reaction is written. If the chemical equation is multiplied by an integer, the equilibrium constant is raised to the power of that integer. For a given reaction, however,

> *the equilibrium constant depends* only *on temperature (and choices of standard states). It is totally independent of other variables, such as pressure or concentrations.*

The constant derived above, $K_{eq} = \mathcal{Q}(a)_{eq}$, is sometimes called the *thermodynamic equilibrium constant* to distinguish it from the pseudoconstant, $\mathcal{Q}(c)_{equil}$. When concentrations or pressures are substituted for activities without activity coefficients, the expression is no longer rigorously constant.

14.4 Calculation of Equilibrium Concentrations

Solution of equilibrium problems typically involves solving a quadratic or cubic equation. For example, for dissociation of acetic acid (CH_3COOH, represented by HOAc),

---

ANSWERS 14.5

A. $\Delta G^\circ = -16.59 \text{ kJ} = -RT \ln K_{eq}$, so

$$K_{eq} = 8.1 \times 10^2 = \frac{P(NH_3)}{[P(N_2)]^{1/2}[P(H_2)]^{3/2}}$$

B. $\Delta G^\circ_{reaction} = \Delta G^\circ_f(N_2O_4) - 2\,\Delta G^\circ_f(NO_2) = 97.74 - 2 \times 51.25 \text{ kJ} = -4.76 \text{ kJ} = -RT \ln K_{eq}$. $K_{eq} = 6.82 = P(N_2O_4)/(P(NO_2))^2$.

C. In expanded form the equation may be written

$$(\Delta G)_{std\ st} - (\Delta G)_{equil} = RT \ln \mathcal{Q}_{std\ st} - RT \ln \mathcal{Q}_{equil}$$

By replacing $\Delta G_{std\ st}$ with ΔG°, ΔG_{equil} with 0, $\mathcal{Q}_{std\ st}$ with 1, and $\mathcal{Q}_{equil}$ with K_{eq}, the equation becomes

$$\Delta G^\circ - 0 = 0 - RT \ln K_{eq}$$

which shows that ΔG° is a measure of how far standard-state concentrations, or activities, differ from an equilibrium state, and K_{eq} is a measure of how far the equilibrium concentrations, or activities, differ from their standard-state values of activities equal to 1. The height of a well, above its floor, is equal to (minus) the depth of the well, below its top.

$$\text{HOAc} \rightarrow \text{H}^+ + {}^-\text{OAc}$$

$K_{eq} = 1.85 \times 10^{-5}$. If a solution is prepared to contain 0.10 M HOAc (before dissociation), the concentrations and therefore the activities (neglecting deviations from ideality) are given as

$$\begin{array}{ccc} \text{HOAc} \rightarrow & H^+ + & {}^-OAc \\ 0.1 - x & x & x \end{array}$$

where x is the amount (as concentration) dissociated. Therefore,

$$\frac{x^2}{0.1 - x} = 1.85 \times 10^{-5}$$

and

$$x^2 + 1.85 \times 10^{-5} x - 1.85 \times 10^{-6} = 0$$

which has the solution $x = 1.35 \times 10^{-3}$ (and -1.369×10^{-3}). Hence the concentration of undissociated acetic acid is $0.10 - 1.35 \times 10^{-3} = 0.0986(5)$ M and the concentrations of H^+ and acetate ion are each 1.35×10^{-3} M. The pH of the solution is

$$\text{pH} = -\log a_{H^+} = -\log(1.35 \times 10^{-3}) = 2.9$$

Alternatively, we could have anticipated that x would be small compared to 0.10 and solved the approximate equation

$$\frac{x^2}{0.1} = 1.85 \times 10^{-5}$$

to get $x = 1.3(6) \times 10^{-3}$. Then confirm that $x^2/(0.1 - 1.36 \times 10^{-3}) = 1.85 \times 10^{-5}$ gives $x = 1.35 \times 10^{-3}$.

The decomposition of phosphine, PH_3, to give solid white phosphorus and hydrogen provides an illustration of the general method of solution of equilibrium equations. Given an initial pressure of PH_3 of 5.0 atm, and that $\Delta G^\circ = -36.48$ kJ, the steps are as follows:

1. Write the reaction (even if it is a familiar one).

$$2PH_3 \rightarrow 2P + 3H_2$$

2. Find K_{eq} from ΔG° (if necessary).

$$\Delta G^\circ = -36.48 \text{ kJ} = -RT \ln K_{eq}$$
$$K_{eq} = 2.46 \times 10^6$$

3. Write down $\mathcal{Q}(a)$ and set it equal to K_{eq}.

$$\mathcal{Q}(a) = \frac{(a(P))^2 (a(H_2))^3}{(a(PH_3))^2} = K_{eq} = 2.46 \times 10^6$$

4. Substitute symbols indicating concentrations, pressures, or other values for activities.

$$a(P) = 1 \qquad a(H_2) = P(H_2) \qquad a(PH_3) = P(PH_3)$$
$$\mathcal{Q}(P) = \frac{(a(P))^2 (P(H_2))^3}{(P(PH_3))^2} \approx K_{eq} = 2.46 \times 10^6$$

5. Determine from the stoichiometry of the reaction and the information given what is known about concentrations or pressures and substitute.

$$2PH_3 \rightarrow 2P + 3H_2$$
$$P_o - 2x \qquad a = 1 \qquad P = 3x$$

and $P_o = 5.0$ atm, so

$$2.46 \times 10^6 = \frac{1 \cdot (3x)^3}{(5 - 2x)^2}$$

6. Solve the equation. Multiplying out the equation gives a cubic equation that would not be easily solved without a computer program. However, inspection of the equation shows that the numerator must be large compared to the denominator, so $5 - 2x$ must be small. Therefore, let $x = \frac{5}{2}$ as a starting estimate, except in the denominator.

$$2.46 \times 10^6 = \frac{(3 \cdot 2.5)^3}{(5 - 2x)^2} \quad \text{and} \quad 5 - 2x = \sqrt{\frac{(7.5)^3}{2.5 \times 10^6}} = 1.31 \times 10^{-2}$$

$$x = \frac{1}{2}(5.0 - 1.31 \times 10^{-2}) = 2.49(35)$$

Therefore,

$$P(PH_3) = 5.0 - 2x = 1.31 \times 10^{-2} \text{ atm}$$

$$P(H_2) = 3x = 7.47 \text{ atm}$$

7. Check the answer for consistency with known information. Recalculation, with $x = 2.49$ instead of 2.50, changes the value of x only in the fourth place, so the calculation is internally consistent and the numbers appear reasonable (e.g., no negative pressures or other obviously inappropriate values).

Omitting steps from the calculation routine will save some seconds of time and sharply increase the probability of error in the result.

When a slightly soluble salt dissolves in water, the activity of the solid is 1 and the equilibrium constant is the product of activities of the ions in solution. To a good approximation, the activities are equal to concentrations at these low concentrations. For example, when MgF_2 dissolves in water,

$$MgF_2 \rightarrow Mg^{2+} + 2F^-$$

and

$$K_{eq} = \frac{a_{Mg^{2+}}(a_{F^-})^2}{a_{MgF_2}} \approx C_{Mg^{2+}}(C_{F^-})^2$$

or, representing the solubility of MgF_2 by S, $K_{eq} \approx S \cdot (2S)^2$.

The equilibrium constant for a slightly soluble salt is often called the solubility product constant *and is represented by* K_{sp}.

EXAMPLES 14.6

A. For MgF_2 dissolving in water, $\Delta G° = 46$ kJ. Find $K_{eq} = K_{sp}$.
B. Find the approximate concentrations of Mg^{2+} and F^- ions, neglecting deviations of γ from 1. □

The Debye–Hückel equation considers the electrostatic interaction of charged particles in a dielectric medium. The activity coefficients depend on the *ionic strength,* μ, which is the sum of concentrations of all ions, each multiplied by the square of its charge number, with the sum multiplied by $\frac{1}{2}$.

$$\mu = \frac{1}{2}\sum_i m_i z_i^2 \tag{14-23}$$

The concentrations, m_i, are given as molality of the ith ion (not distinguishable from molarity for dilute aqueous solutions) and z_i is the number of electronic charges on the ion.

For dilute solutions, in which the ions may be considered isolated from each other, the activity coefficient of an ion is given by

$$\log \gamma_i = -Az_i^2\mu^{1/2} \tag{14-24}$$

The constant A depends on the temperature and the solvent. For water at 25°C, $A = 0.51$.

EXAMPLES 14.7

A. Find the activity coefficients of Mg^{2+} and F^- in a solution saturated with MgF_2 (for which the concentrations of Mg^{2+} and F^- were calculated above to be 1.29 and 2.58 mM and $K_{sp} = 8.65 \times 10^{-9}$). Recalculate the concentrations of the ions, including the effects of the activity coefficients.
B. Find the activity coefficients of Mg^{2+} and of F^- if 0.010 M NaCl is added to the saturated MgF_2 solution. What effect will this have on the solubility of MgF_2?
C. What will happen to the solubility of MgF_2 (measured by the concentration of Mg^{2+} ion) if 0.010 M NaF is added to the solution? □

14.5 Temperature Dependence of Equilibria

Equilibrium constants depend explicitly on the temperature, through the RT factor, and implicitly because $\Delta G°$ is temperature dependent.

$$\frac{d(\ln K)}{dT} = -\frac{d}{dT}\left(\frac{\Delta G°}{RT}\right) = -\frac{1}{RT}\left(\frac{\partial \Delta G°}{\partial T}\right)_P - \Delta G° \frac{d}{dT}\left(\frac{1}{RT}\right)$$

Recalling that

$$\left(\frac{\partial \Delta G}{\partial T}\right)_P = -\Delta S$$

and replacing $\Delta G° + T\Delta S°$ with $\Delta H°$, we find that

$$\frac{d(\ln K)}{dT} = \frac{\Delta H°}{RT^2}$$

or, after integration,

$$\ln \frac{K_2}{K_1} = \frac{\Delta H° \, \Delta T}{RT_1T_2} \tag{14-25}$$

Note the similarity of this equation to the Clausius-Clapeyron equation. The same form arises often. As will be seen in Part III, whenever two states are compared that differ in energy, an exponential dependence on temperature is to be expected.

Temperature and pressure dependence may be predicted qualitatively from Le Châtelier's principle, or equivalently from the pair of equations,

$$\left(\frac{\partial \Delta G}{\partial T}\right)_P = -\Delta S$$

and

$$\left(\frac{\partial \Delta G}{\partial P}\right)_T = \Delta V$$

ANSWERS 14.6

A. $\Delta G^\circ = -RT \ln K_{sp} = 46,000$ J;

$$K_{sp} = 8.65 \times 10^{-9} = a(Mg^{2+}) \times [a(F^-)]^2$$

B. Letting $a(Mg^{2+}) = x$, $a(F^-) = 2x$, so

$$8.65 \times 10^{-9} = x(2x)^2 = 4x^3$$

$$x = \sqrt[3]{2.14 \times 10^{-9}} = 1.29 \times 10^{-3}$$

Recall that the first says that substances tend to decompose, to produce more entities and therefore greater entropy, as the temperature increases. The second says that if a reaction would produce a greater volume of products than reactants, an increase of pressure will inhibit the reaction, tending to drive it backward. The equilibrium constant varies with temperature but not with pressure.

14.6 Electrochemistry

The change in Gibbs free energy at constant temperature and pressure is

{constant T, P} $$dG = w + P\,dV = w'$$

or (14-26)

{constant T, P} $$\Delta G = W'$$

where W' is any work other than $\int P\,dV$ work. In particular, if electrical work is being done by the system,

$$W' = -\mathcal{E}q = -\mathcal{E}It \quad (14\text{–}27)$$

One mole of electrons (transferred) is called an *equivalent*. The charge of one mole of electrons is called a *faraday*, with symbol $\mathcal{F}$.

$$\textit{One}\text{ faraday} = 1\mathcal{F} = N_A e = 6.022 \times 10^{23} \times 1.6022 \times 10^{-19} = 96,485 \text{ C}.$$

The amount of charge, q, transferred in an electrochemical reaction is therefore 96,485 C times the number of equivalents, n.

$$q = n\mathcal{F} \quad (14\text{-}28)$$

and

$$\Delta G = -n\mathcal{F}\mathcal{E} \quad (14\text{-}29)$$

Therefore, the free energy change of a chemical reaction can be measured very accurately by electrochemical measurement of the amount of charge transferred, $q = n\mathcal{F}$, and the electric potential, $\mathcal{E}$.

A positive cell potential, $\mathcal{E}$, corresponds to a spontaneous reaction.

ANSWERS 14.7

A. The ionic strength (assuming concentration values from example 14.6) is

$$\mu = \tfrac{1}{2}[4c + 1 \times (2c)] = 3c = 3 \times 1.29 \times 10^{-3} = 3.87 \times 10^{-3}$$

For the magnesium ion,

$$\log \gamma = -0.51 \times (2)^2(3.87 \times 10^{-3})^{1/2} = -0.1268 \quad \text{and} \quad \gamma = 0.74(68)$$

For the fluoride ion,

$$\log \gamma = -0.51 \times (1)(3.87 \times 10^{-3})^{1/2} = -3.171 \times 10^{-2} \quad \text{and} \quad \gamma = 0.92(96)$$

Inserting these values of γ_i into the equation for solubility makes that equation

$$K_{sp} = 8.65 \times 10^{-9} = 0.75c(Mg^{2+}) \times [0.93c(F^-)]^2$$

Again letting $x = c(Mg^{2+})$, we find that

$$8.65 \times 10^{-9} = 0.6487(4x^3)$$

and

$$x = 1.4(94) \times 10^{-3}$$

This is slightly higher than the original calculation but not sufficiently different to warrant recalculation of the activity coefficients.
B. If 0.010 *M* NaCl is added to the solution, the ionic strength is increased. Neglecting the concentration of ions from the MgF_2, we obtain

$$\mu = \tfrac{1}{2}[(1)^2(0.010) + (1)^2(0.010)] = 0.010$$

and therefore $\mu^{1/2} = 0.10$ and

$$\log \gamma_+ = -0.51 \times (2)^2 0.10 = -0.204$$

$$\gamma_+ = 0.6252$$

$$\log \gamma_- = -0.51 \times (1)^2 0.10 = -0.051$$

$$\gamma_- = 0.8892$$

$$K_{sp} = 8.65 \times 10^{-9} = (0.6252x)(0.8892 \cdot 2x)^2$$

so $x^3 = 4.37 \times 10^{-9}$ and $x = 1.64 \times 10^{-3}$. The effect of the increased ionic strength is to increase the solubility of the MgF_2 slightly.
C. Adding NaF increases the concentration of fluoride ion.

$$MgF_2 \rightarrow Mg^{2+} + 2F^-$$

$$a = 1 \qquad x \qquad 0.010 + 2x$$

If we neglect the $2x$ in comparison to the 0.010, the equation becomes

$$8.65 \times 10^{-9} = 0.6252x[0.8892(0.010)]^2 = 4.94 \times 10^{-5}x$$

so the solubility decreases to

$$x = 1.75 \times 10^{-4} M$$

This effect is called *salting out* one compound by adding a *common ion*.

The entropy change of a reaction is similarly readily determined by measuring the temperature dependence of the potential. From

$$\left(\frac{\partial \Delta G}{\partial T}\right)_P = -\Delta S \qquad \text{and} \qquad \Delta G = -n\mathscr{F}\mathscr{E}$$

we obtain

$$\Delta S = -\left(\frac{\partial \Delta G}{\partial T}\right)_P = n\mathscr{F}\left(\frac{\partial \mathscr{E}}{\partial T}\right)_P$$

Then, from $\Delta G = \Delta H - T\,\Delta S$,

$$\Delta H = -n\mathscr{F}\mathscr{E} + n\mathscr{F}T\left(\frac{\partial \mathscr{E}}{\partial T}\right)_P$$

Summarizing, we obtain

$$\begin{aligned} \Delta G &= -n\mathscr{F}\mathscr{E} \\ \Delta S &= n\mathscr{F}\left(\frac{\partial \mathscr{E}}{\partial T}\right)_P \\ \Delta H &= -n\mathscr{F}\left[\mathscr{E} - T\left(\frac{\partial \mathscr{E}}{\partial T}\right)_P\right] \end{aligned} \tag{14-30}$$

Electrochemical measurements are also directly applicable to equilibrium studies. Free energy is related to activities,

$$\Delta G = \Delta G^\circ + RT \ln \mathscr{Q}(a)$$

and

$$\Delta G = -n\mathscr{F}\mathscr{E} \qquad \text{and} \qquad \Delta G^\circ = -n\mathscr{F}\mathscr{E}^\circ$$

The electrochemical potential for any choice of activities of reactants and products is

$$\mathscr{E} = \mathscr{E}^\circ - \frac{RT}{n\mathscr{F}} \ln \mathscr{Q}(a) \tag{14-31}$$

This is called the *Nernst equation*.

From $\Delta G^\circ = -RT \ln K_{eq}$ and $\Delta G^\circ = -n\mathscr{F}\mathscr{E}^\circ$ we find that

$$\mathscr{E}^\circ = \frac{RT}{n\mathscr{F}} \ln K_{eq} \tag{14-32}$$

Remember that $\Delta G = -n\mathscr{F}\mathscr{E} \leq 0$ at equilibrium, so we cannot set $\Delta G = 0$ (or $\mathscr{E} = 0$) when electrical work is being done.

Cell Potentials and Half-Cell, or Electrode, Potentials

For a device that "uses" electric power, the *cathode* is the negative electrode. For a cell that produces electric power, the cathode is the positive electrode.

The cathode of any electrical device is the electrode at which electrons enter the device.

For a device that uses electric power, the *anode* is the positive electrode. For a device that produces electric power, the anode is the negative electrode. The anode is the electrode at which electrons leave a device.

When zinc is burned in oxygen, the reaction is

$$Zn + \tfrac{1}{2}O_2 \rightarrow Zn^{2+}O^{2-}$$

or

$$Zn \rightarrow Zn^{2+} + 2e$$

$$2e + \tfrac{1}{2}O_2 \rightarrow O^{2-}$$

The zinc is *oxidized*. The oxygen is *reduced*.

When an atom or ion loses electrons, it is said to be oxidized. *When an atom or ion gains electrons, it is said to be* reduced.

The reverse process, in which zinc is changed to the metal, would be

$$Zn^{2+}O^{2-} \rightarrow Zn + \tfrac{1}{2}O_2$$

or

$$Zn^{2+} + 2e \rightarrow Zn$$

$$O^{2-} \rightarrow \tfrac{1}{2}O_2 + 2e$$

When one species is oxidized, losing electrons, some other species must be reduced, taking up those electrons. Such reactions are called *oxidation-reduction* reactions, often abbreviated to *redox reactions*.

It follows from the definitions above that when there is reduction, it must occur at the cathode (where electrons are being supplied); oxidation occurs at the anode, supplying electrons.

EXAMPLES 14.8

In the common dry cell, one electrode is the zinc metal case (often enclosed in steel), the other is a carbon rod through the center.

A. Which is the anode and which the cathode?
B. Which is negative and which positive?
C. Where does oxidation occur? □

The electrochemical potential for a cell reaction can be measured very accurately. The number of possible cell reactions is so great, however, that it is helpful to divide a cell reaction into two parts, one part occurring at the anode, the other part at the cathode. These are called *half-cell reactions* or *electrode reactions*.

ANSWERS 14.8

A. The zinc case is the anode; carbon is the cathode (Fig. 14.1).
B. The zinc case is the negative electrode; the carbon rod is positive.
C. Zinc is oxidized to Zn^{2+}. (At the carbon rod, manganese from MnO_2 is reduced to Mn^{3+}.)

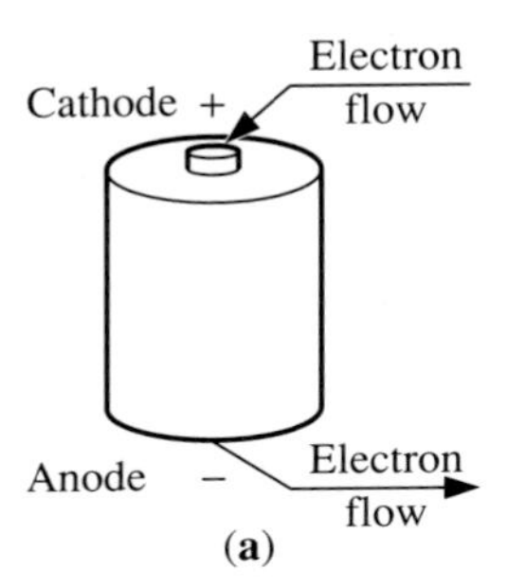

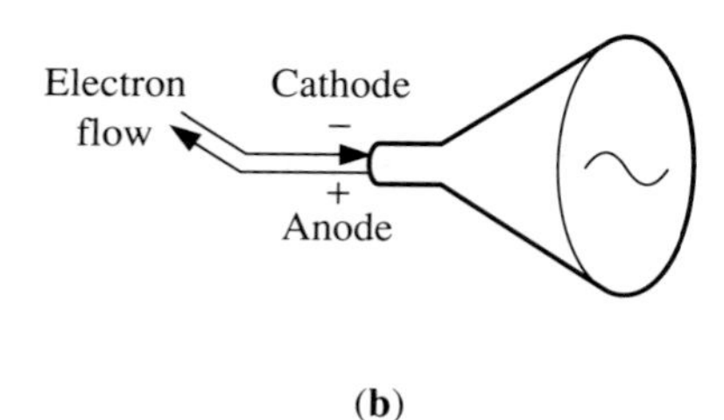

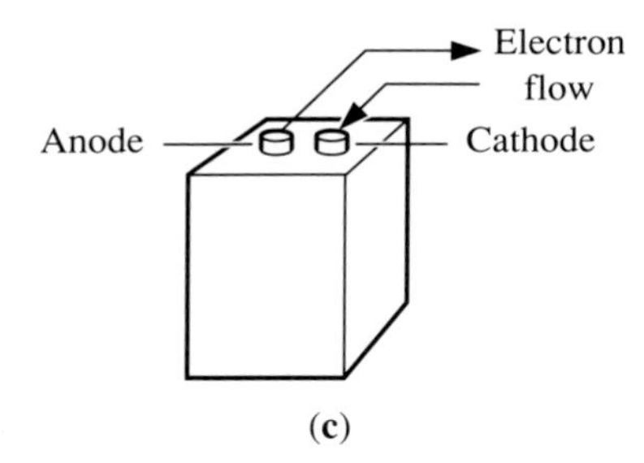

FIGURE 14.1 (a) Electrons leave a dry cell at the electrode that is at a negative potential; this is called the *anode*. (b) Electrons enter a cathode-ray tube at the terminal connected to a negative potential; this terminal is called the *cathode*. (c) For an arbitrary "black box" device, electrons enter at the cathode and leave at the anode; the sign of the potentials at the electrodes depends on whether the device "produces" or "uses" electric power.

It is not possible to carry out one of the electrode reactions without the other, so the potential of an individual half-cell reaction cannot be measured. Therefore, a value is arbitrarily assigned to one half-cell reaction. All other half-cell potentials are measured relative to the assigned half-cell potential.

The standard electrode potential, $\mathcal{E}°$, for

$$2H^+ + 2e \rightarrow H_2$$

is conventionally assigned the value of zero volts.

The cell reaction

$$2H^+ + 2Cl^- \rightarrow H_2 + Cl_2$$

consists of the two half-cell reactions:

$$\text{At the cathode: } 2H^+ + 2e \rightarrow H_2 \qquad \mathcal{E}^\circ = 0.000 \text{ V}$$

$$\text{At the anode: } 2Cl^- \rightarrow Cl_2 + 2e \qquad \mathcal{E}^\circ = \mathcal{E}_A^\circ$$

and $\mathcal{E}^\circ = \mathcal{E}_C^\circ + \mathcal{E}_A^\circ = -1.359$ V. Hence the value -1.359 V is assigned to this anode half-cell reaction. Other half-cell reactions may then be combined with either of these to obtain additional assignments. Some values of half-cell potentials are listed in Table 14.2.

Electrochemical cell reactions are often written in abbreviated form. The reaction

$$Zn + Cu^{2+} \rightarrow Zn^{2+} + Cu$$

would be written

$$Zn/Zn^{2+} \;//\; Cu^{2+}/Cu$$

Table 14.2 Standard Electrode Potentials, 25°C[a]

Reaction	$\mathcal{E}^\circ$	Reaction	$\mathcal{E}^\circ$
$Li^+ + e \rightarrow Li$	−3.045	$Cu^{2+} + e \rightarrow Cu^+$	0.153
$K^+ + e \rightarrow K$	−2.925	$SO_4^{2-} + 4H^+ + 2e \rightarrow H_2SO_3 + H_2O$	0.20
$Na^+ + e \rightarrow Na$	−2.71	$AgCl + e \rightarrow Ag + Cl^-$	0.222
$Mg^{2+} + 2e \rightarrow Mg$	−2.37	$Hg_2Cl_2 + 2e \rightarrow 2Hg + 2Cl^-$	0.2677
$Al^{3+} + 3e \rightarrow Al$	−1.67	$Cu^{2+} + 2e \rightarrow Cu$	0.337
$Mn(OH)_2 + 2e \rightarrow Mn + 2OH^-$	−1.47	$Cu^+ + e \rightarrow Cu$	0.522
$Zn^{2+} + 2e \rightarrow Zn$	−0.763	$I_2 + 2e \rightarrow 2I^-$	0.534
$Cr^{2+} + 2e \rightarrow Cr$	−0.74	$MnO_4^- + e \rightarrow MnO_4^{2-}$	0.54
$2CO_2(g) + 2H^+ + 2e \rightarrow H_2C_2O_4(aq)$	−0.49	$MnO_4^- + 2H_2O + 3e \rightarrow MnO_2 + 4OH^-$	0.57
$Fe^{2+} + 2e \rightarrow Fe$	−0.440	$O_2 + 2H^+ + 2e \rightarrow H_2O_2$	0.682
$Cr^{3+} + e \rightarrow Cr^{2+}$	−0.41	$Fe^{3+} + e \rightarrow Fe^{2+}$	0.771
$Cd^{2+} + 2e \rightarrow Cd$	−0.402	$Hg_2^{2+} + 2e \rightarrow 2Hg$	0.789
$AgI + e \rightarrow Ag + I^-$	−0.151	$Ag^+ + e \rightarrow Ag$	0.7991
$Sn^{2+} + 2e \rightarrow Sn$	−0.136	$Hg^{2+} + 2e \rightarrow Hg$	0.854
$Pb^{2+} + 2e \rightarrow Pb$	−0.126	$2Hg^{2+} + 2e \rightarrow Hg_2^{2+}$	0.919
$O_2 + H_2O + 2e \rightarrow HO_2^- + OH^-$	−0.076	$NO_3^- + 4H^+ + 3e \rightarrow NO + 2H_2O$	0.96
$Cu(NH_3)_4^{2+} + 2e \rightarrow Cu + 4NH_3$	−0.05	$Cr_2O_7^{2-} + 14H^+ + 6e \rightarrow 2Cr^{3+} + 7H_2O$	1.33
$Fe^{3+} + 3e \rightarrow Fe$	−0.036	$Cl_2 + 2e \rightarrow 2Cl^-$	1.359
$2H^+ + 2e \rightarrow H_2$	0.0000	$MnO_4^- + 8H^+ + 5e \rightarrow Mn^{2+} + 4H_2O$	1.52
$AgBr + e \rightarrow Ag + Br^-$	0.073	$Ce^{4+} + e \rightarrow Ce^{3+}$	1.61
$Hg_2Br_2 + 2e \rightarrow 2Hg + 2Br^-$	0.14	$MnO_4^- + 4H^+ + 3e \rightarrow MnO_2 + 2H_2O$	1.67
$Sn^{4+} + 2e \rightarrow Sn^{2+}$	0.15	$H_2O_2 + 2H^+ + 2e \rightarrow 2H_2O$	1.77

[a] Standard reduction potentials in volts. *Standard potentials* may be given either as reduction potentials or as oxidation potentials (with opposite direction, and sign). *Standard electrode potentials* are reduction potentials.

indicating that a zinc anode is in contact with a zinc ion solution and a copper cathode is in contact with a copper ion solution. The two solutions must be electrically connected by a *salt bridge*, represented by the symbol //. In this notational system, the anode is *always* written on the left.

Anode Left *W A Y S*

EXAMPLES 14.9

A. From the *standard electrode potentials* (i.e., standard reduction potentials) given in Table 14.2, find $\mathcal{E}^\circ$ for the silver/silver chloride cell,

$$Ag\ //\ HCl\ //\ AgCl\ //\ H_2, Pt$$

B. Find $\mathcal{E}^\circ$ for the cadmium/calomel cell,

$$Cd\ //\ CdCl_2\ //\ Hg_2Cl_2\ //\ Hg$$

C. Find $\mathcal{E}^o$ for the cell

$$Hg\ //\ Hg_2Cl_2\ //\ CdCl_2\ //\ Cd$$

□

Addition of cell reactions provides an apparent paradox that is illuminating. First, we know that the electrochemical potential cannot depend on the amount of a reaction. (The voltage produced by a dry cell does not double if you keep a flashlight on twice as long.) Therefore, $\mathcal{E}$ and $\mathcal{E}^\circ$ must be the same for

$$Zn \rightarrow Zn^{2+} + 2e \qquad \mathcal{E}^\circ = 0.763V$$

as for

$$\tfrac{1}{2}Zn \rightarrow \tfrac{1}{2}Zn^{2+} + e \qquad \mathcal{E}^\circ = 0.763V$$

Furthermore, cell potentials are additive when half-cell reactions are added together.

$$\begin{aligned} Cl_2 + 2e &\rightarrow 2Cl^- & \mathcal{E}^\circ &= 1.359V \\ Zn &\rightarrow Zn^{2+} + 2e & \mathcal{E}^\circ &= 0.763V \\ Zn + Cl_2 &\rightarrow ZnCl_2 & \mathcal{E}^\circ &= 2.122V \end{aligned}$$

On the other hand, the following values from Table 14.2 seem *not* to be additive.

$$\begin{aligned} Fe &\rightarrow Fe^{2+} + 2e & \mathcal{E}^\circ &= 0.440V \\ Fe^{2+} &\rightarrow Fe^{3+} + e & \mathcal{E}^\circ &= -0.771V \\ Fe &\rightarrow Fe^{3+} + 3e & \mathcal{E}^\circ &= 0.036V \end{aligned}$$

and $0.036\ V \neq 0.440 + (-0.771) = -0.331\ V$.

The problem arises because it was implicitly assumed that cell potentials should be additive. Cell potentials are not state functions, so that there is no inherent reason why they should add. Kirchhoff's law and Hess's law apply to state functions only.

Free energy is a state function and therefore additive. Rewrite the three equations to show free-energy changes.

$$Fe \rightarrow Fe^{2+} + 2e \qquad \Delta G_1 = -n_1 \mathcal{F} \mathcal{E}_1$$
$$Fe^{2+} \rightarrow Fe^{3+} + e \qquad \Delta G_2 = -n_2 \mathcal{F} \mathcal{E}_2$$
$$Fe \rightarrow Fe^{3+} + 3e \qquad \Delta G_3 = -n_3 \mathcal{F} \mathcal{E}_3$$

Then, by Hess's law,

$$\Delta G_3 = \Delta G_1 + \Delta G_2$$
$$-n_3 \mathcal{F} \mathcal{E}_3 = -n_1 \mathcal{F} \mathcal{E}_1 - n_2 \mathcal{F} \mathcal{E}_2$$

and, as expected,

$$-3(0.036) = -2(0.440) - 1(-0.771)$$

Half-cell potentials may be safely added to get cell reactions because electrical neutrality requires that n, the number of equivalents, is the same for each half-cell reaction that is added. Adding half-cell reactions, with constant n, is equivalent to adding ΔG values.

A more general approach to electrochemical processes should take into consideration actual electrical potentials and charge motions at points within the circuit. The instantaneous potentials on the anode and cathode are ϕ_A and ϕ_C. Then, for no reaction ($d\lambda = 0$),

$$d\mathcal{E}_{\text{cell}} = d\mathcal{E}_{\text{electrostatic}} = \phi_A \, d\mathcal{q}_A + \phi_C \, d\mathcal{q}_C = w' \tag{14-33}$$

Let $d\mathcal{q}_o$ be the (positive) charge transferred, from cathode to anode, outside the cell, and let $d\mathcal{q}_i$ be the (positive) charge transferred, from anode to cathode, inside the cell. Then, quite generally,

$$d\mathcal{q}_C = d\mathcal{q}_i - d\mathcal{q}_o = -d\mathcal{q}_A \tag{14-34}$$

However, under the usual operating conditions of a cell, $d\mathcal{q}_i = d\mathcal{q}_o$ and $d\mathcal{q}_A = 0 = d\mathcal{q}_C$.

If the electrochemical process is reversible,

$$\phi_C - \phi_A = \mathcal{E} \tag{14-35}$$

the electrochemical potential of the cell (which is concentration- and temperature-dependent). For steady-state, reversible operation,

$$w'_{\text{rev}} = dG = -\mathcal{A} \, d\lambda$$

More generally,

$$w' = -\mathcal{A} \, d\lambda + \phi_A \, d\mathcal{q}_A + \phi_C \, d\mathcal{q}_C \tag{14-36}$$

ANSWERS 14.9

A. For $Ag + Cl^- \rightarrow AgCl + e$, $\mathcal{E}^\circ = -0.222$ V and for $HCl + e \rightarrow \frac{1}{2} H_2 + Cl^-$, $\mathcal{E}^\circ = 0.000$ V. Thus, for the cell, $\mathcal{E}^\circ = -0.222$ V. [Note that this is not the same as for the cell reaction $Ag \rightarrow Ag^+ + e$. The concentration of Ag^+ ion is much lower in the presence of Cl^-, so the standard-state concentrations are very different.]
B. For $Cd \rightarrow Cd^{++} + 2e$, $\mathcal{E}^\circ = 0.4020$ V, and for $Hg_2Cl_2 + 2e \rightarrow 2Cl^- + 2Hg$, $\mathcal{E}^\circ = 0.2677$ V, so for the cell, $\mathcal{E} = 0.402 + 0.2677 = 0.6697$ V.
C. The reaction shown is the reverse of the previous reaction, with Hg as the anode and Cd as the cathode. Reversing the reaction changes the sign of $\mathcal{E}$ and $\mathcal{E}^\circ$, so $\mathcal{E} = -0.6697$ V.

Let n be the number of equivalents (i.e., moles of electrons) transferred by the reaction per unit of reaction, defined by λ. Then the charge transferred internally is

$$d\mathcal{q}_i = \frac{dn}{d\lambda}\mathcal{F}\,d\lambda$$

As long as there is a complete circuit,

$$d\mathcal{q}_C = d\mathcal{q}_i - d\mathcal{q}_o = -d\mathcal{q}_A$$

and therefore

$$w'_{rev} = -\mathcal{A}\,d\lambda - \phi_A(d\mathcal{q}_i - d\mathcal{q}_o) + \phi_C(d\mathcal{q}_i - d\mathcal{q}_o) \leq (\phi_A - \phi_C)d\mathcal{q}_o = w'$$

Subtracting terms involving external current gives

$$-\mathcal{A}\,d\lambda - (\phi_A - \phi_C)\,d\mathcal{q}_i \leq 0$$

But $\phi_C - \phi_A = \mathcal{E}$ for reversible operation, so

{reversible} $$\mathcal{E} = \mathcal{A}\left(\frac{d\lambda}{d\mathcal{q}_i}\right) = \frac{\mathcal{A}}{(dn/d\lambda)\mathcal{F}} \qquad (14\text{-}37)$$

or

{reversible} $$dG = -\frac{dn}{d\lambda}\mathcal{F}\mathcal{E}\,d\lambda \qquad (14\text{-}38)$$

SUMMARY

For the representative chemical reaction,

$$a\text{A} + b\text{B} \rightarrow c\text{C} + d\text{D}$$

$$\Delta G = \Delta G^\circ + RT \ln \frac{a_C^c a_D^d}{a_A^a a_B^b} = \Delta G^\circ + RT \ln \mathcal{Q}$$

$\mathcal{Q}$ measures the deviation of the substances from their standard states. At equilibrium,

$$\mathcal{Q}_{\text{equil}} \equiv \left(\frac{a_C^c a_D^d}{a_A^a a_B^b} \right)_{\text{equil}} = e^{-\Delta G^\circ / RT} = K_{\text{equil}}$$

ΔG°, and therefore K_{eq}, depends only on the reaction and on temperature (and on choice of standard states). $\Delta G^\circ = -RT \ln K_{\text{eq}}$.

The *activity* is $a_i \equiv f_i / f_i^\circ$. An activity coefficient may be defined as $\gamma_i = a_i / c_i$ or $\gamma_i = a_i / P_i$. Often $| \gamma_i | \approx 1$, but for ions, γ may be significantly less than one. It may be estimated by the Debye–Hückel approximation,

$$\log \gamma_i = -A z_i^2 \mu_i; \quad \mu_i = \tfrac{1}{2} \sum m_i z_i^2$$

(m_i = molality, z_i = charge, $A = 0.51$ for water, at 25°C.)

Temperature dependence of equilibrium constants is given by

$$\frac{\partial \ln K}{\partial T} = \frac{\Delta H^\circ}{RT^2}$$

Electrochemistry: $\Delta G = -n\mathcal{F}\mathcal{E}$ and $\Delta G^\circ = -n\mathcal{F}\mathcal{E}^\circ$.

$$\Delta S = -\left(\frac{\partial \Delta G}{\partial T} \right)_P = n\mathcal{F} \left(\frac{\partial \mathcal{E}}{\partial T} \right)_P .$$

Nernst equation: $\mathcal{E} = \mathcal{E}^\circ - \dfrac{RT}{n\mathcal{F}} \ln \mathcal{Q}$.

The *cathode* is the electrode where electrons enter a device.

Cations (positive) go to the cathode; anions to the anode.

Oxidation occurs (if at all) at the anode; reduction at the cathode.

The short-hand notation for cell reactions puts the anode on the left.

QUESTIONS

14.1. Knowing that $\mathcal{E}$ should not depend on n, why does n appear explicitly in the Nernst equation?

$$\mathcal{E} = \mathcal{E}^\circ - \frac{RT}{n\mathcal{F}} \ln \mathcal{Q}(a)$$

14.2. For all cells, whether they produce electric power or are driven by an external voltage source, a useful generalization is:

Cations move toward the cathode; anions toward the anode.

Cations are positive ions; anions are negative ions. But the cathode is sometimes positive and sometimes negative. Why should the direction of motion be consistent? [Consider the causes of the motion, which are not necessarily the same.]

14.3. An *absolute activity*, λ, is sometimes defined by the equation $\lambda \equiv e^{\mu / kT}$ or equivalently, $\mu = kT \ln \lambda$. (This λ should not be confused with the parameter that measures progress of a process.) How is λ related to the activity, a?

PROBLEMS

14.1. For the reaction

$$CO + \frac{1}{2}O_2 \rightarrow CO_2$$

a. Estimate the value of ΔS at room temperature. Explain the basis of your estimate.
b. How would you expect the equilibrium point of this reaction to shift if the temperature is increased? Explain.
c. How would you expect the equilibrium point of this reaction to shift if the pressure is increased? Explain.

14.2. For the reaction $CO + 2H_2 \rightarrow CH_3OH$, $\Delta G° = -13.47$ kJ at 700 K. At this temperature, find the percentage decomposition of methanol if the initial pressure of (pure) methanol is 5 atm.

14.3. **a.** Find ΔH, ΔG, and ΔS for each of the following reactions at room temperature.

$$4FeO + O_2 \rightarrow 2Fe_2O_3$$

$$2Fe_3O_4 + \frac{1}{2}O_2 \rightarrow 3Fe_2O_3$$

$$3FeO + \frac{1}{2}O_2 \rightarrow Fe_3O_4$$

b. Which form of iron oxide, FeO, Fe_3O_4, or Fe_2O_3, is most stable (in the presence of an oxygen atmosphere) at this temperature? Explain.
c. How would raising the temperature change the relative stabilities (on the basis of the above results)? Explain.

14.4. For the reaction at 25°C

$$3H_2(1\text{ atm}) + N_2(1\text{ atm}) \rightarrow 2NH_3(1\text{ atm})$$

a. Find ΔH.
b. Find ΔE.
c. Find ΔG.
d. Find ΔS.
e. Is the reaction as written, at constant T and P, spontaneous?
f. Does the reaction as written take up or give off thermal energy?
g. Find the pressure, P, of NH_3 for which $\Delta G = 0$.

14.5. What electrochemical measurements would you make to determine ΔH, ΔS, and ΔG of a chemical reaction? Show how you could apply the measured values to obtain the thermodynamic quantities required.

14.6. The standard electrode potential of the silver/silver ion electrode is 0.7996 V. The standard potential of the silver iodide electrode is -0.1519 V. Find the solubility product of silver iodide.

14.7. A scientifically and commercially important reaction, by which solar cell silicon can be produced, is

$$SiH_4 \rightarrow Si + 2H_2$$

The standard free energy of formation of SiH_4 is -39.3 kJ/mol. Ignoring other possible products (such as SiH_3, SiH_3^+, etc.), how much H_2 would you expect to find, at equilibrium, if the pressure of SiH_4 is 0.10 atm?

14.8. From tabulated values for the standard electrode potentials of the half-cell reactions at 25°C for the reaction

$$2Fe^{2+} + Hg_2^{2+} \rightarrow 2Hg + 2Fe^{3+}$$

find

a. $\mathscr{E}°$.

b. The equilibrium constant.

14.9. For the electrochemical reaction

$$Zn(s) + 2AgCl(s) \rightarrow ZnCl_2(0.555\ m) + 2Ag(s)$$

where (s) indicates a solid and m indicates molality, the measured cell potential is $\mathscr{E} = 1.015$ V and $(\partial\mathscr{E}/\partial T)_P = -4.02 \times 10^{-4}$ V/K (at 0°C). For this reaction, find

a. ΔG.

b. ΔS.

PART **III**

Statistical Mechanics: Thermodynamics at the Molecular Level

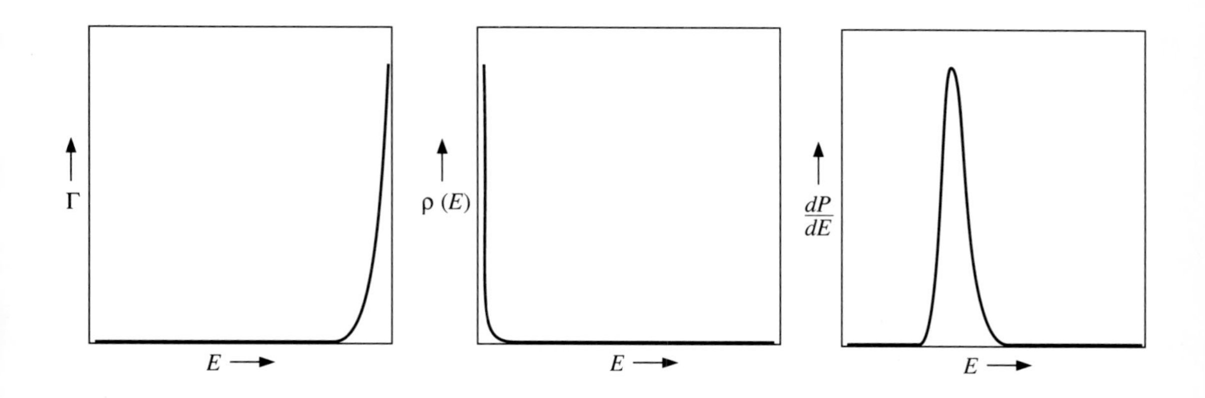

CHAPTER 15

Statistics and Probabilities

Classical thermodynamics avoids entanglement in details that depend on the structure of matter. It is not necessary to have a proper model of the molecular structure of a substance to apply thermodynamic arguments. Properties of individual substances, such as heat capacities, heats of fusion and vaporization, and Joule–Thomson coefficients, are regarded as empirical constants, to be obtained by experiment and applied to the thermodynamic calculations.

Classical thermodynamics is accepting and nonjudgmental concerning the value of such parameters; it does not attempt to tell us whether the values are reasonable or unreasonable. We have seen that this feature is responsible for the reliability and the simplicity of thermodynamics.

The model independence of classical thermodynamics is, at the same time, an important limitation, for it provides no check on our molecular models; it cannot tell us whether they are correct or not. A more detailed theory is required to examine how macroscopic properties depend on molecular properties. That theory is called *statistical mechanics*.

15.1 Rationale for a Statistical Approach

Although statistical mechanics is based on classical mechanics and/or quantum mechanics, it differs in major respects. Rather than following precise motions of individual atoms or molecules, or summing over quantum states occupied by individual particles, statistical mechanics seeks only a probabilistic interpretation, relying on the large numbers of particles present and on statistical arguments of what outcomes are most likely.

The statistical, or probabilistic, approach is necessary. Consider the difficulties of dealing with a small, simple system, such as 1 g of helium gas in the approximation that it is an ideal gas. The equations of motion are those of N point particles (no rotations or internal degrees of freedom) undergoing elastic collisions. However, 1 g (He) $= \frac{1}{4}$ mol, so N is 1.5×10^{23}. For each particle there are three position coordinates and three velocity or momentum coordinates. Hence there are 9×10^{23} variables and 9×10^{23} simultaneous equations of motion.

Two basic difficulties are immediately apparent. First, it would be necessary to supply 9×10^{23} initial values for the variables. We could not put them into our computer by hand. Even at a rate of one number (e.g., six characters, to obtain any semblance of accuracy) per second, it would require nearly 10^{24} seconds to enter the numbers. The age of the universe is no more than 10^{18} seconds.

Second, there is no existing computer, or array of computers, that could handle 9×10^{23} numbers, much less one that could handle that many simultaneous equations. Even if such a computer existed, it would take a very long time to solve the equations. For example, at 10^{10} operations per second (a rough limit to computer speed if each byte must travel 1 cm at the speed of light) it would require 9 million years just to read the initial values from memory. By contrast, the variables in the helium sample would all have changed significantly within a microsecond or less. The computer would be totally useless for predictions even if the computations could be performed and the results somehow read out and analyzed.

The problems are compounded by quantum limitations. Consider a far more restricted sample of 15 billiard balls undergoing elastic collisions on the surface of a table that absorbs no energy. The minimum uncertainty in the initial positions and velocities (speeds and directions) of the 15 hard spheres is given by Heisenberg's inequality. Because the initial conditions are slightly uncertain, the velocities after each collision are increasingly uncertain. After as few as 10 collisions of the balls, the uncertainties are so great that effectively nothing is known or predictable about where the balls are or where they are going. For 9×10^{23} variables, the point at which uncertainty would overtake available information would come at least as quickly.

It might appear that the decision to deal with statistics and probabilities is fundamentally a "sour grapes" approach. Perhaps because we cannot get exact information on the motions of atoms and molecules, we tell ourselves we did not really want that information in the first place. Actually, the justification for choosing the statistical approach is far better than that.

The kind of information about the sample of helium that we would find useful is not, for example, where some atom named Ann is at this moment, but values for variables such as volume, pressure, enthalpy, entropy, free energy, heat capacity, and Henry's law constants, as well as the average energy per atom, which provides the temperature. At equilibrium, each of these is constant, and therefore totally independent of the furious comings and goings of Ann and Bill and Carol and Doug and the other helium atoms. The information inaccessible to us, by limitations of computers and measurements, is quite irrelevant to our search for meaningful descriptions of the equilibrium state of the helium.

Are statistical arguments sufficiently accurate? A calculation of the entropy of 1 g of helium gas gives the value of 31.5 J/K at room temperature. The probability that the entropy might fluctuate from this value by 3×10^{-5} J/K, or one part in a million, is[1] about $10^{-10^{18}}$.

[1]As shown in Section 16.3, $S = k \ln P$, where P is a measure of probability, k is Boltzmann's constant, and P and S are maximum values at equilibrium. Hence $S - S' = k \ln(P/P')$, or $P'/P = e^{-(S-S')/k}$, so if $S - S' = 3 \times 10^{-5}$ J/K, $P'/P = e^{-3\times10^{-5}/1.38\times10^{-23}} = e^{-2.17\times10^{18}} \approx 10^{-(10^{18})}$. To quote W. J. Moore, "It is evident, therefore, that anyone who sees a book flying spontaneously into the air is dealing with a poltergeist and not an entropy fluctuation (probably!)."

15.2 Probabilities and Ensembles

It is often possible to assign an inherent probability, or a priori probability (i.e., "before all else"), to certain events. A proper toss of a proper coin gives 50% probability of heads and 50% probability of tails. (An unlikely "edge" landing would be eliminated from consideration and the coin retossed.) A properly shuffled and dealt deck of cards gives a probability of $\frac{1}{4}$ for any card being a spade and a probability of $\frac{1}{13}$ of any card being an ace.

These probabilities are *not* predictions of what will happen in a real experiment. If a coin is tossed 10 times, the probability of getting 5 heads and 5 tails (in any order) is less than 25%. Bridge hands are seldom evenly matched by suit or by point value. Even for very large numbers of events or chances, the discrepancy between the "expected" number of heads and the experimental number will usually be very large.

For large numbers of random events, the deviation from an average outcome typically increases in proportion to the square root of the number of events. (This is the dilemma of the gambler who relies on the "law of averages.") It is only the *percentage* deviation that becomes small for large numbers.

> *The greater the number of events or chances, the* greater *will be the expected sum of deviations from the theoretical prediction, but the* smaller *will be the percentage deviation between theory and experiment.*

Sometimes the language can become very cumbersome if distinctions must be made between real experiments and the elementary theory of statistics for large numbers. What is meant when we ask: What is the probability of getting equal numbers of heads and tails when a coin is tossed 10 times? Whatever our answer, it is not likely to agree with an experiment in which a coin is tossed 10 times.

The difficulties are largely removed by introducing an *ensemble* of systems, or tests. Rather than letting one person toss one coin 10 times, let that person repeat the same experiment (tossing the coin 10 times) M times, or let M people each toss one coin 10 times, then average the results. All M people and coins are postulated to be equivalent for this limited purpose, so the average of the M sets of tosses should be the same, whether consecutive or simultaneous.[2]

Let M, the number of equivalent samples contributing to the ensemble, approach infinity. Then the expected averages approach the predictions of elementary theory. Turning this around, we specify the meaning of probability in terms of the ensemble.

> *Probability theory predicts average values obtained from an ensemble of M equivalent systems or experiments, where M is very large, approaching infinity.*

Wherever we speak of probability, it will mean the expected average value obtained from such a large ensemble. It will sometimes be called a *background ensemble* to emphasize the distinction from the relatively small number of actual trials under consideration. Results from a single system or test may or may not agree with the expectations for the background ensemble.

[2]The ensemble of M tossers and coins typically might provide a better measure, in practice, since there is no problem of wear on the coin, fatigue of the tosser, or other irrelevant progressive effects that could complicate the experiment.

There is no significant arbitrariness in the ensemble for tossing a coin N times. On the other hand, a prediction of how many aces will appear in a given hand is significantly less valuable than a prediction of how many aces will appear when there are two or more kings, or when there are at least six spades. Each condition affects the others. To make such predictions, different ensembles can be constructed in which the number of spades in a hand is fixed, or in which the number of face cards is constant, and so on.

Special ensembles, in which one or more variables are fixed, are necessary in the analysis of complex physical systems. One time the total energy and the number of particles will be fixed. Another time the average translational kinetic energy and the number of particles will be held constant while examining the probabilities for other variables. Yet another time the number of particles and the volume may be allowed to change, holding constant the average energy.

If an experiment consists of N trials, each yielding one value, there are N independent variables. Usually, the particular values obtained are of less interest than certain properties of the distribution of values. When the number of times each possible value was obtained (or the number of values obtained within a selected interval about each nominal value) is plotted against the values, the graph is called a distribution function.

An example of a possible distribution of grades in a class of 200 students (N) is shown in Table 15.1 and in Fig. 15.1. The grade (interval) appearing most often in this sample is 83–84, which would therefore be called the *modal value*, or the *mode*. Approximately half the students (95) have grades below 81–82, half (88) are above; this grade value is the *median*. The average, or *mean* value, is 79–80 (actually 79.53 if each grade interval is represented by its midpoint; e.g., 79.5). The number of grade intervals is $n = 30$. (Some "empty" intervals are not listed.)

The *variance* is a measure of the spread of the distribution from the mean value. It is the average of the squares of the deviations from the mean,

Table 15.1 Grade Distribution

j	Grade	Number of Students	j	Grade	Number of Students	
0	39–40	2	17	73–74	8	
1	41–42	0	18	75–76	13	
2	43–44	1	19	77–78	15	
3	45–46	0	20	79–80	16	*Mean*
4	47–48	1	21	81–82	17	*Median*
7	53–54	2	22	83–84	20	*Mode*
9	57–58	3	23	85–86	17	
10	59–60	2	24	87–88	13	
11	61–62	1	25	89–90	13	
12	63–64	3	26	91–92	10	
13	65–66	5	27	93–94	8	
14	67–68	6	28	95–96	5	
15	69–70	7	29	97–98	1	
16	71–72	10	30	99–100	1	

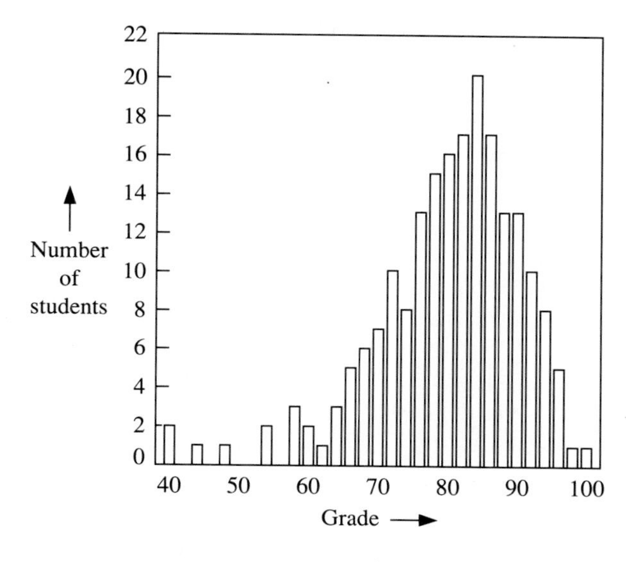

FIGURE 15.1 Distribution function for the grades from Table 15.1. The mode is 83–84 ($j = 22$), the median is 81–82 ($j = 21$), and the mean is 79–80 ($j = 20$). The standard deviation is 10.5.

$(1/N)\Sigma(x_i - \overline{x})^2$. The variance for this set of grades is 110; the variance for the relative grades (j values) is 27.8. The square root of the variance is often considered a better measure of the spread of values. It is called the *standard deviation* and is 10.5 for these grades, or 5.27 for the relative grades (j).

When values, or functions of values, are to be summed, the sum may be taken over all the experimental results (e.g., over the 200 students) or over the possible values (from the lowest value, 39–40, to the last, or highest value, 99–100). Either method gives the same result, but the notations can easily be confused. To distinguish between them, we adopt the following notational convention:

Let N be the number of trials.

Let n be the number of possible values, or intervals (not counting zero).

Let i be any one of the trials, with value x_i; $i = 1$ to N.

Let j designate one of the values, or intervals, representing the value x_j that occurs $P_n(j)$ times; j goes from 0 to n.

Let $P_n(j)$ be the (relative) probability of any of the N trials falling in the jth value interval; $\sum P_n(j) = 1$.

Then the mean value, or average, is

$$\overline{x} = \frac{1}{N}\sum_{i=1}^{N} x_i = \sum_{j=0}^{n} P_n(j)x_j \tag{15-1}$$

The sum of the deviations is zero.

$$\sum_{i=1}^{N}(x_i - \overline{x}) = \sum_{j=0}^{n} P_n(j)(x_j - \overline{x}) = 0 \tag{15-2}$$

The variance is

$$\text{variance} = \frac{1}{N}\sum_{i=1}^{N}(x_i - \overline{x})^2 = \sum_{j=0}^{n}(x_j - \overline{x})^2 P_n(j) \tag{15-3}$$

and the standard deviation is

$$\sigma = \sqrt{\frac{1}{N}\sum_{i=1}^{N}(x_i - \overline{x})^2} = \left[\sum_{j=0}^{n}(x_j - \overline{x})^2 P_n(j)\right]^{1/2} \tag{15-4}$$

In contrast to the average deviation, $(1/N)\sum_{i=1}^{N}(x_i - \overline{x})$, which should go to zero as N becomes large, the standard deviation, σ, is not zero for the background ensemble and may remain nearly constant as N increases. One convenient measure of reliability is called the *standard deviation of the mean*. It is equal to the standard deviation divided by the square root of the number of trials.

$$\begin{aligned}\text{standard deviation of the mean} &= \frac{\sigma}{\sqrt{N}} \\ &= \sqrt{\frac{1}{N^2}\sum_{i=1}^{N}(x_i - \overline{x})^2} \\ &= \frac{1}{N}\left[\sum_{j=0}^{n}(x_j - \overline{x})^2 P_n(j)\right]^{1/2}\end{aligned} \tag{15-5}$$

15.3 Random Walk

Random small errors, or uncontrolled variations contributing to a measured value, may be visualized as consecutive and cumulative. The first contribution may lower the value, the next may lower it again or raise it, and so on, for each of the many contributions. The effect is typically represented by a very drunk person trying to walk home, with equal probability that each step is toward or away from the intended destination and independent of the direction of the preceding step. Therefore the process by which an experimental value changes with N is usually called *random walk*.

If the number (N) of opportunities for error increases, we might expect the total error, $\sum(x_i - x_o)$, to increase. Yet intuitively we expect the experimental values to be more accurate as N becomes large.

As N increases, the likelihood of the sum of contributions becoming large increases, but not linearly with N. The cumulative effect increases with the square root of N. The average deviation is the sum of deviations ($\sim \sqrt{N}$) divided by the number of events (N), and therefore is proportional to $\sqrt{N}/N = 1/\sqrt{N}$. Hence as N increases, the sum of deviations probably increases, but the average deviation probably decreases.

Random walk problems occur frequently. Any time successive contributions are independent, they may be considered as displacements along orthogonal axes of a multidimensional space. The magnitude of the net displacement is then obtained by Pythagorean addition of the component displacements.

$$D^2 = \sum_i d_i^2 \quad \text{and therefore} \quad |D| = \sqrt{d_1^2 + d_2^2 + d_3^2 + \cdots}$$

If all of the contributions are of the same magnitude, or more generally if ϵ is the root mean square of the contributions, then

$$|D| = \sqrt{N\varepsilon^2} = \varepsilon\sqrt{N} \tag{15-6}$$

Brownian motion is a familiar example of random walk. The collision of a single molecule with a visible dust particle is not sufficient to produce detectable motion of the particle. The sum of many molecular collisions, on the other hand, can give a significant acceleration to the particle, but in a completely unpredictable direction and of unpredictable magnitude. The dust particle therefore moves about randomly in the fluid. On the average, the displacement of the particle from an initially observed position increases with the square root of the number of molecular collisions, or the square root of the number of detectable displacements, and therefore in proportion to the square root of time.

Random walk is also an appropriate model for the addition of random phases or phase shifts. Elementary theory would predict that if two musical instruments play the same note, equally loudly, the resultant heard by any observer might be anywhere from zero, if the interference is destructive, to four times the volume of one instrument, if they are in phase. In practice, the phase of each instrument is not fixed, but changes quite rapidly as the instrument is played. The phases of N instruments therefore vary such that the amplitude increases with the square root, $\sqrt{N}$. The loudness, or intensity, is the square of the amplitude, so N instruments are (on the average) $(\sqrt{N})^2 = N$ times as loud as one instrument.

15.4 Binomial Distribution

The possible consequences of N steps of a random walk, and their relative probabilities, are given by the binomial distribution. Let a statistical experiment consist of N events, each of which has a known a priori probability, p, for "success," and therefore a probability $q = 1 - p$ for "failure." For the coin toss, $p = \frac{1}{2}$ for getting heads (or for getting tails). When one die is tossed, each of the six numbers has an equal probability of appearing on top, so $p = \frac{1}{6}$ for getting any one specific number, such as 4. The probability of not getting 4 is then $\frac{5}{6}$. For any single test, $p + q = 1$.

The individual events under consideration are assumed to be independent. If the probability of an event occurring once in one trial is p, the probability of the event occurring twice in two trials is $p \times p$, or p^2, and the probability the event will occur N times, in N tests, is p^N. The number of possible values is equal to the number of trials plus one; i goes from 1 to N (N values) and j goes from 0 to $n = N$ (hence $N + 1$ values).

The probability the event will occur in the first test but not in the second is pq; the probability it will not occur in the first test but will in the second is qp, which is the same. Therefore, the probability of getting the event to occur just once in two tests is $pq + qp = 2pq$ (not equal to $2p$). In N tests, the probability of the event occurring just once is Npq^{N-1} and the probability of the event occurring $N - 1$ times is $Np^{N-1}q$. The nonoccurrence may be in any of the N tests.

Evaluation of such probabilities is entirely equivalent to the mathematical evaluation of $(p + q)^n$. For example,

$$(p + q)^2 = pp + pq + qp + qq = p^2 + 2pq + q^2 = 1$$

and

$$\begin{aligned}(p + q)^3 &= ppp + ppq + pqp + qpp + pqq + qpq + qqp + qqq \\ &= p^3 + 3p^2q + 3pq^2 + q^3 = 1\end{aligned}$$

In general,

$$(p+q)^n = p^n + n\,p^{n-1}q + \cdots + \frac{n!}{j!(n-j)!}p^j q^{n-j} + \cdots$$

$$= \sum_{j=0}^{n} a_{n,j}\,p^j q^{n-j} = \sum_{j=0}^{n} \binom{n}{j} p^j q^{n-j} = 1 \tag{15-7}$$

The coefficients,

$$\binom{n}{j} = a_{n,j} = \frac{n!}{j!(n-j)!} \tag{15-8}$$

are called the *binomial coefficients*.[3] You may recognize $\binom{n}{j}$ as the number of possible combinations of n distinguishable items taken j at a time.

EXAMPLES 15.1

A. If a coin is tossed n times, the probability of j heads is $\binom{n}{j} p^j q^{n-j}$. Let $n = 4$. Find the probabilities for

a. $j = 0$. b. $j = 1$.
c. $j = 2$. d. $j = 3$.
e. $j = 4$.

B. Find the probability of getting doubles (the same number on each die) when two dice are thrown

a. Once.
b. Twice.
c. Three times.

Also find the mean value of doubles for each set. □

The distributions calculated in the examples above are presented graphically in Fig. 15.2. Although it would require n values to describe each distribution fully, three numbers are of particular interest: the most probable, or *modal* value; the average, or *mean* value; and the width of the curve, as measured by the *standard deviation*.

For the binomial distribution the average value is

$$\bar{j} = \sum jP_n(j) = \sum_{j=0}^{n} j\,a_{n,j}\,p^j q^{n-j} = n\,p \tag{15-9}$$

and the standard deviation is

$$\sigma_{n,j} = \left[\sum_{j=0}^{n} (j - n\,p)^2 P_n(j)\right]^{1/2} = \sqrt{n\,pq} \tag{15-10}$$

[3]For convenience, 0! is defined to be equal to 1, so that one expression describes all values of j. Alternatively, the definition of $n!$ may be taken as $n! = \int_0^\infty e^{-t} t^n\,dt$. This gives $0! = 1$ and $n! = \prod_{i=1}^{n} i$ for integral values of n.

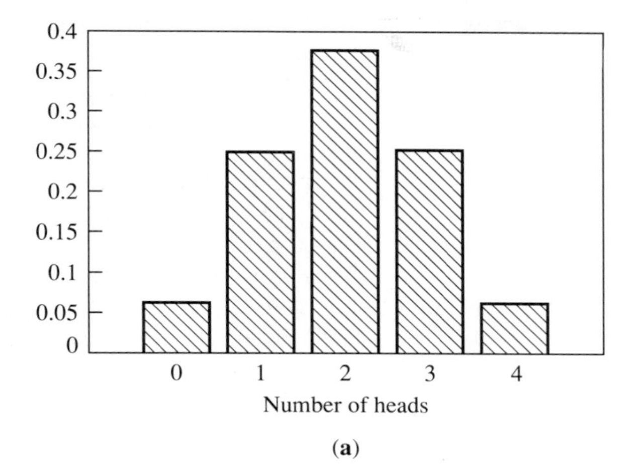

(a)

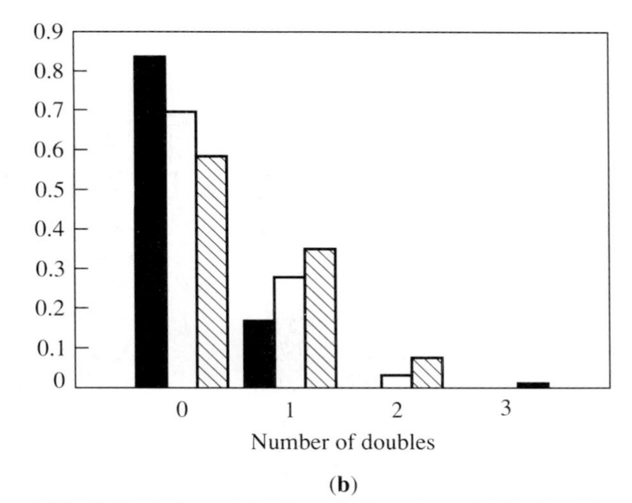

(b)

FIGURE 15.2 Distribution functions: (a) Probability of heads in four coin tosses; (b) probability of throwing doubles in one throw (left bar, 0 and 1), two throws (white bar, 0 to 2), and three throws (right bar, 0 to 3) of two dice.

Remember that the probabilities calculated above and presented in Fig. 15.2 are for background ensembles, the results that would be obtained for M sets, each involving N trials, as M approaches infinity. The modal value, mean, and standard deviation prove especially important in appraising experimental results, where N trials produce a distribution curve that is not likely to agree with any of the calculated probability distributions.

When the number of trials, N, per set of experiments (i.e., one of the M sets contributing to the ensemble) becomes large, the distribution tends to become symmetric, even for $p \neq \frac{1}{2}$. For a symmetric distribution, the modal value, or peak of the distribution, coincides with the mean value or center of the distribution. That condition will be satisfied for most of the distributions important in statistical mechanics, so the modal value will not usually be considered separately from the mean.

ANSWERS 15.1

A. a. $j = 0;\ P = \binom{4}{0} p^0 q^4 = \frac{4!}{0!4!}\left(\frac{1}{2}\right)^4 = \frac{1}{16}.$

b. $j = 1;\ P = \binom{4}{1} p^1 q^3 = \frac{4!}{1!3!}\left(\frac{1}{2}\right)^4 = \frac{4}{16} = \frac{1}{4}.$

c. $j = 2;\ P = \binom{4}{2} p^2 Q^2 = \frac{4!}{2!2!}\left(\frac{1}{2}\right)^4 = \frac{4 \times 3}{2}\frac{1}{16} = \frac{3}{8}.$

d. $j = 3;\ P = \binom{4}{3} p^3 q^1 = \frac{4!}{3!1!}\left(\frac{1}{2}\right)^4 = \frac{4}{16} = \frac{1}{4}\left(\approx j = 1\right).$

e. $j = 4;\ P = \binom{4}{4} p^4 q^1 = \frac{4!}{4!0!}\left(\frac{1}{2}\right)^4 = \frac{1}{16}\left(\approx j = 0\right).$

B. a. One throw: The probability of getting any specific number, such as 4, with one die is $\frac{1}{6}$. To get two 4's, in two throws, the probability is $\left(\frac{1}{6}\right)^2$. Therefore, the probability of getting doubles is the sum over all six numbers, or $6 \times \left(\frac{1}{6}\right)^2 = \frac{1}{6} = p$, and $q = \frac{5}{6}$. Mean value $= \frac{1}{6}$.

b. Two throws: Probability of no double in two throws is

$$P = \binom{2}{0} p^0 q^2 = \frac{2!}{0!2!}\left(\frac{1}{6}\right)^0\left(\frac{5}{6}\right)^2 = \frac{25}{36}$$

Probability of getting one double in two throws is

$$P = \binom{2}{1} p^1 q^1 = \frac{2!}{1!1!}\left(\frac{1}{6}\right)^1\left(\frac{5}{6}\right)^1 = \frac{10}{36}$$

Probability of getting two doubles in two throws is

$$P = \binom{2}{2} p^2 q^0 = \frac{2!}{2!0!}\left(\frac{1}{6}\right)^2\left(\frac{5}{6}\right)^0 = \frac{1}{36}$$

Note that the sum of the probabilities is 1, as required, but $P(0) \neq P(2)$. The probability of one or more doubles is not $\frac{1}{6} + \frac{1}{6}$, but the mean value is

$$\frac{1}{3} = 0 \times \frac{25}{36} + 1 \times \frac{10}{36} + 2 \times \frac{1}{36}$$

c. Three throws: Probability of no double is

$$P = \binom{3}{0} p^0 q^3 = \frac{3!}{0!3!}\left(\frac{1}{6}\right)^0\left(\frac{5}{6}\right)^3 = \frac{125}{216}$$

Probability of one double in three throws is

$$P = \binom{3}{1} p^1 q^2 = \frac{3!}{1!2!}\left(\frac{1}{6}\right)^1\left(\frac{5}{6}\right)^2 = \frac{75}{216}$$

Probability of two doubles in three throws is

$$P = \binom{3}{2} p^2 q^1 = \frac{3!}{2!1!}\left(\frac{1}{6}\right)^2\left(\frac{5}{6}\right)^1 = \frac{15}{216}$$

Probability of three doubles in three throws is

$$P = \binom{3}{3} p^3 q^0 = \frac{3!}{3!0!}\left(\frac{1}{6}\right)^3\left(\frac{5}{6}\right)^0 = \frac{1}{216}$$

The mean value is

$$0 \cdot \frac{125}{216} + 1 \cdot \frac{75}{216} + 2 \cdot \frac{15}{216} + 3 \cdot \frac{1}{216} = 3 \cdot \frac{1}{6} = \frac{1}{2}$$

If you must throw doubles (e.g., to "Get Out of Jail" in a *Monopoly* game), the first answer shows it will take an average of six turns. The second answer shows that it will take an average of three sets of two throws. The last answer shows that it will take an average of two sets of three throws.

15.5 Poisson Distribution

If the chance of success in any event, p, is very low, the probability of observing success may still be of moderate magnitude if the number of chances is very large. The chance of any one card dealt to you being an ace of spades is quite small ($\frac{1}{52}$), but by the time you have 13 cards dealt, the probability has increased to $\frac{1}{4}$. The chance of a single ^{14}C nucleus disintegrating in any 1-second interval is about 6×10^{-12}. However, even a microgram of carbon containing 1% ^{14}C contains 5×10^{14} unstable nuclei, giving an expectation of 3×10^3 disintegrations per second.

Calculations of the binomial coefficients would be unreasonably tedious for very large N. Fortunately, the probabilities of interest are given to an adequate accuracy by a simple expression. In some number of trials, $N = n$, or some time interval, Δt, let j be the number of successes observed. If the average number of successes is $\overline{j} \equiv \mu$, the probability of observing j successes is

$$P_\mu(j) = \frac{\mu^j}{j!}e^{-\mu} \qquad (15\text{-}11)$$

This is called the *Poisson distribution*.

The binomial distribution has two independent parameters, N and p. The Poisson distribution is the limiting form of the binomial distribution,[4] for a fixed product Np, as $N \to \infty$ (and therefore $p \to 0$). The Poisson distribution has only one remaining independent parameter, which is best taken as $\overline{j} = \mu = Np$.

EXAMPLE 15.2

Show that the maximum of the Poisson distribution occurs at $j = \mu$. [Note that the derivative of $\ln j!$ is the derivative of the sum, $\sum_{k=1}^{j} \ln k$, and for very large j, $dk = \Delta k = 1$ is an adequate approximation. Also, $d\,a^p = a^p(\ln a)\,dp$.] □

[4] See Derivation 15.2 at the end of the chapter.

ANSWER 15.2

$0 = \frac{d}{dj} P_\mu(j); \quad d\left(\frac{\mu^j}{j!} e^{-\mu}\right) = e^{-\mu}\left[\frac{\mu^j}{j!}(\ln \mu)\, dj + \mu^j \frac{-1}{(j!)^2}\, dj!\right] =$

$e^{-\mu}\frac{\mu^j}{j!}\{(\ln \mu)\, dj - d(\ln j!)\} = 0$, so $(\ln \mu)\, dj = d(\ln j!)$. But $d(\ln j!) =$

$d\sum_{k=1}^{j} \ln k = \ln(k = j) = \ln j$.

(All terms in the sum $\sum_{k=1}^{j} k$ remain constant when $dj = \Delta j = 1$, except for addition of the last term, $\ln j$.) Hence $\ln \mu = \ln j$ and $\mu = j$ at the maximum.

The Poisson distribution is noticeably asymmetric when the average number of counts is small (e.g., $\mu \leq 3$). For larger numbers (e.g., $\mu \approx 10$) the distribution becomes symmetric and closely resembles the Gaussian distribution (Fig. 15.3).

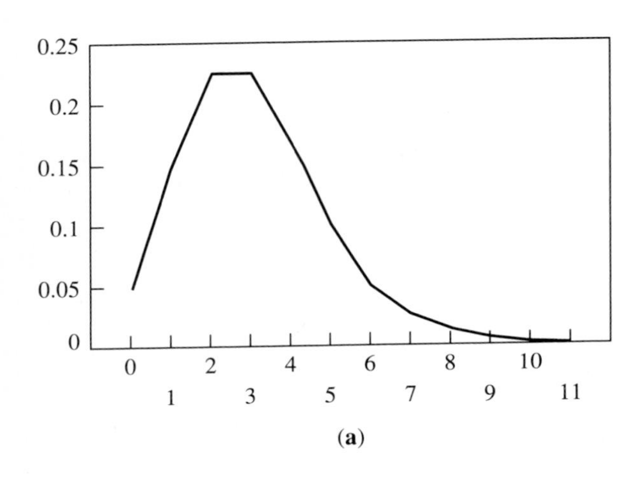

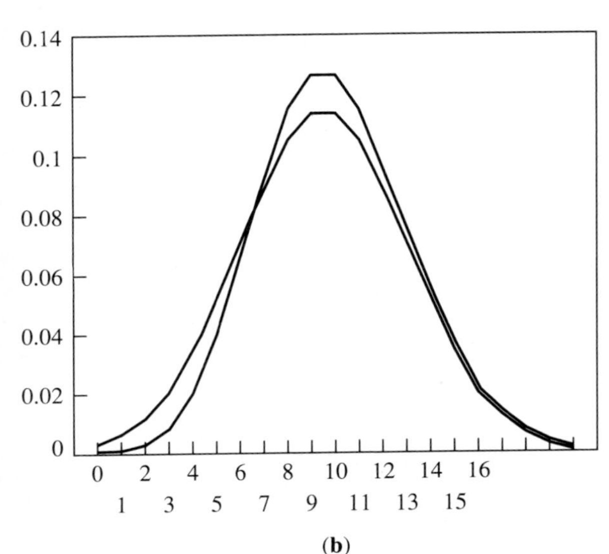

FIGURE 15.3 Poisson distributions: (a) $\mu = 3$; (b) $\mu = 10$, with a Gaussian curve matched roughly to the Poisson distribution. The Gaussian is higher at low values and lower at the peak in this match.

15.6 Gaussian Distribution

We have seen that the binomial distribution, considered above, is fundamental when events, or "successes," are counted. If the probability of success in any one trial is $p \leq 1$, the probability, $P_N(j)$, of j successes in N trials is

$$P_N(j) = a_{N,j}\, p^j q^{N-j} = \binom{N}{j} p^j q^{N-j} = \frac{N!}{j!(N-j)!} p^j q^{N-j} \tag{15-12}$$

where $q = p - 1$. Unless $p = \frac{1}{2}$, the distribution is asymmetric, but the asymmetry becomes less noticeable as N increases. For large $n = N$, where the binomial coefficients would be tedious to calculate, the binomial distribution approaches the simpler Poisson distribution.

$$P_\mu(j) = \frac{\mu^j}{j!} e^{-\mu}$$

where

$$\mu \equiv \bar{j} = \sum_j j P_N(j) = Np = \sum_{j=0}^{N} j P_\mu(j) \tag{15-13}$$

When the outcome of an event is a measurement rather than a "yes" or "no," the outcomes need not differ by fixed increments. The number of possible outcomes is then greatly increased. If p and q are taken as the a priori probabilities for positive and negative contributions, in a random walk, then Npq will be large.

Let

$$P(N, p, j) = \frac{N!}{j!(N-j)!} p^j q^{N-j} \qquad (j = 0 \rightarrow N)$$

To take advantage of the symmetry, define a new variable, $k = j - \bar{j}$; k is the deviation from the mean. Then as we allow the number of points, N, to become infinite, we replace $P(N, p, j)$ with $P(k)$, with $-\infty < k < \infty$.

Expand $\ln P(k)$ in a Taylor series abut the midpoint, $k = 0$.

$$\ln P(k) = \sum_{i=0}^{\infty} \frac{1}{i!} k^i \frac{d^i}{dk^i} \ln P(k) \tag{15-14}$$

with the derivatives evaluated at $k = 0$. Inserting values of the derivatives gives

$$\ln P(k) = \ln P(0) + \frac{1}{2} k^2 \frac{-1}{Npq} + \frac{1}{6} k^3 \frac{q^2 - p^2}{(Npq)^2} + \frac{1}{24} k^4 \frac{-2(p^3 + q^3)}{(Npq)^3} + \cdots \tag{15-15}$$

Neglecting terms beyond the second gives

$$P(k) = P(0) e^{-k^2/2Npq} \tag{15-16}$$

which is the *Gaussian distribution function*.

It is easily shown that terms beyond the second are less than $k^i/(Npq)^{i-1}$, so near the center of the distribution, where $k \ll Npq$, these higher-order terms will be very small and the series converges. For values of k not near the center, the exponential term causes $P(k)$ to be vanishingly small, so that the accuracy of its evaluation is of no consequence.

Summarizing, we find that the assumptions that lead to a Gaussian distribution are

1. The distribution is symmetric about the maximum.
2. The maximum occurs at the average value (these two conditions express the assumption of random errors).
3. The distribution may be treated as a continuous function, for all values of the variable (i.e., $N \to \infty$).

The Gaussian distribution is conveniently written in terms of a continuous variable, x, as

$$P(x) = \mathcal{A} e^{-(x-x_o)^2/2\sigma^2}$$

with $\sigma = \sqrt{Npq}$ the standard deviation and x_o the mean value (see Derivation 7). Requiring that

$$\int_{-\infty}^{+\infty} P(x)\, dx = 1$$

gives the customary form

$$P(x) = \frac{1}{\sigma\sqrt{2\pi}} e^{-(x-x_o)^2/2\sigma^2} \tag{15-17}$$

Because the only independent parameters of the Gaussian curve are the mean value, x_o, and the standard deviation, σ, all Gaussian distributions have exactly the same shape, differing only in placement (i.e., mean value) and in scale of the width. The inflection point ($d^2P/dx^2 = 0$) occurs at $x - x_o = \pm\sigma$. The points at which the Gaussian curve drops to $1/e$ of its maximum value are at $\pm\sqrt{2}\sigma$ from the center. The mean deviation (average value of $|x - x_o|$) is $\sqrt{2/\pi}\sigma = 0.80\sigma$. Also, 68.3% of all values fall within the range of one standard deviation; 95.4% of all values fall within two standard deviations; and 99.7% of all values fall within three standard deviations.

It is important to keep in mind the assumption that the *only* reason for the variance of a Gaussian distribution is assumed to be random contributions. The distribution of molecular speeds in a gas is not a Gaussian curve (although the distribution of velocities is a symmetrical, Gaussian function).

A collection of scores on an examination need not resemble the Gaussian curve, even if N is very large, unless the differences between scores are determined by random effects. An examination score for N students, with n possible values for each score, is not equivalent to N measurements of a single experimental value. The results may, however, resemble a Gaussian distribution, at least if the knowledge of the students is determined by random effects or if the measurement/scoring process is uncertain because of random errors.

On the other hand, most experiments attempt to measure a single quantity and *are* subject to random errors. If the average speed was measured N times for nitrogen gas at a predetermined temperature, those average speeds would be expected to fall on a Gaussian curve, very narrow compared to the Maxwell–Boltzmann distribution. A large number of examples have been accumulated for which the experimental distribution closely resembles a Gaussian distribution. It is therefore commonly called the *error curve* and the *normal distribution*, or simply the *bell curve*.

Lippmann probably best expressed the status of the Gaussian curve when he commented to Poincaré, "Everybody believes in the exponential law of errors: the experimenters, because they think it can be proved by mathematics; and the mathematicians, because they believe it has been established by observation."[5]

In applications of statistics to molecules, the numbers are usually so great that often only the Gaussian function is a practical description, and the value of N is so great that the Gaussian curve should give a very accurate representation of the ensemble of real systems.

15.7 Experimental Distributions

The calculations and graphs considered so far have all been based on the ensembles of M sets of N trials. Because M is allowed to approach infinity, there is no uncertainty in the distribution curves. Means, modal values, standard deviations, and individual probabilities are exact values. Everything is known (in principle) about each of the distribution curves.

Experimentalists face a totally different question. When an experimental distribution curve is drawn from a finite set of N trials, it cannot be expected to agree precisely with any of the infinitely many possible theoretical distribution curves. The problem of the experimentalist is to ascertain, with as much assurance as possible, which of the theoretical distribution curves the experimental set has come from. As the number of trials, N, increases, differences between the experimental curve and the appropriate background ensemble curve are expected to decrease, so it should become easier to guess what is the correct background ensemble, but in principle one can never be sure.

It is most likely that as N increases, the modal, mean, and standard deviation values of the experimental curve will agree closely with the values for the background ensemble. These descriptors are therefore very important. However, they are uncertain for the experimental distribution and may change as more trials are accumulated.[6]

Often a set of N experimental values should each give the same value, except for random errors of measurement that are equally likely to be positive or negative. Other times the differences between experimental values are caused by random influences not part of the experimental procedure and irrelevant to the quantity under investigation. In either case, let x_o be the value that would be observed in the absence of errors or such random contributions. Then, because the deviations from the true value are random, their sum should vanish and $\sum_{i=1}^{N}(x_i - x_o) = 0$. This is the condition for x_o to be the mean value of the distribution; $\overline{x} = x_o$.

[5]Poincaré, *Calcul des probabilités,* p. 149; quoted by Whittaker and Robinson (ref. 28), p. 179.

[6]The convergence of the experimental distribution to the background ensemble distribution need not be monotonic as judged by the common parameters. Apparent convergence to a particular mean value, for example, may disappear with increasing N. The values may eventually converge to some other mean value.

Therefore, if the errors or other contributions to deviations from the true value are random, the mean value of the distribution is the true value, insofar as the experimental distribution curve resembles the background ensemble distribution. If there is a nonrandom, or systematic, error in the measurement, the mean value is the true value plus the systematic error (positive or negative).

The sum of deviations from the mean necessarily vanishes for all distributions, and therefore gives no information on the width or shape of the curve. Two possible descriptors for the width are the sum of absolute values of deviations and sum of squares of deviations. The former, $\sum_{i=1}^{N} |x_i - x_o|$, proves to be less valuable than the latter, $\sum_{i=1}^{N}(x_i - x_o)^2$, which gives the variance,

$$\text{variance} = \frac{1}{N}\sum_{i=1}^{N}(x_i - x_o)^2 = \sum_{j=0}^{n}(x_j - x_o)^2 P_n(j) \tag{15-3}$$

and the standard deviation,

$$\sigma = \sqrt{\frac{1}{N}\sum_{i=1}^{N}(x_i - x_0)^2} = \left[\sum_{j=0}^{n}(x_j - x_0)^2 P_n(j)\right]^{1/2} \tag{15-4}$$

For small N, there are arguments favoring dividing the sum by $N - 1$, rather than by N, because the condition $\sum(x_i - x_o) = 0$ leaves only $N - 1$ independent values of x. For values of N greater than 10 the difference between dividing by N and by $N - 1$ is usually negligible.

SUMMARY

A statistical treatment provides all the information about complex systems that is meaningful, or possible.

Probability theory predicts average values for an *ensemble* of M equivalent systems or experiments, where M approaches infinity.

Ensembles are often "prepared" subject to special restrictions, such as constant total energy or constant average energy or constant volume.

The sum of deviations from the mean increases with N, the number of trials, but the percentage deviation decreases as $1/\sqrt{N}$.

For N trials and n possible values, if $P_n(j)$ is a probability that a trial x_j falls in the jth interval, then

$$\overline{x} = \frac{1}{N}\sum_{i=0}^{N} x_i = \sum_{j=0}^{n} P_n(j)x_j$$

$$\text{variance} = \frac{1}{N}\sum_{i=1}^{N}(x_i - \overline{x})^2 = \sum_{j=0}^{n}(x_j - \overline{x})^2 P_n(j)$$

$$\text{standard deviation} = \sigma = \sqrt{\frac{1}{N}\sum_{i=1}^{N}(x_i - \overline{x})^2} = \left[\sum_{j=0}^{n}(x_j - \overline{x})^2 P_n(j)\right]^{1/2}$$

and standard deviation of the mean $= \sigma/\sqrt{N}$.

Random walk: $\sum(x_i - x_o) \sim \sqrt{N}$; $\overline{(x_i - x_o)} \sim 1/\sqrt{N}$;

mean deviation $= |D| = \sqrt{N\varepsilon^2} = \varepsilon\sqrt{N}$.

Binomial distribution: $(p+q)^n = \sum_{j=0}^{n} a_{n,j}\, p^j q^{n-j}$.

$$a_{n,j} = \binom{n}{j} = \frac{n!}{j!(n-j)!}; \qquad \bar{j} = np; \qquad \sigma_{n,j} = \sqrt{npq}$$

Poisson distribution: $P_\mu(j) = \dfrac{\mu^j}{j!}e^{-\mu}$; $\quad \bar{j} = Np = \mu \qquad (N \to \infty)$.

Gaussian distribution: $P(k) = P(0)e^{-k^2/2Npq}$; $\quad \sigma = \sqrt{Npq}$

$$P(x) = \frac{1}{\sigma\sqrt{2\pi}}\, e^{-(x-x_o)^2/2\sigma^2}$$

Inflection points at $x - x_o = \pm\sigma$; $\quad \overline{|x - x_o|} = \sqrt{\frac{2}{\pi}}\,\sigma$.

Probability (or area) between $\pm\sigma = 68.3\%$; between $\pm 2\sigma = 95.4\%$; between $\pm 3\sigma = 99.7\%$.

DERIVATIONS AND PROOFS

15.1. Show that

$$\frac{n!}{j!(n-j)!} + \frac{n!}{(j+1)!(n-j-1)!} = \frac{(n+1)!}{(j+1)!(n-j)!}.$$

From this result, Pascal's triangle follows. When the binary coefficients are written in centered, horizontal rows (by n), each coefficient is the sum of the coefficients above it, to the right and left. The first few rows are

1
1 1
1 2 1
1 3 3 1
1 4 6 4 1

15.2. Show (by supplying the missing arguments) that

$$P(n, p, j) \equiv \frac{n!}{(n-j)!j!} p^j (1-p)^{n-j}$$

$$= \frac{n(n-1)(n-2)\cdots(n-j+1)p^j(1-p)^{n-j}}{j!}$$

so that if $n \to \infty$ and $p \to 0$, such that $np = \mu$, a constant, and thus $j << n$, then

$$P(n, p, j) \to \frac{n^j p^j (1-p)^n}{j!} = \frac{\mu^j}{j!}(1-p)^{\mu/p}$$

and therefore, because $\lim_{p\to 0}(1-p)^{-1/p} = e$,

$$P(n, p, j) = P_\mu(j) = (\mu^j/j!)e^{-\mu}.$$

15.3. Show by comparison of the Poisson distribution with the binomial distribution that the standard deviation of the Poisson distribution is $\sigma = \sqrt{Npq} = \sqrt{\mu}$.

15.4. Show that the Poisson distribution is not sharply peaked at $j = \mu$, but rather that if $\bar{j} = \mu$, then $P_\mu(\bar{j}-1) = P_\mu(\bar{j})$.

15.5. Show that if $\delta x_i = x_i - a$ for some set of values $\{x_i\}$, the value of a that makes $X = \sum_{i=1}^{N} \delta x_i^2$ a minimum is $a = (1/N)\sum x_i = \bar{x}$.

15.6. Show that if $\delta x_i = x_i - \bar{x}$, then $\sum \delta x_i{}^2 = \sum(x_i^2 - \bar{x}^2)$.

15.7. For large N the binomial distribution is symmetric, so the value of j at the maximum gives the average value of j.

a. Show that if $P(N, p, j) \equiv \dfrac{N!}{(N-j)!j!}p^j(1-p)^{N-j}$,

with N large but p not necessarily small, the maximum obtained by setting $(d \ln P)/dj = 0$ occurs at $\bar{j} = Np$.

b. Show that, at $j = \bar{j} = Np$, $\dfrac{d \ln P}{dj} = 0$ and $\dfrac{d^2 \ln P}{dj^2} = \dfrac{-1}{Npq}$, using results from part (a), and therefore, if $\ln P$ is expanded, to give

$$\ln P = \ln P(\bar{j}) + \frac{d \ln P}{dj}\delta j + \frac{1}{2!}\frac{d^2 \ln P}{dj^2}(\delta j)^2 + \cdots$$

the expansion leads to

$$P(j) = P(\bar{j})e^{-(\delta j)^2/2Npq}$$

It is then easy to show that $Npq = \sigma^2$, as in Derivation 15.3. Normalization gives $\int P(j)\,dj = 1 = \int_{-\infty}^{\infty} P(\bar{j})e^{-(\delta j)^2/2\sigma^2}\,dj$ and

$$P(\bar{j}) = \frac{1}{\sigma\sqrt{2\pi}}$$

15.8. Show that the Gaussian curve,

$$P(x) = \frac{1}{\sigma\sqrt{2\pi}}e^{-(x-x_o)^2/2\sigma^2}$$

has inflection points at $x_o \pm \sigma$.

15.9. Show that $P(x_o \pm \sqrt{2}\sigma) = (1/e)P(x_o)$.

15.10. Show that, if x_o is known and the ratio for two values, $P(x_2)/P(x_1)$, is known, then σ may be determined for a Gaussian distribution.

15.11. The binomial distribution for tossed coins also applies to many other one-dimensional random-walk problems, such as electric or magnetic dipoles orienting in an external field and polymer segments aligning along a stretch direction. If N is the total number, the number aligned in each direction may be written as $\frac{1}{2}(N+n)$ and $\frac{1}{2}(N-n)$, where $n = N - 2j$ and $|n| << N$.

a. Show that the logarithm of

$$\left(\frac{1}{2}\right)^N \frac{N!}{\left[\frac{1}{2}(N+n)\right]!\left[\frac{1}{2}(N-n)\right]!}$$

can be expanded, using $\ln x! = x \ln x - x + \ln \sqrt{2\pi x}$ and $\ln(1+x) = x - x^2/2 + \cdots$ to give $-n^2/2N - \frac{1}{2}\ln(\pi N/2)$.

b. Show that this leads to $P(n) = (2/\pi N)^{1/2} e^{-n^2/2N}$, which is a Gaussian distribution. What is the standard deviation? Why is it not $\sqrt{Npq}$?

PROBLEMS

15.1. How many different sets of bridge hands are possible? (That is, how many ways can 52 cards be divided into groups of 13?)

15.2. A particle of dust experiences random bombardment by air molecules that cause it to undergo random displacements of root-mean-square magnitude 100 μm. What total displacement of the dust particle would you predict after 1 million of these random displacements?

15.3. A coin is suspected of being unsymmetrically weighted, such that the probability of getting heads is 0.7 for each throw.

a. With a coin so weighted, what is the expectation of getting (exactly) 7 heads out of 10 throws?

b. With a normal coin, what is the expectation of getting the same result of 7 heads out of 10 throws?

15.4. Stu Sharpe likes to play with trigonal pyramid dice (four triangular faces, with numbers 1–4) because he thinks the chances of getting the right combination are higher.

a. What is the chance of getting 1 + 1 with one throw of a pair of these?

b. What is the chance of getting 7 with one throw of a pair of these?

15.5. A certain sample of 1 g of carbon should give 5 counts per minute. Find the probabilities that the actual count in the first minute will be

a. 0. **b.** 1. **c.** 2.
d. 3. **e.** 4. **f.** 5.

What factor gives $P_\mu(j)$ from $P_\mu(j-1)$?

15.6. In a trial run, for standardization purposes, a carbon dating experiment was expected to give 30 counts. It gave only 25. What is the probability this was a correct measurement?

15.7. With modern laboratory instrumentation, counting techniques are possible where previously only collective effects could be detected. For example, in Raman spectroscopy, photographic images or currents from photomultiplier tubes are replaced by counting photons arriving at the detector. This allows measurement of much weaker optical signals than previously possible. In one segment of a certain Raman rotational band structure, two adjacent peaks have a known intensity ratio of $\frac{5}{4}$. If there are, on the average, 50 photons/s at the center of the stronger peak:

a. What is the probability of getting only 40 counts in one second of counting at that peak? Is this larger or smaller than the probability of getting 50 counts in 1 s at the weaker peak?

b. What is the probability of getting only 8 counts in 0.20 s of counting at the stronger peak?

c. What is the probability of getting only 4 counts in 0.10 s of counting?

15.8. A certain wavelength is reported as $12.015 \pm 0.007\,\mu\text{m}$, where the uncertainty is the standard deviation. (The experiment gives values centered about 12.015 but in a Gaussian distribution with $\sigma = 0.007\,\mu\text{m}$.)

a. What is the peak value, $\rho(0) \equiv \rho(\lambda = 12.015)$, for probability normalized to 1; $\int \rho(x)\,dx = 1$. Does $\rho(0)$ have units?

b. What is the probability that another measurement will give the value 12.015 μm (i.e., between 12.0145 and 12.0155 μm)?

c. If the Gaussian function is approximated by a triangular function with the peak value calculated in part a, what is the base length of the triangle to give a probability of 1 that all values fall within the triangle?

d. How does the half-width (width of the triangle at half the height) compare with σ, the standard deviation of the Gaussian fit?

e. What fraction of the distribution would be represented by a triangular function of the same half-width as the Gaussian curve? (That is, how does the area of a triangular function, of the peak height calculated in part a and with half the width at half the height equal to σ, compare to the area under the Gaussian curve?)

15.9. The following numbers are selected from an ensemble of Gaussian distributions, normalized so that $\int_{-\infty}^{\infty} \mathcal{F}(x)\,dx = 1$.

x	7.0	9.0	11.0	13.0	15.0	17.0
$\mathcal{F}(x)$	0.00876	0.0648	0.176	0.176	0.0648	0.00876

a. What value would you assign as x_0?

b. What value would you assign as σ?

c. What value of $\mathcal{F}(x)$ would you anticipate for $x = 10.5$?

15.10. A certain lottery sells 2×10^6 tickets at $2.50 each and awards a single grand prize of $2,000,000.

a. What is the actuarial value of one $2.50 ticket?

b. If you buy a ticket, what is the probability that you will win with it?

c. If you buy a ticket for this month's lottery and another for next month's lottery, what is the probability of winning twice?

d. It has been estimated that the probability someone will win a lottery twice in the same year is about 10%. (See P. Diaconis and F. Mosteller, *Journal of the American Statistical Association* 84, 853 (1989).) How can you reconcile this result with your answer to part (c)?

15.11. An investigator suspects there will be a correlation between the order in which students arrive for an exam and their scores on the exam. There are 100 questions on the exam. Analysis shows the probability of a positive result arising solely from chance, for the given class size, is 1.0%. When the tests are given, the results "confirm the null hypothesis"; that is, no correlation is found.

A second investigator, on examining the findings, decides there is additional information in the experimental results. A comparison is made, question by question, to see if there is a correlation between arrival time and test question score. What result would you predict for this more detailed analysis?

Comment on the validity of the generalization: "Statistical analysis of any experiment must be based on the intent of the experiment." (For a discussion related to experiments in the physical sciences, see E. B. Wilson, Jr., *An Introduction to Scientific Research* (New York: McGraw-Hill Book Co., Inc., 1952.)

15.12. During World War II psychologists found a difference in performance of pilots that could be attributed to *field dependence* or *field independence,* a measure of ability of the individual to interpret optical signals independently of extraneous information. More recently, the concept has been extended to learning styles. In one version, there are approximately 30 independent tests with two (or more) outcomes per test, permitting precise identification of the abilities and/or preferences of the individual. Assuming 2 classifications under each of 30 tests, find

a. The number of classes into which students are to be placed.

b. The probable number of students per classification. (Ideally, instruction would be tailored to each class of students.)

15.13. Even valid correlations may require some care in interpretation of cause and effect. Can you suggest causes for the following observations?

a. Studies of birth rates in Holland have shown a positive correlation between children born per year and the stork population.

b. Tests of reading ability in a random sample of students drawn from an elementary school show a positive correlation between reading level and shoe size.

CHAPTER 16

Information and Thermodynamics

Science lives on the flow of information: from experiments and observations, to and from storage, and between individuals and groups. As early as the nineteenth century, a connection was recognized between information and entropy, which is a measure of uncertainty. Within the last few decades especially, the theory of information encoding and transmission has been explored to provide significant new insights into thermodynamics and statistical mechanics, as well as communication theory.

The groundbreaking paper of Shannon provided some fundamental formulas and their interpretations. Subsequent work by Brillouin, Jaynes, and others has broadened the field and increased the number of applications. Their findings[1] will serve as an introduction to application of probability theory to thermodynamics.

16.1 Communication of Information

When Paul Revere's compatriot climbed to the top of the tower to signal movement of the British troops, it was not enough to flash a light. Whether the troops were moving by land or by sea was necessary information for the colonists who planned an appropriate countermove. To distinguish between the three possibilities—no movement, movement by land, or movement by sea—three possible signals were required. No light indicated a continued absence of movement. One light indicated movement by land; two lights indicated movement by sea. When Revere saw the second light he had sufficient information to begin his historic ride.

Information can be transmitted from one person to another by various types of languages, such as spoken English, written English, sign or pictographic language, Morse code, or a bidding convention in the game of bridge. The number of signals, or symbols, required differs from one language to another. For example, "man" requires one symbol in sign language or Chinese, three symbols (plus a space or terminator) in written English, and eight symbols (two dots, four dashes, and two spaces, or blanks, plus another space or terminator) in Morse code.

The number of symbols required is related to the size of the language. Morse code has only three possible signals, including the blank. Written English has 27

[1]See the Bibliography for references.

characters, including the blank (and neglecting punctuation, which is also important). To understand basic Chinese, as might appear in a newspaper, requires a knowledge of 2000 or more characters. Quite generally, the more symbols available, the more information can be carried by each one, so fewer are needed per message.

The information content of a signal is an inverse function of the probability of that signal. When more choices of signal are available, each becomes less probable and carries more information.

A signal that is constrained to be continuously on or continuously off cannot carry information. The smallest amount of information is represented by two alternatives—signal or no signal, or *on* or *off*. Such a signal carries one *bit* of information.

Elementary information theory must distinguish between what may be called "hard" information and "soft" information. Hard information carries no intrinsic value beyond the information transmitted, which can be counted in bits. There is no "good" or "joyful" or "sad" message. Such labels belong to soft information and are beyond the realm of information theory as it has been developed.

Nevertheless, the state of the receiver of information must be considered in evaluating the amount of information transmitted. In particular, if the receiver can predict part or all of the message, the message must be considered to have a lower content than if it is unpredictable. We will find an analog in the description of a system and its preparation in a specific state, allowing prediction of the results of certain measurements before those measurements are performed (or performed again).

16.2 Quantitative Measure of Information

Many transmission systems are *synchronous* or resemble a synchronous system. In each equal short time interval, one signal is sent and received. Consider a system in which the choices are limited to two, which might be *on* or *off,* or *high* or *low,* or some other scheme. Then equal information is carried in each time interval. Over n intervals, there are 2^n choices, so the information content must be a function of 2^n. The total information, however, must be n times the information sent in a single time interval. Hence

$$f(2^n) = nf(2)$$

Such behavior is characteristic of logarithms.

$$\ln(2^n) = n\ln(2)$$

Information is measured by a logarithmic function of the number of choices.

The simplest definition of information content of a message would be $I =$ constant $\times \log_\alpha \mathscr{Q}$ or, without loss of generality,

$$I = K\ln\mathscr{Q} \tag{16-1}$$

where $\mathscr{Q}$ is the number of choices from which the message was selected and K is an arbitrary constant that determines the units for I, taking into account the base of the logarithm. If $K = 1/\ln 2$ (or if the constant is 1 and the base of the logarithm is 2), I is measured in bits.

If there are $\mathscr{Q}$ choices, of equal a priori probability, the probability of any one of those choices is

$$P = \frac{1}{\mathcal{Q}} \tag{16-2}$$

so an alternative definition of information content is

$$I = -K \ln P \qquad \text{or} \qquad I \text{ (bits)} = -\log_2 P \tag{16-3}$$

EXAMPLES 16.1

A. How many bits of information are carried by a signal whose probability is $P = 1$?
B. How many bits are carried by a signal whose probability is $P = 0.01$?
C. How many bits are carried by 10 successive signals, each with $P = 0.01$?
□

When two events are independent, the probabilities are multiplicative.

$$P_{1,2} = P_1 \times P_2 \tag{16-4}$$

and

$$I_{1,2} = -K \ln P_{1,2} = I_1 + I_2 = -K \ln P_1 - K \ln P_2 \tag{16-5}$$

The total amount of information is the sum of the information in all time units, provided that the signal in each time unit may be considered independent of the signal in any other time unit. When the last condition is not satisfied (as in transmitting English, where certain combinations allow the next letter to be predicted with ease), the information content is lowered.

It need not be assumed that all character types have equal probabilities. In English a space (probability 0.2) and the letter e (probability 0.105) are far more likely than j, q, or z (probabilities approximately 0.001). Frequencies of occurrence of letters, in English, are given in Table 16.1.

EXAMPLES 16.2

A. Find the information carried by a letter e in English. Assume no redundancy in the English.
B. Find the information carried by a letter q.
C. Estimate the information carried by a letter u when it follows the letter q in English.
□

In a very long representative sample of English there are n_a occurrences of a, n_b occurrences of b, and so on. $\sum n_i = N$. Then the probability of the ith letter is $p_i = n_i/N$ and $\sum p_i = 1$.

To find the number of possible distinct messages that could be constructed from these N characters, we note first that there are $N!$ permutations possible for N items. However, some of these permutations consist of permutations of the letters a among themselves; these arrangements are indistinguishable, so $N!$ must be divided by $n_a!$. Correcting, similarly, for permutations of b, of c, and so on, we find that

$$\mathcal{Q} = \frac{N!}{\prod_i n_i!}$$

The probability of a particular message that fits these criteria is $P = 1/\mathscr{Q}$,

$$P = \frac{\prod_i n_i!}{N!} \tag{16-6}$$

The information content of the message is

$$I = -K \ln P = -K \ln\left(\frac{\prod_i n_i!}{N!}\right)$$

This can be simplified by means of Stirling's approximation,[2]

$$\ln n! = n \ln n - n$$

to an adequate approximation for present purposes. Set

$$\ln\left(\frac{\prod_i n_i!}{N!}\right) = \sum_i \ln n_i! - \ln N! = \sum_i n_i \ln n_i - \sum_i n_i - (N \ln N - N)$$

Therefore, because $\sum n_i = N$,

$$I = -K \sum_i n_i \ln \frac{n_i}{N} \tag{16-7}$$

Table 16.1 Probabilities of Letters in English

Symbol	p	$-\log_2 p$	$-p \log_2 p$	Symbol	p	$-\log_2 p$	$-p \log_2 p$
space	0.2	2.32	0.464	*c*	0.023	5.44	0.125
e	0.105	3.25	0.341	*f,u*	0.0225	5.47	0.123
t	0.072	3.80	0.273	*m*	0.021	5.57	0.117
o	0.0654	3.93	0.257	*p*	0.0175	5.84	0.102
a	0.063	3.99	0.251	*y,w*	0.012	6.38	0.0766
n	0.059	4.08	0.241	*g*	0.011	6.51	0.0716
i	0.055	4.18	0.230	*b*	0.0105	6.57	0.0690
r	0.054	4.21	0.227	*v*	0.008	6.97	0.0557
s	0.052	4.27	0.222	*k*	0.003	8.38	0.0251
h	0.047	4.41	0.207	*x*	0.002	8.97	0.0179
d	0.035	4.84	0.169	*j,q,z*	0.001	9.97	0.0001
l	0.029	5.11	0.148				

[2]Factorials may be approximated by the series

$$\ln n! = \left(n + \frac{1}{2}\right)\ln n - n + \frac{1}{2}\ln 2\pi + \left(\frac{B_1}{1\cdot 2}\frac{1}{n} - \frac{B_2}{3\cdot 4}\frac{1}{n^3} + \frac{B_3}{5\cdot 6}\frac{1}{n^5} - \cdots\right)$$

where the B_i are Bernoulli numbers. For small n, the factorial can be evaluated directly. For large n only the first terms are significant, so $n! \approx \sqrt{2\pi n}(n + \frac{1}{2})/e^{n+1/2}$ or, adequate for most purposes, $\ln n! = n \ln n - n$. See *Am. J. Phys* **51,** 776 (1983).

ANSWERS 16.1

A. $I = -\log_2 1 = 0.$
B. $I = -\log_2 0.01 = -\ln 0.01/\ln 2 = -\log 0.01/\log 2 = 6.64$ bit.
C. $I = -10\log_2 0.01 = -10\ln 0.01/\ln 2 = 66.4$ bit.

ANSWERS 16.2

A. $I = -\ln 0.105/\ln 2 = 3.25$ bit.
B. $I = -\ln 0.001/\ln 2 = 9.97$ bit.
C. Probability of q being followed immediately by u is probably about 99% in English prose (notwithstanding the exceptions in this example), so $I \approx -\ln 0.99/\ln 2 = 0.015$ bit.

Dividing by N, to find the information per character, and replacing n_i/N with p_i gives *Shannon's formula:*

$$\frac{I}{N} = -K\sum_i p_i \ln p_i \tag{16-8}$$

The average information per character (in appropriate units) is equal to the sum, over all characters, of the product of probability times ln *(probability).*

EXAMPLES 16.3

A. Based on Shannon's formula, which carry more information in a typical English passage, the characters e or x?
B. Shannon's formula does not make specific provision for redundancy. What is the ratio of the number of characters in the original form of the message, "*Ths sntnc hs hd ll th vwls rmvd*" compared to the reduced form given here? □

16.3 Information and Entropy

Let there be N molecules of an ideal gas in volume V at a fixed temperature. Select, arbitrarily, a certain-size volume element, δV, small enough that the probability of more than one molecule occupying the same volume element may be neglected. Then the molecules are distributed among $V/\delta V$ locations. Imagine that it were possible to know the precise *microstate* of the gas (with respect to position, only). That is, assume that it is known precisely which of the volume elements are occupied. To find the information contained in this specification, consider that there are $V/\delta V$ choices of location for each molecule, so for N molecules there are $(V/\delta V)^N$ choices:

$$I = K\ln\left(\frac{V}{\delta V}\right)^N = KN\ln\frac{V}{\delta V} \tag{16-9}$$

except that the molecules are indistinguishable, so we should divide by the permutations of the molecules among the occupied volume elements. Hence

$$I = K\ln\left[\frac{(V/\delta V)^N}{N!}\right]$$

In practice, we know nothing of the microstate. It would be better to think of this expression as a measure of our *ignorance* rather than of information known to us. It is information *required* but not available to us. To emphasize this difference between a fully specified message that carries information $I = K \ln \mathcal{Q}$ and an unspecified microstate that would require information $K \ln \mathcal{Q}$ that we do not have, we will represent the latter situation by a script I, $\mathcal{I}$.

> *The information contained in a fully specified microstate over $\mathcal{Q}$ states is $I = K \ln \mathcal{Q}$; the information required, or our* ignorance, *of a thermodynamic system in which the particles are distributed in an unknown way among $\mathcal{Q}$ states, is $\mathcal{I} = K \ln \mathcal{Q}$.*

Consider now two containers, one with N_A molecules of gas A and the other with N_B molecules of gas B, at the same temperature and volume. Initially, the ignorance (or information required) is just the sum for the two containers.

$$\mathcal{I}_i = K\left\{ \ln\left[\frac{(V/\delta V)^{N_A}}{N_A!}\right] + \ln\left[\frac{(V/\delta V)^{N_B}}{N_B!}\right]\right\}$$

If the two containers are now connected, the gases expand without change of temperature, so the momentum distribution is unchanged. However, each gas now occupies a volume twice as great as before. Our ignorance (or the information required) has increased to

$$\mathcal{I}_f = K\left\{ \ln\left[\frac{(2V/\delta V)^{N_A}}{N_A!}\right] + \ln\left[\frac{(2V/\delta V)^{N_B}}{N_B!}\right]\right\}$$

Most terms drop out in the difference, $\mathcal{I}_f - \mathcal{I}_i$, leaving

$$\begin{aligned}\Delta \mathcal{I} &= K(N_A \ln 2V - N_A \ln V + N_B \ln 2V - N_B \ln V)\\ &= K(N_A \ln 2 + N_B \ln 2)\end{aligned}$$

Comparison of this result with the increase in uncertainty calculated for the same problem in Section 9.6,

$$\Delta S = R(n_A + n_B) \ln 2 \qquad (16\text{-}10)$$

shows that the increase in our ignorance, $\Delta \mathcal{I}$, is exactly equal to the increase in uncertainty, or entropy, ΔS, if $K = k = 1.38 \times 10^{-23}$ J/K.

> *Entropy may be defined by the equation*
>
> $$S = k \ln \mathcal{Q} \qquad (16\text{-}11)$$
>
> *where k is Boltzmann's constant and $\mathcal{Q}$ is the number of states accessible to the system.*

An equivalent result is obtained, in somewhat different fashion, by considering a crystal lattice with 1 mol, N_A, of lattice sites. If each site is occupied by a gold atom, the structure is completely known. There is no ignorance; no further information is required to specify what atoms are where, because each gold atom is indistinguishable from other gold atoms. There is the greatest possible order to the arrangement, with no uncertainty (within the classical model).

ANSWERS 16.3

A. Each e carries much less information than an x, but there are 50 times as many, so collectively the e's contribute 19 times as much to the information content (neglecting redundancy).
B. There are 45 characters in the original sentence, with all vowels restored, compared with 32 in the shortened version (including the terminators), giving a ratio of 1.4. Total redundancy in English is typically on the order of 3/1.

Now imagine an identical set of lattice sites filled with gold and silver atoms in equal numbers. Each site can be filled with equal probability with silver or with gold. There are two choices for each site, independent of the choices for any other lattice site. Then our ignorance is the information required to specify the structure in detail, which is

$$\mathcal{I} = K \ln 2^{N_A} = K N_A \ln 2$$

If we choose $K = k = R/N_A$, the Boltzmann constant,

$$\mathcal{I} = R \ln 2 \tag{16-12}$$

This is precisely the amount of uncertainty, ΔS, introduced if $\frac{1}{2}$ mol of silver and $\frac{1}{2}$ mol of gold are permitted to diffuse into each other to produce a random solid solution.

Imagine that we start with a carefully prepared system of fully specified microstate. For this system, which is not yet at equilibrium, our ignorance is zero and $I = k \ln \mathcal{Q}$ is a maximum. With time, the system decays toward equilibrium. Information, I, decreases and ignorance, $\mathcal{I}$, and hence entropy, S, increase.

$$\Delta I < 0 \qquad \Delta \mathcal{I} = \Delta S > 0 \qquad \Delta I + \Delta \mathcal{I} = 0$$

By intervening we could reorder the system, in whole or in part. For example, we can recompress a gas back to a smaller volume, or separate salt from water, or isolate silver and gold from a solid solution. This would represent an increase in information and a decrease in ignorance, or entropy.

$$\Delta I > 0 \qquad \Delta \mathcal{I} = \Delta S < 0 \qquad \Delta I + \Delta \mathcal{I} = 0$$

However, we know that the decrease of entropy of the system requires an increase of entropy elsewhere, in the surroundings. Our increase in information of the system has resulted in an even greater increase in entropy, or ignorance, of system plus surroundings.

$$\left\{ \begin{matrix} \text{system +} \\ \text{surroundings} \end{matrix} \right\} \qquad \Delta S = \Delta \mathcal{I} = -\Delta I > 0 \tag{16-13}$$

The universe becomes less determined, or more random, with the passage of time. Our increase in information of physical systems is smaller than our increase in ignorance.[3]

[3]It is therefore true, if perhaps irrelevant, that the net effect of every experiment is to decrease our knowledge of the universe.

Previously, we recognized that the amount of information carried by a signal is proportional to the probability. $I = -K \ln P = +K \ln \mathcal{Q}$ with $P = 1/\mathcal{Q}$. Now we take a different approach. If we do not know the "message," or microstate, we can still count the number of possible microstates, which we are calling $\mathcal{Q}$.

Assume now that the a priori probability of each microstate is the same (consistent with an assumed set of external parameters). Then the probability of the system being in *any* of a set of microstates with specific values of macroscopic variables of E, H, G, F, S, and so on, is proportional to the *number* of microstates that correspond to that thermodynamic macrostate. That is, the more states are available, the greater the probability that at least one of those states will be the state that is occupied. We are therefore interested in a probability $P \sim \mathcal{Q}$. This probability is related to our ignorance of the system, $\mathcal{I} = K \ln P$.

The statistical-mechanical definition of entropy may be expressed as

$$S \sim \ln P \tag{16-14}$$

Entropy is proportional to the logarithm of a probability P of a macrostate, where P, in turn, is proportional to the number of microstates accessible to the system contained within that macrostate: $P \sim \mathcal{Q}$.

Note particularly that, following custom, we have now changed from looking at the probability of a microstate ($P \sim 1/\mathcal{Q}$) to looking at the collective probability of the microstates that constitute a given macrostate. Each of the microstates is assumed equally probable, so the probability of a macrostate increases with increasing number of microstates contributing to the macrostate.

16.4 The Partition Function

Consider a system containing N particles, each of which may be assigned some energy value, ϵ_k, where the ϵ_k are not necessarily all different. Compare the energy states to apartments or office suites of a building (not all on different floors), each of which may have an arbitrary number of occupants. Then for even a moderate number of particles and energy states, there will be a very large number of ways in which the particles can be distributed among the states. All the arrangements, however, must satisfy two conditions.

First, the number of particles is predetermined and not subject to change. If the number of particles in the energy state ϵ_k is represented by N_k, called the *occupation number,* then

$$\sum_k N_k = N \tag{16-15}$$

Second, the total energy is fixed, so

$$\sum_k N_k \epsilon_k = E \tag{16-16}$$

An analogy would be a shopping trip with a fixed amount of money to spend (E). Each item has a price (ϵ_k). A few high-priced items could consume the entire budget, or a large number of low-priced items could be purchased. The limitation is that the sum over the number of items (N_k), each multiplied by its price (ϵ_k), cannot differ from the total amount to be spent on the shopping trip (E).

A particular set of occupation numbers, $\{N_k\}$, defines a *distribution,* which is sufficient to determine the (macroscopic) state of the system. There are, however, many different *arrangements* of the particles consistent with the specified distribution.

For simplicity, imagine a linear representation of all N particles according to the energy state they occupy. First are all particles in states of the lowest energy, next are those of the next highest energy, and so on. Now a permutation of the particles does not change the distribution; there are $N!$ ways of ordering the particles, which give $N!$ equivalent arrangements. For present purposes, however, we are concerned only with the number, N_k, in each energy state, not with the order in which the particles were placed in the states. The total number of arrangements is therefore divided by the number of possible orders within each state, $N_k!$. The number of different ways that N particles can be arranged in the energy states, ϵ_k, is thus defined by the number of possible sets of occupation numbers, N_k. The number of arrangements is

$$n(N_k) = \frac{N!}{\prod_k N_k}!$$

There is no a priori reason to prefer any one arrangement of the particles over any other, so it is reasonable to assume that each arrangement is equally likely. The probability of finding the particles distributed in a particular way (i.e., with a specific set of occupation numbers, $\{N_k\}$) is therefore proportional to the number of equivalent arrangements that fit that set of occupation numbers. The probability for a set $\{N_k\}$ is proportional to $n(N_k)$.

$$P \sim \frac{N!}{\prod_k N_k}! \tag{16-17}$$

(A correction may be required for indistinguishable particles, but it will not affect the equations derived here. See Section 20.3.)

Taking the logarithm of each side gives

$$\ln P = \ln N! - \sum_k \ln N_k! + \text{ constant}$$

We are interested in finding the most probable state of the system, or the maximum value of P. At the maximum, the rate of change of P with the N_k is zero.

$$\delta P = 0$$

so

$$\delta \ln P = \delta \ln N! - \delta \sum_k \ln N_k! = 0$$

Because N is fixed, $\delta \ln N! = 0$. Therefore,

$$\delta \sum_k \ln N_k! = \sum_k \delta \ln N_k! = 0$$

Substituting Stirling's approximation in the form

$$\ln m! = m \ln m - m$$

gives

$$\begin{aligned}\sum_k \delta \ln N_k! &= \sum_k \delta(N_k \ln N_k) - \sum_k \delta N_k = 0 \\ &= \sum \ln N_k \delta N_k + \sum N_k \delta \ln N_k - \sum \delta N_k \qquad (16\text{-}18) \\ &= \sum \ln N_k \delta N_k\end{aligned}$$

because $\sum_k N_k \, \delta \ln N_k = \sum N_k \, \delta N_k / N_k = \sum \delta N_k$ (and the last terms, $\sum \delta N_k - \sum \delta N_k$, are zero, anyway). The three conditions are then

$$\begin{aligned}\sum_k \ln N_k \, \delta N_k &= 0 \qquad \text{(maximum } P) \\ \delta N = \sum_k \delta N_k &= 0 \qquad \text{(constant } N)\end{aligned} \qquad (16\text{-}19)$$

and

$$\delta E = \sum_k \epsilon_k \, \delta N_k = 0 \qquad \text{(constant } E)$$

The first equation arises from the assumption that the probability is a maximum, which is equivalent to the second law of thermodynamics. The second equation expresses the constancy of the number of particles, an assumption made implicitly in classical thermodynamics. The third equation is the condition of constant energy, for the system that is not exchanging energy with its surroundings, and is therefore equivalent to the first law of thermodynamics.

The three conditions are to be satisfied simultaneously, which can be accomplished with a method due to Lagrange. Because each of the expressions is equal to zero, their sum must also be zero. However, simply adding the three equations together replaces three conditions with one. A more general result is to take an arbitrary linear combination.

$$\alpha \sum \delta N_k + \beta \sum \epsilon_k N_k + \sum \ln N_k \, \delta N_k = 0 \qquad (16\text{-}20)$$

(If a third constant were inserted, it could be removed by dividing through by that constant.) The equation as written replaces three equations with one equation and two constants—still a total of three conditions to be satisfied. That is, there are implied equations of the form α = some value and β = some value, which specify the state that is being described by the equation.

The equation must be valid for any change in the N_k values. Over time it is expected that particles will exchange energies, so individual particles move between energy states and the occupation numbers can change. The δN_k are arbitrary,[4] so the equation must vanish term by term.

[4]Without the constants α and β, the δN_k would not all be independent. The Lagrange multipliers allow us to assume that all δN_k are independent, and therefore that the coefficients must vanish independently.

$$\alpha\, \delta N_k + \beta \epsilon_k\, \delta N_k + \ln N_k\, \delta N_k = 0 \qquad (\text{all } \delta N_k)$$

and therefore

$$\ln N_k + \alpha + \beta \epsilon_k = 0$$

or

$$N_k = e^{-\alpha} e^{-\beta \epsilon_k} \tag{16-21}$$

It is convenient at this point to group together any of the ϵ_k that are equal, or *degenerate*. If there are g_k energy states with energy ϵ_k, then

$$n_k = g_k e^{-\alpha} e^{-\beta \varepsilon_k} \tag{16-22}$$

where now the ε_k are all different. The different ε_k are called *energy levels*. Each energy level, ε_k, has g_k different component *energy states*, ϵ_k. The g_k values may be different for different energy levels.

By comparison with specific problems that have been solved, such as the barometric equation and the Maxwell–Boltzmann distribution law, the constant β may be recognized as $1/kT$. (A more general argument will be given in Section 19.3.) The constant α has not yet been determined. However, we *can* find relative populations, n_k/N.

$$\sum n_k = N = e^{-\alpha} \sum g_k e^{-\beta \varepsilon_k} \tag{16-23}$$

and therefore

$$\frac{n_k}{N} = \frac{n_k}{\sum n_k} = \frac{e^{-\alpha} g_k e^{-\beta \varepsilon_k}}{e^{-\alpha} \sum g_k e^{-\beta \varepsilon_k}}$$

or, letting

$$Z \equiv \sum_k g_k e^{-\beta \varepsilon_k} \tag{16-24}$$

$$\frac{n_k}{N} = \frac{g_k e^{-\beta \varepsilon_k}}{\sum_k g_k e^{-\beta \varepsilon_k}} \equiv \frac{g_k e^{-\beta \varepsilon_k}}{Z} \tag{16-25}$$

The function Z is called the *state sum* (*Zustandsumme*) or the *partition function*. It is written as a sum of exponential terms over all energy levels, multiplied by the degeneracies, g_k, which give the number of states per energy level. It is also called the *sum over states*.

Because the partition function is proportional to the total number of molecules, it serves as a normalization factor to permit determination of n_k/N at any temperature ($T = 1/k\beta$), provided that the ε_k and g_k are known. When Z and N are known, α can be found from the condition

$$e^{-\alpha} = \frac{N}{Z} \tag{16-26}$$

or $\alpha = \ln Z - \ln N$.

16.5 Evaluation of Thermodynamic Properties

The primary goal of statistical mechanics is to calculate the thermodynamic properties of a system from the detailed molecular model of the system. If the energy levels, ε_k, and degeneracies, g_k, are known, the partition function can be evaluated. The thermodynamic properties of the system are calculated from the partition function and values of external variables.

The average energy per particle is easily found.

$$\bar{\varepsilon} = \frac{\sum n_k g_k \varepsilon_k}{\sum n_k}$$

The energy per mole is $\mathrm{E}_k = N_A \varepsilon_k$ and $\mathrm{E} = N_A \bar{\varepsilon}$. Using

$$\frac{\mathrm{E}_k}{R} = \frac{\varepsilon_k}{k}$$

we obtain

$$\mathrm{E} = \frac{\sum g_k \mathrm{E}_k e^{-\mathrm{E}_k/RT}}{\sum g_k e^{-\mathrm{E}_k/RT}} \equiv \frac{\sum g_k \mathrm{E}_k e^{-\mathrm{E}_k/RT}}{Z} \qquad (16\text{-}27)$$

The derivative of the sum over states, with respect to temperature, is

$$\frac{dZ}{dT} = \frac{1}{RT^2} \sum g_k \mathrm{E}_k e^{-\mathrm{E}_k/RT}$$

Therefore, the energy may be written

$$\mathrm{E} = \frac{RT^2(dZ/dT)}{Z} = RT^2 \frac{d(\ln Z)}{dT} \qquad (16\text{-}28)$$

From the expression for energy, $\mathrm{C}_v = (\partial E/\partial T)_V$ is found.

$$\begin{aligned} \mathrm{C}_v &= \frac{\partial}{\partial T}\left[RT^2 \frac{\partial(\ln Z)}{\partial T}\right] \\ &= RT^2 \frac{\partial^2 \ln Z}{\partial T^2} + 2RT \frac{\partial \ln Z}{\partial T} \\ &= \frac{R}{T^2} \frac{\partial^2 \ln Z}{\partial (1/T)^2} \end{aligned} \qquad (16\text{-}29)$$

Let S_o be the molar entropy at 0 K. Then the entropy at the temperature T is

$$\mathrm{S} = \mathrm{S}_o + \int_0^T \frac{\mathrm{C}_v}{T} dT = \mathrm{S}_o + \int_0^T \frac{\partial}{\partial T}\left(RT^2 \frac{\partial \ln Z}{\partial T}\right) \frac{dT}{T}$$

which integrates to give

$$\mathrm{S} = \mathrm{S}_o + [R \ln Z]_0^T + \left[\frac{\mathrm{E}}{T}\right]_0^T$$

EXAMPLE 16.4

Show that

$$\int \frac{\partial}{\partial T}\left(RT^2 \frac{\partial \ln Z}{\partial T}\right)\frac{dT}{T} = R \ln Z + \frac{\mathrm{E}}{T}$$

using $\mathrm{E} = RT^2[d(\ln Z)/dT]$. □

A reference level for E is chosen such that as $T \to 0$, $\mathrm{E} \to 0$ (see Problem 9). Then $\lim_{T\to 0}(\mathrm{E}/T) = 0$, because $\mathrm{E} \sim e^{-\mathrm{E}_k/RT}$ approaches zero more rapidly than does T. Therefore,

$$\mathrm{S} = \mathrm{S}_o + \frac{\mathrm{E}}{T} + R \ln Z - R \ln Z_o \qquad (16\text{-}30)$$

The first and last terms are independent of temperature. It is conventional to set

$$\mathrm{S}_o = R \ln Z_o = R \ln(g_o e^{-\beta\varepsilon_o}) = R \ln g_o$$

Then

$$\mathrm{S} = \frac{\mathrm{E}}{T} + R \ln Z \qquad (16\text{-}31)$$

and at 0 K,

$$\mathrm{S} = \mathrm{S}_o = R \ln g_o \qquad (16\text{-}32)$$

Combining the value of entropy with the earlier value for energy,

$$\mathrm{E} = RT^2 \frac{d(\ln Z)}{dT}$$

and values of the parameters P, V, and T gives sufficient information to calculate the thermodynamic potentials,

$$E = NkT^2 \frac{d(\ln Z)}{dT}$$

$$S = \frac{E}{T} + Nk \ln Z$$

$$H = E + PV \qquad (16\text{-}33)$$

$$G = H - TS$$

$$F = E - TS = -NkT \ln Z$$

For an ideal gas, the pressure is known to be

$$P = \frac{NkT}{V}$$

and therefore

$$H = E + NkT$$

{ideal gas}
$$G = NkT - NkT \ln Z \qquad (16\text{-}34)$$
$$F = -NkT \ln Z$$

To evaluate Z, and therefore E, S, and other thermodynamic properties, it is only necessary to know the energy states of a system. For an ideal gas, the energy is a sum of energies of individual molecules. Rotational energy states may be calculated from the moments of inertia. Vibrational energy states are calculated from observed frequencies of vibration, and translational states from the mass of a molecule.

The equations can be put in more convenient form by limiting the range of systems. In particular, the partition function of a gas is the product of three parts, representing translation, rotation, and vibration.[5]

$$Z = Z_{\text{tr}} \cdot Z_{\text{rot}} \cdot Z_{\text{v}} \qquad (16\text{-}35)$$

If the gas is ideal (Section 20.4),

$$Z_{\text{tr}} = V\left(\frac{2\pi mkT}{h^2}\right)^{3/2} = \left(\frac{2\pi M}{h^2}\right)^{3/2} \frac{(kT)^{5/2}}{\sqrt{N_A}P} \qquad (16\text{-}36)$$

EXAMPLES 16.5

A. Evaluate the translational partition function, and $\ln Z_{\text{tr}}$, for 1 mol of nitrogen molecules ($M = 28.01340$ g/mol) at 300 K and 1 atm.
B. Evaluate the translational partition function, and $\ln Z_{\text{tr}}$, for 1 mol of methane molecules, CH_4 ($M = 16.04303$ g/mol) at 300 K, 1 atm. □

For a linear molecule, in the approximation of a rigid rotor (Section 18.2),

$$Z_{\text{rot}} = \sum_{J=1}^{\infty} (2J + 1)e^{-BJ(J+1)hc/kT} \qquad (16\text{-}37)$$

but at room temperature this is adequately approximated by the high-temperature value,

$$Z_{\text{rot}} = \frac{kT}{shcB} \qquad (16\text{-}38)$$

where $B = h/(8\pi^2 cI_b)$ is called the *rotational constant*, I_b is the moment of inertia about the center of mass, and s is the symmetry constant, which is 1 if the molecule is asymmetric or 2 if it is symmetric about its center point. (The

[5]See Chapter 18 for more explicit information on the quantized states. The assumption made throughout the discussion is that the energies, or frequencies, of each distinctly different type of energy storage mode—electronic, vibrational, rotational, and so on— are quite different in magnitude and therefore the energies can be separated. This is called the *Born–Oppenheimer approximation* (see Problem 16.8).

ANSWER 16.4

$$\frac{\partial}{\partial T}\left(RT\frac{\partial \ln Z}{\partial T}\right) = RT\frac{\partial^2 \ln Z}{\partial T^2} + R\frac{\partial \ln Z}{\partial T}$$

Therefore,

$$\frac{1}{T}\frac{\partial}{\partial T}\left(RT^2\frac{\partial \ln Z}{\partial T}\right) = 2R\frac{\partial \ln Z}{\partial T} + RT\frac{\partial^2 \ln Z}{\partial T^2}$$

$$= \frac{\partial}{\partial T}\left(RT\frac{\partial \ln Z}{\partial T}\right) + R\frac{\partial \ln Z}{\partial T}$$

so

$$\int \frac{1}{T}\frac{\partial}{\partial T}\left(RT^2\frac{\partial \ln Z}{\partial T}\right) dT = \int d\left(RT\frac{\partial \ln Z}{\partial T}\right) + \int R\, d(\ln Z)$$

$$= RT\frac{\partial \ln Z}{\partial T} + R \ln Z$$

The first term is E/T. The integral must be evaluated between limits $(0 \rightarrow T)$.

ANSWERS 16.5

A. $Z_{tr} = R(2\pi k/N_A h^2)^{3/2}\left[M^{3/2}T^{5/2}P^{-1}\right] = 4.941 \times 10^{31}\left[M^{3/2}T^{5/2}P^{-1}\right]$ in SI units. At 300 K, 1 atm, therefore, $Z_{tr}(N_2) = 3.56 \times 10^{30}$; $\ln Z_{tr} = 70.348$.
B. $Z_{tr}(CH_4) = 1.54 \times 10^{30}$; $\ln Z_{tr} = 69.512$.

---

symmetry constant is discussed in Sections 18.2 and 20.5. It is often represented by σ, which could be confused here with the wavenumber, $\sigma = \nu/c$.)

For a nonlinear rigid molecule, in the high-temperature approximation,

$$Z_{rot} = \frac{1}{s}\sqrt{\frac{\pi}{ABC}\left(\frac{kT}{hc}\right)^3} \qquad (16\text{-}39)$$

with A, B, and C the rotational constants, inversely proportional to I_a, I_b, and I_c, respectively.

EXAMPLES 16.6

A. The rotational constant for nitrogen is 2.010 cm^{-1} = 201 m^{-1}. Evaluate the rotational partition function, and $\ln Z_{rot}$, for N_2 at 300 K.
B. Methane is a spherical rotor, with $A = B = C = 5.25$ cm^{-1} and s = 12. Evaluate the rotational partition function, and $\ln Z_{rot}$, for CH_4 at 300 K. □

In the approximation of a harmonic oscillator (Sections 18.1 and 20.2), each vibrational mode contributes $(1 - e^{-\eta_i})^{-1}$ to Z, where $\eta_i \equiv hc\sigma_i/kT$ and σ_i, the wavenumber, is the classical frequency of vibration divided by the speed of light. Grouping together factors for degenerate modes (of degeneracy g_i) gives

$$Z_v = \prod_i (1 - e^{-\eta_i})^{-g_i} \qquad (16\text{-}40)$$

EXAMPLES 16.7

A. Nitrogen has a single vibrational mode, nondegenerate, with $\sigma = 2359.61$ $\text{cm}^{-1} = 235,961\ \text{m}^{-1}$. Evaluate the vibrational partition function of N_2 at 300 K.
B. Methane has four vibrational frequencies: $\sigma_1 = 2914.2\ \text{cm}^{-1}$ $(g = 1)$, $\sigma_2 = 1526.9\ \text{cm}^{-1}(g = 2)$, $\sigma_3 = 3020.3\ \text{cm}^{-1}(g = 3)$, and $\sigma_4 = 1306.2\ \text{cm}^{-1}(g = 3)$. Evaluate the four contributions to the partition function and their product. □

Provided that there are no internal rotations, the thermodynamic functions are the sum of the terms evaluated above.

$$\begin{aligned} H &= E_o + H_{tr} + H_{rot} + H_v \\ C_p &= C_{p_{tr}} + C_{p_{rot}} + C_{p_v} \\ S &= S_{tr} + S_{rot} + S_v \end{aligned} \tag{16-41}$$

The translational contributions are

$$\begin{aligned} H_{tr} &= \frac{5}{2}RT \\ C_{p_{tr}} &= \frac{5}{2}R \end{aligned} \tag{16-42}$$

and (including the correction for indistinguishability, Section 20.3)

$$\begin{aligned} S_{tr} &= R\left\{\frac{5}{2} + \ln\left[\frac{V}{N}\left(\frac{2\pi mkT}{h^2}\right)^{3/2}\right]\right\} \\ &= R\left[\frac{5}{2} + \ln\left(\frac{2\pi M}{Nh^2}\right)^{3/2}\frac{(kT)^{5/2}}{P}\right] \end{aligned} \tag{16-43}$$

By substituting values in SI units, this may be reduced to

$$S_{tr} = R\left(\frac{3}{2}\ln M + \frac{5}{2}\ln T - \ln P\right) + 172.298$$

known as the Sackur–Tetrode equation.

EXAMPLES 16.8

A. Evaluate the translational contribution to the entropy of N_2 at 300 K and 1 atm.
B. Evaluate the translational contribution to the entropy of CH_4 at 300 K and 1 atm.
□

For linear molecules at room temperatures,

$$\begin{aligned} H_{rot} &= E_{rot} = RT \\ C_{p_{rot}} &= C_{v_{rot}} = R \end{aligned} \tag{16-44}$$

and

$$S_{rot} = R\left(1 + \ln\frac{kT}{hcBs}\right)$$

ANSWERS 16.6

A. $Z_{rot}(N_2) = \dfrac{1.38 \times 10^{-23} \times 300}{2 \times 6.63 \times 10^{-34} \times 3 \times 10^8 \times 201} = 51.868$; $\ln Z_{rot} = 3.949$.

B. $Z_{rot}(CH_4) = \dfrac{1}{12}\sqrt{\dfrac{\pi}{(525)^3}}\left(\dfrac{kT}{hc}^3\right) = 36.969$; $\ln Z_{rot} = 3.610$.

ANSWERS 16.7

A. $\eta_i = \dfrac{6.63 \times 10^{-34} \times 3 \times 10^8 \times 2.35961 \times 10^5}{1.38 \times 10^{-23} \times 300} = 11.317$; $Z = 1.000\,012$.

B. $\eta_1 = 13.9764$, $\eta_2 = 7.32294$, $\eta_3 = 14.4852$, $\eta_4 = 6.26448$; $Z_1 = 1.000\,00$, $Z_2 = 1.001\,32$, $Z_3 = 1.000\,00$, $Z_4 = 1.007\,06$.

ANSWERS 16.8

A. $S_{tr} = 150.44$ J/mol · K for N_2.
B. $S_{tr} = 143.49$ J/mol · K for CH_4.

For nonlinear molecules at room temperatures,

$$H_{rot} = E_{rot} = \frac{3}{2}RT$$
$$C_{p_{rot}} = C_{v_{rot}} = \frac{3}{2}R \qquad (16\text{-}45)$$

and

$$S_{rot} = R\left\{\frac{3}{2} + \ln\left[\frac{1}{s}\sqrt{\frac{\pi}{ABC}}\left(\frac{kT}{hc}\right)^3\right]\right\}$$

EXAMPLES 16.9

A. Evaluate the rotational contribution to the entropy of N_2 at 300 K.
B. Evaluate the rotational contribution to the entropy of CH_4 at 300 K. □

The vibrational contributions are

$$H_v = E_v = RT\sum_i g_i \frac{\eta_i e^{-\eta_i}}{1 - e^{-\eta_i}}$$
$$C_{p_v} = C_{v_r} = R\sum_i g_i \frac{\eta_i^2 e^{-\eta_i}}{(1 - e^{-\eta_i})^2} \qquad (16\text{-}46)$$
$$S_v = R\sum_i g_i \frac{\eta_i e^{-\eta_i}}{1 - e^{-\eta_i}} + R\sum_i g_i \ln(1 - e^{-\eta_i})$$

EXAMPLES 16.10

A. Evaluate the vibrational contribution to the entropy of N_2 at 300 K.
B. Evaluate the vibrational contribution to the entropy of CH_4 at 300 K. □

The entropy and heat capacity are usually the most sensitive tests of the assignment of experimental values to a molecule. The energy term does not include electronic or nuclear energies, which are independent of temperature near or below room temperature, nor does it include spin energies, which are negligible under normal conditions. The entropy also omits the effects of spin (except for the molecular symmetry constant), discussed in Section 20.3.

The derivation presented here has the advantage of great simplicity. The conclusions reached are fundamentally correct. Unfortunately, there are some loose ends and unanswered questions. If the results are to be applied in the context of classical mechanics, the definition of an energy state or energy level must be clarified. If the equations are to be consistent with quantum mechanics, more attention must be given to lowest-energy states, to specification of states of large systems, to the degree of distinguishability of particles, and ultimately to rules governing the simultaneous occupation of a quantum state by more than one particle.

Before undertaking more rigorous derivations that can lead to additional insights into these questions, it will be necessary to consider some background problems in classical mechanics and the modifications of classical mechanics that lead to quantum mechanics.

SUMMARY

Information $= I = K \ln \mathcal{Q}$, where $\mathcal{Q} =$ number of choices.
Probability $= P = 1/\mathcal{Q}$; $I = -K \ln P$.
One "bit" = a choice of two options; I (bits) $= -\log_2 P$.
For independent events, $I = \sum I_i$; $\quad P = \prod P_i$.

Shannon's formula: n_i events of ith type; $\sum n_i = N$ events; $p_i = \dfrac{n_i}{N}$; $\dfrac{I}{N} = -K \sum_i p_i \ln p_i$ gives the average information per event.

Information, I, is needed to fully specify the microstate of a system. If that information is *not* known, it is a measure of our *ignorance*, $\mathcal{I}$, of that system.

An increase in ignorance, $\mathcal{I}$, is an increase in entropy, S, and a decrease in information, I. $\Delta\mathcal{I} = \Delta S = -\Delta I$ for the system. For system plus surroundings, $\Delta\mathcal{I} > |\Delta I|$.

If there are $\mathcal{Q}$ possible states, each of equal a priori probability, then $P \sim \mathcal{Q}$ and $S \sim \ln P = \ln \mathcal{Q} + constant$.

For a fixed number of particles, $\sum_k N_k = N$, and fixed energy, $\sum_k N_k \epsilon_k = E$, the maximum value of the probability, $P \sim N!/\prod_k N_k!$, is found if $N_k = e^{-\alpha} e^{-\beta\epsilon_k}$, where the constant β can be shown to be $\beta = 1/kT$ and the constant α (a measure of the free energy) serves as a normalization constant.

Grouping the ϵ_k into levels ε_k, each of degeneracy g_k and population n_k, let

$$Z \equiv \sum_k g_k e^{-\beta\varepsilon_k} \qquad \text{and} \qquad \frac{n_k}{N} = \frac{g_k e^{-\beta\varepsilon_k}}{Z}$$

Average energy is $\bar{\varepsilon} = \dfrac{\sum n_k g_k \varepsilon_k}{\sum n_k}$ and the molar energy is

$$\mathrm{E} = N_A \bar{\varepsilon} = \frac{\sum g_k \mathrm{E}_k e^{-\mathrm{E}_k/kT}}{Z} = NkT^2 \frac{d(\ln Z)}{dT}$$

ANSWERS 16.9

A. $s_{rot} = 41.15$ J/mol · K for N_2.
B. $s_{rot} = 42.49$ J/mol · K for CH_4.

ANSWERS 16.10

A. $s_v = 1.044 \times 10^{-3}$ J/mol · K for N_2.
B. The contributions are 9.2×10^{-5}, 6.9×10^{-2}, 1.7×10^{-4}, and 2.8×10^{-1} for each of the modes, giving a total contribution of $s_v = 0.35$ J/mol · K for CH_4.

Other thermodynamic functions are

$$S = \frac{E}{T} + Nk \ln Z$$

$$F = -NkT \ln Z$$

For an ideal gas, $PV = NkT$, so

$$H = E + NkT$$

$$G = NkT(1 - \ln Z)$$

$$F = -NkT \ln Z$$

$$Z_{tr} = V\left(\frac{2\pi mkT}{h^2}\right)^{3/2}$$

$$H_{tr} = \frac{5}{2}RT$$

$$c_{p_{tr}} = \frac{5}{2}R$$

$$s_{tr} = R\left\{\frac{5}{2} + \ln\left[\frac{V}{N}\left(\frac{2\pi mkT}{h^2}\right)^{3/2}\right]\right\}$$

For a linear molecule,

$$Z_{rot} = \sum_{J=1}^{\infty}(2J + 1)e^{-BJ(J+1)hc/kT}$$

which becomes, at high temperatures (e.g., room temperature)

$$Z_{rot} = \frac{kT}{shcB}$$

$$H_{rot} = E_{rot} = RT$$

$$C_{p_{\text{rot}}} = C_{v_{\text{rot}}} = R$$

$$S_{\text{rot}} = R\left(1 + \ln \frac{kT}{hcBs}\right)$$

For a polyatomic (nonlinear) molecule at high T,

$$Z_{\text{rot}} = \frac{1}{s}\sqrt{\frac{\pi}{ABC}\left(\frac{kT}{hc}\right)^3}$$

where $A, B,$ and C are rotational constants and s is the symmetry number.

$$A = \frac{h}{8\pi^2 cI_a}, \qquad B = \frac{h}{8\pi^2 cI_b}, \qquad C = \frac{h}{8\pi^2 cI_c}$$

$$H_{\text{rot}} = E_{\text{rot}} = \frac{3}{2}RT; \; C_{p_{\text{rot}}} = C_{v_{\text{rot}}} = \frac{3}{2}R$$

and

$$S_{\text{rot}} = R\left\{\frac{3}{2} + \ln\left[\frac{1}{s}\sqrt{\frac{\pi}{ABC}\left(\frac{kT}{hc}\right)^3}\right]\right\}$$

For a harmonic oscillator, with $\eta_i \equiv \dfrac{hc\sigma_i}{kT} = \dfrac{h\nu_i}{kT}$,

$$Z_v = \prod_i \left(1 - e^{-\eta_i}\right)^{-g_i}$$

$$H_v = E_v = RT\sum_i g_i \frac{\eta_i e^{-\eta_i}}{1 - e^{-\eta_i}}$$

$$C_{p_v} = C_{v_r} = R\sum_i g_i \frac{\eta_i^2 e^{-\eta_i}}{(1 - e^{-\eta_i})^2}$$

$$S_v = R\sum_i g_i \frac{\eta_i e^{-\eta_i}}{1 - e^{-\eta_i}} + R\sum_i g_i \ln\left(1 - e^{-\eta_i}\right)$$

PROBLEMS

16.1. **a.** How many ways can four distinguishable objects (e.g., a penny, nickel, dime, and quarter) be arranged in a straight line?

b. How many bits of information could be carried by a message consisting of the four objects?

c. If a fifth object (e.g., a half dollar) is added, how many bits of information could be carried by the arrangements?

d. Is the information per object constant? (Compare parts b and c.)

16.2. Compare the number of characters in the following message: "Buick Century '85, 4 DR, AC, PS, PB, PDL, cruise, tilt, new tires" with the same message fully written out. What is the ratio? How would the same message appear if it were contained in an advertising brochure describing this car?

16.3. The local postmaster was under intense scrutiny by the local occupation forces, but he nevertheless succeeded in conveying messages to couriers by filling mailboxes selectively with letters visible through the box windows (as "mail" or "no mail" in each box).

a. If he had available five rows of 20 boxes (and adequate letters or fliers), how many bits of information could be displayed at a time?

b. To how many letters of English (at an assumed 3 bit/character average) would this be equivalent?

16.4. What is the increase in ignorance, or entropy, if 1 mol of gas A, 1 mol of gas B, and 1 mol of gas C, each at the same temperature, pressure, and initial volume, are allowed to mix isothermally within the same total volume (three times the original volume/mol)?

16.5. Some molecules, such as CO, can fit into the crystal correctly or turned end for end. There is no reason why the energy difference for such an error cannot be arbitrarily small. There is, however, a minimum ΔS if 1 mol of molecules is frozen into a crystal without control of this orientational degree of freedom, as compared to the crystal in proper alignment. How large is this ΔS in J/mol · K?

16.6. For a polar molecule in an electric field, the energy is given classically as $U = -\mathbf{p} \cdot \mathbf{E} = -pE \cos \vartheta$, but for many systems it is sufficient to consider only the two quantum states in which the moment is aligned with the field or against the field. The problem is then much like a one-dimensional random walk. Of the N dipoles, $\frac{1}{2}(N + n)$ are aligned with the field and $\frac{1}{2}(N - n)$ are aligned against the field. The probability is given by the binomial distribution,

$$P(n) = \frac{N!}{[(N + n)/2]![(N - n)/2]!} p^n q^{N-n}$$

In the absence of a field, $p = q = \frac{1}{2}$, but the field causes one direction to be favored, usually only slightly. The coefficient (involving factorials) is the *statistical weight* of a state of given n. Show by applying Stirling's formula in the form

$$\ln x! = x \ln x - x + \frac{1}{2} \ln 2\pi x,$$

that the statistical weight is $c - n^2/2N$, where c is a constant (independent of n).

16.7. Show that the entropy at absolute zero can be set equal to $s_o = R \ln g_o + C$, where C is any constant, and satisfy all the necessary conditions on s_o. [In practice, some additive terms to the entropy, such as the effect of spin degeneracy, are therefore often ignored without any significant consequences.]

16.8. The Born–Oppenheimer approximation divides the total energy of a molecule into electronic, vibrational, and rotational parts, plus other lower-energy terms: $E = E_{\text{el}} + E_{\text{vib}} + E_{\text{rot}} + \cdots$. Show that to this approximation, the partition function can be factored. $Z = Z_{\text{el}} \cdot Z_{\text{vib}} \cdot Z_{\text{rot}} \cdots$.

16.9. Show that any additive term to the energy, such as the "zero-point" vibrational energy, $\frac{1}{2}\hbar\omega$, contributes only a constant factor to the partition function and may therefore be ignored.

16.10. What is the (approximate) value of 5000!? (Express in conventional scientific notation.)

16.11. Find the total entropy of N_2 at 25°C and 1 atm.

16.12. Find the total entropy of CH_4 at 25°C and 1 atm.

16.13. Select a small molecule (other than N_2 or CH_4; your instructor may suggest an appropriate size or other limitation). Look up the physical constants in the literature and calculate the standard entropy, enthalpy, and free energy, at 25°C. Compare your calculated values with values from thermodynamic tables.

CHAPTER 17

Classical Mechanics and Phase Space

Application of statistical methods at the molecular level requires a detailed model of how atoms and molecules move and interact. For many purposes, classical mechanics is quite adequate. Ultimately, however, all conclusions must be checked by comparing them with calculations based on quantum mechanics. The transition is greatly facilitated by certain choices of variables and equations of motion in the classical treatment. Because readers will have different backgrounds in classical and quantum mechanics, a summary of the important ideas, sufficient for the adaptation required to statistical mechanics, is presented in this and the following chapter.

17.1 Phase Space

To specify the state of any particle it is necessary to know the spatial coordinates and the velocities (i.e., velocity components) and how the particle interacts with its surroundings. However, interactions depend on spatial coordinates, or possibly on velocities. Therefore, if spatial and velocity coordinates are known for one instant of time, the future and the history of the particle can in principle be determined, according to classical mechanics.

It is convenient to describe motions in terms of momentum coordinates rather than velocities. For one-dimensional motion, plot p_x along one axis and $q(= x)$ along another. The particle coordinates then trace out a line with time in this p–q plane. For three-dimensional motion of a particle, there are six orthogonal axes, in six-dimensional space. For N particles, a space of $6N$ dimensions is required.

The $6N$-dimensional space of position and momentum coordinates of N particles is called *phase space*. The location of a single point in this space defines all $6N$ coordinate values and, by the rules of classical mechanics, determines the future trajectory of the point through phase space. The changes in position and momentum coordinates with time describe the "phase" of the motion, in the same sense that the location of the moon, relative to the earth and sun, determines the phase of the moon or the position and velocity of a harmonic oscillator determine the phase of the oscillator. Because early analyses were concerned with gases, the space of $6N$ position and momentum coordinates is also called Γ space (for Γ*as*).

EXAMPLES 17.1

A. What would be the trace in phase space for a one-dimensional harmonic oscillator?
B. What would be the trace in phase space for a ball thrown upward (e.g., with $v_o = 10$ m/s) and subsequently falling back to its original location? □

The probability for the $6N$ coordinates of a system falling on any point in phase space must be vanishingly small. Probabilities must therefore be expressed for the system coordinates to fall within a small volume element, $dp_1\, dp_2 \cdots dp_{3N}\, dq_1\, dq_2 \cdots dq_{3N} \equiv dp\, dq$. That is, the point is in $dp\, dq$ if it falls simultaneously in the ranges $p_1 \to p_1 + dp_1$, $q_1 \to q_1 + dq_1$, $p_2 \to p_2 + dp_2$, $q_2 \to q_2 + dq_2$, and so on.

The probability of the coordinates falling in the phase space volume element $dp\, dq$ is given by a *probability density function,* $\rho = \rho(p_1, p_2, \ldots, p_{3N}, q_1, q_2, \ldots, q_{3N})$. The increment of probability for the element $dp\, dq$ is

$$\delta P = \rho\, dp\, dq \tag{17-1}$$

The integral over phase space must include all possibilities, so

$$\int \rho\, dp\, dq = 1 \tag{17-2}$$

The probability is dimensionless. However, the distribution function and the elements of phase space, $dp\, dq$, are not dimensionless; each depends on the units selected. It is convenient, therefore, to introduce a small "standard" volume element,

$$\Delta p_o\, \Delta q_o \equiv dp_{1o}\, dp_{2o} \cdots dq_{1o}\, dq_{2o} \ldots \equiv \delta \mathscr{R} \tag{17-3}$$

In classical mechanics the size of $\delta \mathscr{R}$ is arbitrary. The size of $dp\, dq$ is unchanged by changes in coordinates (for contact, or canonical, transformations of coordinates), but depends on the units in which p and q are expressed.

The volume element, $\Delta p_o\, \Delta q_o \equiv \delta \mathscr{R}$ has units of *action,* pq, raised to an appropriate power ($3N$, for N particles). Then $\rho\delta \mathscr{R}$ and $dp\, dq/\delta \mathscr{R}$ are each dimensionless quantities, independent of units.

From quantum mechanics it is known that a product $\Delta p_{io}\, \Delta q_{io}$ cannot meaningfully be determined if it is much less than Planck's constant, h. Therefore the standard volume element, $\Delta p_o\, \Delta q_o \equiv \delta \mathscr{R}$ is taken as h^{3N}. This choice leads to a smooth transition from classical descriptions to quantum descriptions of a given system.

In quantum mechanics, densities in continuous phase space are replaced by sums over discrete eigenstates of small systems. However, in large systems the discrete states are so close together that they are quasi-continuous and the exact quantum state cannot be determined, even in principle. It then becomes necessary to apply statistical probability considerations to the discrete quantum states. This leads to a *density matrix* (Section 19.5).

ANSWERS 17.1

A. For any harmonic oscillator, $p^2/2m + \textit{k}q^2/2$ = constant, or $p^2 + aq^2 = c$. As p and q change with time, an ellipse is traced out in phase space (Fig. 17.1a).
B. Momentum (upward) decreases linearly with time, $p = p_o - \int f\,dt$, while position increases, then decreases, with time quadratically; $h = v_o t - \frac{1}{2}at^2$. The point in phase space follows a parabola (Fig. 17.1b); $q = (q_o + p_o^2/2m^2g) - p^2/2m^2g$.

According to classical mechanics, even a complex system has specific values for every q_i and p_i at any instant of time. The coordinates of a complex system therefore occupy *a single point* in phase space, called the *phase point*.

> *The phase point moves about with time, following the laws of mechanics, describing coordinate changes of real particles subject to the laws of mechanics.*

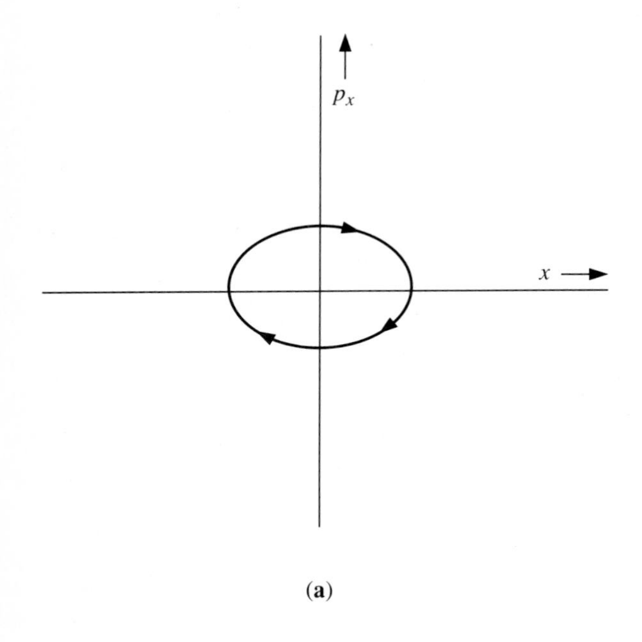

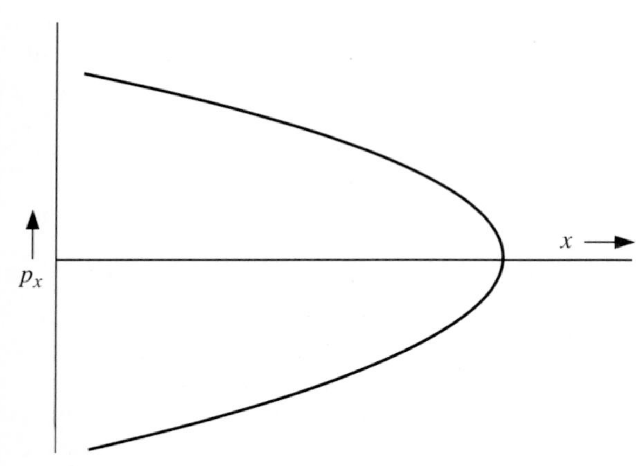

FIGURE 17.1 (a) A harmonic oscillator traces out an ellipse in phase space. (b) A body thrown vertically upward follows a parabolic path in phase space.

The motion of the phase point is called a *phase trajectory.* According to classical mechanics, the values of the p_i and q_i at any instant determine the values at all later times. Therefore, through each point in phase space there is exactly one trajectory.

Although classical mechanics assumes that the system has a unique phase point, the location of the phase point of a complex system at any time cannot be determined, for reasons considered in Section 15.1. In principle, it would be possible to find average positions for a system over a long period of time. In practice, even this is unrealistic, however, because no system is completely isolated. Over long periods of time interactions of the system with its surroundings influence the state of the system.

To circumvent this difficulty, Gibbs considered a collection, or ensemble, of identical systems. Each system is represented by a single point in phase space. Because the number, M, of such systems may be taken to be very large (approaching infinity), statistical methods should apply to the ensemble.

The initial state of the system is not fully known, so the initial states of the systems in the ensemble are not known exactly. Unless each system could be prepared identically, the trajectories of the phase points must differ. The points cannot "collide"; phase trajectories cannot intersect. Even over a long time, the behavior of each system must differ from other systems within the ensemble, as long as the systems are regarded as isolated. How the systems should be prepared becomes an issue.

An alternative, which we follow here, is to divide an original large complex system into many subsystems, each typically containing many particles. For example, if the original system is $\frac{1}{6}$ mmol, divide the system into 10^8 subsystems, each containing 10^{12} particles. Each subsystem should be large enough that the energy of interaction of the subsystem with its surroundings is small compared to the energy of the subsystem. Then the state of each subsystem may be considered independent of the state of any other subsystem, over any period of time that is not too long.

The number of equivalent molecules in each subsystem is usually very large, but it need not be. Individual molecules of a hot polyatomic gas could be treated as subsystems if the pressure is low enough that the molecules interact with each other very little. Each molecule would transfer energy between its various internal modes, subject to conservation laws.

The subsystems form an ensemble, for which statistical averaging is meaningful.

The description of the system by a single phase point is replaced by a probability density, in phase space, for the phase points of the subsystems.

The probability that one of the subsystems of the ensemble falls in the phase space volume element $dp\,dq$ is given by the *probability density function,* $\rho = \rho(p_1, p_2, \ldots, p_{3N}, q_1, q_2, \ldots, q_{3N})$. The increment of probability for the element $dp\,dq$ is $\delta P = \rho\,dp\,dq$.

Our elementary statistical treatment of a macroscopic system assumes that the intensive properties are independent of the size of the system or the number or size of the subsystems. (This is only approximately true when the theory is extended to small numbers of degrees of freedom.) The number of phase points for the system is therefore of no particular consequence. Rather than setting $\int \rho\,dp\,dq = M$, where M is the number of subsystems in the ensemble, it is often more convenient to change the normalization factor so that $\int \rho\,dp\,dq = 1$, retaining, nevertheless, the model of a cloud of phase points, streaming in time through phase space.

The interpretation is then that any particular real subsystem has a fractional probability of being in any given region of phase space, but the system has unit probability of being somewhere in phase space (i.e., unit probability of being represented by the M points distributed quasi-continuously through phase space).

The probability density of phase points considers only the subsystems of the ensemble. It must not be confused with the probability of individual coordinates of particles within a subsystem. All $6N$ values of the p_i and q_i of the N particles of a subsystem are represented by a single phase point in $6N$-dimensional phase space.

17.2 Formulations of Classical Mechanics[1]

Newton originally expressed classical mechanics in the form $\mathbf{f} = \dot{\mathbf{p}}$, where $\dot{\mathbf{p}}$ is the time derivative of the momentum. It is in this form, or the equivalent statement, $\mathbf{f} = m\mathbf{a}$, that mechanics is taught in introductory courses. A simultaneous attempt by Leibniz to develop mechanics in terms of mv^2 led to the development, nearly two centuries later, of the general concept of energy, which proves more helpful in certain kinds of problems.

For many complex problems of mechanics, such as celestial mechanics, vibrations of many-particle systems, and motions with constraints, alternative formulations of classical mechanics are more convenient than Newton's method. These allow greater flexibility in choices of coordinates and easier incorporation of constraints or considerations of energy.

Five equivalent statements of classical mechanics are of particular importance and are likely to be encountered in contemporary physics. Each is briefly defined below and three elementary problems are solved to show how the equations apply. Of course, the results are the same from any of the formulations. However, in complex problems, one method may be significantly easier to apply than another.

Newtonian Formulation

$$\mathbf{f} = \dot{\mathbf{p}} \equiv \frac{d\mathbf{p}}{dt} = \frac{d(mv)}{dt} = m\mathbf{a} \qquad \mathbf{p} = m\dot{\mathbf{q}} \tag{17-4}$$

$\mathbf{p}$ and $\mathbf{q}$ are vectors in a Cartesian coordinate frame.

1. Free particle:

$$\mathbf{f} = 0, \text{ therefore } \dot{\mathbf{p}} = 0; \ \mathbf{p} \text{ is constant, } \mathbf{q} \text{ is undetermined.}$$

2. Harmonic oscillator: A force, f, acts on a mass, m.

$$f = -kq = \dot{p}; \ p = m\dot{q}, \text{ therefore } \dot{p} = m\ddot{q}$$

Therefore, $-kq = m\ddot{q}$; $\ddot{q} = -\frac{k}{m}q$. Define

$$\omega \equiv \sqrt{\frac{k}{m}}$$

[1] The intent of this section is to serve as a review for those who have previously encountered these alternative formulations and as an introduction to the terminology and notation for readers unfamiliar with them. The notations of the Lagrangian, Hamiltonian, and Poisson brackets appear in other sections.

then

$$q = q_0 e^{\pm i\omega t} \qquad \text{or} \qquad q = q_0 \cos \omega t$$

3. Rigid rotor: For an object moving in a circular path at speed v at a distance r from the center of rotation,[2] define:

$$\text{angular speed and angular velocity}: \omega \equiv \dot{\vartheta} = \frac{v}{r} \quad \text{and} \quad \mathbf{v} = \boldsymbol{\omega} \times \mathbf{r}$$

$$\text{angular acceleration}: \alpha \equiv \ddot{\vartheta} \equiv \dot{\omega} = a/r \quad \text{and} \quad \mathbf{a} = \boldsymbol{\alpha} \times \mathbf{r}$$

$$\text{angular momentum}: \mathcal{L} \equiv \mathbf{r} \times \mathbf{p} \equiv \mathbf{r} \times (m\mathbf{v}) \quad \text{and}$$

$$\mathcal{L} = mvr = mr^2\omega = I\omega$$

where $I \equiv mr^2$ is the moment of inertia.

$$\text{torque}: \boldsymbol{\tau} = \mathbf{r} \times \mathbf{f} = \mathbf{r} \times (m\mathbf{a}) = \mathbf{r} \times (m\boldsymbol{\alpha} \times \mathbf{r})$$

and for $\dot{r} = 0$,

$$\tau = rf_{\perp} = mr^2\alpha = I\alpha = I\dot{\omega} = \dot{\mathcal{L}}$$

where $f_{\perp}$ is the force component perpendicular to **r**.

If $\tau = 0$, then $\boldsymbol{\alpha} = 0$, $\boldsymbol{\omega}$ is constant, and $\mathcal{L}$ is constant. The energy is $E = \frac{1}{2}mv^2 = \frac{1}{2}mr^2\omega^2 = \frac{1}{2}I\omega^2 = \mathcal{L}^2/2I$.

Lagrange Formalism

Let $L = L(q, \dot{q}) = T - V$, where $T = T(q, \dot{q})$ is the kinetic energy, $V = V(q)$ is the potential energy (which we assume here to be defined), and q may be a generalized coordinate (including, for example, an angle or a change in a length). The function L is called the *Lagrangian*.

Define a generalized momentum, *conjugate* to q_i,

$$p_i = \frac{\partial L}{\partial \dot{q}_i} \tag{17-5}$$

and a generalized force,

$$f_i = \frac{\partial L}{\partial q_i} \tag{17-6}$$

Then *Lagrange's equation* says that

$$\frac{d}{dt}\frac{\partial L}{\partial \dot{q}_i} - \frac{\partial L}{\partial q_i} = 0 \tag{17-7}$$

which may be recognized as $\dot{p}_i - f_i = 0$.

[2]We choose the same origin for ϑ and **r**. Then ω and $\mathcal{L}$ will be parallel. See, for example, D. Halliday and R. Resnick, *Physics,* 3rd ed. (New York: John Wiley & Sons, Inc., 1978), Secs. 13.2–13.3 (where they have written $\omega \equiv \dot{\phi} \neq \dot{\vartheta}$).

1. Free particle:

$$T = \tfrac{1}{2}m\dot{q}^2, \qquad V = 0 \qquad L = \tfrac{1}{2}m\dot{q}^2$$

$$f = \frac{\partial L}{\partial q} = 0 \qquad \frac{d}{dt}\frac{\partial L}{\partial \dot{q}} = \frac{d}{dt}p = \dot{p} = 0$$

2. Harmonic oscillator:

$$T = \tfrac{1}{2}m\dot{q}^2 \qquad V = \tfrac{1}{2}\,kq^2$$

$$L = \tfrac{1}{2}m\dot{q}^2 - \tfrac{1}{2}\,kq^2$$

$$f = \frac{\partial L}{\partial q} = -\,kq \qquad \frac{d}{dt}\frac{\partial L}{\partial \dot{q}} = \frac{d}{dt}(m\dot{q}) = m\ddot{q}$$

$$m\ddot{q} + kq = 0 \qquad \ddot{q} = -\frac{k}{m}q \qquad \omega \equiv \sqrt{\frac{k}{m}}$$

$$q = q_0 e^{\pm i\omega t} \qquad \text{or} \qquad q = q_0 \cos \omega t$$

3. Rigid rotor:

$$T = \tfrac{1}{2}I\omega^2 \qquad V = 0$$

$$L = T - V = \tfrac{1}{2}I\omega^2 \equiv \tfrac{1}{2}I\dot{\vartheta}^2$$

$$p_\vartheta = \frac{\partial L}{\partial \dot{\vartheta}} = I\dot{\vartheta} \equiv I\omega \equiv \mathscr{L}$$

$$f_\vartheta(\equiv \tau) = \frac{\partial L}{\partial \vartheta} = 0 \qquad \frac{d}{dt}\frac{\partial L}{\partial \dot{\vartheta}} - \frac{\partial L}{\partial \vartheta} = 0 \qquad \dot{\mathscr{L}} = 0$$

Hamiltonian

Define

$$H = H(p, q) \equiv \sum_i p_i \dot{q}_i - L(q, \dot{q}) \tag{17-8}$$

$$dH = \sum (p_i\, d\dot{q}_i + \dot{q}_i\, dp_i) - \sum \left(\frac{\partial L}{\partial q_i} dq_i + \frac{\partial L}{\partial \dot{q}_i} d\dot{q}_i \right)$$

Substituting

$$\frac{\partial L}{\partial \dot{q}_i} = p_i \qquad \text{and} \qquad \frac{\partial L}{\partial q_i} = \frac{d}{dt}\frac{\partial L}{\partial \dot{q}_i} = \dot{p}_i$$

gives

$$dH = \sum (p_i\, d\dot{q}_i + \dot{q}_i\, dp_i) - \sum (\dot{p}_i\, dq_i + p_i\, d\dot{q}_i)$$
$$= \sum (\dot{q}_i\, dp_i - \dot{p}_i\, dq_i)$$

However,

$$dH = \sum \left(\frac{\partial H}{\partial p_i} dp_i + \frac{\partial H}{\partial q_i} dq_i \right)$$

Therefore, by comparison of coefficients, we obtain Hamilton's equations,

$$\frac{\partial H}{\partial p_i} = \dot{q}_i \qquad \text{and} \qquad \frac{\partial H}{\partial q_i} = -\dot{p}_i \tag{17-9}$$

H is called the *Hamiltonian*. For "conservative" systems, $H = T + V$, where $T = T(p, q)$ and $V = V(q)$.

1. Free particle:

$$T = \frac{p^2}{2m} \qquad V = 0$$

$$H = \frac{p^2}{2m} \qquad \frac{\partial H}{\partial p} = \frac{p}{m} = \dot{q} \qquad p = m\dot{q}$$

$\dfrac{\partial H}{\partial q} = 0 = -\dot{p}$, and therefore p is constant

2. Harmonic oscillator:

$$T = \frac{p^2}{2m} \qquad V = \frac{1}{2}kq^2$$

$$H = \frac{p^2}{2m} + \frac{1}{2}kq^2 \qquad \frac{\partial H}{\partial p} = \frac{p}{m} = \dot{q} \qquad p = m\dot{q}$$

$$\frac{\partial H}{\partial q} = kq = -\dot{p} = -m\ddot{q} \qquad \ddot{q} = -\frac{k}{m}q \qquad \omega \equiv \sqrt{\frac{k}{m}}$$

$$q = q_o e^{\pm i\omega t} \qquad \text{or} \qquad q = q_o \cos \omega t$$

3. Rigid rotor:

$$T = \frac{\mathcal{L}^2}{2I} \qquad V = 0 \qquad H = \frac{\mathcal{L}^2}{2I}$$

$$\dot{q} \equiv \dot{\vartheta} \equiv \omega = \frac{\partial H}{\partial p} \equiv \frac{\partial H}{\partial \mathcal{L}} = \frac{\mathcal{L}}{I}$$

$$\dot{\mathcal{L}} \equiv \dot{p} = -\frac{\partial H}{\partial q} \equiv \frac{\partial H}{\partial \vartheta} = 0$$

Hamilton's Principle of Least Action

From an arbitrary point a, at time t_1, a system progresses to point b, at time t_2. It may follow any path through phase space between (a, t_1) and (b, t_2). The Lagrangian is $L(q_i, \dot{q}_i; t) = T(q_i, \dot{q}_i; t) - V(q_i; t) = \mathcal{L}(p_i, q_i, t) = \mathcal{T}(p_i, q_i, t) - \mathcal{V}(q_i, t)$ after conversion to position and momentum coordinates.[3] Hamilton's

[3]The Lagrangian expressed in position and momentum coordinates, $\mathcal{L}(p_i, q_i, t)$, must be distinguished from the angular momentum, $\mathcal{L}$, which may be one of the momentum variables in the Lagrangian.

principle states that there exists some path, or consecutive contiguous values of (p_i, q_i), such that for fixed $t_2 - t_1$, the integral

$$\int_{t_1}^{t_2} L(q_i, \dot{q}_i, t)\, dt = \int_{t_1}^{t_2} \mathscr{L}(p_i, q_i, t)\, dt$$

is an extremum—a maximum, a minimum, or constant for small variations in path. The variation of the integral vanishes for arbitrary small variations of p_i and q_i from that path.

$$\delta \int_{t_1}^{t_2} \mathscr{L}(p_i, q_i, t)\, dt = 0$$

$$= \int_{t_1}^{t_2} \mathscr{L}(p_i + \delta p_i, q_i + \delta q_i, t)\, dt - \int_{t_1}^{t_2} \mathscr{L}(p_i, q_i, t)\, dt \tag{17-10}$$

The path through phase space actually followed by the system will be the extremum. Note that this is not the same, in general, as a path through three-dimensional real space.

Hamilton's principle, also known as the *principle of least action,* is more valuable as a starting point for theoretical derivations than for calculations. Nevertheless, some simple examples may be given to illustrate the meaning.

1. Free particle: $q = \dot{q}t$, $V = 0$, $T = p^2/2m$. The values of p and q will not be the same at all points along a varied path, but they must be the same as the actual path at the endpoints. Therefore, if $p \to p + \delta p$ for the first half of the path, then $p \to p - \delta p$ along the second half, in order that q and t will be the same at the end point. Compare

$$\int_{t_1}^{t_2} L\, dt = \int_0^{\Delta t} \frac{p^2}{2m}\, dt = \frac{p^2}{2m}\Delta t$$

with

$$\int_0^{(1/2)\Delta t} \frac{(p + \delta p)^2}{2m}\, dt + \int_{(1/2)\Delta t}^{\Delta t} \frac{(p - \delta p)^2}{2m}\, dt$$

which gives

$$\frac{p^2}{2m}\Delta t + \frac{(\delta p)^2}{2m}\Delta t$$

a quantity that is larger than the original integral for any δp.

2. Harmonic oscillator: The motion described by $q = q_0 \cos \omega t$, $p = m\dot{q} = -m\omega q_0 \sin \omega t$ gives the minimum value of the integral. As a trial variation we could let $q = q_0 \cos \omega t + \delta q_0 \sin^2 \omega t$, a variation that disappears at the end points of the motion. Then

$$\dot{q} = -\omega q_0 \sin \omega t + 2\delta q_0 \sin \omega t(\omega \cos \omega t)$$

Setting $T = \frac{1}{2}m\dot{q}^2$ and $V = \frac{1}{2}\,k q^2 = \frac{1}{2}m\omega^2 q^2$ and comparing integrals of the Lagrangian between $t = -\pi/\omega$ and $t = +\pi/\omega$ gives the variation $\frac{1}{8}m\omega\pi(\delta q_0)^2$ for small t, which is greater than zero.

Hamilton's principle, in mechanics, is equivalent to Fermat's law in optics, although the latter is written to give a minimum time rather than constant time.

3. Rigid rotor: The rigid rotor is most easily treated by noting that $L = \mathscr{L}^2/2I$ is of identical mathematical form to $L = p^2/2m$. Therefore, the solution of $\delta \int_{t_1}^{t_2} (\mathscr{L}^2/2I)\, dt = 0$ must be identical to the solution of $\delta \int_{t_1}^{t_2} (p^2/2m)\, dt = 0$. Having shown that $\dot{p} = 0$ for the free particle, it follows that $\dot{\mathscr{L}} = 0$ for the rigid rotor.

Poisson Brackets

Let $Y = Y(p_i, q_i; t)$ be an arbitrary function. Then

$$\dot{Y} \equiv \frac{dY}{dt} = \left(\frac{\partial Y}{\partial t}\right)_{p_i, q_i} + \sum_i \left(\frac{\partial Y}{\partial q_i}\frac{dq_i}{dt} + \frac{\partial Y}{\partial p_i}\frac{dp_i}{dt}\right)$$

Making the substitutions

$$\frac{dq_i}{dt} = \dot{q}_i = \frac{\partial H}{\partial p_i} \quad \text{and} \quad \frac{dp_i}{dt} = \dot{p}_i = -\frac{\partial H}{\partial q_i}$$

gives

$$\dot{Y} \equiv \frac{dY}{dt} = \left(\frac{\partial Y}{\partial t}\right)_{p_i, q_i} + \sum_i \left(\frac{\partial Y}{\partial q_i}\frac{\partial H}{\partial P_i} - \frac{\partial Y}{\partial p_i}\frac{\partial H}{\partial q_i}\right) \tag{17-11}$$

Define

$$\{A, B\} \equiv \sum_i \left(\frac{\partial A}{\partial q_i}\frac{\partial B}{\partial p_i} - \frac{\partial B}{\partial q_i}\frac{\partial A}{\partial p_i}\right) \tag{17-12}$$

called the *Poisson bracket of A and B*. Then

$$\dot{Y} = \left(\frac{\partial Y}{\partial t}\right)_{p_i, q_i} + \{Y, H\} \tag{17-13}$$

The partial derivative, $(\partial Y/\partial t)_{p_i, q_i}$, vanishes as long as the definition of Y in terms of positions and momenta does not change with time. It follows, then, that if the Poisson bracket of Y and the Hamiltonian, H, vanishes, $\{Y, H\} = 0$, then $\dot{Y} = 0$; Y is a constant of the motion.

Note that $\{H, H\} = 0$, so provided that H does not explicitly depend on the time [i.e., $(\partial H/\partial t)_{p_i, q_i} = 0$], H is a constant of the motion.

1. Free particle:

$$T = \frac{p^2}{2m} \qquad V = 0 \qquad H = \frac{p^2}{2m}$$

Then, because p and q are independent variables, $\partial q/\partial p = 0 = \partial p/\partial q$ and

$$\dot{q} = \{q, H\} = \frac{\partial q}{\partial q}\frac{\partial H}{\partial p} - \frac{\partial H}{\partial q}\frac{\partial q}{\partial p} = \frac{\partial H}{\partial p} = \frac{p}{m} \qquad \therefore \dot{q} = \frac{p}{m}$$

$$\dot{p} = \{p, H\} = \frac{\partial p}{\partial q}\frac{\partial H}{\partial p} - \frac{\partial H}{\partial q}\frac{\partial q}{\partial p} = 0 \qquad \therefore \dot{p} = 0$$

p is a constant of the motion; q is not.

2. Harmonic oscillator:

$$T = \frac{p^2}{2m} \qquad V = \frac{1}{2} kq^2$$

$$H = \frac{p^2}{2m} + \frac{1}{2} kq^2$$

$$\dot{q} = \{q, H\} = \frac{\partial q}{\partial q}\frac{\partial H}{\partial p} - \frac{\partial H}{\partial q}\frac{\partial q}{\partial p} = \frac{\partial H}{\partial p} = \frac{p}{m} \qquad \therefore \dot{q} = \frac{p}{m}$$

$$\dot{p} = \{p, H\} = \frac{\partial p}{\partial q}\frac{\partial H}{\partial p} - \frac{\partial H}{\partial q}\frac{\partial p}{\partial p} = -\frac{\partial H}{\partial q} = -kq \qquad \therefore \dot{p} = -kq$$

$$p = m\dot{q} \quad \text{so} \quad \dot{p} = m\ddot{q} = -kq \qquad \ddot{q} = -\frac{k}{m}q \qquad \omega \equiv \sqrt{\frac{k}{m}}$$

$$q = q_0 e^{\pm i\omega t} \quad \text{or} \quad q = q_0 \cos \omega t$$

3. Rigid rotor:

$$T = \frac{\mathcal{L}^2}{2I} \qquad V = 0 \qquad H = \frac{\mathcal{L}^2}{2I}$$

$$\dot{\vartheta} = \{\vartheta, H\} = \frac{\partial \vartheta}{\partial \vartheta}\frac{\partial H}{\partial \mathcal{L}} - \frac{\partial H}{\partial \vartheta}\frac{\partial \vartheta}{\partial \mathcal{L}} = \frac{\partial H}{\partial \mathcal{L}} = \frac{\mathcal{L}}{I} \qquad \therefore \dot{q} \equiv \dot{\vartheta} \equiv \omega = \frac{\mathcal{L}}{I}$$

$$\dot{p} \equiv \dot{\mathcal{L}} = \{\mathcal{L}, H\}, = \frac{\partial \mathcal{L}}{\partial \vartheta}\frac{\partial H}{\partial \mathcal{L}} - \frac{\partial H}{\partial \vartheta}\frac{\partial \mathcal{L}}{\partial \mathcal{L}} = 0$$

Note that for any pair of coordinates and momenta,

$$\{q_i, q_j\} = 0 \qquad \{p_i, p_j\} = 0$$

$$\{q_i, p_j\} = \delta_{ij} = \begin{cases} 1 \text{ if } i = j \\ 0 \text{ if } i \neq j \end{cases} \tag{17-14}$$

$$\{A, B\} = -\{B, A\}$$

The symbol δ_{ij} is called the *Kronecker delta*.

The Poisson bracket, in classical mechanics, is closely related to the commutator of operators in quantum mechanics.

17.3 Three-Dimensional Rotors

The simple rigid rotor was solved, in different ways, in the preceding section. Rotations in three-dimensional space involve additional considerations. They appear in so many applications of statistical mechanics that some additional attention to the equations is appropriate here.

The value of I, the moment of inertia, is in general different for each axis through a body. For freely rotating objects, axes passing through the center of mass of the body are of primary interest, and attention is limited to such axes. A plot of I against angle of orientation of the axis in three-dimensional space yields an ellipsoid, called the *momental ellipsoid*. There are three mutually perpendicular directions, or three orthogonal axes of the momental ellipsoid, for which I is a maximum, a minimum, or invariant for small changes of direction. These three axes are called the *principal axes* of the body.

Transformation to axes coinciding with the principal moments of inertia has the effect of diagonalizing the inertial tensor, I. That is, cross terms between components of the angular velocity such as $\omega_x I_{xy} \omega_y$ are eliminated. Angular momentum and rotational energy of a rigid body for an arbitrary rotation have particularly simple forms when written in terms of the moments of inertia and angular momenta about the principal axes.

$$\mathscr{L} = I\omega = I_a\omega_a + I_b\omega_b + I_c\omega_c$$

$$E = \frac{1}{2}I_a\omega_a^2 + \frac{1}{2}I_b\omega_b^2 + \frac{1}{2}I_c\omega_c^2 \tag{17-15}$$

$$= \frac{\mathscr{L}_a^2}{2I_a} + \frac{\mathscr{L}_b^2}{2I_b} + \frac{\mathscr{L}_c^2}{2I_c}$$

Symmetry is particularly important for rotating bodies. If rotation of a body by an angle of $2\pi/n$ about some axis produces a configuration of the body indistinguishable from the original (considering equivalent atoms to be indistinguishable), the axis is called an *n-fold axis*. An n-fold rotation axis is usually represented by the symbol C_n. For example, a sawhorse has a twofold axis of rotation, C_2. A square table has a fourfold axis of rotation and, therefore, also a twofold axis, because two consecutive fourfold rotations (i.e., two consecutive rotations of 90°) are equivalent to a twofold rotation (or 180°); $C_4^2 = C_2$. Although objects may have other symmetry elements (such as planes of symmetry), we need not be concerned with those at this time.

EXAMPLES 17.2

A. Identify the rotation axes of the following molecules.
   a. Water, H_2O.
   b. Ammonia, NH_3 (pyramidal).
   c. Ethylene, $H_2C{=}CH_2$ (planar).

B. Which of the following have a threefold or higher rotation axis? Which have more than one noncoincident threefold or higher axis?
   a. Water, H_2O.
   b. Ammonia, NH_3.
   c. Benzene, C_6H_6 (planar, hexagonal).
   d. Methane, CH_4 (tetrahedral; C at center of a pyramid). □

If an object has a threefold or higher axis of rotational symmetry, two of the principal moments of inertia are equal to each other and independent of direction about the symmetry axis. The momental ellipsoid is then an ellipsoid of revolution and the object is called a *symmetric rotor* or a *symmetric top*.

When there is more than one noncoincident threefold or higher axis of rotational symmetry, the momental ellipsoid is a sphere. All three moments of inertia are then equal and independent of direction and the object is called a *spherical rotor* or a *spherical top*. If there is no threefold or higher axis of rotational symmetry, the object is called an *asymmetric rotor* or an *asymmetric top*. (Note that an asymmetric top may still have a high degree of symmetry, as shown by ethylene.)

The total rotational energy of a symmetric rotor, with $I_b = I_c$, may be rewritten as

---

ANSWERS 17.2

A. (a) Water has a single C_2 axis. (b) Ammonia has a single C_3 axis (and therefore also C_3^2 , a rotation by 240°). (c) Ethylene has three mutually perpendicular C_2 axes.
B. (a) Water has no C_3 or higher axis. (b) Ammonia has a single axis of symmetry C_3 or higher. (c) Benzene has a number of C_2 axes lying in the plane (six axes, three passing through C atoms and three passing through C−C bonds). It has one C_6 axis (which is therefore also a $C_3 = C_6^2$ and a $C_2 = C_6^3$), perpendicular to the plane of the molecule. (d) Methane has a C_3 axis coincident with each C−H bond. Of these four molecules, only methane has more than one noncoincident threefold or higher axis.

$$E = \frac{\mathcal{L}_a^2}{2I_a} + \frac{\mathcal{L}_b^2 + \mathcal{L}_c^2}{2I_b} = \frac{\mathcal{L}_a^2 + \mathcal{L}_b^2 + \mathcal{L}_c^2}{2I_b} + \mathcal{L}_a^2\left(\frac{1}{2I_a} - \frac{1}{2I_b}\right)$$

$$= \frac{\mathcal{L}^2}{2I_b} + \mathcal{L}_a^2\left(\frac{1}{2I_a} - \frac{1}{2I_b}\right) \qquad (17\text{-}16)$$

Molecules may have angular momenta arising from various sources. The molecule as a whole will rotate, at least in the gaseous state. Degenerate vibrations often contribute angular momentum. Electrons may have angular momentum in atoms and in molecules. Many nuclei also have intrinsic angular momenta. To a satisfactory approximation, each of these angular momenta is independent of the others. The total angular momentum of an atom or molecule is the vector sum of the angular momentum contributions from all sources.

The classical model for adding angular momenta is that the angular momentum vectors that form a resultant rotate, or precess, about the resultant, with a speed that increases with the energy of interaction. The strongest couplings should be added first. These precess most rapidly. Weaker couplings then precess more slowly about the new resultant.

17.4 Vibrations of Polyatomic Molecules

A mass on a spring or a diatomic molecule can be solved as a single equation in one unknown. However, many of the generalizations that might be drawn from the one-dimensional problem are no longer valid when there are more than two particles in the system. Polyatomic molecules require solution of a set of simultaneous equations. If this is attempted without simplifying techniques, it will typically be next to impossible. For example, benzene has 12 atoms, giving 36 position coordinates and yielding a 36th-order equation to be solved.

To illustrate the theory without getting heavily involved in the details of solutions, the method will be applied to a heteronuclear diatomic molecule, such as HCl. Let x_1 and x_2 be the position coordinates of the two atoms. Then the kinetic and potential energies are

$$T = \tfrac{1}{2}m_1\dot{x}_1^2 + \tfrac{1}{2}m_2\dot{x}_2^2 \qquad \text{and} \qquad V = \tfrac{1}{2}\,k(x_1 - x_2)^2$$

We introduce mass-weighted coordinates.

$$q_1 = \sqrt{m_1}\,x_1 \qquad \text{and} \qquad q_2 = \sqrt{m_2}\,x_2$$

Then

$$T = \frac{1}{2}\dot{q}_1^2 + \frac{1}{2}\dot{q}_2^2 = \frac{1}{2}\sum \dot{q}_i^2$$

$$V = \frac{1}{2}k(x_1 - x_2)^2 = \frac{1}{2}k\left(\frac{q_1}{\sqrt{m_1}} - \frac{q_2}{\sqrt{m_2}}\right)^2$$

$$= \frac{1}{2}\frac{k}{m_1}q_1^2 + \frac{1}{2}\frac{k}{m_2}q_2^2 - \frac{k}{\sqrt{m_1 m_2}}q_1 q_2 = \frac{1}{2}\sum_{i,j=1}^{2} f_{ij}q_i q_j$$

with $f_{11} = k/m_1$, $f_{22} = k/m_2$, and $f_{12} = f_{21} = -k/\sqrt{m_1 m_2}$.

The Lagrangian is

$$L = T - V = \tfrac{1}{2}\sum \dot{q}_i^2 - \tfrac{1}{2}\sum f_{ij}q_i q_j$$

which gives the equations

$$\ddot{q}_i + \sum_j f_{ij}q_j = 0$$

Anticipating that the motion will be harmonic, we write

$$q_i = q_i^{\circ} \cos \omega t \qquad \text{and therefore} \qquad \ddot{q}_i = -\omega^2 q_i$$

Let $\lambda \equiv \omega^2$. The equations then have the form

$$-\lambda q_i + \sum_j f_{ij}q_j = 0$$

The equations are now conveniently written in matrix form.

$$[f_{ij} - \lambda\delta_{ij}][q_i] = 0 \tag{17-17}$$

where δ_{ij} is the Kronecker delta, or

$$\begin{bmatrix} f_{11} - \lambda & f_{12} \\ f_{21} & f_{22} - \lambda \end{bmatrix}\begin{bmatrix} q_1 \\ q_2 \end{bmatrix} = 0 \qquad \text{or} \qquad \mathbb{A}\mathbb{X} = 0$$

Then either $\mathbb{X} \equiv 0$ (i.e., all displacements are zero) or the determinant of $\mathbb{A}$ is zero; $|\mathbb{A}| = 0$. (Otherwise, $\mathbb{A}^{-1}$ exists and multiplying on the left by $\mathbb{A}^{-1}$ gives $\mathbb{A}^{-1}\mathbb{A}\mathbb{X} = \mathbb{X} = \mathbb{A}^{-1}0 = 0$.) Hence

$$\begin{vmatrix} f_{11} - \lambda & f_{12} \\ f_{21} & f_{22} - \lambda \end{vmatrix} = 0 \qquad \text{or} \qquad |\mathbb{A} - \lambda\mathbb{I}| = 0 \tag{17-18}$$

where $\mathbb{I}$ is the unit matrix (ones along the principal diagonal, otherwise only zeros). This determinantal equation is often called the *secular equation* from its history in applications involving time dependence of motions.

A secular equation for an $n \times n$ determinant is equivalent to a polynomial equation of order n. In this instance,

$$\lambda^2 + c_1\lambda + c_2 = 0$$

Furthermore, the coefficients are easily identifiable properties of the matrix $\mathbb{A}$; c_1 is minus the trace (sum of elements on the principal diagonal), c_2 is the sum of the 2×2 minors (determinants of submatrices along the principal diagonal), and so on. Thus for this 2×2 matrix (for which the 2×2 minor is the entire determinant),

$$\lambda^2 - (f_{11} + f_{22})\lambda + (f_{11}f_{22} - f_{21}f_{12}) = 0$$

When the n values of λ satisfying the equation have been found, each may be inserted, in turn, into the secular equation to solve for the coordinates, q_i, or more precisely, for the ratios of the coordinates.

For the diatomic molecule, with values of the f_{ij} substituted, the equation is

$$\lambda^2 - \left(\frac{k}{m_1} + \frac{k}{m_2}\right)\lambda + \left(\frac{k^2}{m_1 m_2} - \frac{k^2}{m_1 m_2}\right) = 0$$

The two solutions are

$$\lambda = \frac{k}{m_1} + \frac{k}{m_2} \equiv \frac{k}{\mu} \text{ and } \lambda = 0$$

Substitution of the first value of λ into the matrix equation gives two equations, each yielding the result

$$q_1 = -\sqrt{\frac{m_2}{m_1}}\, q_2$$

Substitution of $\lambda = 0$ gives $x_1 = x_2$, representing a simple translation of the molecule without vibration (i.e., with $\lambda \equiv \omega^2 = 0$). Only the ratios of displacements are found. For the classical vibrations of harmonic oscillators, any amplitude of vibration is possible.

The method outlined above opens the door to solution of much larger systems. Again we will illustrate only with the diatomic molecule. Choose coordinates that correspond more closely to the expected answer.

$$q_1 = x_2 - x_1 = \Delta r \qquad \text{and} \qquad q_2 = x_2 + x_1$$

The kinetic energy expression is now a more complicated quadratic form.

$$T = \tfrac{1}{2}m_1\dot{x}_1^2 + \tfrac{1}{2}m_2\dot{x}_2^2 = \tfrac{1}{8}m_1(\dot{q}_2 - \dot{q}_1)^2 + \tfrac{1}{8}m_2(\dot{q}_2 + \dot{q}_1)^2$$

We choose to write this in terms of coefficients of a new matrix that will be labeled $\mathbb{G}^{-1}$ (i.e., G inverse). Hence

$$T = \tfrac{1}{2}\sum_{i,j}(g^{-1})_{ij}\dot{q}_i\dot{q}_j \tag{17-19}$$

The potential energy is of the same form as previously, although the coefficients differ in value.

$$V = \tfrac{1}{2}\, k q_1^2 = \tfrac{1}{2}\sum_{i,j} f_{ij}q_iq_j \tag{17-20}$$

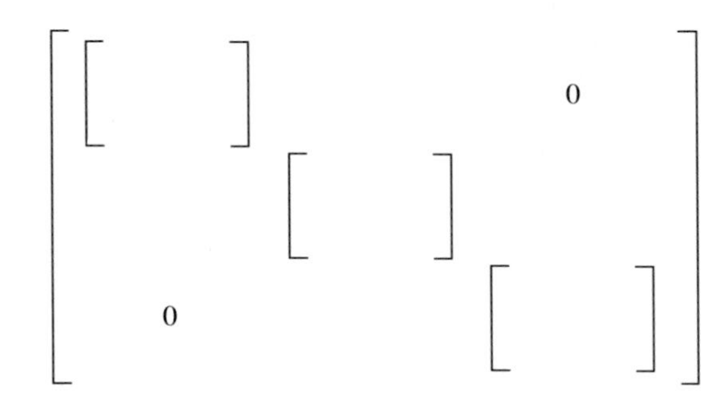

FIGURE 17.2 The $\mathbb{GF}$ matrix product factors into blocks along the principal diagonal, mixing only coordinates of the same symmetry.

Then the Lagrangian equation leads to the determinantal equation

$$\left|\mathbb{F} - \mathbb{G}^{-1}\lambda\right| = 0 \qquad \text{or} \qquad \left|\mathbb{GF} - \lambda\mathbb{I}\right| = 0 \tag{17–21}$$

which may be solved, as before, to give the $\lambda_i \equiv \omega_i^2$ and hence the relative displacements of the atoms.

Solution of the equations of motion would still be difficult or impossible except for simplifications arising from the symmetry of the molecule. If the set of coordinates $\{q_i\}$ exhibits the symmetry of the molecule, there is at least one symmetry operation, such as a rotation or a reflection, that leaves the molecule unchanged, and hence the energy unchanged, but which changes some q_i to $-q_i$ and leaves some q_j of different symmetry unchanged. Hence any cross term, mixing q_i and q_j, would give a change in sign of this energy contribution just by carrying out a symmetry operation (equivalent to the arbitrary renumbering of identical atoms). This cannot be. There can be no cross terms between coordinates of different symmetry.

If molecular coordinates are all selected in accordance with the symmetry of the molecule, they are called *symmetry coordinates*. Such symmetry coordinates can easily be written down by inspection of the molecule, making use of the principles of group theory. The $\mathbb{F}$ and $\mathbb{G}$ matrices for symmetry coordinates factor into blocks along the principal diagonal (Fig. 17.2). Only vibrations of the same symmetry can mix.

Group theory analysis[4] tells what symmetry properties the symmetry coordinates should have and how many there are of each symmetry. It also allows a prediction of which interacting sets of symmetry coordinates will give rise to infrared absorption or emission and which to Raman scattering or other interactions with radiation.

For example, in benzene the 36 vibrations are divided into sets of 2, 0, 1, 1, 0, 2, 2, 2, and pairs of 1, 3, 4, and 2, plus three rotations and three translations (of zero vibrational frequency). There are five quadratic equations, one cubic, and one quartic.

Solution of the secular equation for the λ_i allows solution for the relative values of the q_i. These mixtures of symmetry coordinates, determined by the values of the masses, bond lengths, bond angles, and force constants, are called the *normal coordinates*. They represent the independent modes of oscillation. If a molecule is set into one of the normal modes, it will remain in that mode to the harmonic-oscillator approximation.

[4]For discussions of group theory and its applications to molecular energy levels, see references 44–46.

In general, the normal modes of oscillation cannot easily be predicted. All atoms move in all vibrations except when constrained by symmetry requirements (such as the C in CO_2 in the symmetric stretch). In certain instances, however, a symmetry coordinate will be a C−H stretch or some other frequency well separated from other vibration frequencies. The normal mode is then dominated by this particular motion.

Two generalizations given by Lord Rayleigh are helpful in predicting qualitative changes in vibration spectra when changes are made in the molecule. First, if any force constant in a vibrating system is increased, every vibration will increase in frequency or, in the limit, remain constant. Second, if any mass is increased in a vibrating system, every vibration will decrease in frequency or, in the limit, remain constant. (Sometimes, however, the changes in form are so great that frequencies cross each other, masking this simple effect.) Quantitative predictions of frequency shifts, even on isotopic substitution, can only be made through a complete normal coordinate analysis, although there are certain *sum rules* relating one frequency shift to another.

Vibrational modes in crystals are similar, with important differences. The symmetry is taken as that of the unit cell, typically involving more than one molecular unit. The symmetry elements of unit cells are called *space groups,* in contrast to the *point groups* that describe free molecules. In addition to the rotations, reflections, and rotation–reflection operations of point groups, space groups have translations, screw axes, and glide planes that carry an atom from one unit cell into the position of an equivalent atom in an adjacent cell.

Because adjacent unit cells interact with each other, vibrations in crystals are generally not sharply defined frequencies but are continuum bands of frequencies. Nevertheless, the absorption or emission spectrum may show sharply defined bands because the *density of states* favors one part of a band over all others.

It is, in part, because of the difficulty of calculating or identifying the various normal modes of oscillation of molecules that the methods of statistical mechanics have been so important. When an assignment of normal modes has been made, it is possible to calculate the heat capacity and therefore the thermodynamic functions at room temperature. Comparing these calculated values with experimental values is the most sensitive test of the proper assignment of normal modes of vibration.

In all subsequent discussions of vibrations (unless specifically stated otherwise) we will assume each vibration to be one of the normal modes, at a sharply defined frequency, $\nu \equiv \omega/2\pi$.

17.5 Liouville's Theorem

Although there is no way to determine the instantaneous values of the $6N$ position and momentum coordinates of a subsystem, there is no reason to doubt that each subsystem exists in a state representable, in classical mechanics, by such a point or within an infinitesimal volume element, $dp\ dq = dp_1\ dp_2 \cdots dp_{3N}\ dq_1\ dq_2 \cdots dq_{3N}$. There are many such points for any subsystem that are consistent with whatever constraints might be imposed, so the subsystem moves through phase space with time, tracing out a path.

The paths of subsystems can only be identical *or* mutually exclusive. Different subsystems, similarly but not identically prepared, must have different trajectories.

EXAMPLES 17.3

A. If a system containing 10^{20} molecules is divided into 10^{12} subsystems, each containing 10^8 particles:

a. What is the dimension of phase space for the system?

b. How many points are in this phase space?

B. Why cannot the trajectory of one subsystem cross that for another? □

The sequence of points followed by any subsystem is unknown. Models for describing the motion of a point in phase space have evolved with time. An early assumption was that the system would pass through every point of phase space, compatible with the energy of the system, before it returned to its original position in phase space. This assumption cannot be justified. It was called the *ergodic hypothesis*.

A modification was proposed that the system passes arbitrarily close to any point in phase space, consistent with its energy, over time. This also has not been provable.

A third model proposed was that the relative probabilities of different phase space points being occupied over a long period of time (during which the system is only slightly influenced, in random fashion, by external influences) will normally be the same as the relative probabilities at one instant for the large number of subsystems constituting the ensemble considered.

The density of subsystem phase points in phase space at any instant may usually be replaced by the time-average density in phase space for any one subsystem.

This model is based on the assumption that over a long period of time, the initial state of each subsystem is "forgotten" and each subsystem may be considered statistically equivalent to each other subsystem.

It is also assumed, as a postulate, that different accessible quantum states, or different regions of phase space, of equal volume, have equal a priori probabilities of being occupied. It was this model that was employed in Section 16.4 when the number of accessible states contributing to a macrostate was taken as a measure of the probability of that macrostate.

The probability of any macrostate of a system is assumed to be proportional to the number of microstates contributing to that macrostate, or proportional to the volume of phase space consistent with that state.

This is the *fundamental postulate of statistical mechanics*. It assumes no special preparation for the system, or its subsystems, which is often expressed by saying the systems have *random phases*.

Given the density function

$$\rho = \rho(q_1, q_2, \ldots, p_1, p_2, \ldots, t)$$

normalized such that

$$\int \rho(p, q, t)\, dp\, dq = 1$$

ANSWERS 17.3

A. (a) Each subsystem has 10^8 particles, requiring 3×10^8 position coordinates and 3×10^8 momentum coordinates, so phase space must be of dimension 6×10^8. (b) There are 10^{12} points in this phase space, one point for each subsystem.
B. If, at any instant, two subsystems had the same values for all position coordinates and for all momentum coordinates of the constituent particles, the subsystems would be identically "prepared" and would subsequently follow the same paths, because each system has the same Hamiltonian. Similarly, identical subsystems could not reach the same point in phase space (same values for all q_i and p_i) unless they had previously followed the same path in phase space.

---

for the ensemble, the average value of any quantity $Y(p, q)$ for the ensemble is

$$Y = \int Y(p, q)\rho(p, q, t)\ dp\ dq$$

The number of subsystems in the ensemble is constant with time, so the number of phase points is constant and subject to the laws for continuity of flow of a fluid.

The time rate of change of fluid density is

$$\frac{d\rho}{dt} = \left(\frac{\partial\rho}{\partial t}\right)_{q_i, p_i} + \sum_i \left(\frac{\partial\rho}{\partial q_i}\frac{dq_i}{dt} + \frac{\partial\rho}{\partial p_i}\frac{dp_i}{dt}\right) = 0 \qquad (17\text{-}22)$$

The *equation of continuity* for fluid flow is

$$\left(\frac{\partial\rho}{\partial t}\right)_{q_i, p_i} + \nabla \cdot (\rho\mathbf{v}) = 0 \qquad (17\text{-}23)$$

where the divergence, $\nabla \cdot (\rho\mathbf{v})$, gives the net flow out of a volume element and the partial derivative gives the change in density at a fixed point (within the volume element).

The divergence is

$$\nabla \cdot (\rho\mathbf{v}) = \sum \left(\frac{\partial(\rho\dot{q}_i)}{\partial q_i} + \frac{\partial(\rho\dot{p}_i)}{\partial p_i}\right)$$

$$= \sum \left[\rho\left(\frac{\partial\dot{q}_i}{\partial q_i} + \frac{\partial\dot{p}_i}{\partial p_i}\right) + \left(\frac{\partial\rho}{\partial q_i}\dot{q}_i + \frac{\partial\rho}{\partial p_i}\dot{p}_i\right)\right]$$

Using Hamilton's equations, we find that the first two derivatives become

$$\frac{\partial}{\partial q_i}\frac{\partial H}{\partial p_i} - \frac{\partial}{\partial p_i}\frac{\partial H}{\partial q_i} = \frac{\partial^2 H}{\partial q_i \partial p_i} - \frac{\partial^2 H}{\partial p_i \partial q_i} = 0$$

Hence the equation of continuity becomes

$$\left(\frac{\partial\rho}{\partial t}\right)_{q_i, p_i} + \sum_i \left(\frac{\partial\rho}{\partial q_i}\dot{q}_i + \frac{\partial\rho}{\partial p_i}\dot{p}_i\right) = 0$$

But this is just the time rate of change of fluid density, so

$$\frac{d\rho}{dt} = \left(\frac{\partial\rho}{\partial t}\right)_{q_i, p_i} + \sum_i \left(\frac{\partial\rho}{\partial q_i}\dot{q}_i + \frac{\partial\rho}{\partial p_i}\dot{p}_i\right) = 0 \qquad (17\text{-}24)$$

which is called *Liouville's theorem*.

The first term, $\partial\rho/\partial t$, describes compression or expansion of the fluid at a point (assuming no sources or sinks of points); the terms in the sum describe changes in density along a flow line. For an ensemble that has come to equilibrium, the probability distribution function of the phase points does not change with time (although the individual phase points move); $(\partial\rho/\partial t)_{q_i, p_i} = 0$. Therefore,

$$\sum_i \left(\frac{\partial\rho}{\partial q_i}\dot{q}_i + \frac{\partial\rho}{\partial p_i}\dot{p}_i\right) = 0$$

Substituting derivatives of the Hamiltonian, Liouville's theorem may be written

$$\left(\frac{\partial\rho}{\partial t}\right)_{q, p} = -\sum \left(\frac{\partial\rho}{\partial q_i}\frac{\partial H}{\partial p_i} - \frac{\partial\rho}{\partial p_i}\frac{\partial H}{\partial q_i}\right)$$

or in the more compact notation of Poisson brackets,

$$\left(\frac{\partial\rho}{\partial t}\right)_{q, p} = -\{\rho, H\} \qquad (17\text{-}25)$$

and each side vanishes for an equilibrium ensemble.

The position and momentum coordinates must satisfy Hamilton's equations. Then Liouville's theorem is satisfied for any canonical transformation of coordinates. Description of a phase space of position coordinates and velocities would not, in general, produce the favorable behavior resulting from position and momentum coordinates.

The fundamental postulate of statistical mechanics tells us that, unless other restrictions interfere, all points in phase space are equally probable. Hence the first approximation to a distribution function would be a uniform distribution over phase space. From Liouville's theorem we see that if at some instant of time the distribution function were uniform over all of phase space, it would remain uniform.

A much more probable situation is a distribution function that depends on some variable, such as the energy. For generality, call the variable α. Then we can imagine a distribution function that is uniform over all parts of phase space (a "surface") for which α has a fixed value. Then, if α is a constant of the motion, $\partial\alpha/\partial t = 0$, the distribution function will remain constant, in time, over the surface of constant α.

For an isolated system, E is constant. Then over a surface in phase space of thickness δE, at E, the distribution function may be taken to be uniform. An ensemble subject to this condition is called a *microcanonical ensemble*. It was the only type of system considered in statistical mechanics until the time of Gibbs. However, as pointed out by ter Haar and others, completely isolated systems are of no interest; it is impossible to perform experiments with or on them. Gibbs introduced other ensembles, considered in Chapter 19.

The time and coordinate independence of the volume element, $dp\ dq$, is important in selecting a meaningful minimum volume element. In particular, if the minimum volume, for a system of N particles, is chosen as

$$\delta \mathscr{R} = \prod_i dp_{io}\ dq_{io} = \prod_i h = h^{3N} \tag{17-26}$$

where h is Planck's constant, then the volume element $dp\ dq/\delta \mathscr{R}$ is independent of coordinates and independent of units, and classical statistical mechanics becomes consistent with quantum statistical mechanics, where the minimum size of the volume element is dictated by Heisenberg's principle.

SUMMARY

Phase space for N particles is a $6N$ dimensional space of the variables $x_1, y_1, z_1, x_2, \ldots, p_{x1}, p_{y1}, p_{z1}, p_{x2}, \ldots = pq$.

Each point in phase space represents a microstate—specific values for each of the 3 position and 3 momentum coordinates of each of the N particles. This is often called Γ space.

As the system changes with time, the system point moves about in phase space, following a phase trajectory.

The probability of the system falling within a volume of size $\Delta p_0 \Delta q_0 = dp_{10}\, dp_{20} \cdots dq_{10} dq_{20} \cdots \equiv \delta\mathfrak{R}$ about any point, pq, is expressed by the distribution function $\rho(p, q)$, with $\int \rho\ dp\ dq = 1$. We choose $\delta\mathfrak{R} = h^{3N}$ to satisfy quantum requirements.

An ensemble of systems gives a density of system points in phase space, described by a probability density, with increment of probability $= \delta P = \rho\ dp\ dq$.

The *fundamental postulate* of statistical mechanics is that the probability of any macrostate is proportional to the number of microstates contributing to that macrostate, or proportional to the volume of phase space consistent with that state. The systems are assumed to have *random phases*.

Liouville's theorem tells us that the density of points in phase space is constant with time.

Classical mechanics may be based on Newton's laws or Lagrange's formalism or equations for the Hamiltonian or Hamilton's principle or Poisson brackets. Each has advantages for specific types of problems. Critical equations are

$$L = L(q, \dot{q}) = T - V; \quad p_i = \frac{\partial L}{\partial \dot{q}_i}, \quad \frac{d}{dt}\frac{\partial L}{\partial \dot{q}_i} - \frac{\partial L}{\partial q_i} = 0$$

$$H = H(p, q) = T + V; \quad \frac{\partial H}{\partial p_i} = \dot{q}_i, \quad \frac{\partial H}{\partial q_i} = -\dot{p}_i$$

$$L(q_i, \dot{q}_i, t) = \mathscr{L}(p_i, q_i, t); \quad \delta \int_{t_1}^{t_2} \mathscr{L}(p_i, q_i, t)\, dt = 0$$

$$\{A, B\} \equiv \sum_i \left(\frac{\partial A}{\partial q_i}\frac{\partial B}{\partial p_i} - \frac{\partial B}{\partial q_i}\frac{\partial A}{\partial p_i} \right)$$

Then the time dependence of any quantity Y is given by $\dot{Y} = \{Y, H\} + \left(\frac{\partial Y}{\partial t}\right)_{p_i, q_i}$.

Rotational energy is

$$E_{\text{rot}} = \frac{\mathcal{L}_a^2}{2I_a} + \frac{\mathcal{L}_b^2}{2I_b} + \frac{\mathcal{L}_c^2}{2I_c}$$

The angular momentum is a vector sum of its components, $\mathcal{L} = \mathcal{L}_a + \mathcal{L}_b + \mathcal{L}_c$ and therefore $L^2 = L_a^2 + L_b^2 + L_c^2$, so if $I_b = I_c$ (a symmetric rotor),

$$E_{\text{rot}} = \frac{L^2}{2I_b} + L_a^2\left(\frac{1}{2I_a} - \frac{1}{2I_b}\right)$$

Vibrational energy may be expressed in terms of normal coordinates

$$E_{\text{vib}} = \tfrac{1}{2}\sum \dot{q}_i^2 + \tfrac{1}{2}\sum_{i,j=1}^{2} f_{ij} q_i q_j$$

The $3N - 6$ equations must be solved simultaneously but can be broken into independent sets by using the symmetry of the molecule.

PROBLEMS

17.1. Leibniz and his contemporaries took mv^2 as the measure of the *vis viva* of an object. Is the factor of $\frac{1}{2}$ now introduced completely arbitrary? Could we equally well define kinetic energy as mv^2? If so, what changes would be required in other parts of mechanics and thermodynamics? If not, what difficulties would be encountered?

17.2. A free particle is represented in six-dimensional phase space by a single point with the coordinates q_i and $p_i (i = x, y, z)$. How does this point move with time?

17.3. Consider a system consisting of 1 mol He (4 g), which is partitioned into 10^{11} subsystems of equal size. Consider only the translational degrees of freedom.

a. What is the dimension of phase space for this ensemble?

b. How many points occupy this space?

c. What is the smallest meaningful volume in phase space (including units)?

17.4. The kinetic energy of a homonuclear diatomic molecule is the sum of the kinetic energies of the atoms. $T = \frac{1}{2}m\dot{x}_1^2 + \frac{1}{2}m\dot{x}_2^2$. The potential energy is $V = \frac{1}{2}\,k(x_1 - x_2)^2$.

The coordinates $q_1 = x_1 + x_2$ and $q_2 = x_1 - x_2$ are orthogonal, so if the Lagrangian is expressed as $L(q_1, \dot{q}_1, q_2, \dot{q}_2)$ the equations for q_2 and $\dot{q}_2$ may be solved independently of q_1 and $\dot{q}_1$.

a. Make this substitution, of q_1 and q_2 for x_1 and x_2, and from the defining equations for Lagrange's equation, find p_2, the momentum conjugate to q_2. Give a simple explanation, or alternative analysis, of why the result is as found.

b. Write the Hamiltonian, $H(p_1, q_1, p_2, q_2)$. Use the Poisson bracket, with the Hamiltonian, to find $\dot{p}_1$, $\dot{p}_2$, $\dot{q}_1$, and $\dot{q}_2$. Interpret your answers. [Do not neglect to apply your general understanding of physics to verify equations at each step.]

17.5. Show that Hamilton's principle applied to the harmonic oscillator function variation $q = q_o \cos \omega t + \delta q_o \sin^2 \omega t$ over the range of a full period, $t = -\pi/\omega$ to $t = +\pi/\omega$, gives the variation $(\pi/8)m\omega^2(\delta q_o)^2$.

17.6. For the rigid rotor ($H = \mathscr{L}^2/2I$), show by means of Poisson brackets which of the following are constants of the motion.

a. ϑ. **b.** $\mathscr{L} = I\omega$. **c.** $E = \frac{\mathscr{L}^2}{2I}$.

17.7. Statistical mechanics describes a wide variety of systems, from ideal gases and ideal crystals to liquids, solar systems, and stars. An important question is the extent to which it may be assumed that the phase trajectory of a system will in due time pass through, or pass arbitrarily close to, each accessible point in phase space. Identify at least one familiar system for which you would not expect all regions of phase space compatible with the energy of the system to be visited, and explain why.

17.8. Which of the following molecules have at least one threefold or higher axis of rotation, as required for a symmetric rotor?

a. O_3 (ozone, nonlinear).

b. CO_2 (linear).

c. H_3C-CH_3 (ethane; each CH_3 pyramidal; angle of rotation, about the C$-$C bond, of one CH_3 with respect to the other may be taken as arbitrary).

d. HCl.

17.9. Which of the following molecules have more than one noncoincident threefold or higher axis of rotation, as required for a spherical rotor?

a. CO_2.

b. C_2H_6 (ethane: Take angle of rotation of $-CH_3$ groups to give the most symmetrical arrangement).

c. SF_6 (octahedral: F's are at corners of a regular octahedron, equivalent to two pyramids with square faces placed together).

17.10. A symmetric rotor for which the largest moment of inertia is perpendicular to the unique axis (like a baseball bat) is called a *prolate* rotor. If the largest moment of inertia is along the unique axis (like a Frisbee), the rotor is called *oblate*. If a prolate rotor is set spinning about its axis of symmetry, then loses energy, it will change spontaneously to a tumbling rotation about an axis perpendicular to the symmetry axis. Satellites are therefore designed as oblate rotors, which can spin stably about their symmetry axes. Explain, on the basis of conservation laws, why loss of even a small amount of rotational energy will make the rotation about the symmetry axis unstable for a prolate rotor.

CHAPTER 18

Fundamentals of Quantum Mechanics

Statistical mechanics was developed historically on the foundation of classical mechanics. Some puzzling uncertainties and discrepancies were apparent, however, which could not be explained until quantum mechanics was introduced between 1900 and 1925. Because statistical mechanics is fundamentally concerned with the energies and interactions of atoms and molecules, quantum mechanics must be incorporated into the model from the beginning to obtain a consistent theory. Fortunately, the relevant portions of quantum mechanics[1] are by and large the simplest—particularly the quantization of energy states.

18.1 Quantization of Energy States

Only a few quantum mechanical problems are relevant to the evaluation of thermodynamic functions of simple systems by statistical mechanics. Important problems include: a particle in a box (i.e., a particle confined by a potential well); a harmonic oscillator; angular momentum and rigid rotors, in one or three dimensions; and the specialized problems of electronic states in atoms, molecules, and solids. The first two problems are summarized in this section. Angular momentum, quantum mechanical rotors, and some properties of electronic states are treated in the following section.

Particle in a Box

When a particle is confined to a box of length L, with impenetrable walls, deBroglie's method[2] suffices to determine the possible energy values. The particle

[1]In this chapter we present the notation and specific formulas of quantum mechanics required for statistical mechanics applications, with a brief discussion of the relationship of quantum mechanics to classical mechanics. More explicit information on operator notation is given in the Appendix. For proofs and derivations, see standard quantum mechanics textbooks.

[2]For the derivation (omitted in most textbooks) see L. deBroglie, *Matter and Light* (New York: Dover Publications, Inc.). In summary, deBroglie argued that if electrons are to exhibit solutions characterized by integers, they must be described by oscillations, and hence waves. Identifying the group speed, u, of a particle wave with the particle speed, the phase speed $v = \lambda\nu$ is related to u by $uv = c^2$. Combining Einstein's two energy equations, $E = mc^2 = h\nu$ yields $\lambda = h/p$ if $p = mu$.

has a wavelength of

$$\lambda = \frac{h}{p} \tag{18-1}$$

where p is the momentum. Assuming the potential energy is constant, we may set $\mathscr{V} = 0$ throughout the box and the total energy is the kinetic energy.

$$E = \frac{p^2}{2m} = \frac{h^2}{2m\lambda^2} \tag{18-2}$$

The particle behaves as a standing wave within the box, as depicted in Fig. 18.1, and therefore

$$\frac{\mathrm{n}\lambda}{2} = L \qquad or \qquad \lambda = \frac{2L}{\mathrm{n}}$$

with integral values of n. Therefore,

$$E = \frac{h^2\mathrm{n}^2}{8mL^2} \tag{18-3}$$

For any "box" significantly larger than molecular dimensions, the energy levels are so closely spaced that the particles may be considered to obey classical mechanics. Translational degrees of freedom of an ideal gas are therefore treated as classical degrees of freedom.

EXAMPLES 18.1

A. Find the spacing between the ground state (n = 0) and first excited state (n = 1) for a speck of dust of $m = 10^{-3}$ g in a box of length 1 mm.
B. Find the spacing between the ground state (n = 0) and first excited state (n = 1) for a He atom in a box of length 1 mm.
C. Find the spacing between the ground state (n = 0) and first excited state (n = 1) for an electron in a box (i.e., a molecule) of length 5 Å. □

Harmonic Oscillator

An oscillator with potential energy function $\mathscr{V} = \frac{1}{2} k q^2$ and kinetic energy $\mathscr{T} = \frac{1}{2} m\dot{q}^2$ has a classical frequency

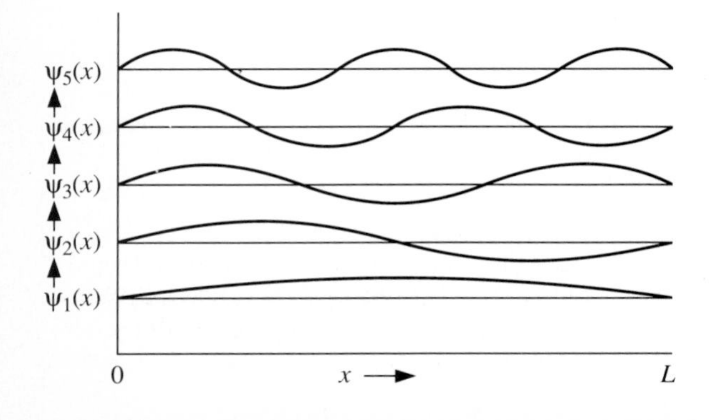

FIGURE 18.1 Amplitude of the wave function, $\Psi_n(x)$, plotted against position, x, for several of the lowest-energy states of a particle in a box. The particle density, or probability of finding the particle in an interval Δx, is proportional to the square of the wave function shown.

$$\nu = \frac{\omega}{2\pi} = \frac{1}{2\pi}\sqrt{\frac{k}{m}} \tag{18-4}$$

The quantum mechanical solution has a time-dependent wave function with the factor

$$e^{i\omega t} \qquad \text{with} \qquad \omega = \sqrt{\frac{k}{m}}$$

so we refer to $\omega/2\pi$ as the frequency of oscillation of the quantum mechanical system.

The classical oscillator may have any energy. The quantum oscillator is limited to the energy values

$$E = \left(\mathrm{v} + \frac{1}{2}\right)\hbar\omega \tag{18-5}$$

with quantum number $\mathrm{v} = 0, 1, 2, \ldots$. The wave functions, Ψ_i, and probability functions, $|\Psi_i^2| = \Psi_i^*\Psi_i$, for some vibrational states are shown in Fig. 18.2. (Ψ_i^* is the complex conjugate of Ψ_i.)

EXAMPLES 18.2

A. Find the spacing between the ground state ($\mathrm{v} = 0$) and the first excited state ($\mathrm{v} = 1$) of a 1000-Hz oscillator.
B. Find the spacing between the ground state and first excited state of a nitrogen molecule, with wavenumber $\sigma = \nu/c = 2.360 \times 10^5\ \mathrm{m}^{-1}$. □

18.2 Angular Momentum: Rotations and Spins

The product of lever arm and momentum, $\mathbf{r} \times \mathbf{p}$, is called the *moment of momentum,* analogous to the torque, which is the moment of a force, $\mathbf{r} \times \mathbf{f}$. Consistent with this definition, the moment of momentum may be written

$$\mathscr{L} = \mathbf{r} \times \mathbf{p} = (\mathbf{i}x + \mathbf{j}y) \times (\mathbf{i}p_x + \mathbf{j}p_y) = (\mathbf{i} \times \mathbf{j})x\,p_y + (\mathbf{j} \times \mathbf{i})y\,p_x = (x\,p_y - y\,p_x)\mathbf{k} \tag{18-6}$$

or a sum of such terms for the particles of a system.

More often, however, this product is called the *angular momentum,* because it is often (but not always!) associated with circular motion. The time rate of change of angular momentum is equal to the torque.

$$\boldsymbol{\tau} = \dot{\mathscr{L}} \equiv \frac{d\mathscr{L}}{dt} \tag{18-7}$$

Angular momentum is one of the few quantities in mechanics that is conserved (along with volume, energy, and linear momentum). The total angular momentum along any axis cannot change; it is simply passed from one object to another. Therefore, the angular momentum of any rotating free body is constant not only in magnitude, but also in direction. We assume that the origin, or the

ANSWERS 18.1

A. $E_1 - E_o = h^2/8mL^2 = 5.5 \times 10^{-56}$ J $= 3.4 \times 10^{-37}$ eV.
B. $E_1 - E_o = 8.3 \times 10^{-36}$ J $= 5.2 \times 10^{-17}$ eV.
C. $E_1 - E_o = 2.4 \times 10^{-19}$ J $= 1.5$ eV.

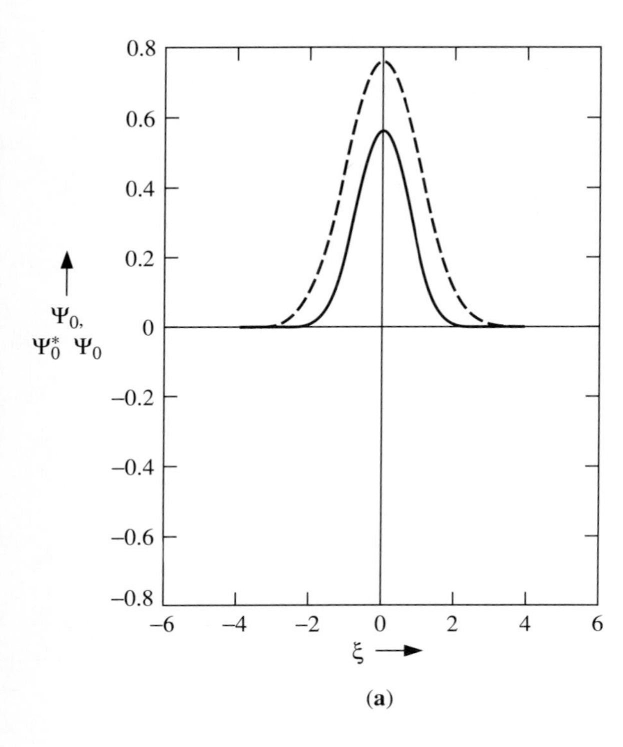

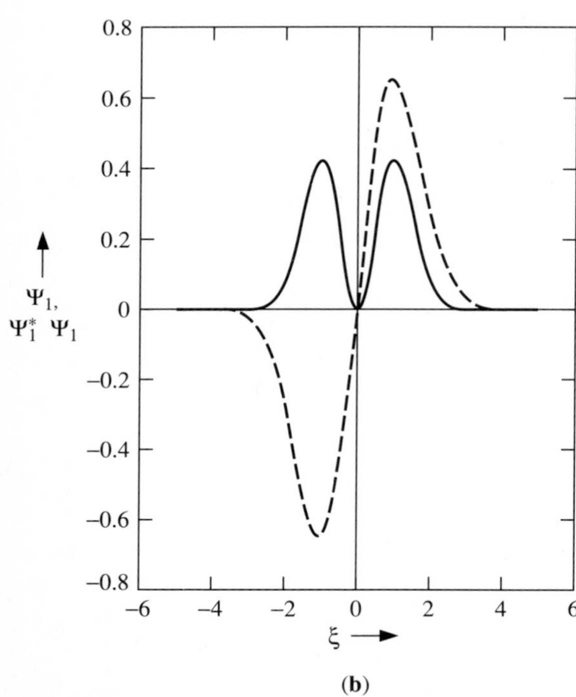

FIGURE 18.2 a, b Harmonic oscillator wave functions: (a) v = 0; (b) v = 1. The dashed lines represent amplitude, Ψ_i; the solid lines represent probability, $\Psi_i^* \Psi_i$.

ANSWERS 18.2

A. $\Delta E = h\nu = 6.6 \times 10^{-31}$ J $= 4.1 \times 10^{-12}$ eV.
B. $\nu = 7.08 \times 10^{13}$ Hz and $\Delta E = 4.69 \times 10^{-20}$ J $= 0.29$ eV.

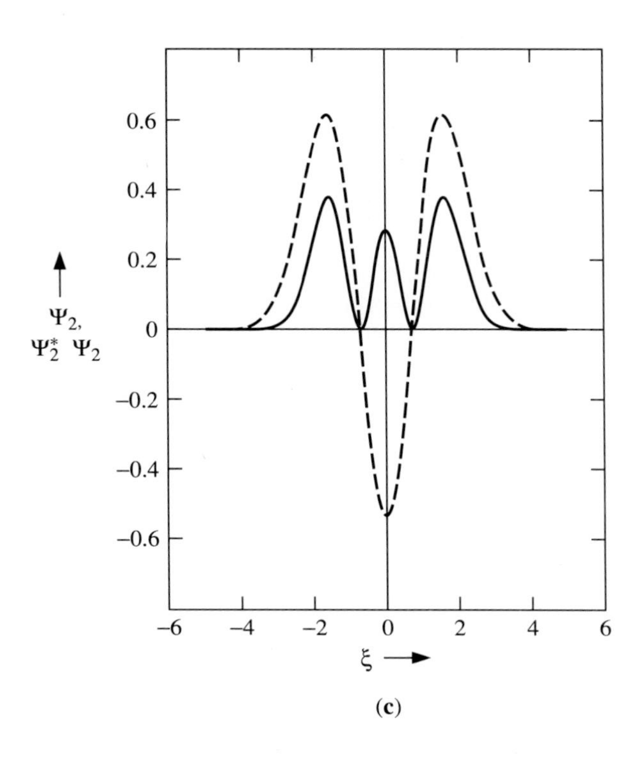

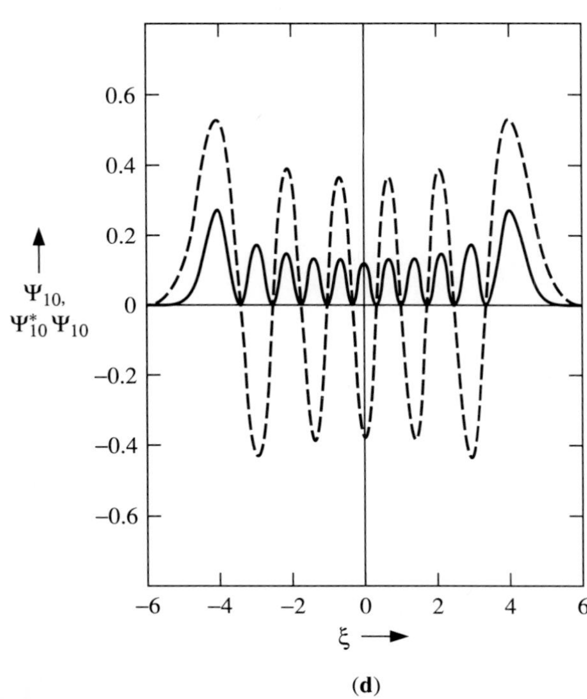

FIGURE 18.2 c, d Harmonic oscillator wave functions: (c) v = 2; (d) v = 10. The dashed lines represent amplitude, Ψ_i; the solid lines represent probability, $\Psi_i^*\Psi_i$.

reference point, for measurement of $\boldsymbol{\omega}$ and $\mathscr{L}$ is the center of mass of the rotating body, in order that rotations are unambiguously separated from other motions.

A one-dimensional rotor is characterized by a moment of inertia, I; by an angle of rotation, ϕ, about an axis uniquely defined by the system or by the experimental apparatus; and by an angular momentum vector, $\mathscr{L} = I\boldsymbol{\omega}$ directed along the axis of rotation according to the right-hand rule, with $\omega = d\phi/dt$. The wave function is of the form

$$\Phi = Ae^{im\phi} \tag{18-8}$$

with m $= 0, \pm 1, \pm 2, \ldots$ or sometimes, when some or all of the angular momentum is the nonclassical *intrinsic angular momentum,* called *spin,* m $= \pm\frac{1}{2}, \pm\frac{3}{2}, \pm\frac{5}{2}, \ldots$.

The rotation is about the unique axis, so $\mathscr{L}$ is directed along that unique axis and is constant in direction and magnitude, barring external intervention. The quantum number m is replaced by other symbols, such as K or s, for certain types of rotors.

About any axis, the kinetic energy of a mass point in rotational motion is $\mathscr{T}_{\text{rot}} = \frac{1}{2}mv^2$. Because $v = r\omega$, $\mathscr{T}_{\text{rot}} = \frac{1}{2}mr^2\omega^2 = \frac{1}{2}I\omega^2$, with $I = mr^2$. About the z axis, for example, $r^2 = x^2 + y^2$, so

$$I = \frac{1}{2}\sum m_i(x_i^2 + y_i^2) \qquad \text{or} \qquad I = \frac{1}{2}\int (x^2 + y^2)\,dm$$

The energy of the quantized rotor is

$$E = \mathscr{T}_{\text{rot}} = \frac{1}{2}I\omega^2 = \frac{(I\omega)^2}{2I} = \frac{\mathscr{L}^2}{2I} \tag{18-9}$$

and $\mathscr{L}$ is limited to the values

$$\mathscr{L} = \text{m}\hbar \qquad (\text{m} = 0, \pm 1, \ldots)$$

so

$$E = \frac{\text{m}^2\hbar^2}{2I} = hcB\text{m}^2 \tag{18-10}$$

if $B = h/8\pi^2 cI$.

The partition function is

$$Z = \left[1 + 2\sum_{\text{m}=1}^{\infty} e^{-\text{m}^2\hbar^2/2IkT}\right]\int d\tau \tag{18-11}$$

Treating the rotation as classical (almost always an excellent approximation), we have

$$Z = 2\int_0^{\infty} e^{-\mathscr{L}^2/2IkT}\,d\mathscr{L}\int d\tau$$

The angle coordinate varies over 2π radians and the volume element, $\int d\tau$, is divided by h. Therefore,

$$Z = \sqrt{2\pi I kT} \times \frac{2\pi}{h}$$

Some molecules have symmetry that reduces the number of distinguishable states (Section 20.5), so we divide also by a *symmetry number,* s

$$Z = \frac{(2\pi I kT)^{1/2}}{s\hbar} \qquad (18\text{-}12)$$

EXAMPLES 18.3

A. Considering a 1-mg speck of dust as a sphere ($I = \frac{2}{5}mr^2$) of density 2 g/cm^3, and therefore $r = 0.5$ mm, find the spacing between the ground state (m = 0) and the lowest rotational state (m = 1), treating the sphere as a one-dimensional rotor.
B. Treating the nitrogen molecule ($B \equiv h/8\pi^2 cI = 2.010$ cm^{-1}) as a one-dimensional rotor, find the spacing between the ground state and lowest rotational state.
C. Rotational spacings increase with increasing quantum number. What value of the quantum number m would give a rotational energy for the speck of dust equal to the energy of the first rotational state of N_2? □

The more general problem of rotations in three-dimensional space leads to wave functions called the Legendre polynomials and the associated Legendre functions. The angular momentum is then

$$\mathscr{L} = \sqrt{\ell(\ell+1)}\hbar \qquad (18\text{-}13)$$

with quantum number $\ell = 0, 1, 2, \ldots$ (or $\frac{1}{2}, \frac{3}{2}, \frac{5}{2}, \ldots$).

The projection of $\mathscr{L}$ on the unique axis (defined by the system or experiment) is m$\hbar$, as for the one-dimensional rotor. It is helpful to think of $\mathscr{L}$ as not fully defined experimentally in direction. The vector $\mathscr{L}$ swings around the fixed axis, with a constant projection on the axis. The projection of $\mathscr{L}$ on axes perpendicular to the unique axis is not experimentally observable without changing the rotational state of the system.

Rotational energy is written, classically, in terms of the moments of inertia and angular momenta about the principal axes as

$$E = \frac{1}{2}I_a\omega_a^2 + \frac{1}{2}I_b\omega_b^2 + \frac{1}{2}I_c\omega_c^2$$

$$= \frac{\mathscr{L}_a^2}{2I_a} + \frac{\mathscr{L}_b^2}{2I_b} + \frac{\mathscr{L}_c^2}{2I_c} \qquad (18\text{-}14)$$

The partition function, in the classical limit, is

$$Z = \iiint \exp\left(-\frac{\mathscr{L}_a^2/2I_a + \mathscr{L}_b^2/2I_b + \mathscr{L}_c^2/2I_c}{kT}\right) d\mathscr{L}_a\, d\mathscr{L}_b\, d\mathscr{L}_c \iiint d\tau$$

ANSWERS 18.3

A. $\Delta E = h^2/8\pi^2 I$, $I = 2 \times 10^{-6} \times 5 \times 10^{-8} = 10^{-13}$, $\Delta E = 5.6 \times 10^{-56}$ J $= 3.5 \times 10^{-37}$ eV.

B. $\Delta E = hcB = 4.0 \times 10^{-23}$ J $= 2.5 \times 10^{-4}$ eV. (The actual spacing for N_2 is twice this value, given by the three-dimensional rotor formula.)

C. $\dfrac{h^2 \mathrm{m}^2}{8\pi^2 \times 10^{-13}} = 4.0 \times 10^{-23}$ J; $\mathrm{m}^2 = 7.2 \times 10^{32}$; $\mathrm{m} \approx 2.7 \times 10^{16}$.

The spatial coordinates give a factor of 4π as the direction of **J** varies over a sphere and a factor of 2π for rotation of the molecule about **J**. Hence

$$\{\text{asymmetric}\} \quad Z = \sqrt{2\pi I_a kT}\,\sqrt{2\pi I_b kT}\,\sqrt{2\pi I_c kT} \times \frac{8\pi^2}{h^3}$$

$$= \frac{(2\pi kT)^{3/2}(I_a I_b I_c)^{1/2}}{\pi \hbar^3} = \left[\frac{\pi}{ABC}\left(\frac{kT}{hc}\right)^3\right]^{1/2} \tag{18-15}$$

where we have defined the rotational constants,

$$A \equiv \frac{\hbar^2}{2hcI_a} = \frac{h}{8\pi^2 c I_a}$$

$$B \equiv \frac{\hbar^2}{2hcI_b} = \frac{h}{8\pi^2 c I_b} \tag{18-16}$$

and

$$C \equiv \frac{\hbar^2}{2hcI_c} = \frac{h}{8\pi^2 c I_c}$$

Asymmetric rotors are usually approximated as symmetric rotors for analysis of spectra. The differences are then treated by perturbation methods.

The total rotational energy of a symmetric rotor, with $I_b = I_c$, is

$$E = \frac{\mathcal{L}_a^2}{2I_a} + \frac{\mathcal{L}_b^2 + \mathcal{L}_c^2}{2I_b} = \frac{\mathcal{L}_a^2 + \mathcal{L}_b^2 + \mathcal{L}_c^2}{2I_b} + \mathcal{L}_a^2\left(\frac{1}{2I_a} - \frac{1}{2I_b}\right)$$

If we choose the symbols J and K for the quantum numbers for total angular momentum and projection of angular momentum, respectively, the square of the total angular momentum is

$$\mathcal{L}^2 = \mathcal{L}_a^2 + \mathcal{L}_b^2 + \mathcal{L}_c^2 = \mathrm{J(J+1)}\hbar^2 \qquad \mathrm{J} = 0, 1, 2, \ldots$$

and the angular momentum about the unique axis is

$$\mathcal{L}_a = \mathrm{K}\hbar \qquad \mathrm{K} = 0, \pm 1, \pm 2, \ldots, \pm \mathrm{J}$$

Both J and K are integers for a quantized molecular rotor, and $|\mathrm{K}| \leq \mathrm{J}$. The angular momentum, $\mathcal{L}$, may have 2J +1 possible orientations in space, and if $\mathrm{K} \neq 0$, two orientations of $\mathcal{L}_a$, as shown in Fig. 18.3.

The energy of the symmetric molecular rotor is therefore

$$E = \frac{\mathcal{L}^2}{2I_b} + \mathcal{L}_a^2\left(\frac{1}{2I_a} - \frac{1}{2I_b}\right) = \frac{J(J+1)\hbar^2}{2I_b} + K^2\hbar^2\left(\frac{1}{2I_a} - \frac{1}{2I_b}\right) \quad (18\text{-}17)$$

Energies are most often expressed as *term values,* equivalent to E/hc, corresponding to the *wavenumber,* $\sigma = \nu/c = 1/\lambda = \Delta E/hc$. The term values of a symmetric rotor are

$$F(J, K) = \frac{E}{hc} = \frac{J(J+1)\hbar^2}{2hcI_b} + K^2\hbar^2\left(\frac{1}{2hcI_a} - \frac{1}{2hcI_b}\right)$$

or

$$\begin{Bmatrix}\text{symmetric}\\ \text{rotor}\end{Bmatrix} \qquad F(J, K) = J(J+1)B + K^2(A - B) \quad (18\text{-}18)$$

The constants A and B are identical for a spherical rotor, so the last term vanishes.

$$\begin{Bmatrix}\text{spherical}\\ \text{rotor}\end{Bmatrix} \qquad F(J, K) = F(J) = J(J+1)B \quad (18\text{-}19)$$

Linear molecules have a very small value for I_A, so $A \gg kT/hc$ and therefore K = 0 at ordinary temperatures (Section 6.4). Hence the term values for a linear molecule also reduce to

$$\{\text{linear}\} \qquad F(J, K) = F(J) = J(J+1)B \quad (18\text{-}20)$$

(Note, however, that the spherical rotor and linear rotor give the same equation for different reasons.) The partition function is

$$\begin{aligned} Z &= \sum_{J=0}^{\infty}(2J+1)e^{-[J(J+1)\hbar^2/2I_b]/kT}\int d\tau \\ &= 2\int_0^{\infty} e^{-\mathcal{L}^2\hbar^2/2I_bkT}\, d\mathcal{L}\int d\tau \end{aligned} \quad (18\text{-}21)$$

in the classical limit, where J may be regarded as large and continuous. The spatial coordinates, integrated over a sphere, give 4π, which is divided by h^2 to give $Z = (2\pi I_b kT/h^2)4\pi$. Including the symmetry number, this reduces to

$$Z = \frac{8\pi^2 I_b kT}{sh^2}$$

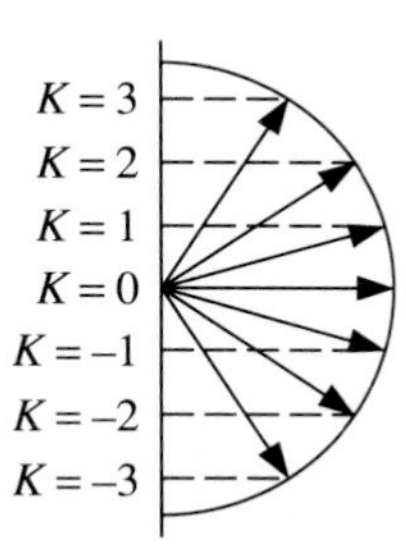

FIGURE 18.3 Possible orientation of the angular momentum with respect to an applied field for a rigid rotor with quantum number of angular momentum J = 3. The magnitude of $\mathcal{L}$ is $\sqrt{3 \cdot 4}\,\hbar$; the maximum projection is exactly $3\hbar$.

or

$$\{\text{linear}\} \qquad Z = \frac{kT}{shcB} \tag{18-22}$$

Angular momentum arises in atoms, and some molecules, because of motions of the electrons and also because electrons have an intrinsic angular momentum (not describable as a classical rotation) called *spin,* of magnitude $s = \sqrt{\frac{1}{2}(\frac{1}{2}+1)}\hbar$ and quantum number $s = \pm\frac{1}{2}$. Any particle with half-integral spin ($s = \frac{1}{2}, \frac{3}{2}, \frac{5}{2}$, and so on, in units of $\hbar$) is called a *fermion*. Any particle with integral spin (0, $1\hbar$, $2\hbar$, and so on) is called a *boson*. Electrons are fermions.

The total angular momentum is still represented by the quantum number J. However, the composition of J is less simple when electronic angular momenta are included than for molecular rotations.

The most common model for describing electrons in atoms is the *one-electron wave function* approximation. Each electron is treated individually, usually by describing it as if it were present by itself and therefore in a state similar to electron states in hydrogen atoms, modified by the differences in nuclear charge. The mathematical function that describes the time-average spatial distribution of the electron, or electron charge, is called an *orbital* (*not* to be confused with an *orbit,* which is a time-dependent path through space). When this is combined with a description of the intrinsic angular momentum, or "spin," the product function is called a *spinorbital*.

Two approximations are common for describing how the values of J are determined. In Russell–Saunders coupling, the orbital angular momenta, ℓ_i, of the electrons couple to give a resultant, $\mathscr{L}$ (with an integral quantum number L), and the electron spin angular momenta, $\mathbf{s}_i$, couple to give a resultant $\mathbf{S}$ (with integral or half-integral quantum number S). Then the orbital and spin angular momenta add vectorially (Fig. 18.4) to give the resultant, $\mathbf{J}$ (with quantum number, J, integral or half-integral).

$$\mathbf{L} + \mathbf{S} = \mathbf{J} \tag{18-23}$$

$$|L - S| \leq J \leq |L + S|$$

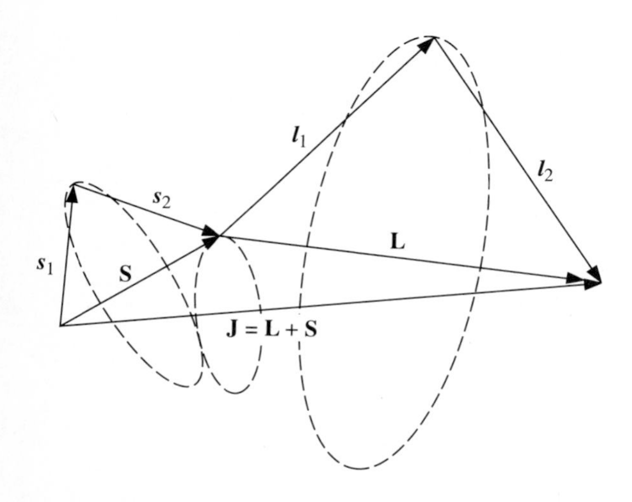

FIGURE 18.4 Russell–Saunders coupling (schematic).

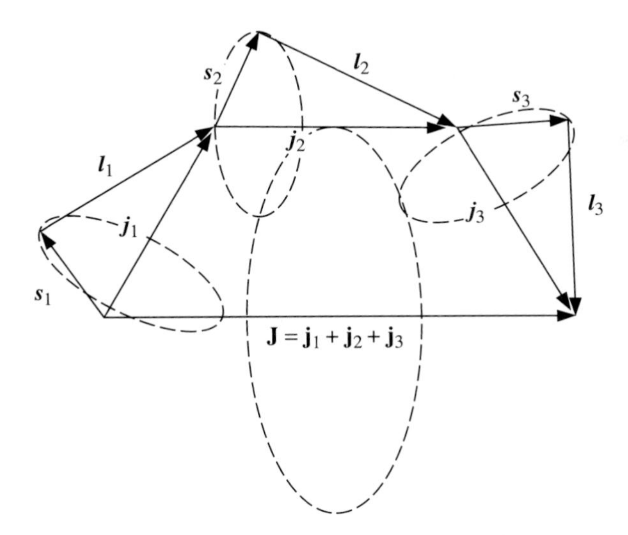

FIGURE 18.5 *j-j* coupling (schematic).

In heavier atoms, the orbital angular momenta of individual electrons add to the spin angular momenta. Representing values for single electrons by lowercase letters (the usual convention),

$$|\mathbf{s}| = \sqrt{s(s+1)}\,\hbar \qquad \text{and} \qquad |\ell| = \sqrt{\ell(\ell+1)}\,\hbar$$

Then addition of angular momenta, $\mathbf{j}_i = \ell_i + \mathbf{s}_i$, is paralleled by the addition of quantum numbers:

$$\mathrm{j}_i = \ell_i + \mathrm{s}_i \qquad \text{and} \qquad \mathrm{J} = \sum \mathrm{j}_i \tag{18-24}$$

This model is called *j–j coupling* (Fig. 18.5).

Recall that classically, the process of summing ℓ and s separately, then adding the resultants to obtain j, would not differ from adding ℓ to s, then adding the resultant j values. There is a difference, however, in quantum mechanics, because each time a sum is formed the magnitude of that sum must be integral (or half-integral).[3]

The classical model for adding angular momenta (Section 17.3) is that the angular momentum vectors that form a resultant rotate, or precess, about the resultant with a speed that increases with the energy of interaction. The strongest couplings are added first and precess most rapidly; weaker couplings then precess more slowly about the new resultant.

Electronic angular momenta in molecules add by rules similar to those for atoms, except that typically it is the angular momentum about a fixed axis of the molecule that is quantized. It is therefore represented by $\mathbf{\Lambda}$ (projection of $\mathcal{L}$) and by $\mathbf{\Sigma}$ (projection of $\mathbf{S}$). The rules for coupling these projected angular momenta to the molecular rotation depend on individual cases, especially on the relative magnitudes of the angular momenta that are coupling.

The energy of an atom or molecule depends on the rotational quantum numbers but not, in general, on the direction of the angular momentum in space. For a symmetric rotor, the resultant degeneracy is $2\mathrm{J} + 1$, the number of possible

[3]The summation of angular momenta described here is the conventional approximation. It is somewhat oversimplified, for it leads to more quantum numbers than degrees of freedom for most systems.

directions for J. In addition, states of the same value of K (for $K \neq 0$), but with **K** pointed in opposite directions, are degenerate.

All rotational states of a spherical rotor with the same value of J are degenerate, corresponding both to the direction of **J** in the molecule and in space. Therefore, a spherical rotor has a rotational degeneracy of $(2J + 1)^2$, with no K degeneracy.

Nuclear spins have little interaction with other energy states of atoms or molecules, so a nucleus with spin $|\mathbf{i}| = \sqrt{i(i + 1)}\hbar$ contributes an additional degeneracy of $2i + 1$. This degeneracy is the same for all states involving the nucleus. We will see later that the presence of nuclear spin can have a significant effect on the properties of the molecule. The properties depend on whether the nuclei are fermions (half-integral spin) or bosons (integral spin). The total wave function must be symmetric (for bosons) or antisymmetric (for fermions) to exchange of identical nuclei. Usually, the electronic and vibrational wave functions can be assumed symmetric (in the ground state). Therefore, the product $\psi_{spin}\psi_{rot}$ must be symmetric or antisymmetric, for bosons or fermions, respectively.[4]

For t nuclei of spin i, there are $(2i + 1)^t$ states that may be considered strictly degenerate for most purposes (at least in the absence of very strong magnetic fields). For example, a homonuclear diatomic molecule ($t = 2$) with integral spin has $(2i + 1)^2$ degenerate states. Of these, $(i + 1)(2i + 1)$ states are symmetric in spin. These must therefore be symmetric in ψ_{rot}; they have even J. The other $i(2i + 1)$ states are antisymmetric in spin and thus antisymmetric in ψ_{rot}, with odd values of J.

In a diatomic molecule with nuclei of half-integral spin, $i(2i + 1)$ states are symmetric in spin and therefore antisymmetric in ψ_{rot}, with odd values of J. The remaining $(i + 1)(2i + 1)$ states are antisymmetric in spin and symmetric in ψ_{rot} and thus have even values of J.

These degeneracies are usually normalized by dividing by $(2i + 1)^t$. This ignores the overall spin degeneracy while taking into account, as necessary, the relative populations of symmetric and antisymmetric rotational states.

Symmetry of spin states with respect to exchange of identical nuclei is not always self-evident. Usually, it is necessary to consider linear combinations of states, writing down the component states and their combinations explicitly.

For example, consider a homonuclear diatomic molecule with $i = 1$ (each nucleus). Two unit vectors, of arbitrary direction but constrained to give a sum of integral length, can add together to give the resultant $I = 0$, $I = 1$, or $I = 2$. If $I = 0$, the projection of I on the axis can only be zero; it is a singlet state ($2I + 1 = 1$). If $I = 1$, the projection can be -1, 0, or $+1$; this is a triplet level

[4]Symmetry of a wave function with respect to exchange of equivalent particles is not the same as spatial symmetry, but these symmetries are related. For example, the harmonic oscillator wave functions (Fig. 18.2) of even quantum number v are symmetric about the center point ($q = 0$). Examination of the form of the vibrations (e.g., Figs. 6.1 and 6.2) shows that when a molecule is vibrating, equivalent atoms may or may not undergo "equivalent" displacement (i.e., symmetric with respect to the nonvibrating molecule). The vibrational wave function of the nonvibrating molecule is fully symmetric to exchange of equivalent nuclei, by definition; it is this equivalence to exchange that defines equivalent nuclei in the molecule. When the molecule is vibrating, it may retain this symmetry (with respect to the symmetry operations of the molecule, such as rotations and reflections) or it may not, depending on the form of the vibration. It can readily be shown (by examining the direct products of irreducible representations, or symmetry species, of the molecular point groups) that a doubly excited vibrational state of any nondegenerate vibration has the same symmetry as the ground state and is therefore symmetric to exchange of equivalent nuclei, as well as being symmetric in the vibrational coordinate. Similar arguments lead to the conclusion that even-numbered rotational states are symmetric to exchange of equivalent nuclei.

Table 18.1 Spin States Projected Values on Axis

i_{a_z}	i_{b_z}	I_z	Symmetry
1	1	2	*s*
1	0	1	
1	−1	0	
0	1	1	
0	0	0	*s*
0	−1	−1	
−1	1	0	
−1	0	−1	
−1	−1	−2	*s*

$(2I + 1 = 3)$. If $I = 2$, the projection can be -2, -1, 0, $+1$, or $+2$; it is a quintuplet level $(2I + 1 = 5)$.

We examine the states more closely by writing the individual projections, i_{a_z} and i_{b_z}, and their sum, I_z, as shown in Table 18.1. Note that only three of the nine states treat nuclei *a* and *b* equivalently. Two of these give $I_z = \pm 2$, showing that the quintet level is symmetric with respect to exchange of nuclei. There are $(i + 1)(2i + 1) = 6$ symmetric states and $i(2i + 1) = 3$ antisymmetric states. It follows that the singlet state is also symmetric and the triplet level must be antisymmetric.

Because every state must remain unchanged (if symmetric) or only change sign (if antisymmetric) when equivalent nuclei are interchanged, we must construct the remaining six states from linear combinations of the six nonsymmetric state representations from Table 18.1. The appropriate linear combinations are shown in Table 18.2. For example, the second and fourth entries in Table 18.1 [(1, 0) and (0, 1)] combine to give the first and fourth entries in Table 18.2 [(1, 0) + (0, 1), symmetric, and (1, 0) − (0, 1), antisymmetric]. The states are shown in Table 18.3 grouped into singlet, triplet, and quintuplet levels.

An alternative representation is often faster for identifying the states. Prepare a matrix of spin values, as shown in Fig. 18.6. Here it is easy to see that the three diagonal states (projections 2, 0, and −2) are symmetric and include states from the quintuplet level. It follows that the remaining symmetric state can only be the singlet and the three antisymmetric states must be the triplet level.

Table 18.2 Spin States Combinations

	i_z	Symmetry
$(1, 0) + (0, 1)$	1	*s*
$(1, -1) + (-1, 1)$	0	*s*
$(-1, 0) + (0, -1)$	−1	*s*
$(1, 0) - (0, 1)$	1	*as*
$(1, -1) - (-1, 1)$	0	*as*
$(-1, 0) - (0, -1)$	−1	*as*

Table 18.3 Symmetric and Antisymmetric Spin States

I	Symmetry	Combination	Projection, I_z
0	*s*	(0, 0)	0
1	*as*	(1, 0) − (0, 1)	1
1	*as*	(1, −1) − (−1, 1)	0
1	*as*	(−1, 0) − (0, −1)	−1
2	*s*	(1, 1)	2
2	*s*	(1, 0) + (0, 1)	1
2	*s*	(1, −1) + (−1, 1)	0
2	*s*	(−1, 0) + (0, −1)	−1
2	*s*	(−1, −1)	−2

If the nuclear spins are half-integral, the particles are fermions and the total wave function must be antisymmetric. For example, if i = $\frac{1}{2}$, as for the hydrogen molecule, then I = 0 (singlet state) or I = 1 (triplet level). The first of these is antisymmetric in the spin, the second is symmetric in the spin (Fig. 18.7). Therefore, for $\psi_{spin}\psi_{rot}$ to be antisymmetric, the singlet state has J = 0, 2, 4, ... and the triplet level has only the rotational states J = 1, 3, 5, ...

Hydrogen, H_2, is an example of a homonuclear diatomic molecule with nuclear spins of $\frac{1}{2}$. There are three symmetric spin states (I = 1 triplet) and only one antisymmetric spin state (I = 0 singlet). The nuclei are fermions, so $\psi_{spin}\psi_{rot}$ must be antisymmetric to exchange (assuming symmetric electronic and vibrational states). Hence the triplet level must have an antisymmetric ψ_{rot}; the singlet state has a symmetric ψ_{rot}. At high temperatures (e.g., room temperature) the triplet level is three times as probable as the singlet state. Because the triplet level is more probable, it is called *ortho* hydrogen; the less probable (singlet) state is called *para* hydrogen. At room temperature, normal hydrogen is a 3:1 mixture of ortho/para.

Note that ortho hydrogen can never fall to the lowest rotational state (J = 0) because this rotational state is symmetric. Therefore, in the presence of a suitable catalyst (a paramagnetic substance that can interact with the nuclear spins through electron coupling), ortho hydrogen converts to para hydrogen, falling into the J = 0 rotational state. The conversion is slow in the absence of catalysts, so it is possible to prepare pure para hydrogen by cooling the gas in the presence of a catalyst, then removing the catalyst before rewarming the gas.

Deuterium, D_2, or molecular heavy hydrogen, has i = 1 and therefore I = 0, 1, or 2. Ortho deuterium (most plentiful at room temperature) consists of

	1	0	−1	
1	2	1	0	quintuplet level (I = 2) is symmetric
0	1	0	−1	triplet level (I = 1) is antisymmetric
−1	0	−1	−2	singlet state (I = 0) is symmetric

FIGURE 18.6 The singlet state and two projections of the quintuplet level fall along the diagonal and are therefore clearly symmetric; the remaining six projections are constructed of linear combinations, half of which are symmetric (and part of the quintuplet level) and the other half antisymmetric (the triplet level).

	$\frac{1}{2}$	$\frac{1}{2}$	
$\frac{1}{2}$	1	0	triplet level (I = 1) is symmetric
$\frac{1}{2}$	0	−1	singlet state (I = 0) is antisymmetric

FIGURE 18.7 Combinations of spins of $\frac{1}{2} + \frac{1}{2}$.

quintuplet and singlet levels, with J = 0, 2, 4, ... Para deuterium is the triplet level, with J = 1, 3, 5, ... At low temperatures the ortho form is favored, in the J = 0 state; at high temperatures the ratio of ortho to para is (5 + 1)/3 or 2/1.

18.3 Uncertainty and Complementarity in Quantum Mechanics[5]

In practice, many molecular properties are best understood and evaluated from the viewpoint of classical mechanics. However, there are important limitations that must be kept in mind to avoid erroneous predictions or interpretations. In this section we emphasize the relationships between classical and quantum mechanics.

Any theory that seeks to provide a mechanical description for small particles must satisfy three principles: the uncertainty principle, the superposition principle, and the correspondence principle. Each is discussed below.

Uncertainty Principle

Classical physics accurately describes systems for which the results of an observation and therefore the phenomenon under investigation may safely be regarded as independent of the observation. Such a physical process may be regarded as happening without regard to whether it is, or can be, observed. Fock has called this "absolutization of physical processes."

Any measurement of position, momentum, or energy of a particle (including a photon) requires that the presence of the particle in some region of space must be detected within some interval of time. The smaller the particle, the more sensitive the detector must be, in general, to give meaningful results. It is easier to locate a locomotive than a baseball, and easier to locate a baseball than an electron.

The fundamental requirement for a particle detector is that the particle must have an effect on the detector. Then Newton's third law requires that the detector must have an equal and opposite effect (measured as force) on the particle. However, this was never considered an important limitation until the present century. It was assumed that detectors could always be constructed that would be sensitive enough that their effect on the particle under observation would be negligible.

If an observation has no significant effect on a system, then it is possible to piggyback, or compound, observations, making one after another. Even if two measurements require quite different experimental conditions of observation, they may be regarded as made simultaneously, or without interference with one another. By compounding of observations, it should be possible, at least in principle, to provide an exhaustive description of the state of any system by performing a sufficient number of observations on the system.

[5]In this section we adopt some of the interpretations and language of V. A. Fock, *Fundamentals of Quantum Mechanics,* translated by E. Yankovsky (Moscow: Mir Publishers, 1978).

The wave–particle duality of light and of particulate matter is a demonstration that the classical description is not suitable for very small objects. At the time quantum mechanics was developed, Heisenberg pointed out a common thread running through all experiments. Regardless of the type of detector—a slit for passage of particles, a photon bounced off the particle, or the sorting of particles in electric and magnetic fields, for example—there is a minimum level of disturbance. The more accurately position is measured, the greater the uncertainty in the momentum after the measurement, or the greater the accuracy of a momentum measurement the more uncertain will be the position. Designating the uncertainties as Δq and Δp gives us

$$\Delta q \, \Delta p \geq \frac{h}{4\pi} \tag{18-25}$$

More generally, p and q may be any pair of conjugate variables, including E and t, energy and time. No experiment can be simultaneously optimized to measure both variables of a conjugate pair, so the product of the uncertainties of the two variables must be limited by the inequality of Heisenberg's uncertainty principle.

There are continuing arguments concerning the interpretation of Heisenberg's uncertainty principle. For present purposes we need only adopt the very conservative interpretation:

Because the results of any experiment always include at least a minimum uncertainty or random error, any theory that seeks to predict results of experiments must incorporate such uncertainty in its predictions.

Any reliable measurement must be objective, relying on observations that are unequivocal. In particular, every measurement requires that the experimenter interact with apparatus that behaves classically. The apparatus, obeying classical mechanics, is a necessary intermediary between a human being and the submicroscopic world.

It is the interaction of a submicroscopic particle with the classical instrumentation that reveals the presence, and therefore the properties, of the particle. Particles may interact with molecules of water to form droplets large enough to be visible, or with atoms of silver to form grains in a visible pattern, or with atoms or molecules to produce photons that generate nerve signals when they strike the retina of the human eye.

Measurement of one property of a small system (e.g., the position of a particle) requires a set of conditions that may or may not be advantageous for measurement of another property of the same system. For example, measurement of the position of an electron and measurement of the momentum of the same electron require quite different apparatus for optimization.

When incompatible experimental conditions are required for the measurement of two distinct properties (which therefore are linked by Heisenberg's uncertainty relationships), compounding of observations is not possible.

Bohr summarized the wave–particle duality in his *complementarity principle*.

A quantum mechanical system may be described as a packet of waves, subject to interference and diffraction, or *as a particle, moving subject to a function that gives only probabilities for positions and momenta.*

Bohr's principle of complementarity is a statement that certain kinds of measurements require incompatible experimental procedures. Because it makes no sense to consider a simultaneous (accurate) measurement of the complementary properties, it is of no particular importance to observe that the properties are different. Wave–particle duality is a statement that we will find what we look for, but not what we are not looking for, in carefully planned and executed experiments.

The interaction of theory and experiment in science is in the prediction of results of experiments. To make predictions—that is, to forecast the results of future measurements—the state of the system on which those future measurements will be made must somehow be delimited. It is of no value to measure a wavelength unless one knows what it is a wavelength of.

There must be a preparatory experiment or an *initial experiment,* perhaps to prepare a monochromatic beam of electrons or photons or to define the locations of the electrons or photons. This preparation also defines the environment into which the system to be measured will be placed (e.g., the properties and position of the crystal or grating or slit) and the instrumentation that will interact with the system to provide information in a classical format.

The details of preparation of the initial experiment are specified in the language of classical mechanics, but the interpretation of results of the experiment involves a theory of quantum mechanics.

The results of an experiment come from a *final experiment,* which requires a choice between mutually incompatible devices, positions, and perhaps other conditions. Different choices may be made for the final experiment for each choice of initial experiment. The purpose of a final experiment is, or may be, to verify the prediction of theory. Fock calls it the *verifying experiment.*

Because the initial state is never fully specified and the final experiment is only a sampling of possible measurements, seemingly identical sequences of experiments may give different results. By statistical analysis, probabilities may be determined for various potential results.

A *total experiment* includes a set of repeated initial experiments and final experiments that will permit comparison of the results with the predictions of a theory. The probability distribution, predicted by a theory, depends on the choice of initial experiment and on the (independent) choice of final experiment. Typically, the final experiment occurs later than the initial experiment, so some allowance must be provided for the time dependence of the quantities under study.

Superposition

One of the critical experiments of quantum mechanics is two-slit diffraction with material particles. When there is a single opening, the intensity observed on the far side of the slit is clearly attributable to particles going through that opening. When a second slit is opened, the classical prediction would be that particles coming through the second slit could only increase intensity at any point. What is observed is that upon opening the second slit, the intensity increases by more than a factor of 2 at some points and decreases at other points, even to the extent of giving zero intensity where the single slit had given significant intensity.

The mathematical description of diffraction of material particles is equivalent to the description of light waves or other waveforms. It is convenient, therefore, to speak of the motion of material particles as being associated with waves, even though there is neither a physical substance nor a force field that undergoes oscillation. Sometimes it is said that "probability" is exhibiting wave motion, but whatever that is intended to mean it would require a radically different definition of probability. Probabilities normally cannot be negative or imaginary.

A state function, $\Psi = \Psi(p, q, t)$, exhibits oscillatory behavior equivalent to a wave amplitude. The probability of finding a particle or small system in a certain volume element of coordinate space $d\tau$ is $\int \Psi^* \Psi \, d\tau$, where Ψ^* is the complex conjugate of Ψ (i.e., $i = \sqrt{-1}$ is replaced by $-i$).

Fourier analysis shows that any state description Ψ can be written as a linear combination of members of a complete set of functions, $\{\Phi_i\}$.

The state function is a linear sum, or superposition, *of the component wave functions,*

$$\Psi = \sum a_i \Phi_i \tag{18-26}$$

and the state of the system, Ψ, at any instant shares the properties of each of the component states, Φ_i, to an extent determined by the magnitudes of the respective coefficients, a.

A probability or intensity is proportional to the square of the absolute value of the wave function.

$$P \sim |\Psi|^2 = \Psi^* \Psi \tag{18-27}$$

Because the terms $a_i \Phi_i$ are oscillatory functions that may be positive or negative, the sum, $\sum a_i \Phi_i$, may vanish when the components are nonzero, or the probability or particle density may be greater than the sum of probabilities for the component states. This allows the intensity for two-slit diffraction to be either greater or less than the sum of the intensities predicted for the two slits separately.

Correspondence Principle

Centuries of testing classical mechanics have shown that the methods and results of classical mechanics are valid for a wide range of experimental conditions. Generally, it is only for very high speeds (when special relativity is required) or very massive objects (requiring general relativity) or very small systems (better described by quantum mechanics) that classical mechanics is inadequate.

It must follow, therefore, that any more powerful description of nature must give the same predictions as classical mechanics for the large fraction of problems for which classical mechanics is known to give correct answers.

The requirement that quantum mechanics must agree with classical mechanics in the classical limit is known as Bohr's correspondence principle.

18.4 The Mathematical Transition to Quantum Mechanics

Specific requirements placed on the mathematics bring the theory of mechanics into agreement with the superposition, correspondence, and uncertainty principles. To achieve linear superposition of states it is sufficient, mathematically, to represent the variables, such as positions, momenta, energies, dipole moments, and time, with *linear operators*.[6] A linear operator may be as simple as multiplication by a number, such as x or t, or multiplication of one matrix by another, or it may be a differential operator, such as d/dx or d/dt.

[6]Operators and their representations are discussed more fully in the Appendix.

Equations of classical mechanics, such as

$$\frac{p^2}{2m} + \frac{1}{2} k q^2$$

are replaced in quantum mechanics by an equivalent form,

$$\frac{\hat{p}^2}{2m} + \frac{1}{2} k \hat{q}^2$$

where $\hat{p}$ and $\hat{q}$ are the linear operators that have been chosen to correspond to p and q. This operator expression then "operates on" an appropriate function (e.g., an algebraic function or a matrix) that represents the state of the system.

A simple but adequate way to build the correspondence principle into a quantum mechanical formalism is to require that each give the same constants of the motion. We have already seen that A will be a constant of the motion in classical mechanics if the Poisson bracket of A with the Hamiltonian, H, vanishes.

$$\dot{A} = \{A, H\}$$

The equivalent condition for linear operators is

$$\dot{\hat{A}} = [\hat{A}, \hat{H}] \tag{18-28}$$

where the square brackets represent the *commutator* of the operators $\hat{A}$ and $\hat{H}$, operating on an appropriate state function.

$$[\hat{A}, \hat{H}] \equiv \hat{A}\hat{H} - \hat{H}\hat{A} \tag{18-29}$$

The correspondence principle will be satisfied automatically, therefore, if the linear operators of quantum mechanics are chosen so that the commutator of the operators is proportional to the Poisson brackets of the corresponding classical variables.

$$[\hat{A}, \hat{H}] \sim \{A, H\} \tag{18-30}$$

Commutators follow the same set of relationships among themselves as do Poisson brackets.

$$\begin{aligned} [q_i, q_j] &= 0 \\ [p_i, p_j] &= 0 \\ [q_i, p_j] &= \delta_{ij} \end{aligned} \tag{18-31}$$

and, in general,

$$[A, B] = -[B, A]$$

The last condition is satisfied for commutators by definition.

$$[\boldsymbol{\alpha}, \boldsymbol{\beta}] = \boldsymbol{\alpha\beta} - \boldsymbol{\beta\alpha} = -(\boldsymbol{\beta\alpha} - \boldsymbol{\alpha\beta}) = -[\boldsymbol{\beta}, \boldsymbol{\alpha}]$$

The other conditions are satisfied by proper choice of operators.

Matrix operators can be constructed that will satisfy the appropriate orthogonality and anticommutative relationships. However, a combination of differential operators with multiplicative operators is probably more familiar.

EXAMPLES 18.4

A. Show that the operator combination $[d/dx, x]$, operating on an arbitrary function, $f(x)$, gives $f(x)$, and therefore the commutator is equal to $+1$.
B. Show that $[d/dx, y]f(x) = 0 = [d/dy, x]f(x)$, and therefore the commutators are each equal to zero. □

We have seen that our mathematical theory must incorporate uncertainty into its predictions. Fortunately, a simple mathematical device provides the necessary uncertainty for us automatically. For compactness and generality, we introduce here the notation of Dirac.[7] Let $\boldsymbol{\alpha}$ be any operator, which operates on a state function—a mathematical function that specifies the state of the system. The state function is represented by a *ket,* $|>$, with or without a label inside.

If the state $\mid i >$ has a definite value of the property represented by the operator $\boldsymbol{\alpha}$, then

$$\boldsymbol{\alpha} \mid i > = a \mid i > \tag{18-32}$$

where a is a constant that is equal to the value of the quantity (represented by $\boldsymbol{\alpha}$) when the system is in the state $\mid i >$. Measurement of the property does not change the state of the system; a second measurement of the same property would yield the same value of a.

More typically there is no definite value of the property represented by $\boldsymbol{\alpha}$ when the system is in a state such as $\mid j >$. Then measurement of the property forces the system into one of the states for which a is determined. However, it is not possible to predict which of the states the system will move into, so the measured value of a cannot be predicted.

$$\boldsymbol{\alpha} \mid j > = \sum_i a_i \mid i >$$

Then the average value of the property is

$$\boldsymbol{\alpha}_{\text{ave}} = < j \mid \boldsymbol{\alpha} \mid j > \tag{18-33}$$

where the premultiplier, $< j \mid$, is called a *bra,* so the product $<|>$ is a bra·ket, or bracket.

The interpretation is that the measurement process selects, for measurement, one of the *characteristic states,* or *eigenstates,* of the operator $\boldsymbol{\alpha}$. Which state is picked determines what value, a_i, is obtained in the measurement. The expected result of a series of measurements is a set of values, with an average value $< j \mid \boldsymbol{\alpha} \mid j >$, called the *expectation value.*

If the initial state is very similar to one of the eigenstates, $\mid j >$, the system will quite probably end up in that state, yielding the corresponding value for the physical quantity. If the initial state has little in common with the eigenstates,

[7]The notation is explained more fully in the Appendix.

the measurement process may give approximately equal probabilities for several eigenstates and the expectation value is an average over all of these states.

Now let $\boldsymbol{\alpha}$ and $\boldsymbol{\beta}$ be two operators that do not commute.

$$[\boldsymbol{\alpha}, \boldsymbol{\beta}] \equiv \boldsymbol{\alpha\beta} - \boldsymbol{\beta\alpha} = i\boldsymbol{\gamma} \tag{18-34}$$

where $i \equiv \sqrt{-1}$ and $\boldsymbol{\gamma}$ is the operator corresponding to a quantity not yet specified. We wish to find the product $(\Delta\boldsymbol{\alpha})(\Delta\boldsymbol{\beta})$, the product of the *dispersions* of $\boldsymbol{\alpha}$ and $\boldsymbol{\beta}$.

The dispersion of x is the mean-square deviation of x from its average value.

$$\Delta x = \text{dispersion of } x = \overline{x^2} - \overline{x}^2 = \overline{(x - \bar{x})^2}$$

Therefore, we wish to find

$$(<|\boldsymbol{\alpha}^2|> - <|\boldsymbol{\alpha}|>^2)(<|\boldsymbol{\beta}^2|> - <|\boldsymbol{\beta}|>^2)$$

By judicious choice of a vector, $\mathbf{V}$, and evaluation of $|\mathbf{V}|^2$, which must be greater than or equal to zero, and with substitution of $i\boldsymbol{\gamma}$ for the commutator of the operators $\boldsymbol{\alpha}$ and $\boldsymbol{\beta}$, we find by the Schwarz inequality[8] that

$$(\Delta\boldsymbol{\alpha})^2 (\Delta\boldsymbol{\beta})^2 \geq \frac{\overline{\boldsymbol{\gamma}}^2}{4} \tag{18-35}$$

Therefore, if we choose operators such that $[\boldsymbol{\alpha}, \boldsymbol{\beta}] = i\hbar$, so $\boldsymbol{\gamma} = \hbar \equiv h/2\pi$, then $(\Delta\boldsymbol{\alpha})^2 (\Delta\boldsymbol{\beta})^2 \geq \hbar^2/4$, or

$$(\Delta\boldsymbol{\alpha})(\Delta\boldsymbol{\beta}) \geq \frac{\hbar}{2} \tag{18-36}$$

The correspondence principle requires only that the commutator be proportional to the Poisson bracket. By choosing the proportionality constant to be $i\hbar$, the condition

$$[\boldsymbol{\alpha}, \boldsymbol{\beta}] = i\hbar\{a, b\} \tag{18-37}$$

satisfies the uncertainty principle, as well.

[8] Let

$$\mathbf{V} \equiv \left(\frac{<|\boldsymbol{\gamma}|>}{2i<|\boldsymbol{\alpha}^2|>}\boldsymbol{\alpha} + \boldsymbol{\beta}\right)|>$$

Then

$$|\mathbf{V}|^2 = <|\left(\frac{-<|\boldsymbol{\gamma}|>}{2i<|\boldsymbol{\alpha}^2|>}\boldsymbol{\alpha} + \boldsymbol{\beta}\right)\left(\frac{<|\boldsymbol{\gamma}|>}{2i<|\boldsymbol{\alpha}^2|>}\boldsymbol{\alpha} + \boldsymbol{\beta}\right)|> \geq 0$$

$$= \frac{<|\boldsymbol{\gamma}|>^2}{4<|\boldsymbol{\alpha}^2|>^2}<|\boldsymbol{\alpha}^2|> + <|\boldsymbol{\beta}^2|> - \frac{<|\boldsymbol{\gamma}|>}{2i<|\boldsymbol{\alpha}^2|>}<|\boldsymbol{\alpha\beta} - \boldsymbol{\beta\alpha}|>$$

Then, since $<|\boldsymbol{\alpha\beta} - \boldsymbol{\beta\alpha}|> = [\boldsymbol{\alpha}, \boldsymbol{\beta}] = i\boldsymbol{\gamma}$,

$$|\mathbf{V}|^2 = \frac{<|\boldsymbol{\gamma}|>^2}{4<|\boldsymbol{\alpha}^2|>} + <|\boldsymbol{\beta}^2|> - \frac{<|\boldsymbol{\gamma}|>^2}{2<|\boldsymbol{\alpha}^2|>} = <|\boldsymbol{\beta}^2|> - \frac{<|\boldsymbol{\gamma}|>^2}{4<|\boldsymbol{\alpha}^2|>}$$

$$<|\boldsymbol{\alpha}^2|><|\boldsymbol{\beta}^2|> \geq \frac{<|\boldsymbol{\gamma}|>^2}{4} \quad \text{or} \quad (\Delta\boldsymbol{\alpha})^2(\Delta\boldsymbol{\beta})^2 \geq \frac{\overline{\boldsymbol{\gamma}}^2}{4}$$

ANSWERS 18.4

A. $\left[\frac{d}{dx}, x\right]f(x) \equiv \left(\frac{d}{dx}x - x\frac{d}{dx}\right)f(x) = f(x) + xf'(x) - xf'(x) = f(x)$.

B. $\left(\frac{d}{dx}y - y\frac{d}{dx}\right)f(x) = yf'(x) - yf'(x) = 0$ and $\left(\frac{d}{dy}x - x\frac{d}{dy}\right)f(x) = 0 \cdot f(x) + x \cdot 0 - x \cdot 0 = 0$.

SUMMARY

A particle confined by a potential to a length L has possible energy states $E =$

$$E(\mathrm{n}) = \frac{h^2\mathrm{n}^2}{8mL^2} \qquad (\mathrm{n}\ \text{integer}).$$

A harmonic oscillator, of classical frequency $\nu = \frac{\omega}{2\pi}$ has energy states $E =$

$$E(\mathrm{v}) = (\mathrm{v} + \tfrac{1}{2})\hbar\omega \qquad (\mathrm{v}\ \text{integer}).$$

A one-dimensional rotor has energy $E = \frac{\mathscr{L}^2}{2I}$ and $\mathscr{L} = \mathrm{m}\hbar$, so $E = E(\mathrm{m}) = \frac{\mathrm{m}^2\hbar^2}{2I}$.

In three dimensions,

$$E_{\mathrm{rot}} = \frac{\mathscr{L}_a^2}{2I_a} + \frac{\mathscr{L}_b^2}{2I_b} + \frac{\mathscr{L}_c^2}{2I_c} \qquad \text{and} \qquad \mathscr{L} = \sqrt{\ell(\ell+1)}\,\hbar$$

(ℓ integer, or sometimes half integer).

Molecular rotations are expressed in terms of the rotational constants,

$$A = \frac{h}{8\pi^2 cI_a}, \qquad B = \frac{h}{8\pi^2 cI_b}, \qquad C = \frac{h}{8\pi^2 cI_c}$$

For a *symmetric rotor*, $\mathscr{L}^2 = \mathrm{J}(\mathrm{J}+1)\hbar^2$ and $\mathscr{L}_a = \mathrm{K}\hbar$ and the term values are $F(\mathrm{J},\mathrm{K}) \equiv \frac{E}{hc} = \mathrm{J}(\mathrm{J}+1)B + \mathrm{K}^2(A - B)$ with $\mathrm{J} = 0, 1, 2\ldots$; $\mathrm{K} = 0, \pm1, \pm2, \ldots \pm \mathrm{J}$.

Linear and *spherical rotors* have term values $F(\mathrm{J}) = \mathrm{J}(\mathrm{J}+1)B$.

Electrons are usually treated in the one-electron wave function approximation, for example, with orbitals or spinorbitals.

In light atoms, angular momenta add in Russell–Saunders coupling: spatial angular momenta add vectorially, $\sum \boldsymbol{\ell}_i = \mathscr{L}$, and spin angular momenta add, $\sum \mathbf{s}_i = \mathbf{S}$. Then $\mathbf{J} = \mathscr{L} + \mathbf{S}$.

In heavier atoms, $j - j$ coupling is more probable:

$$\mathbf{j}_i = \boldsymbol{\ell}_i + \mathbf{s}_i \qquad \text{and} \qquad \mathbf{J} = \sum \mathbf{j}_i.$$

Nuclear spins must be considered because they affect the symmetry of a wave function to exchange of identical particles, thereby determining which states are possible, although the nuclear spins have no significant effect on the energies of the states.

Fermions (particles with half-integral spin) have total wave functions antisymmetric to exchange of identical particles; bosons (particles with integral spin) have total wave functions symmetric to exchange of identical particles.

Quantum mechanical equations are obtained from classical mechanics (for classically meaningful variables) by representing the classical variables with linear operators (to obtain superposition of states), selected such that the commutators of the operators are equal to $i\hbar$ times the Poisson bracket of the classical variable (to obtain correspondence and the uncertainty relationship).

If an experiment is designed to measure wave properties, it will find wave properties; if an experiment is designed to measure particle properties, it will find particle properties. No experiment can be simultaneously optimized for both wave and particle properties.

PROBLEMS

18.1. Show that the Poisson bracket of position and momentum coordinates, x and p_x, is equal to 1, and therefore the commutator of the quantum operators is $[\hat{x}, \hat{p}_x] = i\hbar$.

18.2. Show from classical properties of the momentum and position coordinates $\mathcal{L}$ and ϕ of the one-dimensional rotor that the quantum mechanical commutator is $[\hat{\phi}, \hat{\mathcal{L}}] = i\hbar$.

18.3. Demonstrate by means of the commutators which of the following are constants of the motion in quantum mechanics.

a. For a free particle ($\hat{H} = \hat{p}^2/2m$)

(1) p.

(2) p^2.

(3) x.

b. For a rigid rotor ($\hat{H} = \hat{\mathcal{L}}^2/2I$)

(1) ϕ.

(2) $\mathcal{L}$.

(3) $\mathcal{L}^2$.

c. For a harmonic oscillator ($\hat{H} = \hat{p}^2/2m + \frac{1}{2}k\hat{x}^2$)

(1) x.

(2) p.

(3) p^2.

18.4. One often sees the statement that quantum mechanics converges to classical mechanics for large values of the quantum number.

a. Can you find an example for which this would be an appropriate description (i.e., a system for which the separation of states becomes small as quantum numbers become large)?

b. Give an example for which large values of the quantum number do not approach the continuum characteristic of classical mechanics.

c. Give a more exact statement of the condition for convergence of quantum results to classical behavior, including the appropriate variable(s), applicable to your example in part b.

18.5. The time-dependent wave functions for the harmonic oscillator in the ith state are $\Psi_o = e^{-i\omega_o t}e^{-q^2/2}$, $\Psi_1 = 2qe^{-i\omega_1 t}e^{-q^2/2}$, and $\Psi_2 = (4q^2 - 2)e^{-i\omega_2 t}e^{-q^2/2}$, where $\omega_i = (i + \frac{1}{2})\sqrt{k/m}$ and $q \equiv (mk/\hbar^2)^{1/4}x$.

a. What happens to these wave functions when the time increases by one half period, $\Delta t = 1/(2\nu)$?

b. What happens to these wave functions when the time increases by one period, $\Delta t = 1/\nu$?

c. For which of these functions (if any) is a change of sign equivalent to a change of phase of the motion ($q \to -q$)?

d. What (if anything) do these properties reveal about our ability to describe a quantum oscillator with classical models?

18.6. The free electron has four quantum numbers, as shown by Dirac (and previously predicted empirically), corresponding to three space coordinates plus one time coordinate.

a. How many quantum numbers are assigned to the electron in the hydrogen atom (ground state or excited state)?

b. How many quantum numbers are assigned to the two electrons (individually *and* collectively) in the He atom (ground state or excited state)?

18.7. The Pauli exclusion principle is sometimes said to be the quantum-mechanical statement that two objects cannot occupy the same space at the same time. Test this interpretation by considering bosons, such as deuterons and alpha particles, which have symmetric wave functions.

18.8. Consider n electrons distributed over n states (e.g., in an atom). Then one possible form of the wave function would be $\Psi(1a, 2b, 3c, \ldots)$, showing that electron 1 is in state a, electron 2 in state b, and so on. This wave function is not simply antisymmetric to exchange of particles.

Show that the determinant,

$$\begin{vmatrix} 1a & 2a & 3a & 4a & \cdots \\ 1b & 2b & 3b & 4b & \cdots \\ 1c & 2c & 3c & 4c & \cdots \\ 1d & 2d & 3d & 4d & \cdots \\ \cdots & \cdots & \cdots & \cdots & \cdots \end{vmatrix}$$

gives a combination of products that is antisymmetric to exchange of any two electrons.

CHAPTER 19

Distribution Functions, Energy, and Entropy

The most important equations of statistical mechanics can be derived with very few assumptions about the nature of the system or its parts. We review briefly the division of a system into an ensemble of subsystems and then examine the properties of the distribution functions of the system and its parts.

19.1 Ensembles and Distribution Functions

An isolated system has a fixed total energy. When an isolated system, in internal equilibrium, is divided into subsystems that are nearly isolated, each subsystem also has an energy that is constant, over any period of time that is not too long.

A subsystem may be nearly isolated either because it is large, so that the number of particles interacting with other subsystems is very small compared to the number of particles in the subsystem, or because the particles interact very weakly, or seldom. In a solid, a subsystem must contain many particles; in a gas at low pressure, a subsystem could be a single molecule. In fact, in earlier analyses of ideal gases we worked with a distribution function for the velocities of individual molecules, assuming a uniform distribution function for the coordinates of the molecules. The corresponding phase space for individual *m*olecules is often called μ space, in contrast to the $6N$-dimensional space of a *g*as of N molecules, called Γ *space*.

Because the subsystems are nearly isolated, and in a state that is nearly independent of time, each subsystem should be quite stable. Properties of the full system reflect a statistical average over all the subsystems, and are therefore subject to even smaller fluctuations.

Assume that a system can be divided into subsystems,[1] which are in themselves:

1. Sufficiently large that equilibrium has some meaning within the subsystem, so that it is meaningful to describe the subsystem as in a time-independent state for some reasonable period of time; and

[1]Recall that an alternative analysis considers an ensemble consisting of the system and (imaginary) replicates of the system, with random phases, just as in analyzing an experimental measurement we consider it to be one member of an infinite ensemble of equivalent measurements, differing by random errors, even though the other measurements of that ensemble have not been carried out.

2. Sufficiently large that interactions with neighboring subsystems are small compared to the total energy of the subsystem, at least on a time-average basis; but
3. Sufficiently small, relative to the system, so that the state of each subsystem may be considered independent of the state of any other subsystem at any given time.

Then each subsystem is "nearly isolated." The probabilities and the distribution functions, ρ_i, of different subsystems are multiplicative.

$$\rho_{12} = \rho_1 \rho_2$$

The probability for subsystem A to be in state 1 *and* for subsystem B to be in state 2 is equal to the product of the probability for subsystem A to be in state 1 and the probability for subsystem B to be in state 2.

It follows that the logarithms of the probabilities are additive.

$$\ln \rho_{12} = \ln \rho_1 + \ln \rho_2$$

The probability of a subsystem falling in a small volume of phase space is

$$\delta P = \rho \, dp \, dq$$

The integral over phase space must include all possibilities, so

$$\int \rho \, dp \, dq = 1$$

The probability is dimensionless. However, the distribution function and the elements of phase space, $dp\,dq$, are not dimensionless; each depends on the units selected. We introduced, therefore, a small volume element $\delta \mathscr{R}$, with units of *action* pq, raised to an appropriate power ($3N$ for N particles). Then $\rho\, \delta \mathscr{R}$ and $dp\, dq/\delta \mathscr{R}$ are each dimensionless quantities, independent of units. The introduction of $\delta \mathscr{R}$ also facilitates the transition between classical statistics and quantum statistics if $\delta \mathscr{R}$ is the appropriate power of Planck's constant, h.

The *relaxation time* of a system is the time required for the system to come to equilibrium, or more precisely, to approach to within some fraction, such as $1/e$, of the distance to equilibrium. Distribution functions for the subsystems are independent of time for times that are long compared to the relaxation times of the subsystems, provided either that the system is in equilibrium or the time is short compared to the relaxation time of the system as a whole. The distribution functions are therefore functions only of quantities that are independent of time. That is, ρ_i depends only on the constants of the motion of the subsystem.

The only additive constants of the motion in classical mechanics are the energy (a scalar), momentum (a vector), and angular momentum (a vector). Of these, only the energy is a necessary consideration. This will be demonstrated at the end of this section, but the following arguments are relevant.

All reference frames differing by a uniform translational velocity are equivalent, so there can be no difficulty in selecting a reference frame for any subsystem in which the momentum of the subsystem is zero. In general, it would not be as easy to make the angular momentum vanish, because a rotating reference frame is accelerated and introduces extraneous fictitious forces (centrifugal and Coriolis). In the present situation, however, no subsystem can have an appreciably different angular momentum from its neighbors unless there is extreme turbulence, which would be inconsistent with the assumption of at least quasi-equilibrium. Therefore, we can safely assume that the reference frame can be chosen to make the angular momentum disappear also, and the effects of such a choice must be quite negligible on the equilibrium state.

It follows that the distribution function for the quasi-isolated subsystem is a function only of energy,

$$\rho_i \, \delta\mathscr{R} = \rho_i(E) \, \delta\mathscr{R} \tag{19-1}$$

If the ith subsystem is described by $\rho_i \, \delta\mathscr{R}_i = \rho_i(E_i) \, \delta\mathscr{R}_i$, then

$$\rho \, \delta\mathscr{R} = \prod \rho_i \, \delta\mathscr{R}_i \qquad \text{and} \qquad \ln[\rho \, \delta\mathscr{R}] = \sum_i \ln[\rho_i(E_i) \, \delta\mathscr{R}_i] \tag{19-2}$$

Because the energies of the subsystems are additive, as is $\ln \rho_i \, \delta\mathscr{R}_i$, $\ln \rho_i \, \delta\mathscr{R}_i$ is a linear function of the energy. The most general linear function of energy is

$$\ln[\rho_i \, \delta\mathscr{R}] = -\alpha_i - \beta_i E_i$$

or (19-3)

$$\rho_i \, \delta\mathscr{R} = e^{-\alpha_i} e^{-\beta_i E_i}$$

The signs are chosen to make the constant β positive; ρ_i must go to zero as E_i becomes infinite.

We return now to the question of momentum, $\mathbf{P}_i$, and angular momentum $\mathbf{M}_i$. They also are additive constants of the motion and therefore appear in linear form,

$$\ln[\rho_i \, \delta\mathscr{R}] = -\alpha_i - \beta_i E_i - \boldsymbol{\gamma}_i \mathbf{P}_i - \boldsymbol{\delta}_i \mathbf{M}_i \tag{19-4}$$

However, to be additive functions of E, $\mathbf{P}$, and $\mathbf{M}$, the three coefficients, β, $\boldsymbol{\gamma}$, and $\boldsymbol{\delta}$, must be the same for all subsystems of the entire isolated system. Hence β, $\boldsymbol{\gamma}$, and $\boldsymbol{\delta}$ can be expressed in terms of the seven constants of the motion $(\beta, \gamma_x, \gamma_y, \gamma_z, \delta_x, \delta_y, \delta_z)$ of the entire system. It is therefore sufficient to choose a reference frame for the entire isolated system for which the momentum and angular momentum disappear.

There are no longer any (nonzero) additive constants of the motion, other than energy, for the system enclosed inside a rigid box, and the box cannot affect the statistical properties of the subsystems. The energy is the only additive constant of the motion and the distribution function is

$$\ln[\rho_i \, \delta\mathscr{R}_i] = -\alpha_i - \beta E_i$$

or (19-5)

$$\rho_i \, \delta\mathscr{R}_i = e^{-\alpha_i} e^{-\beta E_i}$$

with β the same for all subsystems.

Only when the observation time is greater than the relaxation time of a single subsystem but smaller than the relaxation time for the ensemble is it necessary to consider partial equilibrium states, in which different subsystems have different values of β, γ, and δ.

19.2 Density of States

Quantum mechanical states of small systems may be continuous, as for classical states, or may be discrete and countable. A continuum of states may be grouped according to whether or not they fall inside certain volume elements of phase space (e.g., of size $\delta\,\mathcal{R}$). The number of states can then be regarded as countable for all systems. It is important to be able to move back and forth between classical and quantum descriptions. Therefore, the equations will be written in parallel.

Let $\Gamma(E)$ represent the number of states of a system (quantum or classical) with energies less than or equal to E. For quantum systems, Γ is the sum of all the discrete quantum states with energies equal to or less than E (although some clarification of the counting process is required, as discussed under the topic of the *density matrix*). The number of classical states with energy in the interval E to $E + dE$ is conveniently defined as the volume of phase space, in units of $\delta\,\mathcal{R}$, corresponding to states of that energy interval,

$$d\Gamma = \frac{dp\,dq}{\delta\mathcal{R}} = \frac{1}{\delta\mathcal{R}}\left(\frac{dp\,dq}{dE}\right)dE \tag{19-6}$$

and Γ is then the integral from $E = 0$ to $E = E$. The increment of probability is

$$\delta P_i = \rho_i\,\delta\mathcal{R}_i\frac{dp\,dq}{\delta\mathcal{R}_i} = \rho_i\,\delta\mathcal{R}_i\,d\Gamma_i \tag{19-7}$$

When the states are discrete, let the number of states be $W(i)$, where i is a parameter that describes the set of states; it is the *statistical weight* for the states of given i. Then

$$\Gamma = \sum W(i) \tag{19-8}$$

where the sum is over the discrete states from the value of i corresponding to $E = 0$ to the value corresponding to E, the energy available to the subsystem. The *probability* of a given state, w (from *wahrscheinlichkeit*), is the a priori probability multiplied by statistical weight.

For example, for N dipoles aligned in a very weak electric or magnetic field, the a priori probability of any particular arrangement of dipoles, with or against the field, is $(\frac{1}{2})^N$. The statistical weight for all those arrangements in which there are n more dipoles aligned with the field than against the field is

$$W(n) = \frac{N!}{\left[\frac{1}{2}(N+n)\right]!\left[\frac{1}{2}(N-n)\right]!} \tag{19-9}$$

Therefore, summing from complete alignment ($n = N$) to the highest accessible state yields

$$\Gamma = \sum_{n=N}^{n} W(n) \tag{19-10}$$

with the upper limit determined by the energy available (hence in principle by the temperature). The probability of the system having an excess of n dipoles aligned with the field is therefore

$$w(n) = \left(\frac{1}{2}\right)^N \frac{N!}{\left[\frac{1}{2}(N+n)\right]!\left[\frac{1}{2}(N-n)\right]!} \tag{19-11}$$

For systems as large and complex as the nearly isolated subsystems of most ensembles, $\Gamma(E)$ increases very rapidly as the energy, E, available to the subsystem increases. As a crude approximation, divide the energy of a subsystem into small units, δE, equal to or larger than the minimum energy quantum for the molecules or to $h/(4\pi\,\delta t)$, where δt is some reasonable time for the observation. Then there are $E/\delta E$ units of energy to be divided among the N particles. Each unit of energy may go to any of the N particles of the subsystem. There is no need to worry about putting too many of the energy units on one particle, provided that the distribution is random.

The number of "equitable" distributions is so great that inequitable distributions are of negligible probability.

The importance of this result can scarcely be overestimated.

Even if an ensemble consists of a single atom or molecule, increasing energy allows excitation of different energy states. The number of states accessible increases with energy.

EXAMPLES 19.1

A. Show, by counting, that $5(= X)$ indistinguishable objects may be placed in $3(= N)$ boxes 21 distinct ways. Show that this is equal to $(5 + 3 - 1)!/(2!5!)$.
B. Show that if $\Gamma = (X + N - 1)!/[(N - 1)!X!]$, then to the approximation that $dX = \delta X = 1$, $d\Gamma/dX \approx (N/X)\Gamma \gg 1$ for large Γ. Rearranging, $d\Gamma/\Gamma \approx N(dX/X)$ and therefore $d\ln\Gamma \approx d\ln X^N$. As energy increases, X becomes large and X^N becomes very large, so Γ increases rapidly. □

The number of states, Γ, is large and increases rapidly with energy, E (Fig. 19.1). By contrast, the distribution function decreases exponentially with energy (Fig. 19.2).

$$\rho_i\,\delta\mathcal{R}_i = e^{-\alpha_i}e^{-\beta E_i} \qquad (\beta > 0)$$

The probability is a product of the rapidly increasing $\Gamma_i = \int dp_i\,dq_i/\delta\mathcal{R}_i$ and the rapidly decreasing $\rho_i\,\delta\mathcal{R}_i$, and is therefore very sharply peaked (Fig. 19.3).

ANSWERS 19.1

A. The possible distributions are (5, 0, 0), (4, 1, 0), (4, 0, 1), (3, 2, 0), (3, 1, 1), (3, 0, 2), (2, 3, 0), (2, 2, 1), (2, 1, 2), (2, 0, 3), (1, 4, 0), (1, 3, 1), (1, 2, 2), (1, 1, 3), (1, 0, 4), (0, 5, 0), (0, 4, 1), (0, 3, 2), (0, 2, 3), (0, 1, 4), and (0, 0, 5), for a total of 21.

$$\frac{(5+3-1)!}{2!5!} = \frac{7!}{2!5!} = \frac{7 \cdot 6}{2} = 21$$

B. $$\frac{d\Gamma}{dX} = \lim_{\delta X \to 0} \left[\frac{(X+N-1+\delta X)!}{(N-1)!(X+\delta X)!} - \frac{(X+N-1!)}{(N-1)!X!} \right] \underset{(\Delta X = 1)}{=} \frac{(X-N-1)!}{(N-1)!X!}$$
$$\times \left(\frac{X+N}{X+1} - 1 \right) = \frac{N-1}{X+1}\Gamma \approx \frac{N}{X}\Gamma \gg 1$$
As X increases, Γ and $d\Gamma/dX$ increase rapidly.

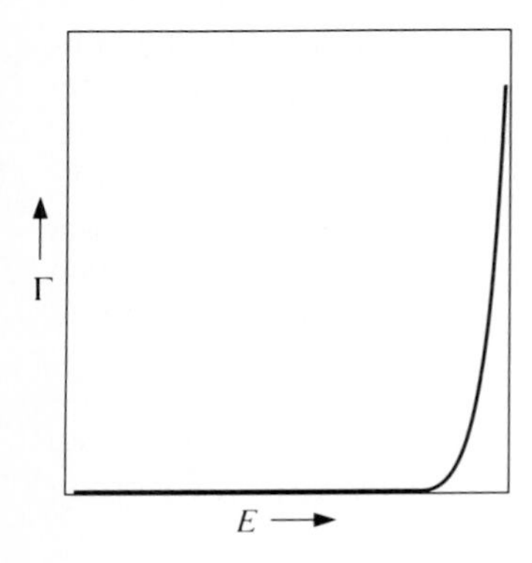

FIGURE 19.1 The number of accessible states, Γ, increases rapidly with energy.

The rate of change of probability with energy is

$$\frac{\delta P}{dE} = \rho(E)\frac{dp\,dq}{dE}$$

Because the average value of E must fall within the narrow spike of probability, $\overline{E}$ may be replaced by the most probable energy, $\hat{E}$, without significant error. Let ΔE be the half width of the spike (sometimes called the *full width, half maximum,* FWHM).

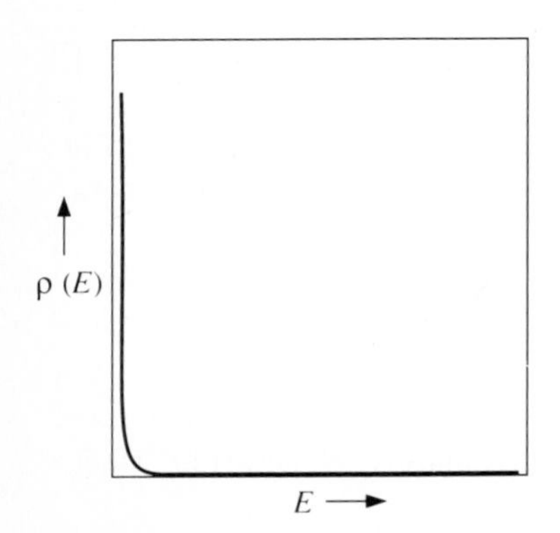

FIGURE 19.2 The distribution function, ρ, decreases exponentially.

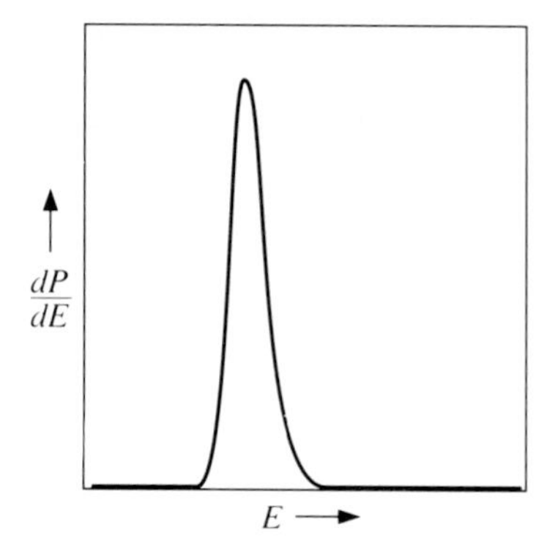

FIGURE 19.3 Probability, $\delta P/dE = \rho(E)d\Gamma/dE$, exhibits a sharp spike about the average energy.

The area of a triangular function is equal to the half width $= \Delta E$, times the height $= \delta P(\hat{E})/dE = \delta P(\overline{E})/dE$:

$$\text{area} = \Delta E \frac{\delta P(\overline{E})}{dE} = 1 \tag{19-12}$$

which is equivalent to the statement that the subsystem must certainly fall somewhere under the curve, or to the requirement that

$$(\Delta p\, \Delta q)\rho(\overline{E}) = 1 = \Delta\Gamma \rho(\overline{E})\, \delta \mathcal{R} \tag{19-13}$$

The importance of this equation is that if $\Delta\Gamma$ (or $\Delta p\, \Delta q$) is small, maximum probability (i.e., the actual state of the system) occurs for a large value of the distribution function, $\rho(\overline{E})$. If the value of $\rho(\overline{E})$ is small, the system is spread over a large number of states.

The range of variation to be expected for the energy is ΔE. The range of variation to be expected for $q_1, q_2, \ldots, p_1, p_2,$, or the volume of phase space likely to be occupied by the subsystem, is $\Delta p\, \Delta q$. The number of states occupied by the subsystem is measured by $\Delta\Gamma$, which is the *statistical weight* of the subsystem distribution.

Note that ΔE is small but much larger than $\delta E = h/\delta t$. Although ΔE depends on the energy (i.e., on where the spike occurs along the energy axis) the dependence is small and may be neglected over the variations in average energy between the subsystems of an ensemble. On the other hand, $\Delta p\, \Delta q/\delta\, \mathcal{R}_i$ and $\Delta\Gamma$, which measure the number of states that are likely to be occupied, are very large, although still very much smaller than the total number of possible states. The probability density decreases strongly with E, and the statistical weight, $\Delta\Gamma = 1/\rho(\overline{E})\, \delta\, \mathcal{R}$, increases sharply with $\overline{E}$.

The more energy that is available to the subsystem, the more the phase points of the subsystem will be spread out.

For the dipole distribution considered previously, the net moment of the subsystem is $M = n\mu$, where μ is the moment of an individual particle. The distribution function is (Chapter 15, Derivation 15.10)

$$w(n) = \left(\frac{2}{\pi N}\right)^{1/2} e^{-n^2/2N} \tag{19-14}$$

which is a Gaussian distribution, centered about $n = 0$ and hence $M = 0$. This is the distribution expected in the absence of a field or at very high temperatures when the a priori probability of each arrangement is the same, as had been assumed.

19.3 Entropy

Define a dimensionless quantity, σ, as

$$\sigma \equiv \ln \Delta\Gamma = \ln \frac{\Delta p\,\Delta q}{\delta\,\mathcal{R}} \quad \text{or} \quad \frac{\Delta p\,\Delta q}{\delta\,\mathcal{R}} = \Delta\Gamma = e^{\sigma} \tag{19-15}$$

The number of states occupied, $\Delta\Gamma$, or number of volume elements of phase space, $\Delta p\,\Delta q/\delta\,\mathcal{R}$, cannot be less than 1, so σ cannot be negative. It will be called the *entropy*.

The statistical weight, or number of states accessible, is multiplicative when subsystems are added, so σ is additive.

$$\sigma = \sum_i \sigma_i \tag{19-16}$$

The entropy of a system is the sum of the entropies of its constituent parts.

The increment of probability is

$$\delta P = \rho\,\delta\mathcal{R}\,\frac{dp\,dq}{\delta\mathcal{R}} = \rho\,\delta\mathcal{R}\,d\Gamma = \rho\,\delta\mathcal{R}\,\frac{d\Gamma}{dE}\,dE$$

Replacing $d\Gamma/dE$ by $\Delta\Gamma/\Delta E$ and replacing $\Delta\Gamma$ with e^{σ} gives

$$\delta P = \frac{\rho\,\delta\mathcal{R}}{\Delta E} e^{\sigma}\,dE \tag{19-17}$$

The energy width, ΔE, is nearly constant, so the probability increases exponentially with $\sigma = \sigma(E)$.

The greater the value of the entropy, consistent with the range allowed by the available average energy, the greater the probability of the state.

The probability, as a function of energy, peaks at $\hat{E}_i = \overline{E}_i$, for subsystems in statistical equilibrium. Therefore, also,

Entropy is a maximum when the subsystem has come to equilibrium.

Removal of a constraint can only increase the volume of phase space available to a subsystem. Therefore, total entropy can only increase (or remain constant) when a constraint is removed. An important example is the preparation of a subsystem in some prescribed state. Allowing the subsystem to then find its own equilibrium state is expected to produce an increase of entropy.

EXAMPLES 19.2

A. For dipoles in an external field of zero strength, the distribution function is

$$w(n) = \left(\frac{1}{2}\right)^N W(n) = \left(\frac{1}{2}\right)^N \frac{N!}{[\frac{1}{2}(N+n)]!\,[\frac{1}{2}(N-n)]!}$$

The entropy of the most probable state is

$$\sigma = \ln \Delta\Gamma = \ln W(0)$$

Show that this can be evaluated, with Stirling's approximation, to give $\sigma \approx N \ln 2 - \frac{1}{2}\ln(N\pi/2)$.

B. Show that the entropy associated with the entire range of states (all values of n) is $\sigma = N \ln 2$, which is negligibly larger than the entropy of the most probable state for large values of N. □

The number of states per energy interval is $d\Gamma/dE$, or $\Delta\Gamma/\Delta E$. The reciprocal, $\Delta E/\Delta\Gamma$, is a measure of the energy difference per state, or the average separation of adjacent energy levels in the vicinity of $\overline{E}$. This *energy splitting*, or average separation, is therefore

$$\frac{\Delta E}{\Delta\Gamma} = \Delta E\, e^{-\sigma(E)} \tag{19-18}$$

It follows that the splitting decreases exponentially with σ.

Because σ is additive, and therefore proportional to the number of particles, the energy splitting decreases exponentially with the size of the system, as measured by the number of particles.

From

$$(\Delta p\, \Delta q)\rho(\overline{E}) = 1 = \Delta\Gamma\, \rho(\overline{E})\, \delta\mathscr{R}$$

or

$$\frac{\Delta p\, \Delta q}{\delta\mathscr{R}} = \Delta\Gamma = \frac{1}{\rho(\overline{E})\, \delta\mathscr{R}}$$

we find that

$$\ln \frac{\Delta p\, \Delta q}{\delta\mathscr{R}} = \ln \Delta\Gamma = -\ln[\rho(\overline{E})\, \delta\mathscr{R}]$$

so that

$$\sigma = -\ln[\rho(\overline{E})\, \delta\mathscr{R}] \tag{19-19}$$

We have seen that the logarithm of the distribution function is a linear function of the energy, the only additive constant of the motion. For the system,

$$\ln[\rho(E)\, \delta\mathscr{R}] = -\alpha - \beta E \tag{19-20}$$

which must be satisfied by the average energy, $\overline{E}$. Hence entropy is a linear function of energy.

$$\sigma = -\ln[\rho(\overline{E})\, \delta\mathscr{R}] = \alpha + \beta\overline{E} \tag{19-21}$$

The average value of a linear function is the function of the average value.

$$\overline{-\alpha - \beta E} = -\alpha - \beta\overline{E}$$

ANSWERS 19.2

A. $W(0) = N!/[(\frac{1}{2}N)!]^2$ so that $\sigma = \ln W(0) = N \ln N - N + \frac{1}{2}\ln(2\pi N) - 2[(N/2)\ln(N/2) - N/2 + \frac{1}{2}\ln(\pi N)] = N \ln 2 - \frac{1}{2}\ln(\pi N) + \frac{1}{2}\ln 2 = N \ln 2 - \frac{1}{2}\ln(\pi N/2)$. For large N, the second term is negligible compared to the first.
B. There are 2^N possible arrangements of N dipoles, so that $\sigma = \ln \Delta\Gamma = \ln 2^N = N \ln 2$. For $N = 100$, $\sigma(0)$ differs by less than 4%; for $N = 10^5$ the difference is less than 0.01%. As summarized by Kittel, this result shows that all accessible phase space has the properties of the most probable condition of the system, to no significant error in the entropy.

Therefore,

$$\sigma = -\ln[\rho(\overline{E})\,\delta\mathscr{R}\,] = -\overline{\ln[\rho(E)\,\delta\mathscr{R}\,]} = -\overline{(-\alpha - \beta E)} = \alpha + \beta\overline{E}$$

Entropy depends only on the average value of the energy and the average value of $\ln[\rho(E)\,\delta\mathscr{R}\,]$.

We repeat a derivation from Section 9.3 with emphasis on the new notation and definitions. The total differential of the energy is

$$dE = \sum\left(\frac{\partial E}{\partial x_i}\right)dx_i$$

Several terms may be specifically recognized or defined:

$$\frac{\partial E}{\partial \sigma} = \theta \text{ which will be called } \textit{temperature} \tag{19-22}$$

$$\frac{\partial E}{\partial N_i} = \mu_i \text{ which is the } \textit{chemical potential} \tag{19-23}$$

For the present we will write any additional terms in the form

$$\frac{\partial E}{\partial x_i} = X_i \text{ which will be called a } \textit{generalized force} \tag{19-24}$$

For example, $\partial E/\partial V = -P$, which is the *pressure*.

Then for a single substance,

$$dE = \theta\, d\sigma + \mu\, dN + \sum X_i\, dx_i \tag{19-25}$$

In general, the terms in the summation will be work terms.

If a system at equilibrium is divided into two parts, then

$$d\sigma = \frac{\partial\sigma_1}{\partial E_1}dE_1 + \frac{\partial\sigma_2}{\partial E_2}dE_2 = 0$$

because σ is a maximum at equilibrium. The total energy is fixed, so $dE_1 = -dE_2$.

$$\frac{d\sigma}{dE_1} = \frac{\partial\sigma_1}{\partial E_1} + \frac{\partial\sigma_2}{\partial E_2}\frac{dE_2}{dE_1} = \frac{\partial\sigma_1}{\partial E_1} - \frac{\partial\sigma_2}{\partial E_2} = 0$$

and

$$\frac{\partial \sigma_1}{\partial E_1} = \frac{\partial \sigma_2}{\partial E_2}$$

The derivative $\partial\sigma/\partial E$ must be the same throughout all parts of a system at equilibrium. From the definition of θ given above,

$$\frac{\partial \sigma}{\partial E} = \frac{1}{\theta}$$

Then, at equilibrium, θ, the temperature of the system, is constant throughout the system.

From

$$\sigma = \alpha + \beta E$$

$$\frac{\partial \sigma}{\partial E} = \beta = \frac{1}{\theta} \qquad (19\text{-}26)$$

This confirms the earlier interpretation of β as proportional to the inverse temperature.

Note that the definitions given here are fully consistent with the equations of Chapter 9, where it was found that

$$\left(\frac{\partial S}{\partial E}\right)_V = \frac{1}{T}$$

and the exponential factor is the same as in Section 7.1,

$$e^{-E/kT} = e^{-\beta E}$$

Thus

$$\beta = \frac{1}{kT} = \frac{1}{\theta} \qquad (19\text{-}27)$$

and from $d\sigma = \beta\, dE$ and $dS = (1/T)\, dE$,

$$\sigma = \frac{S}{k} \qquad (19\text{-}28)$$

The temperature, θ, has units of energy and the entropy, σ, is dimensionless.

The pressure was given above as

$$P = -\left(\frac{\partial E}{\partial V}\right)_\sigma \qquad (19\text{-}29)$$

Then, using

$$\left(\frac{\partial E}{\partial \sigma}\right)_V = \theta$$

$$dE = \theta\, d\sigma - P\, dV$$

and

$$d\sigma = \frac{1}{\theta}\, dE + \frac{P}{\theta}\, dV$$

Therefore,

$$\left(\frac{\partial \sigma}{\partial V}\right)_E = \frac{P}{\theta} \tag{19-30}$$

Again we divide an isolated system into two parts. Then

$$d\sigma = \frac{\partial \sigma_1}{\partial V_1}\, dV_1 + \frac{\partial \sigma_2}{\partial V_2}\, dV_2 = 0$$

$$\frac{d\sigma}{dV_1} = \frac{\partial \sigma_1}{\partial V_1} - \frac{\partial \sigma_2}{\partial V_2} = \frac{P_1}{\theta_1} - \frac{P_2}{\theta_2} = 0$$

at equilibrium. It is necessarily true that $\theta_1 = \theta_2$ at equilibrium, so the equation shows that $P_1 = P_2$ at equilibrium. For arbitrary change in V_1,

$$d\sigma = \left(\frac{P_1}{\theta_1} - \frac{P_2}{\theta_2}\right) dV_1 > 0$$

so if P is positive (as expected) and $P_1 > P_2$, then $dV_1 > 0$. The region at higher pressure expands, compressing the region at lower pressure.

More generally, for $P > 0$ and $\theta > 0$,

$$\left(\frac{\partial \sigma}{\partial V}\right)_E > 0 \tag{19-31}$$

so expansion is the spontaneous process. However, if $P < 0$, as sometimes suggested for special situations, the system contracts spontaneously.

In similar fashion, one can show that for equilibrium, at constant temperature and pressure, μ must be constant, for each substance, throughout the system. If μ is not constant, then dN will be such as to move material from higher μ to lower μ. For any other generalized force, X_i, the condition for equilibrium must be that X_i will be constant throughout the system. If X_i is not constant, the change, dx_i, is determined by the requirement that $d\sigma \geq 0$.

General thermodynamic arguments suggest that any reversible process should be slow, even though there is no time dependence in classical thermodynamics. If a process (that can be reversed) is sufficiently slow, it is thermodynamically reversible.

Let the process be described by a progress variable, λ, which might be the amount of material transferred from one volume or phase to another, the position of a piston, or any other convenient measure of progress. Assume that the entropy is some function of λ, $\sigma = \sigma(\lambda)$, and the system is never too far from equilibrium for $d\lambda/dt$ to change sign when some controlling parameter is changed by a modest amount. Then the time rate of change of entropy is a function of $d\lambda/dt$, which is assumed to be small.

$$\frac{d\sigma}{dt} = f\left(\frac{d\lambda}{dt}\right)$$

The function can be written as a power series in $d\lambda/dt$.

$$\frac{d\sigma}{dt} = f\left(\frac{d\lambda}{dt}\right) = a_o + a_1\left(\frac{d\lambda}{dt}\right) + a_2\left(\frac{d\lambda}{dt}\right)^2 + \cdots \qquad (19\text{-}32)$$

When $d\lambda/dt = 0$, nothing happens; $d\sigma/dt = 0$. Therefore, the first term must be zero. The second coefficient, a_1, must also be zero, because the process can move in either direction: $d\sigma/dt = a_1\,(d\lambda/dt) \geq 0$ but $d\lambda/dt$ may be positive or negative. Hence the lowest nonzero coefficient is a_2.

$$\frac{d\sigma}{dt} = a_2\left(\frac{d\lambda}{dt}\right)^2$$

Then

$$d\sigma = a_2\left(\frac{d\lambda}{dt}\right)^2 dt = a_2\,\frac{d\lambda}{dt}\,d\lambda$$

and

$$\frac{d\sigma}{d\lambda} = a_2\,\frac{d\lambda}{dt}$$

Therefore,

$$\text{as } \frac{d\lambda}{dt} \to 0 \qquad \frac{d\sigma}{d\lambda} \to 0 \qquad (19\text{-}33)$$

and any process (that can be reversed) is reversible if it is sufficiently slow. How slow must it be? Basically, the requirement is that the rate of the process must be slow compared to the relaxation time, or the rate for attainment of equilibrium.

19.4 Approximations to the Distribution Function

The question of the form of the distribution function, ρ, has been avoided. The information contained in ρ necessarily reflects, in part, the extent of our knowledge about the system and its microstates.

The classical distribution function for a system of N particles is a function of the $6N$ variables $p_1, p_2, \ldots, p_{3N}, q_1, q_2, \ldots, q_{3N}$. However, we have insufficient information to follow points in phase space, and need not do so, both because the density of points in phase space remains unchanged (Liouville's theorem) and because the distribution function can be written solely in terms of energy, as shown above.

Nevertheless, in contemplating the form of the distribution function, it may be helpful to attempt to construct a function from consideration of the detailed microstructure of a system. We consider two types of systems, at known temperature, volume, and pressure: an ideal gas and finely powdered NaCl crystals.

For an ideal gas any volume element of real space should be equally likely to be occupied, by any of the particles. The distribution function, ρ, should therefore be *uniform* across the $3N$ q_i variables of phase space.

Not all values of the p_i are equally probable, because of the known energy constraints. However, we saw that *if* the energy were totally unknown, so that all p_i were equally likely for a subsystem, Liouville's theorem tells us that the uniform distribution of phase points in phase space would represent a steady state. If the distribution is uniform at some instant, it must remain uniform for all time, while the system remains suitably isolated.

Knowing nothing about the precise locations of sodium ions or chloride ions in the small crystals of powdered NaCl, we might have to settle for a uniform distribution of the q_i to describe this system, also. In fact, however, each ion is bound to within a small distance of an equilibrium position, and those equilibrium positions may be regarded as fixed with respect to time. Therefore, the distribution function cannot actually be a uniform function of the q_i. Even within the limits of oscillation expected for an individual ion, the distribution function cannot be uniform, for both classically and by quantum mechanical description an oscillator spends a larger fraction of its time at or near the turning points of the motion than at the central equilibrium position. The uniform distribution is unnecessarily crude.

The next step in refinement of the distribution function is to concentrate on a much smaller number of variables than the $6N$ position and momentum coordinates. The *microcanonical distribution* identifies each phase point with a subsystem energy value and specifies the probabilities for phase points according to that single parameter. We have seen above the justification for a distribution function dependent solely on energy.

From a strict Newtonian point of view we would describe the distribution function as a surface of constant energy in phase space. However, there are difficulties with assigning a single value of energy. The distribution function must be infinite, to prevent the integral from vanishing, just as the area under a point in two-dimensional space is zero, or the volume of a point or a two-dimensional surface in three-dimensional space is zero.

It is also unrealistic to assume that the distribution function is a surface of a single, fixed energy because of known quantum mechanical uncertainties, or for a real classical system for which only imperfect information is available (Section 15.1). A better description for the distribution function is a surface of finite thickness, $\delta E \geq h/\delta t$, where δt is a reasonable time period for observation of the system. Then δE is extremely small compared to the energy of the subsystem, but definitely not zero.

The *microcanonical distribution* sets the distribution function equal to a constant value for phase points falling within a thin surface of nearly constant energy ($E = E \rightarrow E + \delta E$) and is zero everywhere else. It is a rectangular function or Dirac delta function,

$$\rho = \text{constant} \times \delta(E - E_o) \qquad (19\text{-}34)$$

The properties of the system do not depend on the width, δE.

Not all points in phase space are accessible to a system or subsystem, because of limitations of energy. The microcanonical distribution divides phase space into a narrow sheet that is accessible and the remainder that is not allowed, for a given average energy of the system or subsystem.

A more difficult question is whether all points in phase space allowed by the energy are equally probable. It is not meaningful to require that the phase point visit all points in phase space; a more modest requirement is to state that over a sufficiently long period of time, the phase point will come arbitrarily close to each point in phase space. Either assumption is known as the *ergodic hypothesis*.

The microcanonical distribution assumes the ergodic hypothesis to be valid for each subsystem. That is, it does not allow for the possibility that a real subsystem may, in practice, not traverse all parts of phase space. The microcanonical distribution was the only form applied in statistical mechanics until Gibbs, at the end of the nineteenth century, proposed more convenient functions.

Gibbs' *canonical distribution* is easily obtained from the equations developed above. From the condition that the distribution function must depend only on constants of the motion, in particular on the energy, and furthermore that it must be a linear function of energy because of the requirement of additivity, we found that

$$\ln[\rho\,\delta\mathcal{R}\,] = -\alpha - \beta E = -\sigma$$
$$\rho\,\delta\mathcal{R} = e^{-\alpha}e^{-\beta E} = e^{-\sigma}$$

It was also shown that $\beta = 1/\theta$ must be constant throughout the system at equilibrium, and β was identified as equal to $1/kT$.

Now define the function F as

$$F = -\frac{\alpha}{\beta} \equiv -\alpha\theta \tag{19-35}$$

Then

$$F = E - \theta\sigma \equiv E - TS \tag{19-36}$$

which is the Helmholtz free energy, or work function.

The canonical distribution function is written in terms of F and E.

$$\rho\,\delta\,\mathcal{R} = e^{-\sigma} = e^{(F-E)/\theta} \equiv e^{(F-E)/kT} \tag{19-37}$$

The canonical distribution is the appropriate statistical distribution for a subsystem of a nearly isolated system, as it allows for fluctuations in the energy. The same distribution may be applied to an isolated body, except that then the fluctuations in energy allowed by the distribution must be ignored.

In all discussions thus far it has been assumed that the number of particles, N_i, in each subsystem, as well as in the system, is held constant. Often, particles move back and forth between system (or subsystem) and surroundings, for example as a result of physical equilibrium or chemical processes. Gibbs defined the *grand canonical distribution* to allow for exchange of particles between subsystems. Let

$$\sigma = \sigma(E, V, N) \qquad \text{or} \qquad S = S(E, V, N)$$

Then

$$d\sigma = \left(\frac{\partial\sigma}{\partial E}\right)_{V,N} dE + \left(\frac{\partial\sigma}{\partial V}\right)_{E,N} dV + \left(\frac{\partial\sigma}{\partial N}\right)_{E,V} dN$$

where we have recognized that

$$\left(\frac{\partial\sigma}{\partial E}\right)_{V,N} = \frac{1}{\theta} \qquad \left(\frac{\partial\sigma}{\partial V}\right)_{E,N} = \frac{P}{\theta} \quad \text{and} \quad \left(\frac{\partial\sigma}{\partial N}\right)_{E,V} = -\frac{\mu}{\theta}$$

and μ is the *chemical potential*.[2]

$$\left(\frac{\partial \sigma}{\partial N}\right)_{E,N} = -\frac{\mu}{\theta} = -\frac{1}{\theta}\left(\frac{\partial G}{\partial N}\right)_{P,\theta}$$

The total differential of the entropy is then

$$d\sigma = \frac{1}{\theta}\, dE + \frac{P}{\theta}\, dV - \frac{\mu}{\theta}\, dN \tag{19-38}$$

Integrating with respect to the three variables E, V, and N gives the *grand canonical distribution function*,

$$\rho\, \delta\, \mathcal{R} = e^{(-PV+\mu N-E)/\theta} \equiv e^{(-PV+\mu N-E)/kT} \tag{19-39}$$

The grand canonical distribution is applicable to equilibrium involving separate phases. It also provides information on the probability of fluctuations in all the variables, including number of particles in elementary systems such as a gas divided into two containers open to each other, or fluctuations within a gas or other substance within a single container.

The canonical and grand canonical distributions may be obtained in a somewhat different manner that helps illustrate their significance. Assume a system of energy E° and temperature T°. Within this system there is a subsystem, or body, of energy E in equilibrium with its surroundings, and therefore at the same temperature, T°. The remainder of the system (i.e., the system minus the body) will be called the *medium;* it has energy $E' = E^\circ - E$ and temperature T°. We wish to know the probability, w_n, that the body exists in a particular quantum state of energy $E = E_n$.

The statistical weight for the medium is a large number, $\Delta\Gamma'$. The statistical weight for the body in the single quantum state is $\Delta\Gamma = 1$. Let $\Delta E'$ be the range of energy for the states included in $\Delta\Gamma'$. Then

$$\int dw = w_n = \text{constant} \times \int \delta(E_n + E' - E^\circ)\, d\Gamma'$$

The statistical weight of the medium, $\Delta\Gamma'$, is a strong function of the energy of the medium, E'.

$$d\Gamma' = \frac{d\Gamma'(E')}{dE'}\, dE' = \frac{1}{\Delta E'} e^{\sigma(E')}\, dE'$$

and

$$\frac{d\Gamma'}{dE'} = \frac{\Delta\Gamma'}{\Delta E'} = \frac{e^{\sigma}}{\Delta E'} \tag{19-40}$$

[2] $\theta\, d\sigma = T dS$ and from Section 13.6, $T dS = dE + P dV - \sum_i \mu_i\, dn_i$, so

$$\left(\frac{\partial S}{\partial n_i}\right)_{E,V} = -\frac{1}{T}\mu_i \equiv -\frac{1}{T}\left(\frac{\partial G}{\partial n_i}\right)_{T,P} \quad \text{and} \quad \left(\frac{\partial \sigma}{\partial n_i}\right)_{E,V} = -\frac{1}{\theta}\left(\frac{\partial G}{\partial n_i}\right)_{\theta,P}$$

For convenience, for $i \geq 1$, let $\sum_i \mu_i\, dn_i \equiv \mu\, dN$.

Therefore,

$$w_n = \text{constant} \times \int \frac{e^{\sigma}}{\Delta E'} \, \delta(E' + E_n - E^\circ) \, dE'$$

$$= \text{constant} \times \left(\frac{e^{\sigma}}{\Delta E'} \right)_{E'=E^\circ - E_n} \qquad (19\text{-}41)$$

because integration over the delta function simply selects out those states for which $E' = E^\circ - E_n$.

Now, because we assume that the body is small compared to the total system, $E_n \ll E^\circ$. Also, $\Delta E'$ is a weak function of energy and $\Delta E' \approx \Delta E^\circ$, so $\Delta E'$ may be replaced with ΔE°, which is not a function of E_n and may be included in the constant coefficient.

Because $E_n \ll E^\circ$, the entropy may be expanded about E°.

$$S'(E') = S'(E^\circ - E_n) = S'(E^\circ) - E_n \frac{dS'(E^\circ)}{dE^\circ} \qquad (19\text{-}42)$$

Recall, however, that $dS/dE = 1/T$, and the temperature is uniform throughout the system. Therefore,

$$\frac{dS'(E^\circ)}{dE^\circ} = \frac{1}{T^\circ} = \frac{1}{T}$$

and thus

$$w_n \sim e^{\sigma(E')} = e^{S'(E')/k} e^{-E_n/kT}$$

or

$$w_n = \mathscr{A} e^{-E_n/kT} \qquad (19\text{-}43)$$

which is Gibbs' canonical distribution. It is the statistical distribution for any macroscopic body that is a small part of a large, isolated system in equilibrium.

It was shown in Section 19.3 that entropy is a function of the average value of the distribution function.

$$\sigma = -\ln[\rho(\overline{E}) \, \delta\mathscr{R}] = -\overline{\ln w_n}$$

Replacing w_n with $\mathscr{A}^{E_n/kT}$ gives $\ln w_n = \ln \mathscr{A} - E_n/kT$, from which we obtain

$$S = -k \ln \mathscr{A} + \frac{\overline{E}_n}{T} \qquad \text{or} \qquad k \ln \mathscr{A} = \frac{\overline{E}_n - TS}{T} = \frac{F}{T} \qquad (19\text{-}44)$$

recognizing that $\overline{E}_n$ is simply the energy of the object (system) and $E - TS = F$, the Helmholtz free energy. Thus the constant is readily incorporated into the exponent, to give

$$w_n = e^{(F - E_n)/kT} \qquad (19\text{-}45)$$

or the equivalent classical expression, for s degrees of freedom,

$$\rho = \frac{1}{h^s} e^{[F-E(p,q)]/kT} \tag{19-45a}$$

Assume now a system of N° particles with energy E° and temperature T°. Proceeding as above, out of this we take a subsystem, defined by a specified volume, V, that is a negligible part of the whole. The subsystem has energy E, temperature T, and N particles, leaving the remainder of the system, which serves as the medium for the subsystem, with energy $E' = E^\circ - E$ and number of particles $N' = N^\circ - N$. The temperatures are assumed equal; $T = T^\circ = T'$. We are content with a "macroscopic" definition of the state of the medium. The subsystem is in some energy state E_n, but the state itself can depend on the number of particles in the system, so we label the energy E_{nN}.

The entropy of the medium is

$$S' = S'(E', N') = S'(E^\circ - E, N^\circ - N)$$

The probability for the system to be in a state such that the subsystem has N particles and energy E_{nN} is

$$w_n \sim e^{S'(E^\circ - E_{nN}, N^\circ - N)/k}$$

where again it has been assumed that $\Delta E'$ is nearly constant.

Because the subsystem is a very small part of the whole, the entropy may be expanded in terms of E_{nN} and dN.

$$S'(E^\circ - E_{nN}, N^\circ - N) = S'(E^\circ, N^\circ) - \frac{1}{T}E_{nN} + \frac{\mu}{T}\, dN + \cdots \tag{19-46}$$

and the probability may be written

$$w_{nN} \sim e^{S'(E')/k} e^{(\mu N - E_{nN})/kT} \qquad \text{or} \qquad w_{nN} = \mathcal{A} e^{(\mu N - E_{nN})/kT} \tag{19-47}$$

For the system,

$$S = -k\,\overline{\ln w_{nN}} = -k \ln \mathcal{A} - \frac{\mu \overline{N}}{T} + \frac{\overline{E}}{T}$$

with $\overline{E} \equiv E$ and $\overline{N} \equiv N$. Multiplying through by T gives

$$kT \ln \mathcal{A} = E - TS - \mu N = F - G = \Omega \equiv -PV$$

Hence

$$w_{nN} = e^{\Omega/kT} e^{(\mu N - E_{nN})/kT} = e^{(\Omega + \mu N - E_{nN})/kT} \tag{19-48}$$

Summing over all states for N particles and over the number of particles gives an expression for Ω.

$$\sum_N \sum_n w_{nN} = 1 = e^{\Omega/kT} \sum_N e^{\mu N/kT} \sum_n e^{-E_{nN}/kT} \tag{19-49}$$

This is Gibbs' *grand canonical distribution*.

One could work only with this grand canonical distribution as a starting point. Then if N is considered constant, $\Omega + \mu N$ is equivalent to $G - PV = F$, the Helmholtz free energy, giving the canonical distribution. The canonical distribution, in turn, differs from the microcanonical distribution by assuming a constant temperature (average energy) but allowing for fluctuations in the energy of any subsystem in equilibrium with a thermal reservoir. If energy fluctuations are to be ignored, the canonical distribution becomes equivalent to the microcanonical distribution.

19.5 The Density Matrix[3]

It has been necessary to distinguish between the two limiting conditions of purely classical variables and discrete quantum states. The former are described by a statistical distribution function $\rho(p, q)$ and the states are measured by a volume element in phase space, $dp\,dq$ or, better, by $dp\,dq/\delta\mathscr{R}$, where $\delta\mathscr{R}$ can be taken as h^s, for s degrees of freedom.

Quantum states may also be continuous, and therefore describable in the same form as classical states. More often we are concerned with discrete states, which are therefore enumerable. An integer (or in some cases, a half-integer) quantum number or a set of quantum numbers is assigned to each discrete quantum state as a label.

A small quantum system is in a specific state at a specific time or, more precisely, has a time-dependent or time-independent probability of being in a specific state. For example, if $\{\Phi_i\}$ is a mathematically complete set of functions, then any state, Ψ, is expressible as a sum of the form

$$\Psi = \sum a_i \Phi_i \tag{19-50}$$

If the state Ψ is a steady state, independent of time, the a_i are time independent and give the relative probabilities of the system being in the individual states, Φ_i. A moving particle, a molecule undergoing a transition from one state to another, or any other system that is changing with time, is described by

$$\Psi(t) = \sum a_i(t)\Phi_i \tag{19-51}$$

The time dependence of the a_i produces a change in $\Psi(t)$ as the system changes from one state to another or through a succession of states.

For any system or particle, there is an inherent uncertainty, measured by Planck's constant, associated with each quantum state, totally unrelated to the statistical uncertainty of the distribution function.

For macroscopic systems, the splitting between energy levels is inversely proportional to the volume and also decreases exponentially with σ, and therefore with N, the number of particles in the system. When the energy separation between a group of states is small compared to the uncertainty in energy, $\delta E = h/\delta t$, a very large number of quantum states have practically the same energy, and may

[3]Density matrices are not required for understanding other parts of this book. This section is intended primarily to inform the reader why and when they are sometimes necessary or helpful and to reconcile the notation of density matrices with other notations. For a more complete explanation of density matrices, see, for example, L. D. Landau and E. M. Lifshitz, *Course of Theoretical Physics:* Vol. 3, *Quantum Mechanics, Non-relativistic Theory,* and Vol. 5, *Statistical Physics:* (Reading, Mass.: Addison-Wesley Publishing Co., Inc., 1958).

therefore be said to be *degenerate*. The elementary description of a system by a single quantum state therefore breaks down for macroscopic systems.

Analysis of the inherent accidental degeneracies of macroscopic systems requires a statistical matrix, called a *density matrix*. Recall that the average value, or expectation value, of any quantity f can be expressed in terms of the *matrix element*

$$f_{ij} \equiv \int \Phi_i \mathbf{f} \Phi_j \, d\tau \tag{19-52}$$

where $\mathbf{f}$ is the operator corresponding to the variable f, the Φ_i are the wave functions belonging to the complete set $\{\Phi_i\}$, and the integration is over the appropriate volume element in the space of the Φ_i. The expectation value is then

$$\langle f \rangle = \sum_i \sum_j a_i^* a_j f_{ij} \tag{19-53}$$

Time independence of the state implies that the a_i are unchanged. For example, the Hamiltonian does not mix the states, leaving the a_i unchanged. Now we wish to allow for mixing of the states on a random, or statistical, basis. A convenient way of handling the mathematics of this mixing has been developed, called the *statistical matrix* or the *density matrix*.

For macroscopic systems, described by quantum states, the product, $a_i^* a_j$, is replaced with an element of a square matrix, w_{ji}, that represents a mixing of the a_i elements.

$$a_i^* a_j \rightarrow w_{ji} \tag{19-54}$$

The expectation value then becomes

$$\langle f \rangle = \sum_i \sum_j w_{ji} f_{ij} = \sum_j (\mathbf{wf})_{jj} \tag{19-55}$$

Hence the expectation value for f is the trace of a matrix, and therefore independent of the representation by which it is expressed. The elements of w_{ji} may be interpreted as average values of the products $a_i^* a_j$, but the density matrix does *not* describe a system moving between specific states, Ψ. Rather, it combines the inherent probability characteristics of quantum states with the statistical uncertainty arising from incomplete knowledge of the state of the macroscopic system. These are not separable for macroscopic quantized systems.

We will not be concerned with density matrices, except to adopt some of the notation, for consistency with other discussions. The diagonal elements of the density matrix, w_{jj}, or simply w_j, are positive and normalized.

$$\sum_j w_j = 1$$

which is related to the corresponding equation,

$$\sum_i |c_i|^2 = 1$$

The diagonal elements, w_j, replace the probability density, ρ, and follow the same general rules. However, just as ρ contains no information about quantized states, w contains no explicit information about the coordinates p and q.

The increment of probability is δw, so the microcanonical distribution becomes

$$\delta w = \text{constant} \times \delta(E - E_o) \prod_i d\Gamma_i$$

Like ρ, w is monotonically decreasing with energy, but

$$W \equiv w \frac{d\Gamma}{dE}$$

is sharply peaked. The normalization conditions give

$$\int w \, d\Gamma = 1 \qquad \text{and} \qquad \int W \, dE = 1$$

The energy width of the distribution is

$$\Delta E = \frac{1}{W(\overline{E})}$$

and therefore

$$\Delta\Gamma = \left(\frac{d\Gamma}{dE}\right)_{E=\overline{E}} \Delta E = \frac{1}{w(\overline{E})} \tag{19-56}$$

The time dependence is reflected in

$$\dot{w} = \frac{i}{\hbar}[w, H] \tag{19-57}$$

with opposite sign to the usual time-dependence equation. Because w plays the role of a distribution function, it should be constant with time, for equilibrium states. The corresponding operator, **w**, therefore commutes with the Hamiltonian. The Hamiltonian is diagonal when the system has a definite energy, so **w** must also be diagonal. We saw above that even though **f** will not in general be diagonal, the expectation value of (**wf**) is given by the trace of this product matrix. For mixed states,

$$\dot{w}_{ij} = \frac{i}{\hbar}(E_j - E_i)w_{ij} \tag{19-58}$$

and the linear dependence on energy is expressed by

$$\ln w_k = -\alpha_i - \beta E_{ki} \tag{19-59}$$

for the kth energy state of the ith subsystem.

SUMMARY

Subsystems are defined sufficiently small that the state of each subsystem may be considered independent of the state of any other subsystem, but sufficiently large that each may come to a time-independent equilibrium, with negligible effects of neighboring subsystems.

The probability distributions for these *nearly isolated* subsystems are independent and therefore multiplicative.

The equilibrium distribution function is independent of time, and therefore can only be a function of constants of the motion, and hence of the energy.

The number of accessible states, or the accessible volume of phase space, for the ith subsystem, is $d\Gamma_i = dp_i\, dq_i/\delta\Re_i$ and the increment of probability is $\delta P_i = \rho_i\, dp_i\, dq_i = \rho_i\, \delta\Re_i\, d\Gamma_i$.

Γ increases rapidly with energy, but ρ decreases rapidly with energy. Therefore, $\delta P_i = \rho_i\, \delta\Re_i\, d\Gamma_i$ tends to be very sharply peaked, with a width $\Delta\Gamma$ and energy spread ΔE.

Energy is distributed randomly, and therefore approximately evenly, among the accessible volumes of phase space, rather than collected in "hot spots." The number of "equitable" distributions is so great that inequitable distributions are of negligible probability.

Define entropy as $\sigma \equiv \ln \Delta\Gamma = \ln[(\Delta p\ \Delta q)/\delta\Re]$, or $\Delta\Gamma = e^{\sigma}$.

Entropy is additive (i.e., extensive).

Probability increases exponentially with σ: $\delta P = (\rho\, \delta\Re/\Delta E)e^{\sigma}\, dE$ for accessible states. Entropy is a maximum for the equilibrium state.

The average separation of energy levels decreases exponentially with σ, and therefore with size of the system (number of particles).

Temperature may be defined as $\partial E/\partial\sigma = \theta$; pressure may be defined as $-\partial E/\partial V = P$; and chemical potential as $\partial E/\partial N_i$, consistent with

$$dE = T\,dS - P\,dV + \sum \mu_i\, dN_i \qquad \text{and} \qquad \theta = kT, \quad \sigma = \frac{S}{k}$$

If a process (that can be reversed) is sufficiently slow, it is thermodynamically reversible.

The microcanonical distribution assigns equal probability to a surface of constant energy. $\rho = \mathit{constant} \times \delta(E - E_0)$, where δ represents the Dirac delta function; $\delta(x) = 0$ unless $x = 0$.

The ergodic hypothesis assigns equal probability to all accessible states (as in the microcanonical ensemble), and hypothesizes that the system phase point will come arbitrarily close to each point in phase space before returning to its initial position.

The canonical distribution allows for variations in energy:

$$\rho\, \delta\Re = e^{-\sigma} = e^{(F-E)/kT}$$

The grand canonical distribution allows changes in number of particles, volume, and energy: $\rho\, \delta\Re = e^{-\sigma} = e^{(-PV+\mu N-E)/kT}$. The density matrix formulation combines the statistical variation of a subsystem with the quantum mechanical uncertainty to provide quantum mechanically correct statistical mechanics for systems, such as solids, with close-lying energy levels.

QUESTIONS

19.1. Why is a system divided into subsystems? How should it be divided?

19.2. How is "volume" in phase space related to the number of quantum states?

19.3. Sketch very briefly the paths, or arguments, that lead to

$$\rho_i\, \delta \mathcal{R} = e^{-\alpha_i} e^{-\beta E_i}$$

or equivalent forms. [How many different paths to this result?]

19.4. Starting from statistical arguments, how are the following quantities defined?
   a. Entropy.
   b. Temperature.
   c. Pressure.

19.5. Show the following.
   a. Entropy increases with N.
   b. Entropy increases with volume.
   c. The average separation of energy levels decreases with V and decreases exponentially with N.
   d. Entropy is a linear function of energy.
   e. Thermal energy flows from "warmer" to "cooler."
   f. Pressures must be equal at equilibrium. If they are not, the higher-pressure region expands.
   g. A process that can be reversed is thermodynamically reversible if it is slow.

19.6. What are the *microcanonical, canonical,* and *grand canonical distributions*? Where would each be useful?

19.7. What is the ergodic hypothesis?

19.8. Why must the density matrix be introduced?

19.9. Using the method of equations 19-25 to 31, show that
   a. At equilibrium (with $T_1 = T_2$) the chemical potential of each species must be the same for systems in contact.
   b. If the chemical potential in phase A is greater ($\mu_{iA} > \mu_{iB}$) for the ith substance, that substance will pass from phase A to phase B.

19.10. What important conclusions, or physical properties of a system, can be drawn from the equation $\delta P = (\rho\, \delta\, \mathcal{R}/\Delta E)e^{\sigma}\, dE$?

19.11. What important conclusions, or physical properties of a system, can be drawn from the equation $\Delta E/\Delta\Gamma = \Delta E\, e^{-\sigma(E)}$?

19.12. Which of the following would be suitable subsystems of a larger sample at room temperature? Justify each choice (i.e., why it is, or is not, suitable).
   a. Single molecules of naphthalene vapor, $C_{10}H_8$, at a very low pressure.
   b. One unit cell of CaF_2 (4 CaF_2 "molecules" in a face-centered cubic lattice).
   c. Half of a 1-g sample of H_2 gas.
   d. 10^{-9} g of liquid water, part of a 1-g sample of water.

19.13. Example 19.2 shows that $\Gamma = 2^N$ and $\Delta\Gamma = 2^N\sqrt{2/N\pi} << \Gamma$ for large N. Is the entropy really about the same for $\sigma = \ln 2^N$ and $\sigma = \ln\left[2^N\sqrt{2/N\pi}\right]$ if N is, for example, Avogadro's number?

CHAPTER 20

Applications of Statistical Mechanics

Only a few examples can be given of the application of statistical mechanics. Illustrations have been chosen that emphasize different aspects of the theory without reliance on understanding of other specialized topics.

20.1 Equipartition Principle

The law of equipartition of energy was given without proof in Section 7.3. The principle can now be derived by looking at a classical distribution function,

$$\frac{\delta N}{N} = Ce^{-\varepsilon/kT} dp_1\, dp_2 \cdots dp_{3N}\, dq_1\, dq_2 \cdots dq_{3N} \tag{20-1}$$

The constant C satisfies the normalization condition for the multiple integral.

$$\int \cdots \int \frac{\delta N}{N} = 1$$

Choose any one of the p_i or q_i and integrate by parts. For example, choosing q_1,

$$u = Ce^{-\varepsilon/kT} \qquad du = \frac{-1}{kT}\frac{\partial \varepsilon}{\partial q_1} dq_1 Ce^{-\varepsilon/kT}$$

$$v = q_1 \qquad dv = dq_1$$

$$\int u\, dv = uv - \int v\, du$$

or

$$1 = \int \cdots \int \left[Ce^{-\varepsilon/kT} q_1\right]_{q_1=a}^{q_1=b} dp_1\, dp_2 \cdots dp_{3N}\, dq_2 \cdots dq_{3N}$$

$$+\int \cdots \int_{q_1=a}^{q_1=b} C e^{-\varepsilon/kT} q_1 \frac{1}{kT} \frac{\partial \varepsilon}{\partial q_1} dp_1\, dp_2 \cdots dp_{3N}\, dq_1\, dq_2 \cdots dq_{3N} \quad (20\text{-}2)$$

Now consider several possibilities:

1. If $\partial\varepsilon/q_1 = 0$, the result is of little interest. The second integral vanishes and the normalization constant contains a factor $1/(b-a)$.

2. If (a) $\partial\varepsilon/\partial q_1 \neq 0$ and
   (b) $q_1 = 0$ *or* $\varepsilon = \infty$ (20-3)
   at a *and* at b

then the first integral vanishes and the second integral gives the average value for $(q_1/kT)(\partial\varepsilon/\partial q_1)$.

$$\left\langle \frac{q_1}{kT} \frac{\partial \varepsilon}{\partial q_1} \right\rangle = 1$$

or

$$\left\langle q_1 \frac{\partial \varepsilon}{\partial q_1} \right\rangle = kT \quad (20\text{-}4)$$

EXAMPLES 20.1

A. What is the average energy, $\langle \varepsilon_i \rangle$, if $\varepsilon_i = \frac{1}{2}A_i\xi_i^2$?

B. A pendulum swinging between $\pm\vartheta$ has kinetic and potential energies.

$$\varepsilon = \tfrac{1}{2}I\omega^2 + mg\ell(1 - \cos\vartheta)$$

For a simple pendulum, $I = m\ell^2$. The speed of the tip of the pendulum is $v = \ell\dot{\vartheta} \equiv \ell\omega$. For small angles,

$$1 - \cos\vartheta = 1 - \left(1 - \tfrac{1}{2}\vartheta^2 + \cdots\right) = \tfrac{1}{2}\vartheta^2$$

Therefore,

$$\varepsilon = \tfrac{1}{2}m\ell^2\omega^2 + \tfrac{1}{2}mg\ell\vartheta^2$$

a. What, if anything, can be concluded from the equipartition principle concerning the average value of the kinetic energy term?

b. What, if anything, can be concluded from the equipartition principle concerning the average value of the potential energy term? □

20.2 Simple Harmonic Oscillators

An important approximation suggested by Born and Oppenheimer provides the basis for analysis of almost all problems involving molecules. Because the energy differences, and therefore the classical frequencies or quantum-mechanical characteristic times, are very different in general for nuclear spins, molecular rotations,

---

ANSWERS 20.1

A. As $\xi_i \to \infty$, $\varepsilon_i \to \infty$, so the equipartition law tells us that $\left\langle \xi_i \frac{\partial(\frac{1}{2}A_i\xi_i^2)}{\partial \xi_i} \right\rangle =$ $\langle A_i\xi_i^2 \rangle = kT$ and $\langle \varepsilon_i \rangle = \frac{1}{2}kT$.
B. For an individual pendulum, given an arbitrary initial energy, the equipartition principle tells us nothing. On the other hand, a pendulum may be in thermal equilibrium with its environment (e.g., by Brownian motion). For an idealized pendulum, for which $\varepsilon = \frac{1}{2}I\omega^2 + \frac{1}{2}mg\ell\vartheta^2$ for *all* ω and ϑ,

$$\left\langle \omega \frac{\partial \varepsilon}{\partial \omega} \right\rangle = \langle I\omega^2 \rangle = kT \qquad \text{and} \qquad \left\langle \vartheta \frac{\partial \epsilon}{\partial \vartheta} \right\rangle = \langle mg\ell\vartheta^2 \rangle = kT$$

Because $T \ll mg\ell/k$ the average amplitude of the pendulum is small ($\vartheta_{\max} \ll 1$) and the real pendulum resembles the idealized pendulum. It follows that

$$\langle \varepsilon \rangle = \tfrac{1}{2}kT + \tfrac{1}{2}kT = kT$$

The same result is found for other oscillators, such as meter needles and torsional pendulums.

---

molecular vibrations, and electronic states, it is possible to approximate the wave function of a molecule by the product

$$\psi = \psi_{\text{electronic}} \times \psi_{\text{vibrational}} \times \psi_{\text{rotational}} \times \psi_{\text{nuclear}}$$

This makes the total energy a sum of the respective contributions,

$$E = E_{\text{electronic}} + E_{\text{vibrational}} + E_{\text{rotational}} + E_{\text{nuclear spin}}$$

and the partition function becomes a product,

$$Z = Z_{\text{el}} \cdot Z_{\text{vib}} \cdot Z_{\text{rot}} \cdot Z_{\text{nucl}} \tag{20-5}$$

The Born–Oppenheimer approximation may be interpreted in either of two ways. First, in classical or quantum mechanics, the coupling of two motions, or energy states, involves a "resonance" denominator. The strength of the coupling depends strongly on the frequency, or energy, separation. Hence for frequencies, or energies, that are quite different, the coupling can be predicted to be quite small.

Alternatively, the coupling effects may be interpreted in terms of average versus instantaneous parameters. For example, a rotating molecule is subject to centrifugal distortion, equivalent to an increase in bond lengths. Molecular vibrations then take place about this new equilibrium bond length but are not otherwise influenced by the fact the molecule is rotating (slowly) as it vibrates. By contrast, during any complete rotation the molecule will typically undergo many vibration cycles. The appropriate moment of inertia for the rotation is not an instantaneous value but an average over a complete vibration cycle. This averaging tends to wash out the interaction of vibration and rotation.

The Born–Oppenheimer approximation will be assumed without further explicit mention. However, it should be remembered that it is only an approximation;

it breaks down severely when energies become comparable, as for certain low-energy electronic transitions that may mix with vibrational states or low-energy vibrational states that may mix with rotational levels within the restrictions set by conservation laws.

The partition function is of the form

$$\sum_i e^{-\varepsilon_i/kT} = \sum_k g_k e^{-(\varepsilon_{\text{transl}}+\varepsilon_{\text{rot}}+\varepsilon_{\text{vib}}+\cdots)/kT}$$
$$= Z_{\text{transl}} \times Z_{\text{rot}} \times Z_{\text{vib}} \times \cdots$$

Therefore,

$$F = -NkT \ln Z$$

may be written

$$F = -NkT \sum_j \ln Z_j \tag{20-6}$$

where the Z_j are Z_{transl}, Z_{rot}, and so on.

The distribution function for an ensemble of classical oscillators is

$$\frac{dN}{N} = \mathscr{A} e^{-[\mathscr{T}(p)/kT + \mathscr{V}(q)/kT]} dp\, dq \tag{20-7}$$

with $\mathscr{T}(p) = p^2/2m$ and $\mathscr{V}(q) = \frac{1}{2} \mathscr{k} q^2$. This is separable, giving two integrals,

$$\int_{-\infty}^{+\infty} a e^{-p^2/2mkT}\, dp = 1 \qquad \text{and} \qquad \int_{-\infty}^{+\infty} b e^{-\mathscr{k}x^2/2kT}\, dx = 1$$

Each is of the standard form,

$$\int_{-\infty}^{+\infty} e^{-\alpha x^2} = \sqrt{\frac{\pi}{\alpha}}$$

The normalization constants, $\sqrt{\alpha/\pi}$, are

$$a = (2\pi mkT)^{-1/2} \qquad \text{and} \qquad b = \sqrt{\frac{\mathscr{k}}{2\pi kT}}$$

The average energy (kinetic + potential) of a classical oscillator is given by the equipartition principle,

$$\langle E \rangle = kT$$

Quantized oscillators have an energy

$$E = \left(\text{v} + \tfrac{1}{2}\right) h\nu_{\text{o}} = \left(\text{v} + \tfrac{1}{2}\right) \hbar\omega_{\text{o}}$$

where $\omega_{\text{o}} = \sqrt{\mathscr{k}/m}$ is the classical angular frequency of the oscillator. The ground-state energy is $\frac{1}{2}\hbar\omega_{\text{o}}$.

The distribution function (or density matrix element) is

$$w_v = a e^{-(v+\frac{1}{2})\hbar\omega_o/kT} = \mathscr{A} e^{-v\hbar\omega_o/kT} \tag{20-8}$$

For an ensemble of oscillators, the expectation value, or average value, for the energy, relative to the ground state, is

$$\langle E \rangle = \frac{\sum_{v=0}^{\infty} v\hbar\omega_o e^{-v\hbar\omega_o/kT}}{\sum_{v=0}^{\infty} e^{-v\hbar\omega_o/kT}} \tag{20-9}$$

which may also be written

$$\langle E \rangle = \frac{kT^2 \frac{\partial}{\partial T} \sum_{v=0}^{\infty} e^{-v\hbar\omega_o/kT}}{\sum_{v=0}^{\infty} e^{-v\hbar\omega_o/kT}} \tag{20-10}$$

Let $\eta \equiv \hbar\omega_o/kT \equiv h\nu_o/kT \equiv hc\sigma/kT$. Then

$$\sum_{v=0}^{\infty} e^{-v\hbar\omega_o/kT} = \sum_{v=0}^{\infty} e^{-\eta v} = \frac{1}{1-e^{-\eta}}$$

and $\partial\eta/\partial T = -\eta/T$.
Therefore,

$$\left\langle \frac{E}{kT} \right\rangle = \frac{T\frac{\partial}{\partial T}(1-e^{-\eta})^{-1}}{(1-e^{-\eta})^{-1}} = \frac{\eta}{e^{\eta}-1} \tag{20-11}$$

EXAMPLES 20.2

A. Show that $(1-e^{-\eta})^{-1} = \sum_{j=0}^{\infty} e^{-\eta j}$.

B. Find the ratio of quantum to classical expectation values, that is, $\left\langle \frac{E}{kT} \right\rangle$, for a harmonic oscillator when $\eta \equiv \hbar\omega_o/kT$ is

a. ≈ 0. b. 0.05.
c. 1.0 d. 4.0. □

Thermodynamic functions are typically expressed as a difference between the value at a chosen temperature and the value for a standard state at 0 K. The PV difference between E and H, between F and G, and between C_v and C_p arises in the translational coordinates and does not enter the equations for molecular vibrations. The contributions of a harmonic oscillation are, per mole:

$$\frac{\mathrm{E}-\mathrm{E_o}}{RT} = \frac{\mathrm{H}-\mathrm{H_o}}{RT} = \frac{\eta}{e^{\eta}-1}$$

$$\frac{\mathrm{C}}{R} = \frac{\eta^2 e^{\eta}}{(e^{\eta}-1)^2}$$

$$\frac{\mathrm{S}}{R} = \int \frac{\mathrm{C}}{RT}\,dT = \frac{\eta}{e^{\eta}-1} - \ln(1-e^{-\eta}) \tag{20-12}$$

$$\frac{\mathrm{F}-\mathrm{F_o}}{RT} = \frac{\mathrm{G}-\mathrm{G_o}}{RT} = \ln(1-e^{-\eta})$$

20.3 Spin and Quantum Statistics

Classical statistics does not take into consideration the effects of particle spin. However, even to obtain the correct classical partition function, it is necessary to be careful about how equivalent particles are treated.

Boltzmann Statistics

Each molecule contributes to the free energy a term

$$F_i = -kT \ln\left(\int e^{-p_i^2/2mkT}\, dp_i \int e^{-\mathcal{V}(q_i)/kT}\, dq_i\right) = -kT \ln z_i$$

If there are N molecules in the same volume,

$$F = -kT \ln Z = -kT \ln\left(\prod z_i\right) = -kT \ln(z_i)^N = -NkT \ln z_i$$

giving a free energy proportional to the number of molecules, as expected.

Now consider N_1 molecules of an ideal gas in volume V_1 at some temperature and pressure, separated by a thin membrane from N_2 molecules of the same gas in an equal volume, V_2, at the same temperature and pressure (so $N_1 = N_2$). Then we would write

$$F = -(N_1 + N_2)kT \ln z_i$$

But if the membrane separating the two gases is now punctured or removed,

$$F = -(N_1 + N_2)kT \ln z_i'$$

and $z_i' = 2z_i$ because the volume over which dq_i is to be integrated has doubled. Therefore, we obtain

$$F = -(N_1 + N_2)kT \ln(2z_i)$$

More generally, if each molecule in a gas is allocated a volume that is $1/N$ of the total gas volume, to calculate the partition function, z_i, of the molecule, then for the total gas of N molecules,

$$F = -NkT \ln(Nz_i) = -kT \ln(Nz_i)^N = -kT(N \ln z_i + N \ln N)$$

Removing a partition between equivalent gas samples causes no perceptible change in the gas and the free energy should be unchanged.[1] An extra term, $N \ln N$, has appeared, however, simply from removal of barriers. It appears the free energy is not additive, because of this extra term, a consequence known as *Gibbs' paradox*. The extra term arose because we expanded the range of integration when the systems were added, which is equivalent to interchanging the identical particles.

To remove the extra term and thereby make the additive thermodynamic functions add properly, divide the partition function by $N!$. Such a division is

[1]The situations with membrane intact and penetrable are not identical, because in the latter case fluctuations in numbers of molecules on the two sides are possible. We neglect this slight difference here.

ANSWERS 20.2

A. Divide $1/(1 - x)$ to get $1 + x + x^2 + \cdots = \sum_{j=0}^{\infty} x^j$.
B. (a) 1.00 (b) 0.9752 (c) 0.5820 (d) 0.0746

equivalent to eliminating the rearrangements of identical particles.[2] Then, to the approximation $N! = N \ln N$,

$$F = -kT \ln \frac{(Nz_i)^N}{N!} = -kT(\ln z_i^N + \ln N^N - N \ln N)$$

or

$$F = -kT \ln z_i^N = -NkT \ln z_i = NF_i \tag{20-13}$$

This remedy is appropriate provided that:

1. The particles are indistinguishable and can interchange positions.
2. The occupation numbers are small, so that the original calculation (before dividing by $N!$) assumed no more than one molecule per state. (Otherwise, it would have been necessary to correct for interchange of identical molecules occupying the same energy state and the $N!$ correction would overcorrect for indistinguishability.)

Systems that satisfy these conditions are said to obey *Boltzmann statistics*. In Boltzmann statistics, the chemical potential, μ, is always large but negative: $\mu \ll 0$.

Quantum Statistics

If the density of particles in available states is not very small (i.e., if the number of available states is not very much larger than the number of particles), the indistinguishability of the particles can be very important in determining the distribution function. The specific behavior of the particles depends on the spin of the nuclei.

Particles of integral spin (in units of $\hbar$), such as photons, deuterons, and alpha particles, are called *bosons*. They are described by wave functions that are symmetric to exchange of identical particles. Particles of odd half-integral spin, $s = n\hbar/2$ with n odd, are called *fermions*. They are described by wave functions that are antisymmetric to exchange of particles. Such particles obey the Pauli exclusion principle; no two equivalent particles may occupy the same state. Fermions are therefore restricted to occupation numbers of 0 or 1.

[2]In the late nineteenth century it was recognized that division by $N!$ was necessary, but no adequate justification could be provided. In the early twentieth century the importance of operational definitions was recognized, especially by Einstein. From this viewpoint it is not possible experimentally to distinguish one atom or molecule from another, so the equations must treat them as indistinguishable. Insistence on operational definitions provides the foundation for quantum mechanics as well as relativity. With the development of quantum mechanics, Gibbs' paradox is resolved by the observation that of all the $N!$ permutations of identical particles, only one combination satisfies the requirement that the total wave function should be symmetric (bosons) or antisymmetric (fermions) to exchange of identical particles.

The effect of quantum statistics on the number of states may be seen by calculating $\Delta\Gamma$ under different assumptions for putting N particles into g boxes. A simple representation is to write down all the particles *and* all the dividers in a row. There are $N + (g + 1)$ items (including the two ends of the row of boxes). Two of the dividers must be on the ends, so there are only $N + g - 1$ particles and dividers to be rearranged. A first approximation solution would be that there are $(N + g - 1)!$ arrangements. Assuming the particles to be distinguishable but the $g - 1$ dividers not to be distinguishable, we obtain

{Boltzmann}
$$\Delta\Gamma = \frac{(N + g - 1)!}{(g - 1)!} \tag{20-14}$$

as the number of states for N distinguishable particles in g boxes.

If the particles are indistinguishable (as must be assumed for quantum mechanical solutions), the general solution, corresponding to Bose-Einstein statistics, is the value given above corrected for permutations of the N particles.

{Bose–Einstein}
$$\Delta\Gamma = \frac{(N + g - 1)!}{N!(g - 1)!} \tag{20-15}$$

However, if no more than one indistinguishable particle can be put in any one box, the problem is to choose N of the g boxes (N necessarily less than, or in the limit equal to, g) that will be occupied. This is the familiar problem of g things taken N at a time, with the solution $\binom{g}{N}$. That is,

{Fermi–Dirac}
$$\Delta\Gamma = \frac{g!}{N!(g - N)!} \tag{20-16}$$

In the limit of $g \gg N$, each of these may be approximated.

Boltzmann :
$$\Delta\Gamma = \frac{(N + g - 1)!}{(g - 1)!} = (N + g - 1)(N + g - 2)\cdots g \approx g^N$$

F–D:
$$\Delta\Gamma = \frac{g!}{N!(g - N)!} = \frac{g(g - 1)\cdots(g - N + 1)}{N!} \approx \frac{g^N}{N!}$$

B–E:
$$\Delta\Gamma = \frac{(N + g - 1)!}{N!(g - 1)!} = \frac{(N + g - 1)(N + g - 2)\cdots g}{N!} \approx \frac{g^N}{N!}$$

This confirms that for "dilute" systems ($N \ll g$), both Fermi–Dirac and Bose–Einstein go over to Boltzmann statistics, with the $N!$, representing indistinguishability of particles, divided out. It was shown above that even Boltzmann statistics must have this $N!$ divisor (or an equivalent correction in the range of integration) to make the extensive thermodynamic functions additive.

The form of the distribution function depends on the statistical weight. For Boltzmann statistics we may take the expression above, in the limit of $g \gg N$. To emphasize the effect of grouping of levels into (nearly) degenerate levels, however, we write the number of ways that N molecules can be distributed among the levels, of energy ε_k, as

$$\Delta\Gamma = \frac{N!}{n_1!\, n_2!\, n_3!\, n_4! \cdots} = \frac{N!}{\prod n_k!}$$

but within each level, of energy ε_k, there are g_k states, with $g_k \gg n_k$. Hence each molecule has g_k "choices." The total statistical weight then is

$$\Delta\Gamma = \prod \frac{g_k^{n_k}}{n_k!}$$

after dividing by $N!$ Taking the logarithm and setting the variation, with the n_k, equal to zero gives

$$\delta \ln \Delta\Gamma = \delta \ln P = 0 = -\sum (\ln n_k - \ln g_k)\, \delta n_k$$

Similarly, the Fermi–Dirac and Bose–Einstein expressions give, from (20-16),

$$\delta \ln \Delta\Gamma = \delta \ln P = 0 = -\sum [\ln n_k - \ln(g_k - n_k)]\, \delta n_k$$

and, from (20-15), neglecting 1 in comparison to n_k and g_k,

$$\delta \ln \Delta\Gamma = \delta \ln P = 0 = -\sum [\ln n_k - \ln(n_k + g_k)]\, \delta n_k$$

Following the mathematics already developed, we may simply combine these equations with the equations $\delta N = \sum \delta n_k = 0$ and $\delta E = \delta \sum n_k \varepsilon_k = 0$ to obtain the three equations,

$$\{\text{Boltzmann}\} \qquad \sum_k \left(\ln \frac{n_k}{g_k} - \alpha - \beta\varepsilon_k \right) \delta n_k = 0$$

$$\{\text{Fermi–Dirac}\} \qquad \sum_k \left(\ln \frac{n_k}{g_k - n_k} - \alpha - \beta\varepsilon_k \right) \delta n_k = 0 \qquad (20\text{-}17)$$

$$\{\text{Bose–Einstein}\} \qquad \sum_k \left(\ln \frac{n_k}{g_k + n_k} - \alpha - \beta\varepsilon_k \right) \delta n_k = 0$$

These equations then give the occupation numbers, n_k, as

$$\{\text{Boltzmann}\} \qquad n_k = \frac{g_k}{e^{\alpha + \beta\varepsilon_k}}$$

$$\{\text{Fermi–Dirac}\} \qquad n_k = \frac{g_k}{e^{\alpha + \beta\varepsilon_k} + 1} \qquad (20\text{-}18)$$

$$\{\text{Bose–Einstein}\} \qquad n_k = \frac{g_k}{e^{\alpha + \beta\varepsilon_k} - 1}$$

The normalization condition, $\sum n_k = N$, gives the value of $e^{-\alpha-\beta\varepsilon_k} = \mathcal{A} e^{-\beta\varepsilon_k}$ as $e^{(\mu-\varepsilon_k)/kT}$. Alternatively, we may start with the grand canonical distribution (Section 19.4),

$$w_{nN} = e^{\Omega/kT} e^{(\mu N - E_n)/kT} \qquad (20\text{-}19)$$

Consider only particles of energy ε_k, summing over the number of particles; $E_k = n_k \varepsilon_k$. The potential $\Omega \equiv -PV$ for energy ε_k may also be labeled Ω_k. The particles must be in one of the states, so

$$\sum_N w_{kN} = 1 = e^{\Omega/kT} \sum_{n_k} e^{(\mu N - E_k)/kT}$$

or

$$\Omega_k = -kT \ln \sum_{n_k} e^{(\mu - \varepsilon_k)n_k/kT} \tag{20-20}$$

and

$$\frac{\partial \Omega_k}{\partial \mu} = -kT \frac{\frac{d}{d\mu} e^{(\mu - \varepsilon_k)n_k/kT}}{e^{(\mu - \varepsilon_k)n_k/kT}} = -kT \frac{n_k}{kT} = -n_k \tag{20-21}$$

which is consistent with the general thermodynamic relationship. These equations are a convenient starting point to obtain the Fermi–Dirac and Bose–Einstein distribution equations.

Fermi–Dirac Statistics

If we apply the restriction that n_k can be only 0 or 1, the sum becomes

$$\Omega_k = -kT \ln \sum_{n_k=0}^{1} e^{(\mu - \varepsilon_k)n_k/kT} = -kT \ln \sum_{n_k=0}^{1} \left[e^{(\mu - \varepsilon_k)/kT} \right]^{n_k}$$

$$= -kT \ln \left[1 + e^{(\mu - \varepsilon_k)/kT} \right]$$

Substituting

$$\overline{n}_k = -\frac{\partial \Omega_k}{\partial \mu}$$

gives

{fermions} $$\overline{n}_k = \frac{e^{(\mu - \varepsilon_k)/kT}}{e^{(\mu - \varepsilon_k)/kT} + 1} = \frac{1}{e^{(\varepsilon_k - \mu)/kT} + 1} \tag{20-22}$$

For the normalization chosen above,

$$\Omega = \sum_k \Omega_k = -kT \sum_k \ln \left[1 + e^{(\mu - \varepsilon_k)/kT} \right] \tag{20-23}$$

The chemical potential, μ, may be positive or negative.

Bose–Einstein Statistics

Other particles, called bosons, may have an unlimited number in any state. Bosons are not subject to conservation laws; they may be created and destroyed at will. It is possible to have any number of these bosons in a given state. From the equations above, modified for the range of n_k, we have

$$\Omega_k = -kT \ln \sum_{n_k=0}^{\infty} [e^{(\mu-\varepsilon_k)/kT}]^{n_k}$$

The sum is a geometric series of the form

$$\sum_{i=0}^{n-1} r^i = \frac{r^n - 1}{r - 1}$$

or

$$\begin{matrix} r < 1 \\ n \to \infty \end{matrix} \qquad \sum_{i=0}^{n-1} r^i = \frac{-1}{r-1} = (1-r)^{-1}$$

The potential Ω_k is a finite quantity for all ε_k. However, the series converges only if the exponential term is less than 1, so $\mu - \varepsilon_k < 0$. Because ε_k may be zero, μ must be negative.

$$\mu < 0$$

Summing the series gives

$$\Omega_k = -kT \ln \sum_{n_k} [e^{(\mu-\varepsilon_k)/kT}]^{n_k} = -kT \ln[1 - e^{(\mu-\varepsilon_k)/kT}]^{-1}$$

or

$$\Omega_k = kT \ln[1 - e^{(\mu-\varepsilon_k)/kT}] \tag{20-24}$$

and the mean occupation numbers are

$$\{\text{bosons}\} \qquad \overline{n}_k = -\frac{\partial \Omega_k}{\partial \mu} = \frac{e^{(\mu-\varepsilon_k)/kT}}{1 - e^{(\mu-\varepsilon_k)/kT}}$$

$$= \frac{1}{e^{(\varepsilon_k-\mu)/kT} - 1} \qquad (\mu < 0) \tag{20-25}$$

The three distribution functions may be compared.

$$\left\{\begin{matrix}\text{Boltzmann} \\ \text{statistics}\end{matrix}\right\} \qquad \overline{n}_k = e^{-\alpha} e^{-\varepsilon_k/kT} = e^{-(\varepsilon_k-\mu)/kT}$$

$$= \frac{1}{e^{(\varepsilon_k-\mu)/kT}} \qquad (\mu \ll 0)$$

$$\left\{\begin{matrix}\text{Fermi–Dirac} \\ \text{statistics}\end{matrix}\right\} \qquad \overline{n}_k = \frac{1}{e^{(\varepsilon_k-\mu)/kT} + 1} \qquad (\mu \gtrless 0) \tag{20-26}$$

$$\left\{\begin{matrix}\text{Bose–Einstein} \\ \text{statistics}\end{matrix}\right\} \qquad \overline{n}_k = \frac{1}{e^{(\varepsilon_k-\mu)kT} - 1} \qquad (\mu < 0)$$

Note that the + sign must go with Fermi–Dirac statistics, to keep $\overline{n}_k$ smaller than for Bose–Einstein statistics.

EXAMPLES 20.3

Consider a system with three energy levels, such that $\varepsilon_k - \mu = \varepsilon_k + |\mu| = 0.05kT, \frac{1}{2}kT$, and $\frac{3}{2}kT$. Find $\overline{n}_k$, the relative populations of each of the three states for the conditions specified below. (Note that the function as written is not normalized.)

A. Assume Boltzmann statistics.
B. Assume Fermi–Dirac statistics.
C. Assume Bose–Einstein statistics.
D. Can you explain why the bosons seem to "attract" or "crowd together" in the low-energy states relative to the other types of systems? □

For either the Fermi–Dirac or Bose–Einstein distributions, the sum over the occupation numbers is equal to the number of particles, N. Also, in each case, if the exponential term becomes very small ($\overline{n}_i << 1$), the distribution becomes equivalent to the Boltzmann distribution. The classical Boltzmann distribution is therefore the limiting case of both Fermi–Dirac and Bose–Einstein statistics.

20.4 Ideal Gas

Any classical system for which the energy is independent of position follows a distribution law of the form

$$\frac{\delta N}{N} = \rho(p, q) = \rho(p) \sim e^{-E(p)/kT}$$

Not only translational motion of an ideal gas, but also, for example, Brownian motion and classical rigid rotors follow such a distribution law.

The distribution over position coordinates (which is uniform) and the distribution over momenta can be normalized separately. The former may be represented as

$$\int_V \left(\frac{dN}{N}\right)_q = 1 = \int_V Ce^{-\mathscr{V}(q)/kT}\, dq = \int_V C\, dq = CV$$

because $\mathscr{V}(q)$, the potential energy, is a constant and may be set equal to zero. The normalization constant, C, is the reciprocal of the volume. The momentum distribution is

$$\int_{p=-\infty}^{p=+\infty} \left(\frac{dN}{N}\right)_p = 1 = \int ae^{-(p_x^2+p_y^2+p_z^2)/2mkT}\, dp_x\, dp_y\, dp_z \tag{20-27}$$

Recalling the definite integral

$$\int_{-\infty}^{+\infty} e^{-\alpha x^2} = \sqrt{\frac{\pi}{\alpha}}$$

integrating over the momentum components gives

$$a\,[\pi(2mkT)]^{3/2} = 1$$

ANSWERS 20.3

A. $n_o = e^{-0.05} = 0.9512, n_1 = e^{-1/2} = 0.6065, n_2 = e^{-3/2} = 0.2231$. ($n_k / \sum n_k = 0.5341, 0.3406, 0.1253$.)

B. $n_o = 1/(1 + e^{0.05}) = 0.4875, n_1 = 1/(1 + e^{0.5}) = 0.3775, n_2 = 1/(1 + e^{1.5}) = 0.1824$. ($n_k / \sum n_k = 0.4654, 0.3604, 0.1741$.)

C. $n_o = 1/(e^{0.05} - 1) = 19.50, n_1 = 1/(e^{0.5} - 1) = 1.54, n_2 = 1/(e^{1.5} - 1) = 0.2872$. ($n_k / \sum n_k = 0.9143, 0.0722, 0.0135$.)

D. We can "fudge" the answer by saying that there is a quantum mechanical force or potential that makes bosons attract each other (like the peculiar "exchange force" and "exchange energy" invoked in early valence theory). However, no such force is introduced in the equations. A better analysis would be to say that the Bose–Einstein distribution is the natural one, in the absence of special restrictions (such as appear for fermions). Although the Boltzmann distribution is "incorrect" (i.e., it does not give the values predicted by the more accurate Bose–Einstein expression for high densities), we can write the Bose–Einstein expression as

$$\frac{1}{e^{(\varepsilon_k - \mu)/kT} - 1} = \frac{e^{-(\varepsilon_k - \mu)/kT}}{1 - e^{-(\varepsilon_k - \mu)/kT}}$$

showing that when $\varepsilon_k - \mu \gg kT$, the denominator approaches 1 and the Bose–Einstein and Boltzmann distributions are indistinguishable. Thus we can use the simpler Boltzmann distribution, derived classically, for the conditions usually met in classical problems.

---

and therefore

$$a = (2\pi mkT)^{-3/2}$$

Distribution functions expressed in terms of speeds or velocity components can be found in much the same manner.

In Section 16.4 the distribution of particles over energy states ϵ_k was found to be

$$n_k = e^{-\alpha} e^{-\beta_k}$$

Degenerate energy states were subsequently grouped, into g_k energy levels, ε_k, and the normalization factor was found to involve the sum over states, or partition function.

$$\frac{n_k}{n} = \frac{g_k e^{-\beta \varepsilon_k}}{Z}$$

At least two faults may be found with the argument. First, there was no attention to the fact that the particles are indistinguishable. That introduces a factor involving $N!$, which must then be removed, as shown in the preceding section, if it makes any difference. (In practice, this constant value drops out of most formulas.) More significant is the assumption that n_k must be assumed to be large, to apply Stirling's approximation. In fact, the number of molecules per elementary volume of phase space is small for an ideal gas.

To avoid difficulties of the earlier derivation, recognize approximate degeneracies from the beginning. Assume that N particles are distributed among the energy states, ϵ_k; then group the energy states. For example, states between ϵ_k and $\epsilon_k + \Delta\epsilon$ are grouped into a level, ε_j, containing g_j states, with $g_j \gg 1$, and N_j molecules in this level. If N and V are sufficiently large, N_j can be made large. Then $N_j \gg 1$ and $g_j \gg 1$, but $g_j \gg N_j$.

The g_j states may now be regarded as boxes into which the N_j molecules are to be placed. Each molecule has g_j possibilities, so the number of options[3] is $(g_j)^{N_j}$. However, it is of no consequence as to which particle is in which box, so the number of options must be divided by the number of permutations, $N!$. The statistical weight or number of possible arrangements is therefore

$$\Delta\Gamma_j = \frac{(g_j)^{N_j}}{N_j!} \qquad \text{and} \qquad \Delta\Gamma = \prod_j \Delta\Gamma_j \tag{20-28}$$

The entropy is

$$S = k \ln \Delta\Gamma = k \sum_j \ln \Delta\Gamma_j$$

$$S = k \sum (\ln g_j^{N_j} - \ln N_j!) = k \sum_j (N_j \ln g_j - \ln N_j!)$$

Now the numbers are sufficiently large that Stirling's formula may be applied to $\ln N_j!$, so that with the substitution of $N_j \ln e$ for N_j,

$$S = k \sum_j (N_j \ln g_j - N_j \ln N_j + N_j) = k \sum_j N_j \ln \frac{e g_j}{N_j}$$

The density of particles per energy state is

$$\overline{n}_j = \frac{N_j}{g_j}$$

with $\overline{n}_j \ll 1$, in general, and the entropy may be written

$$S = k \sum_j g_j \overline{n}_j \ln \frac{e}{\overline{n}_j} \tag{20-29}$$

If we let

$$g_j = \frac{(\Delta p)_j\,(\Delta q)_j}{\delta\,\mathscr{R}} = \Delta\tau_j \to d\tau$$

and

$$\overline{n}_j \to n$$

the sum may be written as an integral,

[3]A previous calculation for putting particles into boxes would give here $\Delta\Gamma_j = (g_j + N_j - 1)!/(g_j - 1)!$. There are N_j factors in $(g_j + N_j - 1)!/(g_j - 1)!$, each between g_j and $g_j + N_j$, so that when $g_j \gg N_j$, their product is approximately $g_j^{N_j}$.

$$S = k \int n \ln \frac{e}{n}\, d\tau \tag{20-30}$$

It has not yet been necessary to assume the gas to be in equilibrium.

Now require the gas to be in equilibrium, with constant number of particles and constant energy.

$$dS = 0 \qquad dN = d\left(\sum g_j \overline{n}_j\right) = 0 \qquad dE = d\left(\sum \varepsilon_j g_j \overline{n}_j\right) = 0$$

Then with the Lagrange multipliers $-\alpha$ and $-\beta$,

$$\frac{dS}{k} - \alpha\, dN - \beta\, dE = 0$$

so

$$\frac{\partial}{\partial \overline{n}_j}\left(\frac{S}{k} - \alpha N - \beta E\right) = 0$$

$$= \frac{\partial}{\partial \overline{n}_j}\left(\sum g_j \overline{n}_j \ln \frac{e}{\overline{n}_j}\right) - \alpha \frac{\partial}{\partial \overline{n}_j}\left(\sum g_j \overline{n}_j\right) - \beta \frac{\partial}{\partial \overline{n}_j}\left(\sum \varepsilon_j g_j \overline{n}_j\right)$$

$$= g_j \ln \frac{e}{\overline{n}_j} - g_j \overline{n}_j \frac{1}{\overline{n}_j} - \alpha g_j - \beta \varepsilon_j g_j$$

$$= -g_j(\ln \overline{n}_j + \alpha + \beta \varepsilon_j) = 0$$

and therefore

$$\overline{n}_j = e^{-\alpha - \beta \varepsilon_j}$$

The equation $\dfrac{dS}{k} - \alpha\, dN - \beta\, dE = 0$, relating S, N, and E, must be the same as the thermodynamic equation for constant T and V,

$$dE = T\, dS + \mu\, dN$$

Therefore, $\alpha = -\mu/kT$ and $\beta = 1/kT$.

For an ideal gas,

$$\mu = \left(\frac{\partial F}{\partial N}\right)_{T,V} = \frac{F}{N} = -kT \sum_j \ln z_j$$

$$G = F + PV = F + NkT = -NkT \sum_j \ln z_j + NkT$$

If the gas is monatomic, the partition function, z_j, for a single molecule is

$$\int_V dq \int e^{-(p_x^2 + p_y^2 + p_z^2)/2mkT}\, dp_x\, dp_y\, dp_z / \delta \mathscr{R} = \frac{(2\pi mkT)^{3/2} V}{\delta \mathscr{R}} \tag{20-31}$$

Setting $\delta \mathscr{R} = h^3$ and $Z = z_j^N/N!$ gives

$$Z = \frac{1}{N!}\left[V\left(\frac{2\pi mkT}{h^2}\right)^{3/2}\right]^N$$

With Stirling's formula this becomes, for the monatomic ideal gas,

$$Z_{\text{transl}} = \left[e\frac{V}{N}\left(\frac{2\pi mkT}{h^2}\right)^{3/2}\right]^N \tag{20-32}$$

It follows that

$$S_{\text{transl}} = -\left(\frac{\partial F}{\partial T}\right)_V = Nk\ln\left[\frac{V}{N}\left(\frac{2\pi mkT}{h^2}\right)^{3/2}\right] + \frac{5}{2}Nk \tag{20-33}$$

This equation, involving Planck's constant and the correction for indistinguishable molecules, was obtained independently by Sakur and by Tetrode in 1915. Confirmation by low-temperature measurements was important in the development of quantum statistics and the third law.

From $E = F + TS$ we obtain

$$E = \frac{3}{2}kT$$

as expected.

20.5 Heat Capacity of Gases and Solids at Low Temperatures

The rotational contribution to the partition function of a heteronuclear diatomic molecule was found in Section 18.2 to be

$$Z = \frac{8\pi^2 I_b kT}{h^2} \equiv \frac{kT}{hcB}$$

in the high-temperature limit. (The symmetry number is 1 for heteronuclear diatomics.)

In the low-temperature limit, rotations cannot be assumed to behave classically. When $kT \ll hcB$, the partition function becomes

$$Z_{\text{rot}} = \sum_{J=0}^{\infty}(2J+1)e^{-hcBJ(J+1)/kT} = 1 + 3e^{-2hcB/kT} + \cdots$$

From this the heat capacity is found to be

$$C_{\text{rot}} = 3Nk\left(\frac{2hcB}{kT}\right)^2 e^{-2hcB/kT} \tag{20-34}$$

EXAMPLES 20.4

A. Show that $\sum_{J=0}^{\infty}(2J+1)e^{-hcBJ(J+1)/kT} = 1 + 3e^{-2hcB/kT}$ for large (negative) values of the exponent.

B. Show that C_{rot} is obtained from the temperature derivative of the energy, $RT^2\,(d\ln Z)/dT$, under the same limiting conditions. □

ANSWERS 20.4

A. The first three terms of the sum are: J = 0: $Z_0 = 1 \times e^0 = 1$; J = 1: $Z_1 = 3e^{2hcB/kT}$; J = 2: $Z_2 = 5e^{-6hcB/kT} \ll 3e^{-2hcB/kT}$ for $2hcB/kT \gg 1$.

B. $$c_v = \frac{\partial}{\partial T}\left(RT^2\frac{d\ln Z}{dT}\right) = \frac{\partial}{\partial T}\left\{RT^2\frac{d}{dT}\left[1+3e^{-2hcB/kT}\right]\right\} = \frac{\partial}{\partial T}\left\{RT^2 \times 3\left[\frac{2hcB}{kT^2}e^{-2hcB/KT}\right]\right\} = R\frac{6hcB}{kT^2}\left(\frac{2hcB}{k}\right)e^{-2hcB/kT} = 3N_Ak\left(\frac{2hcB}{kT}\right)^2 e^{-2hcB/kT}.$$

Plots of energy and c_v against temperature show that the energy falls below expectations at low temperatures (i.e., $E_{rot} < kT$), but then the energy climbs faster than expected as it catches up. Hence the heat capacity, the slope of the curve of energy versus temperature, is greater than the classical limit for some intermediate temperature interval.

For homonuclear diatomic molecules the symmetry number is 2 and the partition function is one-half as large. The value of the spin of the identical pairs of nuclei is also important. For example, if the nuclei have i = 0, the sum must also be zero; I = 0. Hence the spin state is symmetric and because the product $\psi_{spin}\psi_{rot}$ must be symmetric to exchange (assuming symmetric electronic and vibrational states), only the states of J = 0, 2, 4, ... can occur. It is because the odd states are missing that the symmetry number factor, s = 2, gives the correct statistical weight at high temperatures.

If I = 1, the spin function is antisymmetric, so ψ_{rot} must be antisymmetric and J = 1, 3, 5, ... This, also, is consistent with the symmetry number, s = 2.

If the nuclear spins are half-integral, the particles are fermions and the total wave function must be antisymmetric. For example, if i = $\frac{1}{2}$, as for the hydrogen molecule, we saw that I = 0 gives a singlet state and I = 1 gives a triplet state. The first of these is antisymmetric; the second is symmetric. Therefore, the singlet state has only J = 0, 2, 4, ... and the triplet state has only the rotational states J = 1, 3, 5, ...

Deuterium, D_2, or molecular heavy hydrogen, has i = 1 and therefore I = 0, 1, or 2. Ortho deuterium (most plentiful at room temperature) consists of quintuplet and singlet states, with J = 0, 2, 4, ... Para deuterium is the triplet state, with J = 1, 3, 5, ... At low temperatures the ortho form is favored, in the J = 0 state; at high temperatures the ratio of ortho to para is (5 + 1)/3 or 2/1. Because of the limitations of J values, the heat capacity curves are substantially different for H_2, HD, and D_2. Heat capacity versus temperature is shown in Fig. 20.1 for three molecules differing only in nuclear spins: one having only symmetric rotational states (like p-H_2 or o-D_2), one having only antisymmetric rotational states (like o-H_2 or p-D_2), and one having both symmetric and antisymmetric rotational states (like H-D). Heat capacity of the natural mixture of ortho and para hydrogen falls between the values for ortho and para hydrogen. Also shown is a plot of rotational energy versus temperature for p-H_2, for o-H_2, and for H_2 with the symmetry of HD. Note that the heat capacity curves display the slopes of the energy curves.

Einstein Model for Heat Capacities of Solids

The law of Dulong and Petit predicts a molar heat capacity[4] for any solid element of $3R$, but the prediction fails for low temperatures or even near room temperature

[4]The value given by Dulong and Petit was 6.4 cal/mol · deg , which was c_p. For solids, c_v is typically about 0.4 cal/mol · deg less. See Fowler and Guggenheim (ref. 26, p. 142).

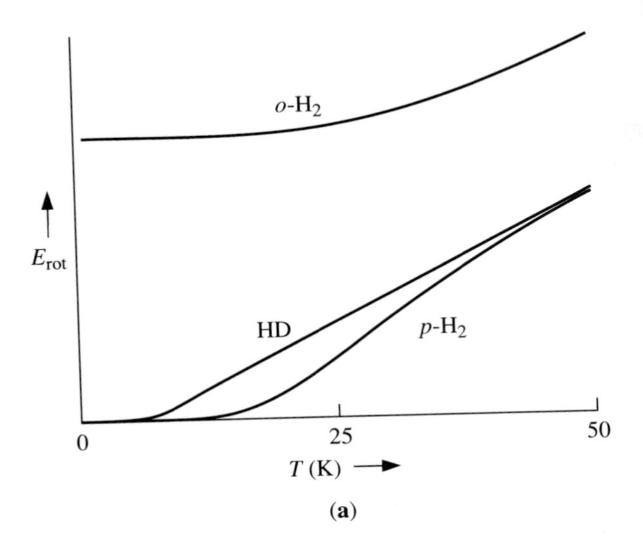

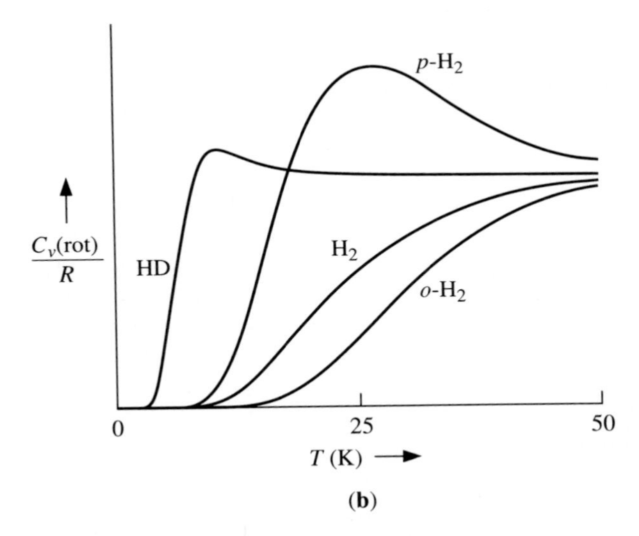

FIGURE 20.1 (a) Energy versus T for J even ($p-H_2$), J odd ($o-H_2$), and for HD symmetry (all J allowed). (b) C_v (rot)/R for $p-H_2$, $o-H_2$, natural H_2 (3/4 ortho), and for HD symmetry (H_2 masses).

for light elements. At very low temperatures, heat capacities of solids approach zero.

Einstein carried out the first analysis of heat capacities of crystals at low temperatures including quantization effects. His model for a crystal of N particles consisted of $3N$ independent harmonic oscillators (i.e., vibrations along x, y, and z axes), each of the same frequency ν_o and energy $\varepsilon_k = h\nu_o$. The energy of $3N$ oscillators is $3N\bar{n}_k\varepsilon_k$ or, as shown in Section 20.2,

$$E = 3NkT\frac{\eta}{e^{\eta} - 1} = \frac{3Nh\nu_o}{e^{h\nu_o/kT} - 1} = \frac{3Nk\Theta}{e^{\Theta/T} - 1} \tag{20-35}$$

where $\Theta \equiv h\nu_o/k \equiv \eta T$.

Taking the derivative of E with respect to temperature gives the heat capacity, C_V.

$$C_V = \frac{\partial E}{\partial T} = 3Nk\frac{(\Theta/T)^2 e^{\Theta/T}}{(e^{\Theta/T} - 1)^2} \tag{20-36}$$

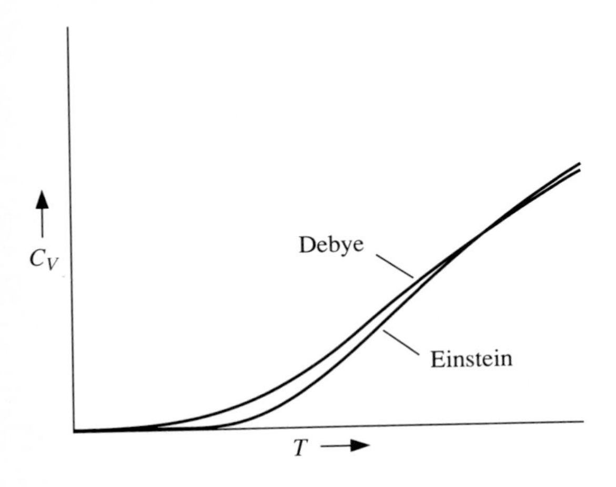

FIGURE 20.2 Einstein and Debye equations for heat capacities of solids at low temperatures. The ratio Θ_E/Θ_D is set at 0.73.

This equation, obtained in 1907, reproduces the most important characteristics of the heat capacity. First, in the high-temperature approximation, $T \gg \Theta$, $e^{\Theta/T} = 1 + \Theta/T$ and $C_V \to 3Nk$, or $3R$ per mole, as obtained empirically by Dulong and Petit. On the other hand, if $T << \Theta$ (low-temperature approximation), the exponential terms, which dominate, give $(e^{\Theta/T})^{-1}$, which goes to zero, so $C_V = 0$ at 0 K, as observed (Fig. 20.2).

Debye Model

Although Einstein's analysis was far better than the purely classical calculations that preceded it, the behavior of his solution at very low temperatures is detectably different from measured values of real solids. Debye improved on the calculation, in 1912, by replacing the set of identical, noninteracting oscillators with an equal number of oscillators that interact and therefore give a continuous range of frequencies, from zero to some maximum.

It is well known from classical mechanics that N oscillators of frequency ν_o, coupled together, will give N frequencies of oscillation. The maximum separation, however, is determined by the largest interaction, which is the nearest-neighbor interaction. Therefore, the N frequencies fall within a finite frequency range, determined by the strength of the coupling. Debye took the range as $\nu = 0$ to $\nu = \hat{\nu}$, an unspecified upper limit.

Possible frequencies in solids are best approached in terms of the vibrational waves[5] that can propagate through the solid. Three polarizations are possible: two orthogonal vibrations transverse to the direction of propagation and one longitudinal mode, along the direction of motion. Let these have an "average" speed $\bar{v}$ (which will actually be found by averaging reciprocals of cubes of the speeds). Each mode has a *wave vector* of magnitude equal to the wavenumber, $\nu/v = 1/\lambda$, and direction equal to the direction of propagation of the wave.

The number of possible modes of frequency ν in any volume V is the number of wave vectors that will fit into the volume (cf. Fig. 7.8). That number is $4\pi(\nu/v)^2\, d(\nu/v)$ times the three polarizations, or

$$\frac{d\Gamma}{V} = \frac{12\pi\nu^2\, d\nu}{v^3} \tag{20-37}$$

[5]It is often convenient to treat such waves as quasi-particles, called *phonons*. We need here only the wave properties.

The contributions of a single harmonic oscillation to the thermodynamic functions, E, S, F, and so on, were found in Section 20.2. The free energy is

$$F = kT \ln(1 - e^{-\eta})$$

with $\eta \equiv \hbar\omega_o/kT$. The free energy for a collection of $3N$ oscillators of frequencies $j\nu_0$, therefore, is

$$F = kT \sum_{j=1}^{3N} \ln(1 - e^{-\hbar\omega j/kT}) + N\varepsilon_I \tag{20-38}$$

where ε_I is an average energy of interaction of the oscillators with each other, which is a function of the volume of the solid.

At low temperatures, the exponential term becomes negligible for $j\omega$ much greater than zero, so the upper limit of the sum may be stated as ∞, without significant change. Then the sum over $3N$ discrete frequencies can be replaced by an integral over frequency, after multiplying by the degeneracy, or number of modes of the same frequency. The free energy is therefore

$$F = kT\frac{12\pi V}{\overline{v}^3}\int_0^\infty \ln[1 - e^{-h\nu/kT}]\nu^2\,d\nu + N\varepsilon_I$$

If we replace frequency with angular frequency, $\nu^2 d\nu = \omega^2 d\omega/(2\pi)^3$,

$$F = kT\frac{3V}{2\pi^2\overline{v}^3}\int_0^\infty \ln[1 - e^{-\hbar\omega/kT}]\omega^2 d\omega + N\varepsilon_I \tag{20-39}$$

The integral is of a form known as the Riemann zeta function and produces values expressed in terms of the gamma function and Bernoulli numbers. Evaluation gives

$$F = -\frac{V\pi^2(kT)^4}{30(\hbar\overline{v})^3} + N\varepsilon_I$$

$$S = \frac{2V\pi^2(kT)^4}{15T(\hbar\overline{v})^3} + N\varepsilon_I \tag{20-40}$$

and

$$E = \frac{V\pi^2(kT)^4}{10(\hbar\overline{v})^3} + N\varepsilon_I$$

and therefore

$$C_v = \frac{2\pi^2 kV}{5(\hbar\overline{v})^3}(kT)^3 \tag{20-41}$$

This shows that the lattice vibration contribution to the heat capacity (which is the only contribution, near 0 K) is proportional to the cube of the temperature at low temperatures, in agreement with experiments.

At higher temperatures η is small, so $1 - e^{-\eta}$ may be replaced by η. Then

$$F = kT \sum_{j=1}^{3N} \ln \eta + N\varepsilon_I = 3NkT \ln \overline{\eta} + N\varepsilon_I$$

The energy is

$$E = 3NkT + N\varepsilon_I$$

and the heat capacity is

$$C_V = 3NkT$$

in agreement with Dulong and Petit, or to a better approximation,

$$C_V = 3NkT - aT \tag{20-42}$$

with the constant, a, usually positive. In these equations N is the number of atoms in the crystal of an element. As shown in Section 6.5, the formula may be applied, with caution, to some ionic compounds.

For low temperatures it was possible to integrate from $T = 0$ to infinity because only very low frequencies contribute significantly to the integral. For intermediate temperatures the equations are not as simple. However, the number of frequencies in a solid is equal to the number of oscillators, $3N$. A continuous spectrum representation of a finite number of frequencies must have some maximum value, $\hat{\nu}$. Define the *Debye temperature*, Θ, from the relation

$$k\Theta = h\hat{\nu} \tag{20-43}$$

Then the properties at intermediate temperatures are expressible as a function of the ratio of the Debye temperature to the actual temperature, Θ/T. Atoms of large mass, or loose bonding to neighbors, have a small Θ; light atoms, or those tightly bound, have a large Θ, requiring higher temperatures for comparable properties.

20.6 Magnetic Moments at Low and Negative Temperatures

An atom in a symmetric electronic state (^{1}S, produced by filled shells or subshells) has no electronic angular momentum and no magnetic moment. Application of an external magnetic field induces a motion of the electrons in accordance with Lenz's law, producing an internal field that opposes the applied field. The substance is therefore repelled by a magnetic field. The effect, called *diamagnetism*, is quite weak, however. It is not easily observed experimentally and is completely obscured by other effects in most materials.

Paramagnetism

Most atoms have unpaired electrons that give a magnetic moment from the intrinsic "spin" and/or have orbital angular momentum, and a corresponding magnetic moment, from electrons in p, d, f, or higher unfilled shells. Assume the spin and orbital effects can be treated together by a quantum number J, with projection on the axis of the applied field of $M\hbar$, where M can have 2J + 1 values, from − J to + J. Then the component of the magnetic dipole along the axis is $\mathcal{g}M\mu_B$, where $\mathcal{g}$ is the magnetogyric (or gyromagnetic) ratio and μ_B is the Bohr magneton; $\mu_B = 9.27 \times 10^{-24}$ J/T. The magnetic energy, per atom, is $-\mu \cdot \mathbf{B} = \mathcal{g}M\mu_B B$. The effect is called *paramagnetism.*

Paramagnetism is also a weak effect. The product $\mu_B B$ amounts to a temperature (ε/k) of less than 1 K for a field of 1 T, so at room temperature, with magnetic fields of normal strength, the magnetic moments are almost entirely random, with only a slight preference for alignment with the field. Nevertheless, the effect can be observed in many substances, including some molecules and many salts in which the electron spinorbitals contributing are sufficiently buried within the chemical unit so that there is negligible interaction between neighboring units. For example, oxygen, O_2, is paramagnetic. When liquid oxygen is poured between the poles of a strong magnet, the liquid is deflected toward the poles, as iron filings would be except much more weakly.

Many nuclei have magnetic moments, also. The mass of the particle enters into the formula for the magnetic moment, however, so the magnetic moments of nuclei are smaller than those of electrons by a factor of nearly 2000.

The average magnetic moment, in the direction of the applied field, B, is

$$\overline{\mu} = \frac{\overline{\varepsilon}}{B} = \frac{1}{B}\frac{\sum_{M=-J}^{M=+J} \varepsilon(M)e^{-\varepsilon(M)/kT}}{\sum_{M=-J}^{M=+J} e^{-\varepsilon(M)/kT}} \tag{20-44}$$

Evaluation of the average magnetic moment is generally difficult, but it may readily be evaluated under certain important limiting conditions. For very large values of J and hence M, the sum may be replaced by an integral over M, which gives the average magnetic moment as

$$\overline{\mu} = J\mathcal{g}\mu_B L(y) \tag{20-45}$$

where $L(y)$ is the *Langevin function,*

$$L(y) = \coth y - \frac{1}{y}$$

shown in Fig. 20.3, and (20-46)

$$y \equiv J\mathcal{g}\mu_B \frac{B}{kT}$$

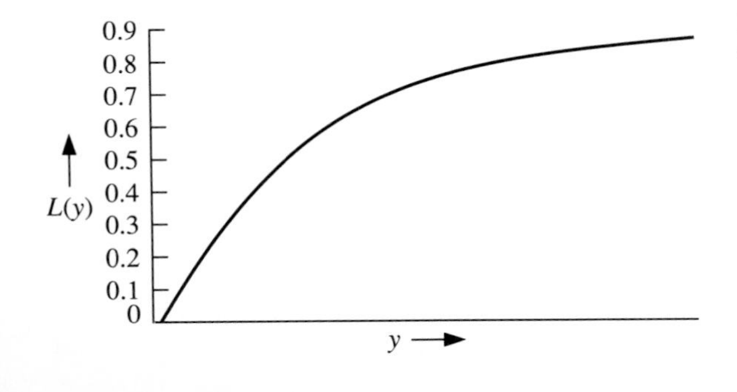

FIGURE 20.3 Langevin function, $L(y)$, versus y.

For low fields, $\overline{\mu}$ is proportional to B, but at high fields $\overline{\mu}$ reaches its saturation value as all moments become aligned. A better linear approximation for low applied fields is obtained by assuming that ε/kT is small and evaluating the sum in powers of ε/kT, which leads, upon substitution of $\sum M^2 = (2J + 1) \cdot J(J + 1)/3$, to

$$\overline{\mu} = J(J + 1)(\mathscr{g}\mu_B)^2 \frac{B}{3kT} \tag{20-47}$$

in agreement with the linear portion of the previous expression except for the substitution of the more precise $J(J + 1)$ for J^2.

Adiabatic Demagnetization

The energy of aligned magnetic dipoles in a field is

$$\varepsilon = -\boldsymbol{\mu} \cdot \mathbf{B} < 0$$

Control of the external field provides an important way of extracting thermal energy from systems at very low temperatures, where substantial alignment can be achieved.

In paramagnetic substances, the interaction is sufficiently weak between the spins and the crystal lattice that the spins may be considered to occupy a "spin lattice" that interpenetrates the crystal lattice and interacts slowly with it. A strong field is applied to a paramagnetic substance that is in good thermal contact with a reservoir at T_1, the lowest temperature readily achieved in the laboratory. Then the lattice modes and the spins are in equilibrium at T_1, with the spins aligned.

When the applied magnetic field is removed, the spins reorient to a random configuration, which is a higher-energy state than when aligned. To reach this higher-energy state, the spin lattice must gain energy, which it can get only from the already cold crystal lattice. The crystal lattice energy, and hence temperature, is thereby lowered. Temperatures below 1 μK are achieved by such adiabatic demagnetization procedures.

Negative Temperatures

It is not possible to reach 0 K or to go below this absolute zero for states in equilibrium. The crystal lattice would not be stable. At constant volume and constant number of particles in any system, the entropy of the system at equilibrium is a function of the internal energy of the system only. For nonrotating systems

$$S = S(E\text{int}) = S\left(E - \frac{p^2}{2m}\right) = \sum_i S_i\left(E_i - \frac{p_i^2}{2m_i}\right)$$

where E is the total energy, E_i is the total energy of the ith subsystem with momentum p_i, and the sum is over all subsystems.

It follows that

$$\frac{\partial S}{\partial p_i} = \frac{\partial S}{\partial E}\frac{\partial E}{\partial p_i} = \frac{1}{T}\left(-\frac{p_i}{m_i}\right)$$

and

$$dS = -\frac{p_i}{m_i T}\, dp_i \geq 0 \tag{20-48}$$

As long as T, p_i, and m_i are positive, $dp_i \leq 0$. Each subsystem therefore moves at constant v or slows down. However, if $T < 0$, then $dp_i > 0$. The body must fly apart. Therefore, negative temperatures are not possible for states in equilibrium.

In practice, however, the interaction of a spin lattice with its surroundings is very weak, so the spins do not equilibrate with momentum over short time scales. The spins do interact with externally imposed magnetic fields. The energy of a spin state in a field is

$$E = E_o + E_{mag} = E_o - \boldsymbol{\mu} \cdot \mathbf{B} \tag{20-49}$$

Assume that the energy per nucleus, $E_n = -\boldsymbol{\mu} \cdot \mathbf{B}$, is small compared to kT. Expanding the partition function for nuclear spins in a magnetic field to second order gives

$$Z_{mag} = \sum_n e^{-E_n/kT} \approx \sum_n \left[1 - \frac{E_n}{kT} + \frac{1}{2}\left(\frac{E_n}{kT}\right)^2\right]$$

For n identical nuclei the sum may be replaced by

$$Z_{mag} = g^N \left[1 - \frac{1}{kT}\overline{E}_n + \frac{1}{2}\left(\frac{1}{kT}\right)^2 \overline{E^2}_n\right] \tag{20-50}$$

where g is the number of possible orientations of a single magnetic moment relative to the lattice or the applied field, so g^N is the number of orientations of N moments or the statistical weight. Taking the logarithm gives the free energy.

$$F_{mag} = -kT \ln Z_{mag} = -kT\left\{\ln g^N + \ln\left[1 - \frac{\overline{E}_n}{kT} + \frac{1}{2}\frac{\overline{E^2}_n}{(kT)^2}\right]\right\}$$

Let

$$x = -\left[\frac{\overline{E}_n}{kT} - \frac{1}{2}\frac{\overline{E^2}_n}{(kT)^2}\right]$$

Then, to second order,

$$\ln(1 + x) = x - \frac{x^2}{2} = -\frac{\overline{E}_n}{kT} + \frac{1}{2}\frac{\overline{E^2}_n - \overline{E}_n^2}{(kT)^2}$$

but $\overline{y^2} - \overline{y}^2 = \overline{(y - \overline{y})^2}$, so

$$F_{mag} = -NkT \ln g + \overline{E}_n - \frac{1}{2}\frac{1}{kT}\overline{(E_n - \overline{E}_n)^2}$$

$$\begin{aligned} S_{\text{mag}} &= Nk \ln g - \frac{1}{2kT^2}\overline{(E_n - \overline{E}_n)^2} \\ E_{\text{mag}} &= \overline{E}_n - \frac{1}{kT}\overline{(E_n - \overline{E}_n)^2} \\ C_{\text{mag}} &= \frac{1}{kT^2}\overline{(E_n - \overline{E}_n)^2} \end{aligned} \tag{20-51}$$

As $T \to 0$, E_n/kT increases rapidly for $E_n > 0$, so only the lowest quantum energy state is occupied. As T increases, E_n/kT decreases and more states are occupied. The magnetic moments become more random until, at $T = \infty$, the distribution would be completely random, or uniform with respect to angle.

Consider now an experiment in which the nuclei are (nearly) aligned at $T \gtrsim 0$. The applied external field is then quickly reversed, so that the nuclei are aligned *against* the field; E_n has changed sign (positive to negative) for each nucleus. The resultant state is not an equilibrium state for any value of the lattice temperature.

However, a change in sign of E is mathematically equivalent to a change in sign of T. If a small, negative temperature is assigned to the spin system, the antialigned state is predicted. Then, with time, the spins decay from this high-energy state toward a random orientation (T decreases toward $-\infty$, statistically equivalent to $+\infty$) as the nuclei give up their excess energy to the lattice. The spin system continues to decay from $+\infty$ toward normal alignment and a normal positive, small temperature.

Negative temperatures provide a meaningful description only for semiisolated nontranslational degrees of freedom, such as nuclear spins in a magnetic field.

EXAMPLES 20.5

There are several thermodynamic and common-sense restrictions on temperature and thermal energy transfer. If T can be negative, is this:

A. Consistent with the general equations developed earlier for entropy change and equilibrium?

B. Consistent with reasonable physical expectations for energy transfer?

C. What requirements are placed on heat capacity, C? □

20.7 Rates of Chemical Reactions and Isotopic Effects

In Chapter 14 it was shown that chemical equilibrium is well described by classical thermodynamics. The equilibrium constant is independent of concentrations or pressures at any temperature, but varies with temperature. At a given temperature, the concentrations or pressures, or more generally the activities, may vary widely, subject only to a product function,

$$\mathcal{Q}(a_i) = \frac{\prod_{\text{products}} a_i}{\prod_{\text{reactants}} a_i}$$

remaining constant at equilibrium. The equilibrium point, however, gives no information about the rate at which equilibrium will be approached. Statistical mechanics provides a bridge, connecting rate theory to equilibrium properties, through the work of Arrhenius, Eyring, and many others.

The Arrhenius Model

A very early model suggested that two atoms or molecules approaching each other could react only if they had sufficient energy. The number having an energy E^*, necessary to react, can be approximated by $e^{-E^*/kT}$. (This approximates the number under the tail of the Maxwell–Boltzmann curve, above E^*, by the height of the curve at E^*, or by such a height multiplied by a constant that is then incorporated into the effective value of E^*.) The energy, E^*, was called the *activation energy.*

An early refinement of the activation energy model recognized that the relative orientation of the molecules, as they approached, could be important. For example, if a small molecule must react at a certain site on a larger molecule, collisions with other points are likely to be ineffective. This geometric dependence was incorporated by multiplying the Maxwell–Boltzmann term by a number smaller than one, called the *steric factor.* In practice, both the activation energy and the steric factor were found experimentally by measuring rates at different temperatures and finding the best fit of the experimental results to Arrhenius' equation,

$$k_f = \mathscr{A} e^{-E^*/kT} \tag{20-52}$$

where k is the (forward) *rate constant* (see below) and $\mathscr{A}$ is the steric factor.

The activation energy, E^*, is interpreted as a barrier that must be crossed by the system before it falls into the lower-energy state representing the product(s) of the reaction. Although the interpretation is almost certainly correct in its broad interpretation, it is of little help in determining what the constants should be.

At equilibrium, reaction does not stop, but the forward rate and reverse rate become equal. If the forward rate *could* be expressed as

$$R_f = k_f a_A^a a_B^b$$

and the reverse rate as

$$R_r = k_r a_C^c a_D^d$$

then it would follow that when $R_f = R_r$,

$$K_{\text{eq}} = \frac{k_f}{k_r} = \left[\frac{a_C^c a_D^d}{a_A^a a_B^b}\right]_{\text{at equil}}$$

However, most reaction rates are *not* proportional to the products of activities of all the reactants, so this is not a useful approach to seeking the rate.[6]

[6]There is a very limited sense in which the last derivation can be said to be valid, although of dubious utility. There are, in general, different paths by which any given reaction can proceed. Usually, one path will be much more rapid than any others and will be the only path of any significance. However, the *principle of microscopic reversibility* states that *at equilibrium,* the forward rate along *any* path must equal the reverse rate along the same path. (Otherwise, adding and removing catalysts, which are effective only for certain paths, would permit shifting equilibria one way, then the other, with no significant expenditure of energy or increase in entropy, but with the possibility of extracting energy from the reaction as it oscillates.) It can be argued that the path in which a molecules of A combine with b molecules of B to form the products is one of the possible paths (even though it may be exceedingly improbable compared to faster paths). Application of the principle of microscopic reversibility then leads to the last equation for K_{eq}. Of course, the rate constants in that equation are of no significance in studying the actual rates of reaction.

ANSWERS 20.5

A. The equation,

$$dS = \left(\frac{1}{T_1} - \frac{1}{T_2}\right) dE_1 > 0$$

predicts that if $T_1 < 0$, then $dE_1 < 0$, so the thermal energy flow would be from the system at negative temperature to the surroundings. Negative temperatures may be considered to be *hotter* than normal temperatures rather than cooler.
B. The nuclei are in a high-energy state, so one should expect energy to flow from the system to the surroundings, in agreement with the formal prediction.
C. $C = Q/\Delta T$, $Q < 0$, and $\Delta T < 0$ as the temperature decreases to $-\infty$ and then decreases from $+\infty$ toward 0. Therefore, C is a positive quantity.

Absolute Rate Theory

Eyring proposed that reaction rates could be treated as a pseudoequilibrium problem. His *absolute rate theory* leads to equations of a mathematical form that can readily be adapted as empirical descriptions of rates, whether or not the detailed model can be determined or verified for specific reactions.

The rate equation is not usually directly predictable from the chemical reaction, because the reaction proceeds through a series of steps, which occur at different speeds. Only the slowest step, called the *rate-determining step,* is important in determining the overall reaction rate.

Most chemical reactions are found to be first-order or second-order processes, meaning that the rate is proportional to one concentration or proportional to a product of two concentrations (or a concentration squared). If the slowest step depends on the concentration of one reactant (perhaps through a sequence of other steps governed by equilibrium constants), the reaction will be first order. If the rate-determining step depends on a product of two concentrations of reactants, the reaction is second order. A few reactions are of zero order, independent of concentrations, and a few are third-order.

Absolute rate theory postulates that chemical reactions typically occur by means of a path that includes formation of an *activated complex*. The activated complex is an intermediate that is either unstable or slightly metastable. The complex is in equilibrium with the reactants or with substances produced, more rapidly, from the reactants. In most cases it can be assumed the complex has a 50% chance of returning to the original state from which it came and a 50% chance of forming the products of the reaction. The rate of reaction should then be proportional to the amount of the activated complex present, which is controlled by an equation of the form of the chemical equilibrium equations. In particular,

$$k_f = \mathcal{A} e^{-\Delta G^*/RT} \tag{20-53}$$

The free energy of the activated complex, ΔG^*, is related to the enthalpy and entropy of the complex,

$$\Delta G^* = \Delta H^* - T\,\Delta S^* \tag{20-54}$$

Often, the entropy can be calculated or estimated from the geometric properties of the assumed activated complex, giving the reaction rate in terms of an enthalpy or energy, not unlike the energy of activation of Arrhenius' theory.

The advantage of this more explicit theory is that it provides a means of treating the inherently nonequilibrium reaction process with equilibrium statistical mechanics. The extent of agreement, then, between predictions and measured rates is a test of whether the partition function is correct, and hence whether the molecular model from which the partition function was constructed is realistic for the reaction.

More advanced analyses of some reactions treat the formation and decomposition of the activated complex as a single dynamic process, in which the memory of the initial momenta of the molecules is not forgotten. The reaction is then traced through the phase space of the reacting species.

Isotope Effects

It is a general rule that isotopic substitution does not change the electronic structure of a molecule. Therefore one would not expect a change in the chemical equilibrium constant or in rate processes. Two types of exceptions are well known, however, both arising from statistical mechanics principles already considered.

First, because isotopic substitution (e.g., deuterium for hydrogen or ^{235}U for ^{238}U) changes the mass of the molecule, transport processes are affected. In idealized cases, $\overline{v} \sim \sqrt{1/M}$, so the molecule containing the heavier isotope moves more slowly. Two examples of the isotopic mass effect on transport rates are of particular historical significance.

When water is electrolyzed, H^+ ions move to the cathode, where they combine to form H_2. The $^1H^+$ ion moves faster than the deuterium ion, $^2H^+ \equiv D^+$, so the residual water becomes enriched in deuterium. This method of separating deuterium from seawater was exploited during World War II, in occupied Norway (before the facility was destroyed) and in the United States, and is now a commercial process.

The percentage difference in mass between ^{235}U and ^{238}U is very small, and smaller yet for the gaseous compound UF_6, but it is predictable that if UF_6 is allowed to diffuse through a membrane, the gas passing through will be slightly enriched in ^{235}U. The gaseous diffusion plant at Oak Ridge, Tennessee, performed multiple successive diffusion steps as the means of preparing enriched uranium.

Natural processes may similarly produce some fractionation. As water evaporates from the ocean, precipitates as rain, and reevaporates from surfaces, rivers, and lakes, the lighter isotopes of hydrogen and oxygen tend to come off first. The isotopic composition of atmospheric and lake water samples is therefore not identical with ocean samples.

The second influence on rate processes is more subtle. Because of differences in mass, vibrational frequencies are slightly changed when isotopic composition is changed. In the extreme case of substituting deuterium for hydrogen, the ground-state vibrational energy may be lowered by as much as 30% (by $1/\sqrt{2}$), although for most vibrations the effect is smaller.

Changes in the ground state and in the spacing of vibrational levels affect the partition function and therefore change rates of reaction. In terms of rate theory, it is the separation between ground state and excited state, $E^* - E_o$, which is sensitive to E_o, that determines rates. Even though the effects are typically small, they are often detectable. When multiple sequential reaction processes are involved, the isotopic effects may lead to measurable changes in isotopic composition.

20.8 Elasticity of Polymeric Elastomers

In Chapter 9 we saw that stretching or compressing a spring may be reversible, isothermal, adiabatic, and isentropic. Reversible compression or expansion of an ideal gas may be isothermal *or* adiabatic; only the adiabatic process is isentropic.

For any coordinate, x_i, we define a generalized force, X_i, conjugate to x_i, by

$$-\frac{X_i}{\vartheta} = \frac{\partial \sigma}{\partial x_i}$$

For an ideal gas, $E \neq E(V)$ and $\sigma(V) = \ln V^N = N \ln V$, so

$$-\frac{X_i}{\vartheta} = \left(\frac{\partial \sigma}{\partial V}\right)_{E,N} = \frac{N}{V} = \frac{P}{\vartheta}$$

Hence (negative) pressure is the generalized force, conjugate to the volume. Pressure arises from the time rate of momentum change at the walls.

Polymers such as rubber share some attributes with crystalline solids and other attributes with an ideal gas. The chains may be modeled as perfectly flexible joints linking segments of fixed length. Hence there is no internal potential energy. However, the stretched elastomer resembles a crystalline solid. Because the unstretched polymer has random orientations of the individual segments, it is noncrystalline in character, with higher entropy than the stretched material. The entropy of the elastomer decreases as the polymer is stretched isothermally.

As a polymer is stretched adiabatically (and reversibly), work is done on the polymer. There is no potential energy change, so the energy added to the polymer must appear as thermal energy; the temperature increases. Thermal entropy is thus increased as the spatial entropy decreases; the overall adiabatic process is isentropic.

Like the coiled spring, the elastomer may be released, permitting it to contract adiabatically (and irreversibly) without doing work on the surroundings. The process is then isoergic but not isentropic. Entropy increases as the polymer goes from the ordered linear, or "crystalline," form back to random orientation with shortening of the chain. There is no change in thermal energy, so temperature is constant.

If, however, the elastomer is allowed to contract reversibly and adiabatically, against an external restraining force, it does work on the surroundings, giving energy to the surroundings. There is still an increase in spatial entropy as the chains become random, but now there is a decrease in thermal energy and therefore in thermal entropy. The total change in entropy is zero for the adiabatic, reversible contraction.

The polymer changes shape, but there is little change in volume, so negligible work is done against the atmosphere. Work is done by, or against, the force associated with the coordinate describing the randomization and ordering of the polymer chains. The overall length of the chain, L, serves as a measure of this randomization process.

Consider a one-dimensional polymer chain consisting of N links, each of length μ. Then each link adds $\pm\mu$ to the length along the axis. If $n = N_+ - N_-$ is the excess of links forward along the axis to links backward along the axis, the overall length is $L = n\mu$. For such a random-walk problem, the entropy is

$$\sigma = \ln \Delta\Gamma = \ln W(n) = \ln \frac{N!}{[\frac{1}{2}(N+n)]!\,[\frac{1}{2}(N-n)]!} \tag{20-55}$$

We have seen that this reduces, with Stirling's formula in the form $\ln x! = x \ln x - x + \ln \sqrt{2\pi x}$, to

$$\sigma = -\frac{n^2}{2N} + \text{constant} = -\frac{L^2}{2N\mu^2} + \text{constant} \tag{20-56}$$

The generalized force, K, conjugate to L is, therefore,

$$-\frac{X_i}{\vartheta} \equiv -\frac{K}{\vartheta} = \frac{\partial \sigma}{\partial L} = -\frac{L}{N\mu^2}$$

Relating L to $L_o \equiv N\mu$ gives an equation of state for the polymer in terms of the generalized force.

$$\frac{L}{L_o} = \frac{n\mu}{N\mu} = \frac{n}{N} = \frac{L/\mu}{N} = \frac{K\mu}{\vartheta}$$

or

$$L\vartheta = K\mu L_o = K\mu^2 N \tag{20-57}$$

From this equation of state we see that $K > 0$. This confirms that work must be done on the system to stretch the system.

$$w = K\,dL > 0 \quad \text{if} \quad dL > 0 \tag{20-58}$$

Note also that $L\vartheta = K\mu^2 N =$ constant, so the length is inversely proportional to the temperature,

$$L \sim \frac{1}{\vartheta} \tag{20-59}$$

The equation for $\sigma(n)$ also confirms that entropy is greatest when the chain is contracted ($n \to 0$). At low temperatures the chain can "relax" to $L \to L_o$, but at higher temperatures the thermal motion of the links causes the chain to contract. The thermal coefficient of expansion of rubber is negative.

If the polymer were to be viewed as a gas, the "pressure" is $-K < 0$. Hence the polymer tends to spontaneously contract.

$$d\sigma = \frac{\partial \sigma}{\partial L}\,dL = -\frac{K}{\vartheta}\,dL > 0 \tag{20-60}$$

20.9 Photons, Electrons, and Condensed Bosons

A gas consisting of elementary particles (i.e., considered structureless for present purposes) of spin s will be described by Bose–Einstein or Fermi–Dirac statistics, depending on the value of s, although for sufficiently dilute systems, such as atomic or molecular gases, Boltzmann statistics is an adequate approximation.

The energy of the particle is kinetic energy of translation, which is at least quasi-classical.

$$\varepsilon = \frac{1}{2m}(p_x^2 + p_y^2 + p_z^2) \tag{20-61}$$

The occupation numbers are

$$\overline{n}_i = \frac{1}{e^{(\varepsilon_i - \mu)/kT} \pm 1} \tag{20-62}$$

with the upper (+) sign for Fermi–Dirac particles and the lower (−) sign for Bose–Einstein particles (or zero, at least compared to the exponential term, for Boltzmann statistics). The number of particles, dN, in an element of phase space, $dp_x\, dp_y\, dp_z\, dV$, is equal to $\overline{n}_k g\, d\tau$, with

$$d\tau = \frac{dp_x\, dp_y\, dp_z\, dV}{h^3}$$

and $g = 2s + 1$. The energy is $\varepsilon = p^2/2m$. Integrating over volume and over the angular coordinates of spherical momentum space gives

$$dN_p = \frac{4\pi g V p^2 dp}{h^3[e^{(\varepsilon-\mu)/kT} \pm 1]}$$

and

$$dN_\varepsilon = \frac{4\sqrt{2}\,\pi g V m^{3/2}\sqrt{\varepsilon}\, d\varepsilon}{h^3[e^{(\varepsilon-\mu)/kT} \pm 1]} \tag{20-63}$$

These are the analogous equations to the Boltzmann distribution for the quantum particles with spin. Integrating dN_p gives the number of particles in the gas, which implicitly determines μ, the chemical potential. Alternatively, integrating $\varepsilon\, dN_\varepsilon$ gives the total energy of the gas,

$$E = \int_0^\infty \varepsilon\, dN_\varepsilon = \frac{4\sqrt{2}\,\pi g V m^{3/2}}{h^3}\int_0^\infty \frac{\varepsilon^{3/2} d\varepsilon}{e^{(\varepsilon-\mu)/kT} \pm 1} \tag{20-64}$$

It was also shown in Section 20.3 that

$$-\Omega_k = \pm kT \ln[1 \pm e^{(\mu-\varepsilon_k)/kT}]$$

with the + sign for fermions and the − sign for bosons. Integrating over volume and the spherical angles of momentum space, as above, gives

$$-\Omega = \pm\frac{4\sqrt{2}\,\pi V g m^{3/2} kT}{h^3}\int_0^\infty \sqrt{\varepsilon}\ln[1 \pm e^{(\mu-\varepsilon)/kT}]\, d\varepsilon$$

which can be integrated by parts to yield

$$-\Omega = PV = \frac{2}{3}\frac{4\sqrt{2}\pi g V m^{3/2}}{h^3}\int_0^\infty \frac{\varepsilon^{3/2}\, d\varepsilon}{e^{(\varepsilon-\mu)/kT} \pm 1} \tag{20-65}$$

Comparison with the energy shows that

$$PV = \frac{2}{3}E \tag{20-66}$$

is a general, exact result for a gas of particles, valid for quantum statistics as well as in the limit of Boltzmann statistics, provided that the energy is given by the classical expression. Additional equations can be derived that are very similar to conventional equations for ideal gases. For example, if $\Delta S = 0$, then $VT^{3/2} =$ constant, $PV^{5/2} =$ constant, and $T^{5/2}/P =$ constant,[7] although $C_p/C_v \neq \frac{5}{3}$ and $C_p - C_v \neq R$. The first-order correction to ideal gas behavior is given by

$$PV = NkT\left[1 \pm \frac{\pi^{3/2}}{2g}\frac{N\hbar^3}{V(mkT)^{3/2}}\right]$$

Therefore the kinetic energy, E, is

$$E = \frac{3}{2}NkT\left[1 \pm \frac{\pi^{3/2}}{2g}\frac{N\hbar^3}{V(mkT)^{3/2}}\right] \approx \frac{3}{2}NkT \tag{20-67}$$

If the energy is given by the relativistic expression

$$\varepsilon = cp \tag{20-68}$$

a similar analysis shows that

$$PV = \frac{1}{3}E \tag{20-69}$$

This result is valid for photons as well as material particles.

Photon Gas

Photons have spin 1 and are therefore bosons, consistent with the observation that photons may be created and destroyed at will. (The number of fermions cannot be changed if one includes antiparticles as a negative number of particles.) An electromagnetic field may be treated as a photon gas. Let σ represent the wavenumber, ν/c. Then the number of modes in the volume element dV is

$$4\pi\sigma^2 d\sigma\, dV \times 2$$

including the factor $g = 2$ for the two polarizations. The number of photons of angular frequency $\omega = 2\pi\sigma c$ is

$$dN_\omega = \frac{V}{\pi^2}\frac{\omega^2\, d\omega}{c^3}\overline{n}_k \tag{20-70}$$

and the increment of energy corresponding to $d\omega$ is

$$dE_\omega = \frac{V\hbar}{\pi^2 c^3}\frac{\omega^3 d\omega}{e^{\hbar\omega/kT} - 1} \tag{20-71}$$

[7]See Landau and Lifshitz, *Statistical Physics* (ref. 13, p. 158).

This is Planck's equation for the energy of blackbody radiation. For low frequencies it gives the Rayleigh–Jeans equation; for high frequencies it gives Wien's equation.

Integrating[8] gives

$$E = \frac{\pi^2 V(kT)^4}{15\hbar^3 c^3} = \frac{8\pi^5 V(kT)^4}{15(hc)^3} \tag{20-72}$$

This is the Stefan–Boltzmann equation,

$$\frac{E}{V} = \frac{4}{c}\sigma_{SB} T^4 \tag{20-72a}$$

with the constant

$$\sigma_{SB} = \frac{2\pi^5 k^4}{15h^3c^2} = 5.67 \times 10^{-8}\mathrm{W/m^2 \cdot K^4}$$

Note that this is energy per volume. Only $\frac{1}{4}$ of this will cross a plane (from one side) per unit of time, so this equation differs from the radiation emitted or absorbed by a plane surface by the factor $c/4$ (cf. effusion equation (7-15), Section 7.3).

Electron Gas

At ordinary temperatures, electrons in solids are condensed into the lowest-available energy states, with very little spillover into higher states. The electrons are therefore said to form a degenerate gas. If we apply analysis techniques as above, the number of electrons may be expressed in terms of the momentum, p_0, of the highest occupied energy level.

$$N = \frac{1}{h^3}\int\int_0^{p_0} g \cdot 4\pi p^2 \, dp \, dV = \frac{gVp_0^3}{6\pi^2\hbar^3} \tag{20-73}$$

Solving for p_0 and for ε_0 gives

$$p_0 = \left(\frac{6\pi^2 N}{gV}\right)^{1/3} \hbar$$

and

$$\varepsilon_0 = \frac{p_0^2}{2m} = \left(\frac{6\pi^2 N}{gV}\right)^{2/3} \frac{\hbar^2}{2m} \tag{20-74}$$

where $g = 2s + 1 = 2$. The mean occupation numbers are

$$\overline{n}_i = \frac{1}{e^{(\varepsilon_i - \mu)/kT} + 1} \tag{20-75}$$

[8]The integral is of the form

$$\int_0^\infty \frac{z^{2n-1}\,dz}{e^z - 1} = \frac{(2\pi)^{2n} B_n}{4n}$$

with $n = 2$ and therefore $B_n = \frac{1}{30}$. (B_n are the Bernoulli numbers.)

As the temperature approaches zero, $\overline{n}_i$ approaches zero for all $\varepsilon_i > \mu$ and $\overline{n}_i$ approaches 1 for all $\varepsilon_i < \mu$. Therefore, μ must be the upper level of energy; $\mu = \varepsilon_0$ is the highest level filled at $T = 0$. This is called the *Fermi level.*

The Fermi level is an energy far above kT at room temperatures. Therefore, one cannot assume electrons to behave classically, and in particular, the average energy of electrons in a metal is not even approximately equal to $\frac{3}{2}kT$. A few values are given in Table 20.1.

At temperatures above zero kelvin but well below $T_o \equiv \varepsilon_o/k$, the Fermi level may be taken as the level with probability of occupancy of $\frac{1}{2}$. The spread of partially filled electron states will be approximately equal to kT. Values of the Fermi level calculated from the equation

$$T_0 = \frac{\varepsilon_0}{k} = \left(\frac{6\pi^2 N}{gV}\right)^{2/3} \frac{\hbar^2}{2mk} \tag{20-76}$$

which treats the electrons as an ideal gas of fermions, are in approximate agreement with experiment. Unlike molecular gases, the electrons in a solid more nearly approach ideal gas behavior at high densities.

Electrons in any but the highest-energy states are trapped in their respective states. Each state contains one electron and can hold no more. To gain energy, therefore, an electron must escape to an empty state which will be near or above the Fermi level. The heat capacity of a degenerate electron gas is therefore quite low and temperature dependent.

$$\mathrm{c} = \frac{Nk\pi^2 T}{2T_0} \sim T \tag{20-77}$$

EXAMPLE 20.6

One definition of the Fermi level is that it is the energy value for which the average occupation number, $\overline{n}_i$, is equal to $\frac{1}{2}$. Find the energy, ε_F, from this definition. □

Properties of Condensed Bosons

Electrons are fermions with the restriction that $\overline{n}_i = 0$ or 1. They cannot all go into the low-energy states. Like filling general-admission seats, the first electrons get the preferential states and later electrons must take less desirable, or higher-energy, states. This simple description in terms of one-electron wave functions is inadequate in some respects. The electrons play "musical chairs," exchanging

Table 20.1 Fermi Level Energies

Metal	ε_0 **(eV)**	$T_0 = \varepsilon_0/k$ **(K)**
Na	3.2	3.8×10^4
K	2.1	2.5×10^4
Cu	7.0	8.2×10^4
Ag	5.5	6.4×10^4
Ba	3.6	4.2×10^4
Al	11.7	13.6×10^4

Source: A. J. Dekker, *Solid State Physics* (Englewood Cliffs, N.J.: Prentice Hall, Inc., 1957; and N. W. Ashcroft and N. D. Mermin, *Solid State Physics* (New York: Holt, Rinehart, and Winston, 1976).

ANSWER 20.6

Set $\overline{n}_k = \frac{1}{2} = 1/[e^{(\varepsilon_F-\mu)/kT}+1]$ and cross multiply to obtain $e^{(\varepsilon_F-\mu)/kT}+1 = 2$. Therefore, $e^{(\varepsilon_F-\mu)/kT} = 1$, $\varepsilon_F = \mu$. The Fermi level is equal to the chemical potential, which is the effective free energy of the electrons; $\mu = (\partial G/\partial N)_{T,P}$.

---

places. However, this exchange does not affect the numbers or the energies of the states occupied at a given instant.

In contrast, bosons seek the lowest energy level, regardless of the presence of other particles in that level. The effect is particularly significant at very low temperatures. Then energy differences are more important than entropy, so a "gas" of bosons may be described as condensing into the lowest-energy state.

A model for boson condensation was first proposed by Einstein. It was shown above that the number of bosons of given energy can be represented as

$$dN_\varepsilon = \frac{4\sqrt{2}\,\pi g V m^{3/2}\sqrt{\varepsilon}\,d\varepsilon}{h^3[e^{(\varepsilon-\mu)/kT} \pm 1]} \tag{20-63}$$

This cannot be quite right, however, for it does not provide for any particles in the lowest energy level, $\varepsilon = 0$. A boson gas at very low temperatures should have *all* particles in the zero energy level.

A suitable resolution of the problem is to divide the energy levels into the zero level and all others, populated, respectively, by N_0 and N_ε particles. The number in energy levels above $\varepsilon = 0$ is given by the expression for dN_ε above.

Then if the total number of particles is N,

$$N = N_0 + N_\varepsilon \tag{20-78}$$

The average occupation numbers are given by

$$\overline{n}_i = \frac{1}{e^{(\varepsilon_i-\mu)/kT} - 1}$$

with $\mu \leq 0$. For the ground state, the degeneracy is one and $\varepsilon = 0$, so

$$N_0 = \overline{n}_0 = \frac{1}{e^{-\mu/kT} - 1}$$

Rearranging this gives

$$e^{-\mu/kT} = \frac{1}{N_0} + 1 \tag{20-79}$$

When $T \approx 0$, N_0 is large, so we conclude that $\mu = 0$ at $T = 0$ and $\mu \approx 0$ for $T \approx 0$.

The population of the levels N_ε above $\varepsilon = 0$ increases rapidly with ε up to $N_\varepsilon = N$, as shown in Fig. 20.4. Subtracting N_ε from N gives N_0, which is also plotted in Fig. 20.4.

Consider now the effect of lowering the temperature of the boson gas. Above some cutoff temperature T_0, there are no particles (or a negligible number) in the ground state. Below T_0, however, the number in the ground state increases sharply. If there is any energy gap between the ground state and the first upper level, particles once in the ground state remain there unless they receive a significant jolt.

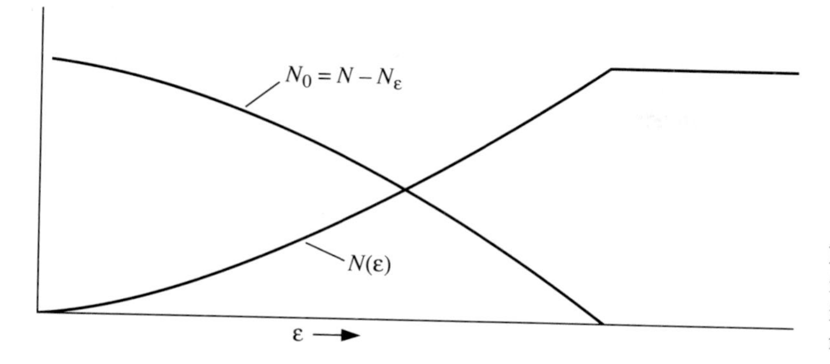

FIGURE 20.4 With increasing temperature, N_ε increases, so $N_o = N - N_\varepsilon$ must decrease.

The condensation does not occur for a photon gas, because the photons are in equilibrium with material oscillators at temperatures well above zero. Similarly, gases of bosons such as ^4He and D_2 that are in thermal equilibrium with warm surroundings are prevented from condensing to the low-energy state.

However, if bosons, such as ^4He, are cooled to very low temperatures, the quantum effects of condensation become quite important. Below the transition temperature of 2.17 K, called the *lambda point* (because a curve of heat capacity versus temperature looks much like a backward λ; Fig. 20.5), normal liquid ^4He (He I) begins to condense into a new liquid phase, called He II. Between 2.17 K and a temperature below 1 K, both He I and He II coexist, with the fraction[9] in the superfluid state increasing as the temperature is lowered.

The interatomic forces are so weak that the low-temperature phase, He II, is not a solid; the zero-point vibration amplitude is greater than the width of the potential well formed by adjacent atoms. The atoms are therefore free to slip past each other. The shallow well of intermolecular attraction is sufficient, however, to produce quantized energy levels. He II is simply He atoms that have condensed into the ground state. The amount of He II dissolved in the He I increases as the temperature decreases below $T_o = 2.17$ K.

The He II atoms share a common momentum state and therefore tend to move collectively. Lacking the energy to change states, they cannot gain or lose momentum and therefore move without viscosity but show infinite thermal conductivity. The liquid is said to be a *superfluid*.

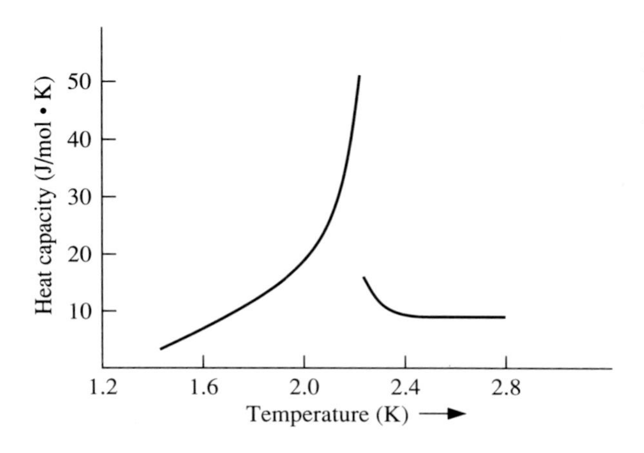

FIGURE 20.5 Heat capacity versus temperature for liquid helium as it undergoes the second-order transition between helium I and helium II near 2.17 K.

[9]It is convenient to think of the liquid as a mixture of two fluids, but this cannot be correct. The fluid state is a superposition of the two states. See L. D. Landau and E. M. Lifshitz (ref. 13, pp. 198–206).

The magnitude of the gap between the low-energy state of He and the upper states (which may be approximated as a continuum) can be judged from the observation that when the speed of fluid flow is increased to about 7 m/s, atoms begin to make transitions out of the superfluid state because of collisions with the walls.

A related but clearly different phenomenon is observed with electrons in solids. As electrons pass through the lattice, they encounter nuclei that are out of their equilibrium lattice positions. This causes scattering of the electrons, both elastically and inelastically. The inelastic collisions convert electron energy acquired from the superimposed electrical field into thermal energy of nuclear motion. Because they are fermions, the electrons cannot condense to a single low-energy state.

In certain solids, however, the interactions of the electrons with the nuclei are strong and such that two electrons, of different spin state, can share a spatial quantum state (all quantum numbers except the spin direction). Two fermions produce a boson, so the two electrons, not necessarily near each other but each interacting with the lattice, appear in many respects as if they were coupled together as an electron pair obeying boson statistics.

The only significant forces are the coulombic attractions of nuclei and electrons and coulombic repulsions of electrons. Nevertheless, the electrons appear to attract each other. The effect is similar in principle, although differing in detail, to the apparent attraction of two protons in the presence of one or more electrons to form H_2^+ or H_2. It is the electron–nuclear attractions that lead to the apparent attraction of particles of like charges.

When the electron pairs act in concert, they have the properties of bosons, which can then condense into the lowest energy state. They cannot be scattered out of this lowest state without significant energy input. The electron pairs therefore exhibit a form of superfluidity, which in this case is called *superconductivity.*

A theory of superconductivity was developed by Bardeen, Cooper, and Schrieffer, which seemed adequate until the recent discovery of high-temperature superconductors. It appears probable that although some major adjustments to the theory of superconductivity will be forthcoming, the basic concept will survive these changes.

SUMMARY

Equipartition principle: $\langle q_i\, \partial\varepsilon/\partial q_i \rangle = kT$ if $\partial\varepsilon/\partial q_i \neq 0$ and if $q_i = 0$ *or* $\varepsilon = \infty$ at the limits of variation of the coordinate (q_i = position or momentum).

The Born–Oppenheimer approximation divides the energy into additive terms: $E = E_{\text{electronic}} + E_{\text{vibration}} + \cdots$.
The partition function then factors:
$Z = Z_{\text{electronic}} \times Z_{\text{vibrational}} \times \cdots$.
The free energy is also additive:
$F = -NkT \ln Z = F_{\text{electronic}} + F_{\text{vibrational}} + \cdots$.

A harmonic oscillator, with $\eta = \hbar\omega_o/kT$, gives an energy contribution $E_{\text{vib}}/nRT = \eta/(e^{\eta} - 1)$ and free energy $F_{\text{vib}}/nRT = G_{\text{vib}}/nRt = \ln(1 - e^{-\eta})$.

When particles are not distinguishable (e.g., equivalent particles in a gas), integration over volume and summing over particles introduces an error of double counting, which may be removed by dividing the partition function by $N!$ if the density is low. Otherwise, one must take into consideration the spins of the particles. Fermions have wave functions antisymmetric to exchange; boson wave functions are symmetric to exchange. The corresponding occupation numbers are

{Boltzmann} $$n_k = \frac{g_k}{e^{\alpha+\beta\varepsilon_k}}$$

{Fermi–Dirac} $$n_k = \frac{g_k}{e^{\alpha+\beta\varepsilon_k}+1}$$

{Bose–Einstein} $$n_k = \frac{g_k}{e^{\alpha+\beta\varepsilon_k}-1}$$

Then

{Boltzmann statistics} $$\bar{n}_k = \frac{1}{e^{(\varepsilon_k-\mu)/kT}} \qquad (\mu \ll 0)$$

$$\Omega = -kT \ln \sum_{n_k}^{\infty} e^{(\mu-\varepsilon_k)n_k/kT}$$

{Fermi–Dirac statistics} $$\bar{n}_k = \frac{1}{e^{(\varepsilon_k-\mu)/kT}+1} \qquad (\mu \gtrless 0)$$

$$\Omega = -kT \sum_k \ln[1 + e^{(\mu-\varepsilon_k)/kT}]$$

{Bose–Einstein statistics} $$\bar{n}_k = \frac{1}{e^{(\varepsilon_k-\mu)/kT}-1} \qquad (\mu < 0)$$

$$\Omega = kT \sum_k \ln[1 - e^{(\mu-\varepsilon_k)/kT}]$$

For an ideal gas, $\mu = -kT \sum_j \ln z_j$, where z_j is the partition function of an individual molecule,

$$z_j = \frac{(2\pi mkT)^{3/2}V}{h^3}$$

Then, for the gas,

$$Z_{\text{transl}} = \left[e\frac{V}{N}\left(\frac{2\pi mkT}{h^2}\right)^{3/2}\right]^N$$

$$S_{\text{transl}} = -\left(\frac{\partial F}{\partial T}\right)_V = Nk \ln\left[\frac{V}{N}\left(\frac{2\pi mkT}{h^2}\right)^{3/2}\right] + \frac{5}{2}Nk$$

$$E_{\text{transl}} = \frac{3}{2}NkT$$

At low temperatures, the rotational contribution to the heat capacity of a heteronuclear diatomic molecule is

$$C_{\text{rot}} = 3Nk\left(\frac{2hcB}{kT}\right)^2 e^{-2hcB/kT}$$

which goes to zero as $T \to 0$. Homonuclear diatomics have odd or even rotational levels missing, depending on whether the nuclear spin states are symmetric or antisymmetric and the spin is integral or half integral.

Einstein considered a solid as a collection of $3N$ independent harmonic oscillators of frequency ν_o and showed that the energy, in terms of $\Theta \equiv h\nu_0/k \equiv \eta T$, would be

$$E = 3NkT\frac{\eta}{e^{\eta} - 1} = \frac{3Nk\Theta}{e^{\Theta/T} - 1}$$

and the heat capacity, $C_v = 3Nk\dfrac{(\Theta/T)^2 e^{\Theta/T}}{(e^{\Theta/T} - 1)^2}$, which goes to zero as $T \to 0$.

Debye allowed for coupling of the oscillators, giving a frequency range $\nu = 0 \to \hat{\nu}$. Then, for $\Theta_D = \dfrac{h\hat{\nu}}{k}(\approx 1.37\Theta_E)$,

$$E = \frac{V\pi^2(kT)^4}{10(\hbar\overline{v})^3} + N\varepsilon_I$$

and therefore, at low temperatures,

$$C_v = \frac{2\pi^2 kV}{5(\hbar\overline{v})^3}(kT)^3 = \frac{12\pi^4 Nk}{5}\left(\frac{T}{\Theta_D}\right)^3$$

These go to zero as $T \to 0$, with $C_v \sim T^3$, and give $C_v = 3Nk$ at high temperature.

Uncoupled magnetic moments (paramagnetism) align with a magnetic field, B, to give a moment

$$\overline{\mu} = \mathrm{J}g\mu_B\left(\coth y - \frac{1}{y}\right) \qquad y \equiv \mathrm{J}g\mu_B\frac{B}{kT}$$

or, for low fields,

$$\overline{\mu} = \mathrm{J}(\mathrm{J}+1)(g\mu_B)^2\frac{B}{3kT}$$

If spins are aligned in a strong magnetic field at a very low temperature, then the field is removed, the spins revert to the higher-energy random orientation, extracting energy from the lattice and lowering the sample temperature. This is called *adiabatic demagnetization*.

If the magnetic field is reversed on a low-temperature, aligned sample, the spin lattice is conveniently described as having a negative temperature, $T < 0$. It then loses energy to the lattice and decays through $T = -\infty$, equivalent to $T = +\infty$, to a positive, low temperature.

An important model for chemical reaction rates considers a reaction to proceed by formation of an unstable (or metastable) activated complex, according to the rules for chemical equilibrium. The activated complex then goes forward to the products or back to reactants. This model couples thermodynamics to reaction rate theory.

Isotopic substitution does not change chemical properties, but may influence rates through several mechanisms. Molecular speeds are mass dependent, so rates of effusion, diffusion, or transport in electrolytic cells vary with mass. Vibrational energy levels are determined by vibrational frequencies, which depend on masses. Although the shifts are small, isotopic substitution may significantly change the separation between vibrational energy levels and energy levels in other potential wells, influencing rates and probabilities of reaction.

A generalized force, X_i, conjugate to the variable x_i, is defined by $X_i = -\theta\, \partial\sigma/\partial x_i$. A one-dimensional polymer chain, of N links, each of length μ, and total length $L = n\mu$, has entropy $\sigma = -L^2/2N\mu^2 + \text{constant}$. This produces a generalized force, $X_i = K = \theta L/N\mu^2 > 0$. Work required to stretch the system is $w = K\, dL$, so there is a "pressure" $-K < 0$. The polymer tends to contract spontaneously: $d\sigma = (\partial\sigma/\partial L)\, dL = -(K/\theta)\, dL > 0$.

If the energy of particles (quantum or classical) is $\varepsilon = p^2/2m$, then $PV = \frac{2}{3}E$. If the energy is $\varepsilon = cp$, then $PV = \frac{1}{3}E$.

The number of modes of oscillation of an electromagnetic field, or the number of photons, is $N_\omega = \dfrac{V}{\pi^2}\dfrac{\omega^2\, d\omega}{c^3}\bar{n}_k$. If $\bar{n}_k = 1$ and $\varepsilon = kT$ for each mode, $\dfrac{dE}{d\omega} = \dfrac{VkT\omega^2}{\pi^2 c^3}$, which is the Rayleigh–Jeans formula. If $\overline{n}_k = \dfrac{1}{e^{\hbar\omega/kT} - 1}$ and $\varepsilon = \hbar\omega$ for each mode, then

$$\frac{dE}{d\omega} = \frac{V\hbar}{\pi^2 c^3}\frac{\omega^3}{e^{\hbar\omega/kT} - 1}$$

which is Planck's equation for blackbody radiation.

Integration over frequency gives the Stefan–Boltzmann equation,

$$\frac{E}{V} = \frac{\pi^2(kT)^4}{15\hbar^3 c^3} = \frac{8\pi^5(kT)^4}{15(hc)^3} = \frac{4}{c}\sigma_{\text{SB}}T^4$$

$$\sigma_{\text{SB}} = \frac{2\pi^5 k^4}{15h^3c^2} = 5.67 \times 10^{-8}\ \text{W/m}^2 \cdot K^4.$$

The highest occupied energy level of a degenerate electron gas is $\varepsilon_o = \left(\dfrac{6\pi^2 N}{gV}\right)^{2/3}\dfrac{\hbar^2}{2m} = \mu$, the Fermi level.

Heat capacity of electrons in a solid is low at room temperatures: $c = Nk\pi^2/2T_o$, which is of the order of $10^{-4}T\dfrac{\text{J}}{\text{mol} \cdot \text{K}}$.

At $T \approx 0$, all bosons should condense into the lowest energy state. The number in energy levels above $\varepsilon = 0$ depends on temperature and is given by

$$dN_\varepsilon = \frac{4\sqrt{2}\pi g V m^{3/2}\sqrt{\varepsilon}\, d\varepsilon}{h^3\left[e^{(\varepsilon-\mu)/kT} - 1\right]}$$

The average occupation numbers are

$$\overline{n}_i = \frac{1}{e^{(\varepsilon_i-\mu)/kT} - 1} \qquad (\mu \lesssim 0)$$

so $N_o = \overline{n}_o = \left[e^{-\mu/kT} - 1\right]^{-1} = N - N_\varepsilon \qquad (N_o = N \text{ at } T = 0)$

This condensation provides an explanation for superfluidity and for superconductivity when electrons are coupled, through nuclear interactions, to form electron pairs that act as bosons.

QUESTIONS

20.1. Ideal gas expansions may be reversible or irreversible, but compressions are normally reversible.
a. Why must this be so? Under what conditions would it not be true?
b. What would be the analogous restriction on expansions or contractions of an elastomer?

20.2. **a.** What is the Born–Oppenheimer approximation?
b. For what types of molecules would you expect to find breakdown of this approximation? Be explicit, giving examples of specific molecules and energy levels if possible.

20.3. The Seebeck effect is the basis of function of thermocouples.
a. Explain how this effect can be related to the temperature dependence of the Fermi level in dissimilar metals.
b. Explain, in terms of the direction of transport of electrons at the metal/metal interface, why each junction should change in temperature (and in which direction).

20.4. The Peltier effect is the reverse of the Seebeck effect; a voltage is applied to a loop made of dissimilar metals. The junctions may (initially) be at the same temperature. Explain, in terms of the direction of transport of electrons at the metal/metal interface, why each junction should change in temperature (and in which direction).

20.5. Give three distinctly different versions of the third law of thermodynamics. Explain what limitations, if any, each has, employing statistical mechanics arguments.

20.6. Write an expression for the rotational term values of
a. F_2.
b. BF_3 (planar).
c. UF_6 (octahedral).

20.7. Identify each of the following molecules as an asymmetric rotor, symmetric rotor, spherical rotor, or linear molecule.
a. H_2O.
b. CH_2Cl_2 (substituted CH_4).
c. H_3C-CF_3.
d. CO_2.
e. $H_3C-C{\equiv}C-CH_3$ (linear chain).

20.8. What assumptions did Einstein make to predict the (approximate) form of $C_v(T)$ for crystalline solids? Specifically,
a. How many oscillators?
b. How many frequencies?
c. How were the frequencies distributed?
d. How did the energy depend on frequency and on temperature?

20.9. What assumptions did Debye make concerning vibrational modes in solids to predict the T^3 dependence at low temperatures?
a. How many oscillators?
b. How many frequencies?
c. How were the frequencies distributed?
d. How did the energy depend on frequency and on temperature?

20.10. Quantum statistics leads to the *occupation numbers*,

$$n_i = \frac{1}{e^{(\epsilon_i - \mu)/kT} \pm 1}$$

a. How can one show, simply from the expected values of n_i, which type of particles get the + sign and which get the − sign?
b. By similar arguments concerning expected values of n_i, what limitations exist on the values of μ for Fermi–Dirac statistics?
c. What limitations exist on μ for Bose–Einstein statistics?
d. What limitations exist on μ for Boltzmann statistics?

20.11. Classical mechanics treats particles as distinguishable in principle. Therefore Boltzmann statistics, in original form, differs from Bose–Einstein or Fermi–Dirac statistics even in the limit where classical physics should be applicable.
a. Why do we believe Boltzmann statistics must be modified for indistinguishability?
b. How is it modified?

DERIVATIONS AND PROOFS

20.1. The energy of a relativistic free particle may be written

$$E = \sqrt{c^2 p_x^2 + c^2 p_y^2 + c^2 p_z^2 + m_0^2 c^4}$$

Find an expression for an average value of the energy of a particle traveling along the x axis, involving p_x and E, using the equipartition principle.

20.2. The Rayleigh–Jeans law for blackbody radiation, valid for low frequencies, may be written

$$\frac{dE}{dV} = \frac{8\pi\nu^2 kT\,d\nu}{c^3}$$

Derive this equation, assuming the equipartition law ($\frac{1}{2}kT$ for each mode of oscillation).

20.3. Derive the law of Dulong and Petit (for c_v of a solid element, in the classical limit), showing
a. Why the principle of equipartition of energy should be applicable to this problem.
b. The results of application of the equipartition principle.

20.4. Show that the contributions of different vibrational modes to c_v are simply additive.

20.5. Show that a vibration of degeneracy g_i contributes g_i times as much to c_v as it would if it were nondegenerate.

20.6. Show that a vibration with $\eta = hc\sigma/kT$ gives a contribution to the heat capacity of

$$\frac{c_v}{R} = g\frac{\eta^2 e^{-\eta}}{(1 - e^{-\eta})^2}$$

PROBLEMS

20.1. A collection of N paramagnetic particles, each of magnetic moment μ, produce a magnetic moment $n\mu$, where $n \ll N$ is the excess number of particles aligned in one direction.

a. In the absence of an applied field, what is the average, or most probable, value of n?
b. In the absence of an applied field, what is the standard deviation of the distribution for n?
c. Write a distribution function for $P(n)$.
d. How would the distribution function be changed if a magnetic field were applied that made the a priori probability for a spin to be aligned with the field, $P = a/b$, slightly greater than $\frac{1}{2}$?

20.2. A paramagnetic system is brought to equilibrium in a strong magnetic field at 4.0 K. The direction of the magnetic field, **H**, is then instantaneously reversed.
a. Write the Boltzmann factor for the distribution function of spins at this point in the experiment. Which factors are positive and which are negative?
b. Apply second-law analysis to show the direction of energy flow between the spin system and its surroundings (the crystal lattice).
c. If the heat capacity of the spin system is $C_s = q/dT$, is C_s positive or negative? Justify your answer.
d. What is the sign of ΔT for the spin system as it approaches equilibrium?

20.3. What happens to the length, L, of an elastomer when it is warmed under constant tension? Why?

20.4. Indicate the sign (> 0, < 0, or $= 0$) for Q, W, ΔE, ΔS, and ΔT for an elastomer
a. If it is stretched reversibly.
b. If it contracts reversibly.
c. If it contracts irreversibly (without constraint).
Assume that the processes are adiabatic.

20.5. The population of any level of energy E_i and degeneracy g_i is $n_i/N = g_i e^{-E_i/kT}/Z$.
a. What fraction of nitrogen molecules ($\sigma = 235{,}961$ m^{-1}, $g = 1$) are in the ground vibrational state (v $= 0$) at 300 K?
b. What fraction of nitrogen molecules are in the first excited state (v $= 1$) at 300 K?

20.6. **a.** What fraction of methane molecules are in the (nondegenerate) ground vibrational state at 300 K? (See Example 16.7.)
b. What fraction are in the first excited vibrational state of σ_4 (130,620 m^{-1}, $g = 3$)?
c. Do your results support the generalization, "Molecules do not vibrate at room temperature"?

20.7. Carbon tetrabromide, CBr_4, has the same symmetry and forms of vibration as methane, but lower frequencies: $\sigma_1 = 267$ cm^{-1}($g = 1$), $\sigma_2 = 123$ cm^{-1}($g = 2$), $\sigma_3 = 681$ cm^{-1}($g = 3$), and $\sigma_4 = 183$ cm^{-1}($g = 3$). Find the population of the ground vibrational state ($v_1 = v_2 = v_3 = v_4 = 0$; $g = 1$) of CBr_4 at 298 K.

20.8. Calculate the expected vibrational contribution to c_v for N_2 at 300 K.

20.9. Consider a gas with molar mass of 30 g/mol at uniform temperature of 27°C,
a. In a uniform gravitational field, $\mathcal{V} = mgh$.
b. In a gravitational field $\mathcal{V} = -Km/r$.
Which of these (if either) satisfies the requirements for application of the equipartition principle?

c. What is the average height (or the height for an average potential energy) of the gas molecules?

20.10. Find the contribution of electrons to the heat capacity at 300 K of

a. K.

b. Al.

20.11. Explain why the Einstein temperature, $\theta_E = h\nu_o/k$, should be expected to be less than the Debye temperature, $\theta_D = h\hat{\nu}/k$, from the way in which these are defined.

20.12. Given $\overline{n}_k = \dfrac{1}{e^{(\varepsilon_k-\mu)/kt}}$ or $\overline{n}_k = \dfrac{1}{e^{(\varepsilon_k-\mu)/kT} \pm 1}$, show from limitations on the values of n_k that

a. $\mu \ll 0$ for the Boltzmann distribution.

b. $\mu < 0$ for the Bose–Einstein distribution.

c. μ may be positive or negative for the Fermi–Dirac distribution.

CHAPTER 21

Perspectives on the First and Second Laws

The first law of thermodynamics is the law of conservation of energy. It is one of the first ideas encountered in the study of motion. The conservation principle could not be formulated, however, until the middle of the nineteenth century, when "heat" was clearly recognized as molecular motion, rather than the mysterious caloric fluid. Conservation laws are now the foundation of modern physics.

Many thermodynamic properties have been considered in this text, such as $E, H, T, P, V, F, G, S, c_v$, and c_p. Which of these quantities is/are subject to a conservation law?

21.1 The Meaning of Conservation

Conservation laws provide continuity between classical physics and relativistic and quantum physics. Because of the fundamental importance of the conservation laws to all of science, and certainly to thermodynamics, and the common misunderstandings associated with some of them, it is worthwhile to pause for a brief look at what is, and is not, meant by "conservation" in the technical sense.

Conservation laws are so powerful that the concept is frequently overused or misused. When a quantity remains constant, for any system, it is likely to be described as being "conserved." With this definition, there are many conservation laws, but few if any that are universally valid. Conservation laws apply not to the system alone but to the system and the surroundings. As a special case, they can be applied to isolated systems, but we cannot experiment with isolated systems.

If a piece of string is cut, the string is "shorter" but the rest of the length is still around. The total length, including both parts, has not changed (Fig. 21.1). Is length a conserved quantity? Is there any one experiment or process that can change total length?

It is not hard to think of processes that will change the total length. If a 6 ft 2×4 is ripped, along its length, to give two 6 ft 2×2 pieces (all nominal widths), then we have gone from 6 ft of board to 12 ft of board, with the loss

FIGURE 21.1 The total length of string is not changed by cutting.

of only a couple of inches in width (Fig. 21.2). Or we can pull taffy to get very long pieces from very short pieces. Total length is not a constant. Length is not a conserved quantity.

The next most complex quantity is area. If a circle is cut from a sheet of paper, does the area change (Fig. 21.3)? Is area conserved?

When a circle is cut from a piece of paper, the circle has a smaller area than the original paper, but the rest of the area is still there. The sum of the areas is the same.

If the change in area "inside" is compensated by a change in area "outside," there is no net change in area. Can you devise an experiment in which total area really will change? (A convenient conceptual measure of area is to ask how much paint would be required to cover the surface.)

Perhaps you found a better example, but certainly one way of changing area would be to warm a copper block so that it expands. A more dramatic way would be to drop a glass brick. The fragments would require much more paint to cover them than did the original brick. Area clearly is not a conserved quantity.

Is volume any different? Can you suggest any experiment that will change the total volume?

An example often quoted as demonstrating conservation of volume is to pour a liquid from one container into another of different shape (Fig. 21.4a). Does this change the volume of the liquid? Does it demonstrate conservation of volume?

A second common example is to immerse a steel ball in water (Fig. 21.4b). How does the steel ball affect the level of the liquid in the container? Does this provide information about conservation of volume?

The volume of liquid is constant as it is poured from one container to another, but only for constant temperature and only if we do not add or subtract liquid. When the steel ball is added to water, the water level rises by an amount corresponding to the volume of the ball. But this really shows nothing more than the additivity of volumes. Even this form of additivity is not fully general. If 50 mL of alcohol is added to 50 mL of water, the volume is less than 100 mL (see

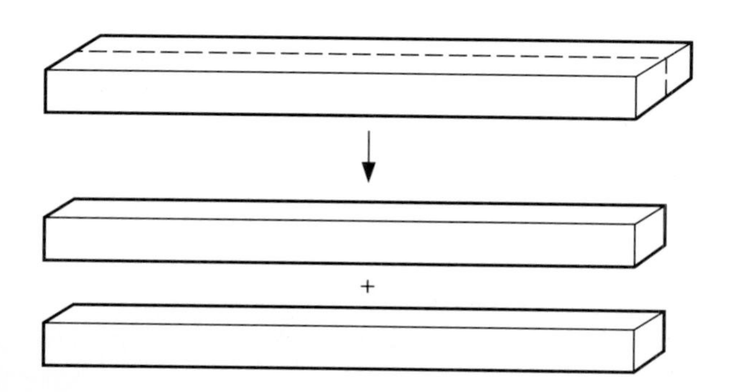

FIGURE 21.2 Cutting the board lengthwise increases the total length.

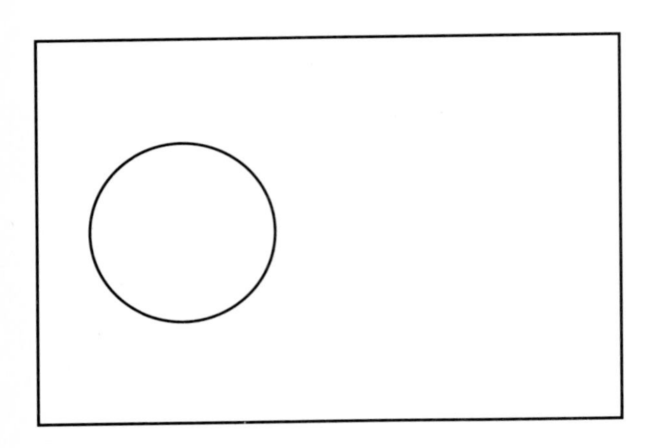

FIGURE 21.3 Does the area change if a circle is cut from a sheet of paper?

Section 5.1). Indeed, if we add 1mL of magnesium sulfate, $MgSO_4$, to 100 mL of water, the volume will be less than 100 mL.[1]

The examples of Fig. 21.4 show that the volume of a system may be constant under restricted conditions but need not always be so. They have little relevancy to the question of whether volume is or is not conserved. Remembering the analogy with length and area, perhaps we are not asking the proper question. Is there a loss of volume somewhere else, or is there a sum of volumes before and after that must be constant?

Consider the example of a large deflated weather balloon stretched across the floor of the room (Fig. 21.5). What happens when you open the valve to inflate the balloon? You will begin to feel very crowded because as the balloon expands, occupying a greater volume, the volume outside the balloon decreases. In other words, if the balloon is called the system, the sum of the volumes of the system and the surroundings is constant.

The system can change its volume, increasing or decreasing, but there is always a concomitant, compensating change in the volume of the surroundings. That is what conservation laws are about. They say that the sum of the amount of the conserved quantity for the system plus the surroundings is the same, for any process whatsoever.

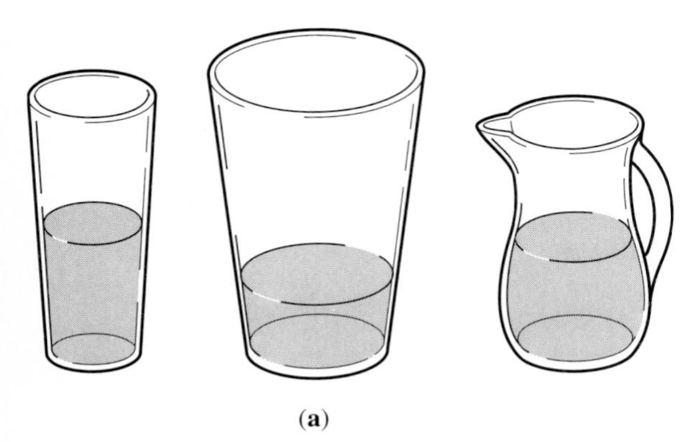

FIGURE 21.4 (a) Does volume of liquid change when the liquid is poured from one container to another?

[1]Water is a likely candidate for such effects because, like ice, the liquid has a relatively open structure produced by hydrogen bonding between water molecules. When ions, such as doubly charged magnesium or sulfate ions, are inserted into water, they attract the oxygen atoms or hydrogen atoms strongly enough to collapse the open water structure around them, producing a shrinkage of volume. Alcohol forms its own hydrogen bonds with water, replacing other water molecules, leading to a volume greater than the original water volume but less than the sum of water and alcohol volumes.

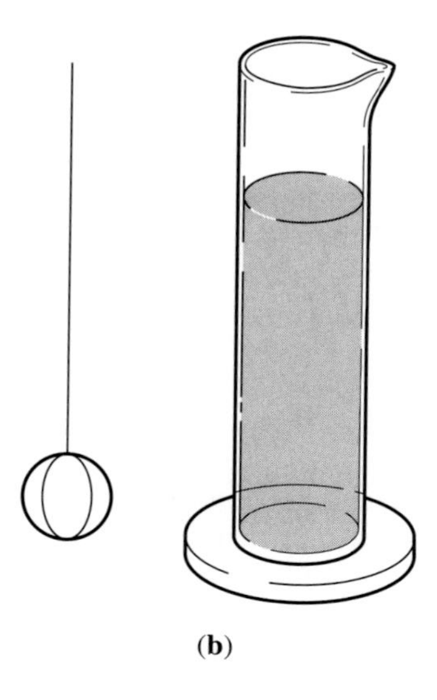

(b)

Figure 21.4 (b) When the steel ball is immersed in liquid, how much does the liquid rise?

Conservation laws live a perilous existence, for even one clear exception is enough to invalidate the law. The conservation laws accepted today are those that have withstood this stringent testing.

Unfortunately, if a conserved quantity is not constant for the system (i.e., it is not a constant of the motion, in the terminology of classical mechanics), it is sometimes said to be not conserved. By contrast, in the technical meaning of conservation that we imply when we equate the first law of thermodynamics to the conservation of energy, thus far no violation has been found for any one of the conservation laws listed in Table 21.1.

"Nonconservative" systems are systems undergoing a process in which energy is transferred to a form that is less obvious. Mechanical energy may "disappear"

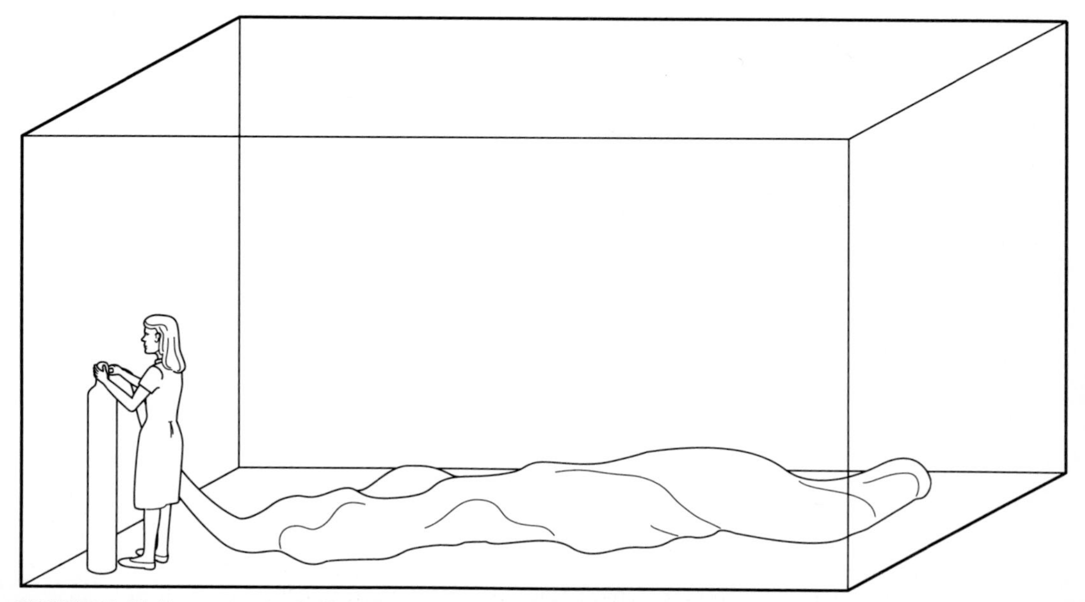

FIGURE 21.5 As the balloon expands, the volume of the surroundings is decreased. The sum of volumes of system and surroundings does not change.

Table 21.1 Conserved Quantities[a]

Quantity	Nature
Volume	Scalar
Energy	Scalar
Mass[b]	Scalar
Momentum	Vector
Angular momentum	Vector
Charge	Scalar algebraic sum: positive − negative

In addition, there are conservation rules that apparently hold for fundamental particles, such as the total number of fermions, the number of leptons, and perhaps the number of hadrons, where the number is taken as the number of particles minus the number of antiparticles.

[a]Cosmologists once thought that there might be a violation of conservation of energy on a scale that is undetectable in laboratory measurements (the steady-state universe model, which has given way to the rival big-bang model), and one might argue that an expanding universe changes total volume, but since these effects are not demonstrable on the scale of laboratory measurements, we need not be concerned about them here.

[b]Rest mass is not conserved. Only relativistic mass is a conserved quantity. This is a direct consequence of conservation of energy and the relation $m = E/c^2$, rather than an independent conservation law. (Of course, $E + m$ is also conserved, because each is separately conserved.) Conservation of mass is listed separately here because it is convenient to consider mass as an independent measure in nonrelativistic problems.

into internal energy or the energy of an applied field, but there is no change in the total energy.

A special case to which conservation laws are often applied is of particular importance.

> Special Case: *The law of conservation of energy states that the energy of an isolated system is constant.*

This special case is often substituted for a more general statement of the law of conservation of energy, but it is much too narrow for the general principle required. An isolated system is generally understood to be isolated in all respects. There can be no transfer of energy, momentum, angular momentum, material, charge, or other quantities across the boundary. It would be a very weak form of conservation of energy, however, to stipulate that energy is constant *only* when momentum is constant for a system, just as it would be too limiting to apply conservation of momentum *only* in problems for which the energy of a system is constant.

The importance of the conservation laws is that they are independent statements. Regardless of how energy, momentum, and other quantities may flow across the boundaries of a system, the change in each conserved quantity *for the system plus the surroundings* is always zero.

The general statement of the first law of thermodynamics is

$$\Delta E_{\text{system}} = -\Delta E_{\text{surroundings}}$$

or

$$\Delta E_{\text{system + surroundings}} = 0 \tag{21-1}$$

or

The energy of the universe is constant.

The first law may be obtained from the first-law equation if several auxiliary equations or conditions are added.

First,

$$Q_{\text{system}} = -Q_{\text{surroundings}}$$

and second, (21-2)

$$W_{\text{system}} = -W_{\text{surroundings}}$$

Third, it must be assumed that there are no other forms of energy transfer (e.g., no mass transfer to or from the system). Then it follows that

$$\Delta E_{\text{system}} = -\Delta E_{\text{surroundings}}$$

However, even with these explicit conditions, the first-law equation is less general than the first law, because W, and therefore Q, is not operationally defined for typical real processes in which friction is present or radiation is transferred between objects not in thermal equilibrium.

A conservation law may only apply to extensive properties of a system. What would it mean to "conserve" temperature when a beaker of water at 25°C is divided into two beakers at 25°C? Intensive properties, such as T, P, c_v, and c_p may be changed at will without changing the surroundings (e.g., by a chemical reaction in a closed, insulated steel bomb).

We have seen that E and V are conserved quantities. The second law tells us that S is not conserved but tends to increase with time for system and surroundings. Because the product PV can change, H is not conserved. Because S increases and T is not constant, F and G are not conserved.

21.2 Definition of Energy

Energy is a difficult concept to define, because it assumes so many different forms.

Energy is recognized as a variety of forms, including kinetic energy, $\frac{1}{2}mv^2$, that can be quantitatively converted into each other under suitable conditions.

A familiar attempt at a definition is

Energy is the ability to do work.

This is particularly weak in two respects. First, as discussed in Part II, conversion of thermal energy to work can never be fully accomplished at achievable temperatures. Perhaps more important, however, work has meaning only as a form of transfer of energy from surroundings to system. The "definition" of energy then becomes a statement that energy is the ability to transfer energy in a certain way, which is of little help in understanding the meaning of energy.

A better but still less than satisfactory definition of energy, would be

Energy is the ability to raise the temperature of a system.

Because a system at a low temperature (containing energy) cannot raise the temperature of another system at a higher temperature, this is also a weak definition.

Even classification systems for energy tend to overlap. For example, a convenient division is

1. Internal energy.
2. Energy of (system) motion, which includes kinetic energy of translation of the center of mass and rotational energy about the center of mass.
3. Potential energy, or energy of position (and/or orientation) of the system.

Alternative classifications include:

a. Chemical energy.
b. Electrical energy.
c. Nuclear energy.
d. Vibrational energy.
e. Electromagnetic energy.
f. Sound energy.

Examination of these lists will reveal that the categories are not sharply defined. They overlap in various ways. For the study of thermodynamics, it is helpful to select a somewhat different classification scheme for the energy of a system and its interaction with its surroundings.

One part of the energy may be called *external energy.* The system as a whole may be moving with *kinetic energy* $\frac{1}{2}mv^2$, it may be rotating with *rotational energy* $\frac{1}{2}I\omega^2$ (or, more generally, $\frac{1}{2}\boldsymbol{\omega}^* I \boldsymbol{\omega}$, where I is the inertia tensor and $\boldsymbol{\omega}$ is angular velocity), and it may be raised or lowered in a gravitational field and subject to electric or magnetic *fields*.

In addition, the system will have energy, which we label *internal energy,* that does not depend on the motion or location of the system as a whole.[2] An equation of state relates internal variables, such as T, P, V, and the internal energy. Energy of motion or energy of position must be subtracted from the total energy to find the internal energy. (In practice, however, this is seldom a significant consideration because we do not pretend to know either the total energy or the total internal energy, and therefore are concerned only with changes of energy with respect to specified variables.)

Internal energy may take many forms. Start with fundamental particles to form *nuclei,* add electrons to form free *atoms,* combine atoms to form *molecules* and/or *ions,* and condense molecules or ions to form a crystalline solid, glass, or liquid *condensed phase*. The atomic or molecular entities of a condensed phase have *potential energy* (usually treated as negative) because of their mutual interactions and, because they are moving, have *kinetic energy* and possibly *rotational energy.*

Some of these energy terms have alternative names. Energy of interaction of atoms to form molecules is usually called *chemical energy.* Part of the molecu-

[2]The *rest mass*, m_o, of the system is determined by the internal energy plus the rotational energy, plus any energy that resides in external fields but is assigned to the system, all divided by c^2.

lar internal energy and part of the potential energy of atomic and/or molecular interaction may be called *vibrational energy.*

When thermal energy is transferred, the transfer is primarily in the form of molecular translational kinetic energy (with some possible rotational energy transfer). However, the kinetic energy undergoes equilibration with other forms of energy in the receiving body after the transfer. The energy contributions are distributed, on a time average, equally among all equivalent particles within the system. Such energy contributions are called *random energy.*

Randomized internal energy arises when the number of particles is very large and the internal energy equilibrates among the particles and modes of storage. The average kinetic energy in each degree of freedom is an observable quantity in many-particle systems. A thermometer measures the average value of $\frac{1}{2}mv^2$ for the particles (as long as the energy is of the form $E_K = \sum p_i^2/2m$ and is not constrained by quantization effects). This average value is directly related to the average values of other randomized energy components that are in dynamic equilibrium with the translational energy.

When the internal energy of a system is increased, by thermal energy transfer, or by work when the energy is subsequently randomized within the system, there will be a temperature increase of the system and/or some or all of the system will undergo a phase change.[3]

We define the increment of randomized internal energy as thermal energy.

Thermal energy, in the interval $T_1 \rightarrow T_2$, includes the increase in random kinetic energy of motion of the molecules in a system, plus the increase in all other forms of energy in dynamic equilibrium with the translational kinetic energy, when the temperature of the system increases from T_1 to T_2.

Thermal energy is dependent on the temperature interval. For water between -1 and $+1$°C, the heat of fusion must be considered a contributor to the thermal energy, but for water between 5 and 6°C, the phase energy at 0°C is irrelevant to the thermal energy.

Contributing influences to the thermal energy generally include translational and rotational kinetic energies, intermolecular potential energy, and vibrational energy (especially very-low-frequency modes). Energy forms that typically do not contribute to the thermal energy include nuclear energy, chemical energy, phase energy for transitions outside the temperature interval, external energy, and internal strain energy.

Not all internal energy need be randomized. Nuclear energy, for example, is concentrated in individual nuclei and not equilibrated near room temperatures. A system consisting of billiard balls on a table may have all the energy of motion in the cue ball at one moment and in some fraction of the balls at another time. The state is dependent on nonrandom external influences.

Similarly, a spring that is compressed has potential energy capable of being transferred isothermally as work. If the spring (or an individual molecule) is oscillating, the vibrational potential energy is quantitatively converted to vibrational kinetic energy and back.

A "rigid" object may be bent, introducing strains that can later restore the object to its original form or that render the object inhomogeneous to chemical

[3]We deliberately ignore here the real possibility that chemical changes or other changes may occur in conjunction with the temperature change. Such complications do not alter the fundamental arguments of this section.

attack. If an electric field is established between two plates of a capacitor, that stored energy may be recovered quantitatively to give an electric current or energy stored in the field of an inductive coil. Such internal energy is not random.

21.3 Potentials

Many of the expressions for work that were given in Table 3.1 are differentials of products of extensive and intensive variables that are called *potentials*. A portion of the total energy may be represented by a sum,

$$E = m\Phi + \mathcal{E}q + \tfrac{1}{2}qV_E + \tfrac{1}{2}V\mathbf{E}\cdot\mathbf{P} + \tfrac{1}{2}\mu_o V\mathbf{H}\cdot\mathbf{M} + \mathcal{T}(L - L_o) + \gamma A + \sum \mu_i n_i \tag{21-3}$$

Then

$$dE = m\,d\Phi + \Phi\,dm + \mathcal{E}\,dq + q\,d\mathcal{E} + \cdots \tag{21-4}$$

Some of these differentials are work terms, some are not. Only those terms that can be written as a product of a generalized force and a displacement represent work.

Some potentials do not contribute to the energy. The product PV has dimensions of energy, but pressure is not an energy density and PV is not energy. The change in PV is

$$d(PV) = PdV + VdP$$

The first part is often (but not always) equal to work done by a system on the surroundings, but VdP is never an energy, or work, term. Similarly,

$$d(TS) = TdS + SdT$$

Although TdS is sometimes equal to Q, there may be a change, TdS, when $Q = 0$, $W = 0$, and therefore $\Delta E = 0$. The change SdT is not, in general, either work or thermal energy transfer.

Potential terms such as PV and TS may be added to or subtracted from the energy, as discussed in Section 11.4. The potentials resulting from such Legendre transforms are no longer energy. For example, $E + PV = H$, which is enthalpy, not energy.

Furthermore, not all energy changes are expressible as differentials of potentials. For example, kinetic energy is quadratic in the usual variables, $\frac{1}{2}mv^2 + \frac{1}{2}I\omega^2$. Frictional forces are a well-known example of forces not derivable from a potential. Friction contributes terms to dE but not to E.

21.4 Maxwell's Demon

The second law, the law of increase of entropy, is comparable in importance to the first law but is far less widely recognized. It is intimately involved with the molecular properties of matter and with the probabilities associated with statistical distributions.

Entropy and temperature are statistical concepts. Each molecule has a given amount of kinetic energy, rotational energy, and other forms; the amounts of these vary with time and from one molecule to another. The average value of the kinetic energy, for a large number of molecules, is constant for a system at equilibrium. From this average kinetic energy, a temperature $T = \frac{2}{3}$ K.E./k can be assigned, but rules governing temperature and entropy are of little value in making predictions concerning the energy or behavior of an individual molecule.

If you could view the world as an enlightened atom, words such as *temperature* and *thermal energy* would be meaningless. You and your friends would undergo seemingly random collisions, so sometimes you would go long distances in one direction and other times you would go only a short distance before changing direction. Kinetic and potential energy would be transferred between you and your neighbors, and each such collision could be described as a work-type interaction, a force acting through a short distance.

Temperature and thermal energy become meaningful only when we are dealing with very large numbers of atoms and molecules. After subtracting out the average motion (i.e., the motion of the center of mass of the macroscopic system), the random motions left give rise to the effects we call temperature. If the random motions are passed from one object to another, across the boundaries of what has been termed the system, the process is thermal energy transfer, Q.

Maxwell, in 1891, emphasized the distinction between the rules for molecules and the statistical rules for collections of molecules by discussing a thought experiment in which temperature differences could be generated without expenditure of energy. He envisioned a gas, in two parts at equilibrium, separated by a barrier. In the barrier is a small hole or door, operated by an intelligent being who can see individual molecules.

When a fast molecule approaches from the left, the door is opened, allowing the molecule to pass through; when a slow molecule approaches from the right, the door is opened, allowing the molecule to pass through. Other times the door is kept closed. In this way the fast molecules are collected on the right side and the slow molecules on the left, so a temperature difference is created between the two sides. It is assumed that through skillful design, the door requires less free energy for its operation than is gained by the selection process.

The intelligent being that can see individual molecules and selectively open the door for their passage has come to be known as a *Maxwell demon*. It has drawn considerable attention, more as a curiosity and a potential flaw in thermodynamic generalizations than as a serious possibility for power generation.

Development of information theory provided new insight into the operation of a Maxwell demon and the power that could, in principle, be obtained if such selection were possible. A necessary prerequisite for the selection process is that the demon must be able to observe the molecules and determine their speeds. Combining some knowledge of measurement theory, as augmented by quantum mechanics, with classical theory, observation of the molecules requires that the demon have a flashlight, or other light source, with which to illuminate the molecules.

Turning on the flashlight entails energy dissipation, or entropy production, but is necessary to acquire the information required for a decision on whether or not to open the door. Calculations of the entropy increase necessary to increase the demon's information show that there is no net gain. The process provides no net decrease in entropy, even apart from losses associated with operation of the door and the demon.

Lest we show too much disappointment at this loss of a possible power source, another calculation[4] has estimated the amount of power that could be produced by operation of the demon, neglecting the energy expenditure by the demon and his flashlight. If a gas acquires a temperature differential of 10 K, the work that could theoretically be extracted, at maximum efficiency, would be 1.7 J per mole of gas. However, if a room full of air (ca. 3000 mol) is to be given a temperature difference of only 2 K by such a demon, the time required would be at least 6×10^{13} s, or 1.9 million years. Thus it would scarcely be worth our time to seek a means of producing a Maxwell demon or attempting to circumvent the limitations imposed by information theory on the effectiveness of such a demon.

21.5 Jaynes' Formalism

If the probabilities of various individual outcomes are known, such as the probabilities of heads in a coin flip or of getting a 6 on a toss of a die, the average values, or expectation values, can be calculated for experiments that depend upon those individual outcomes.

Let $p_i(x_i)$ be the probability of the outcome x_i. Then the average value of any function of x_i, $f(x_i)$, is

$$\langle f(x_i) \rangle = \sum_i f(x_i) p_i \tag{21-5}$$

where it is assumed the probabilities are normalized:

$$\sum_i p_i = 1 \tag{21-6}$$

Suppose, now, that the probabilities, p_i, are not known but a series of measurements has established values considered to be reliable for the average values of a set of functions $f_k(x_i)$. The problem, then, is to solve

$$\sum_i p_i = 1$$

and

$$\langle f_k(x_i) \rangle = \sum_i f_k(x_i) p_i$$

to find the probabilities, p_i.

No rigorous solution of the equations can be produced for the general case, but Jaynes has observed that the most conservative assignment of probabilities is that which leaves the greatest uncertainty in their values.[5]

The uncertainty, or entropy, associated with a set of probabilities p_i, or the amount of information conveyed by a message of N symbols with the probabilities p_i, is

$$S = I = K \sum_i^N p_i \ln p_i \tag{21-7}$$

[4]H. S. Leff, *Am. J. Phys.* **55,** 701 (1987).
[5]E. T. Jaynes (ref. 38).

and therefore this sum is an extremum for variations in the p_i. This assumption concerning assignment of probabilities has been called *Jaynes' formalism*.

Addition of the assumption that $dS = dI = 0$ provides a third equation which, together with the two previous equations, leads directly to

$$p_i = e^{-\alpha} e^{-\sum_k \beta_k f_k(x_i)} \tag{21-8}$$

where α and the β_k are Lagrange multipliers, not yet determined. The mathematical steps are identical to the earlier calculations of distribution functions and occupation numbers involving the energy.

The value of α is obtained from the normalization condition.

$$\alpha = \ln\left[\sum_i e^{-\sum_k \beta_k f_k(x_i)}\right] \tag{21-9}$$

and differentiation of α with respect to the β_k,

$$-\frac{\partial \alpha}{\partial \beta_k} = \frac{\frac{d}{d\beta_k}\sum_i e^{-\sum_k \beta_k f_k(x_i)}}{\sum_i e^{-\sum_k \beta_k f_k(x_i)}} = \sum_i p_i f_k(x_i) = \langle f_k(x_i)\rangle \tag{21-10}$$

provides conditions from which the β_k may, in principle, be determined.

SUMMARY

Conservation is given many meanings, including minimizing loss and describing a constant of the motion. The important meaning, especially in thermodynamics, is that the total quantity, in system and surroundings, remains constant. As a special case, the value is constant for an isolated system. Only extensive properties can be conserved.

In classical mechanics, E, $\mathbf{p}$, and $\mathscr{L}$ are conserved, but also V, charge, (relativistic) mass, and numbers of certain particles are subject to conservation laws, which are among the most important principles of physics.

Energy is not readily definable, except that all forms of energy, including kinetic energy, are quantitatively interconvertible. It is convenient to distinguish (overlapping) types of energy, such as chemical, electrical, nuclear, vibrational, and so on.

It is necessary to distinguish internal energy from its various components, such as (molecular) vibrational, rotational, chemical, and lattice energies, and to distinguish between randomized internal energy, or thermal energy, and nonrandom forms such as strain energy and chemical or nuclear energy that do not change (reproducibly) with temperature.

Energy includes many contributions, many of which are expressible as potentials. Changes of potentials give changes of energy of the system, with corresponding energy transfer to the surroundings that may be as Q or as W. Other energy changes and transfers are not expressible as changes of potentials.

If it were possible to sort molecules (e.g., with Maxwell's demon) according to their energies, an apparent violation of the second law would be possible. More careful analysis, however, shows that the sorting process requires an entropy increase greater than the anticipated entropy decrease, which would in any event be too small to be of practical importance.

Jaynes has shown that if $p_i(x_i)$ is the probability of the outcome x_i, so that the average value of the function $f(x_i)$ is $\langle f(x_i)\rangle = \sum_i f(x_i)p_i$, then it is not possible to solve the final equation to find the $p_i(x_i)$. However, the assumption that leaves the greatest uncertainty in the values of the $p_i(x_i)$ leads to the information content

$$I = K \sum_{i}^{N} p_i \ln p_i$$

which is consistent with other formulations of information theory and statistical mechanics. Subject to this assumption, the p_i may be evaluated from the $f_k(x_i)$.

QUESTIONS

21.1. E. A. Abbott (A Square) has written a delightful book titled *Flatland* about life in a two-dimensional world. In that world there is length and area but no volume. What spatial property, if any, would be conserved there?

21.2. Linking the first-law equation ($\Delta E = Q + W$) with the law of conservation of energy requires that $W_{\text{system}} = -W_{\text{surroundings}}$. For an expanding gas, this means that $\int P\,dV$ for the system is equal and opposite to the same integral for the surroundings.

a. Is it always true that $P_{\text{system}} = P_{\text{surroundings}}$? (Consider carefully how system and surroundings are defined, and how the pressure is exerted.)

b. Is it always true that $dV_{\text{system}} = -dV_{\text{surroundings}}$? (If so, how can you be sure? If not, give a counterexample.)

21.3. There have been many sightings of black crows and some white (albino) crows have been observed, but no yellow crows have been seen. Is the following statement valid?

"If it is a crow, it is not yellow."

Compare the evidence for the first law of thermodynamics (or any other conservation law) with the evidence for the absence of yellow crows.[6]

21.4. The word "tendency" should be used with caution and backed by the precision of a mathematical description whenever possible. We use the term *escaping tendency* for fugacity. The chemical potential could serve to measure the tendency of molecules to move from one system or phase to another, but its range is less appropriate; as $f \to 0$, $\mu = RT \ln f + B(T) \to -\infty$, but the tendency of molecules to escape goes to zero.

a. Is there a property of a system that "tends to escape" in proportion to the pressure (or perhaps $-P$) of a system? If so, what property "escapes"?

b. Is there a property of a system that "tends to escape" in proportion to the temperature of a system? If so, what property "escapes"?

21.5. The product $P\,dV$ is strongly suggestive of work, just as $T\,dS$ is suggestive of q. Give an example for which $dE = T\,dS - P\,dV$ and V changes but $P\,dV \neq w$ and $T\,dS \neq q$.

[6] If you believe yellow crows exist, substitute for "yellow," "orange and purple, with green polka dots."

APPENDIX

Matrix and Operator Notations

An operator is a generalization of ordinary algebra. If we know the state of motion of a particle, such as one undergoing oscillatory motion, we can write

$$\mathscr{T} = \mathscr{T}(p, q) = \frac{p^2}{2m} \qquad \text{and} \qquad \mathscr{V} = \mathscr{V}(p, q) = \mathscr{V}(q) = \frac{1}{2}\omega^2 m q^2$$

The functions $\mathscr{T}$ and $\mathscr{V}$ can be regarded as prescriptions on how to evaluate the kinetic and potential energies. In a similar sense, there is a formalism that allows $\mathscr{T}$ and $\mathscr{V}$ to be *operators* that operate on a description of the state of the system to yield values of the kinetic and potential energies. In this sense we could say that $\mathscr{T}$ is the prescription to square the momentum and divide by twice the mass.

An operator is any quantity that is capable of changing a second kind of quantity: $2x$ differs from x, so 2 is an operator operating on (multiplying) x; $\sqrt{5}$ means to operate on the number 5 with the square-root operator, and so on. *Linear operators* are those, $\boldsymbol{\alpha}$, for which

$$\boldsymbol{\alpha}(u + v) = \boldsymbol{\alpha} u + \boldsymbol{\alpha} v, \qquad \text{and therefore} \qquad \boldsymbol{\alpha}(cu) = c\boldsymbol{\alpha}(u) \qquad \text{(A-1)}$$

$2(u + v) = 2u + 2v$, and $2(7u) = 7 \cdot 2u$, but $\sqrt{u + v} \neq \sqrt{u} + \sqrt{v}$ and $\sqrt{7u} \neq 7\sqrt{u}$. Quantum mechanics is concerned primarily with linear operators. It makes little or no difference what kind of linear operators are chosen, provided only that they satisfy some simple, explicit relationships among themselves, as given in Section 18.4.

There are times when differential operators are most convenient for treating quantum mechanical equations, but other times matrices or other special operators, such as ladder operators and annihilation operators, can simplify arguments and notation. For much of statistical mechanics it is not important which notation is selected. The best notation is then the most general form, which can easily be converted explicitly into any convenient system. In this section we give a brief introduction to matrices and their properties, then show the correspondence between the three most common descriptions of quantum mechanical states.

A.1 Matrix Notation

A matrix is a rectangular array of numbers, of *dimension* $m \times n$, m rows and n columns. The general element of a matrix $\mathbb{A}$ is written a_{ij}, the first subscript denoting the row number and the second the column number; $\mathbb{A} = [a_{ij}]$. Two matrices are equal only if they are identical, element for element.

Every square matrix has a *determinant,* a single number defined as the sum of products of the elements of the matrix (from different rows and columns), with appropriate signs. For any determinant larger than 3×3, the definition is too cumbersome for calculation purposes, so determinants are evaluated by shortcut methods, such as the Laplace expansion, triangularization, or pivotal condensation.

If any two rows, or any two columns, of a matrix are interchanged, the determinant of the matrix changes sign; therefore, if two rows, or two columns, are identical, the determinant is zero. A matrix with zero determinant (including a matrix that is not square) is said to be *singular.*

The sum of two matrices is the matrix whose elements are the sums of the respective matrix elements.

$$\mathbb{A} + \mathbb{B} = [a_{ij}] + [b_{ij}] = [(a_{ij} + b_{ij})] \tag{A-2}$$

Matrices must have the same dimensions to be added.

Multiplication of two matrices is defined only if the number of columns in the first matrix is equal to the number of rows in the second matrix.

$$\text{If} \quad \mathbb{A}\mathbb{B} = \mathbb{C} \quad \text{then} \quad c_{ij} = \sum_k a_{ik} b_{kj} \tag{A-3}$$

For example, if $\mathbb{A}$ is a row matrix and $\mathbb{B}$ is a column matrix,

$$\mathbb{C} = [c_{ij}] = [a_{i1} \quad a_{i2} \cdots] \begin{bmatrix} b_{1j} \\ b_{2j} \\ \vdots \end{bmatrix} = (a_{i1} b_{1j} + a_{i2} b_{2j} + \cdots) \tag{A-4}$$

(The i and j are unnecessary, for the row and column, and may be omitted.) A row times a column gives a scalar, or ordinary number. A row matrix, or column matrix, is often called a *vector.* The matrix product is analogous to a *dot product* or *scalar product*.

$$[2 \quad 3 \quad 0] \begin{bmatrix} 5 \\ 3 \\ 7 \end{bmatrix} = (2 \times 5 + 3 \times 3 + 0 \times 7) = 19$$

The same method applies to matrices of any dimensions $m \times p$ and $p \times n$, giving a product matrix of dimension $m \times n$. The matrix elements of the product are products of the rows and columns of the matrices to be multiplied, as above.

$$[a_{ik}][b_{kj}] = [c_{ij}] \equiv \left[\sum_k a_{ik} b_{kj}\right] \tag{A-5}$$

$$\begin{bmatrix} a_{11} & a_{12} \\ a_{21} & a_{22} \end{bmatrix} \begin{bmatrix} b_{11} & b_{12} \\ b_{21} & b_{22} \end{bmatrix} = \begin{bmatrix} (a_{11}b_{11} + a_{12}b_{21}) & (a_{11}b_{12} + a_{12}b_{22}) \\ (a_{21}b_{11} + a_{22}b_{21}) & (a_{21}b_{12} + a_{22}b_{22}) \end{bmatrix}$$

or

$$\begin{bmatrix} 2 & 3 & 0 \\ 1 & 4 & 6 \end{bmatrix} \begin{bmatrix} 5 & 1 & 3 & 2 \\ 3 & 4 & 1 & 0 \\ 7 & 2 & 3 & 6 \end{bmatrix}$$

$$= \begin{bmatrix} (10 + 9 + 0) & (2 + 12 + 0) & (6 + 3 + 0) & (4 + 0 + 0) \\ (5 + 12 + 42) & (1 + 16 + 12) & (3 + 4 + 18) & (2 + 0 + 36) \end{bmatrix}$$

$$= \begin{bmatrix} 19 & 14 & 9 & 4 \\ 59 & 29 & 25 & 38 \end{bmatrix}$$

In general, $\mathbb{AB} \neq \mathbb{BA}$. If $\mathbb{A}$ is $m \times p$ and $\mathbb{B}$ is $p \times n$ but $m \neq n$, then $\mathbb{AB}$ is defined, but the multiplication in reverse order, $\mathbb{BA}$, is not defined.

The determinant of a product is equal to the product of the determinants.

$$\det(\mathbb{AB}) = \det \mathbb{A} \times \det \mathbb{B} \tag{A-6}$$

The product of a matrix and a scalar is the original matrix with each element multiplied by the scalar.

$$c\mathbb{A} = c[a_{ij}] = [c \times a_{ij}] = \mathbb{A}c \tag{A-7}$$

The *diagonal,* or *principal diagonal,* of a matrix is the set of elements running from the upper left to the lower right corner. The *trace,* or *character,* of a matrix is the sum of the elements along the principal diagonal. A matrix with zero elements everywhere except along the principal diagonal is called a *diagonal matrix.* A diagonal matrix must be square.

The *unit matrix,* $\mathbb{I}$, is a diagonal matrix of any dimensions with all elements along the diagonal equal to 1. $\mathbb{I} = [\delta_{ij}]$. ($\delta_{ij}$ is the Kronecker delta; it is 1 if $i = j$ and 0 if $i \neq j$.) The unit matrix has the same function in matrix algebra as the scalar 1 has in scalar algebra. It may be inserted or suppressed at will.

$$\mathbb{AI} = \mathbb{A} = \mathbb{IA} \tag{A-8}$$

A constant times the unit matrix is called a *constant matrix:*

$$c\mathbb{I} = \mathbb{I}c = [c\,\delta_{ij}] \tag{A-9}$$

The zero matrix, or null matrix, is a matrix of any dimensions that has only zero elements.

The *transpose,* $\tilde{\mathbb{A}}$, of a matrix, $\mathbb{A}$, is obtained by interchanging rows and columns. The transpose of a column matrix is a row matrix, and vice versa. Transposing a square matrix is equivalent to rotating the elements about the principal diagonal.

$$\tilde{\mathbb{A}} = [\tilde{a}_{ij}] = [a_{ji}] \tag{A-10}$$

The *conjugate,* or *complex conjugate,* of a matrix is the matrix with each element replaced by its complex conjugate. $\overline{\mathbb{A}} = [\overline{a}_{ij}]$. A *real matrix* is a matrix with real elements, $\overline{a}_{ij} = a_{ij}$ so $\overline{\mathbb{A}} = \mathbb{A}$. The *transposed conjugate,* $\tilde{\overline{\mathbb{A}}}$, is represented by $\mathbb{A}^*$, which may also represent the complex conjugate of a scalar or the transpose of a real matrix.

The *inverse* of a matrix is a matrix that when multiplied times the original matrix gives the unit matrix. The determinant of the inverse of $\mathbb{A}$ is the inverse of the determinant of $\mathbb{A}$; therefore, an inverse matrix, $\mathbb{A}^{-1}$, can exist only if $\mathbb{A}$ is square and nonsingular (determinant not equal to zero).

From the definitions, it follows that

$$\begin{aligned}(\widetilde{\mathbb{A}\mathbb{B}}) &= \tilde{\mathbb{B}}\tilde{\mathbb{A}} \\ (\mathbb{A}\mathbb{B})^{-1} &= \mathbb{B}^{-1}\mathbb{A}^{-1} \\ (\mathbb{A}\mathbb{B})^* &= \mathbb{B}^*\mathbb{A}^*\end{aligned} \tag{A-11}$$

A.2 Matrix Transformations

If

$$\mathbb{A}\mathbb{X} = \lambda\mathbb{X}, \quad \text{or } [a_{ij}][x_j] = \lambda[x_j] \tag{A-12}$$

λ is called an *eigenvalue,* or *root,* of the matrix $\mathbb{A}$ and the column matrix $\mathbb{X}$ is called an *eigenvector* of $\mathbb{A}$. In general, an $n \times n$ matrix has n eigenvalues (corresponding mathematically to the n roots of an nth-order polynomial or n simultaneous equations) and has n eigenvectors. The eigenvalues, λ_i, and eigenvectors, $\mathbb{X}_i$, satisfy the equation

$$(\mathbb{A} - \lambda_i\mathbb{I})\mathbb{X}_i = 0 \qquad \text{and therefore} \qquad |\mathbb{A} - \lambda_i\delta_{ij}| = 0 \tag{A-13}$$

Simultaneous premultiplication and postmultiplication of a matrix is called a *transformation*. (Row and column matrices are transformed only by post- or premultiplication, respectively.) The most important class of transformations is the *similarity* transformations, where the pre- and postmultipliers are inverses of each other.

$$\mathbb{A}' = \mathbb{H}^{-1}\mathbb{A}\mathbb{H} \tag{A-14}$$

The matrices $\mathbb{A}'$ and $\mathbb{A}$ are said to be *equivalent* under the similarity transformation. Any matrix equation satisfied by $\mathbb{A}$ will be satisfied by $\mathbb{A}'$, or by any other matrix equivalent to $\mathbb{A}$ under a similarity transformation. Therefore, such a transformation leaves all of the eigenvalues of $\mathbb{A}$ unchanged, and among other properties, the trace (which is the sum of the eigenvalues) and the determinant (the product of the eigenvalues) are unchanged.

The eigenvectors of a matrix may be combined to form a square matrix.

$$\mathbb{H} = \begin{bmatrix}\mathbb{X}_1\mathbb{X}_2\mathbb{X}_3\cdots\end{bmatrix} \qquad \text{and} \qquad \mathbb{H}^{-1} = \tilde{\mathbb{H}} = \begin{bmatrix}\tilde{\mathbb{X}}_1 \\ \tilde{\mathbb{X}}_2 \\ \vdots\end{bmatrix} \tag{A-15}$$

Then, because $\mathbb{A}\mathbb{X}_i = \lambda_i \mathbb{X}, \mathbb{A}\mathbb{H} = [\lambda_i \, \delta_{ij}]\mathbb{H} \equiv \mathbb{A}\mathbb{H}$, and $\mathbb{H}^{-1}\mathbb{H} = \mathbb{I}$ (as proved below), so

$$\mathbb{H}^{-1}\mathbb{A}\mathbb{H} = [\lambda_i \, \delta_{ij}] \equiv \quad \mathbb{\Lambda} \tag{A-16}$$

That is, the matrix consisting of the eigenvectors of $\mathbb{A}$ will transform $\mathbb{A}$ by a similarity transformation into a diagonal matrix, $\mathbb{\Lambda}$, which has the eigenvalues of $\mathbb{A}$ as its diagonal elements.

Two matrices, $\mathbb{A}$ and $\mathbb{B}$, can be diagonalized by the same similarity transformation (e.g., with $\mathbb{H}$ and $\mathbb{H}^{-1}$) if and only if $\mathbb{A}$ and $\mathbb{B}$ commute. The transforming matrix that diagonalizes $\mathbb{A}$ (and $\mathbb{B}$) consists of the eigenvectors of $\mathbb{A}$ (and $\mathbb{B}$). It follows that if two matrices, $\mathbb{A}$ and $\mathbb{B}$, can be diagonalized by the same transforming matrix, $\mathbb{H}$, then the columns of $\mathbb{H}$ (and rows of $\mathbb{H}^{-1}$) must be the eigenvectors of both $\mathbb{A}$ and $\mathbb{B}$.

In the quantum mechanical representation, the eigenvectors represent the state of the system, each eigenvector corresponding to an eigenvalue that would be the result of a measurement of the physical property if the system were in that state. If two matrices, $\mathbb{A}$ and $\mathbb{B}$, representing two operators, $\boldsymbol{\alpha}$ and $\boldsymbol{\beta}$, can be diagonalized simultaneously, the set of eigenvectors that constitute the diagonalizing matrix must represent eigenstates of both $\boldsymbol{\alpha}$ and $\boldsymbol{\beta}$. Then $\boldsymbol{\alpha}$ and $\boldsymbol{\beta}$ (and hence $\mathbb{A}$ and $\mathbb{B}$) represent different properties of the same set of states. The corresponding variables, a and b, can be measured simultaneously.

All diagonal matrices commute. If a matrix $\mathbb{A}$ commutes with a diagonal matrix, $\mathbb{D}$, all of whose eigenvalues are different, $\mathbb{A}$ must also be diagonal. All square matrices commute with any constant matrix or zero matrix (of suitable dimensions), but the constant matrix and zero matrix do not have all their eigenvalues (i.e., diagonal elements) different.

A.3 General Theorems of Operators

A few theorems concerning operators are of special importance and recur often in applications of science and mathematics. Of particular interest are the following properties of hermitian operators, several of which have been required above.

In general, $< k|\boldsymbol{\alpha}|m >^* = < m|\boldsymbol{\alpha}^*|k >$. If $\boldsymbol{\alpha} = \boldsymbol{\alpha}^*$, then $\boldsymbol{\alpha}$ is said to be *hermitian*.

1. Eigenvalues of a hermitian operator are real.

$$< i|(\boldsymbol{\alpha}|i >) = \lambda_i < i|i > \qquad \text{and} \qquad (< i|\boldsymbol{\alpha})i > = \lambda_i^* < i|i > \tag{A-17}$$

If $\boldsymbol{\alpha} = \boldsymbol{\alpha}^*$, then $< i|\boldsymbol{\alpha}|i >^* = < i|\boldsymbol{\alpha}^*|i > = < i|\boldsymbol{\alpha}^*|li >$, and $\lambda = \lambda^*$.

2. Eigenvectors of a hermitian operator belonging to different eigenvalues are orthogonal.

$$< j|\boldsymbol{\alpha}|i > = \lambda_i < j|i > = \lambda_j < j|i >$$

Therefore,

$$(\lambda_j - \lambda_i) < j|i > = 0 \tag{A-18}$$

so if $\lambda_j \neq \lambda_i$, then $< j|i > = 0$. It follows that a matrix of eigenvectors is orthogonal to its own transpose. (If two or more eigenvalues are the same, any linear combination of their eigenvectors is an eigenvector, so the corresponding eigenvectors may be adjusted to be orthogonal.) Hence the transpose is the inverse.

Most physical problems of both classical and quantum mechanics are of a type known as Sturm–Liouville problems. The eigenvectors of Sturm–Liouville problems are orthogonal and form a complete set, which means that the solutions of one problem can be expressed, by Fourier's theorem, in terms of the solutions of another problem.

3. The Hamiltonian does not mix eigenvectors belonging to different eigenvalues of a constant of the motion. If Λ is a constant of the motion, with eigenvalues λ,

$$\mathbf{\Lambda}|j>= \lambda_j|j> \tag{A-19}$$

then $[\mathbf{\Lambda}, \mathbf{H}] = \dot{\Lambda} = 0$, so $< |\mathbf{\Lambda H}| > = < |\mathbf{H\Lambda}| >$. Therefore,

$$(< k|\mathbf{\Lambda})\mathbf{H}|j >= \lambda_k < k|\mathbf{H}|j > = < k|\mathbf{H}|j > \lambda_j = < k|\mathbf{H}(\mathbf{\Lambda}|j >)$$

and thus

$$(\lambda_k - \lambda_j) < k|\mathbf{H}|j >= 0, \text{ so } \lambda_k = \lambda_J \text{ or } < k|\mathbf{H}|j >= 0 \tag{A-20}$$

Therefore, if the operator and the Hamiltonian commute, any eigenvalue of the operator is a constant of the motion.

Because eigenvectors belonging to different eigenvalues are orthogonal,

$$\int \Psi_j^* \Psi_i \, d\tau = 0 \qquad (i \neq j) \tag{A-21}$$

Therefore, if for some operator $\boldsymbol{\alpha}$ it is found that

$$\int \Psi_j^* \boldsymbol{\alpha} \Psi_i \, d\tau \neq 0 \qquad (i \neq j) \tag{A-22}$$

the product of $\boldsymbol{\alpha}$ and Ψ_i must be different from Ψ_i and must, in fact, contain a contribution from Ψ_j. It is said, therefore, that $\boldsymbol{\alpha}$ has caused a *transition* from Ψ_i to Ψ_j.

For example, if Ψ_j and Ψ_i represent harmonic oscillator states, an operator representing a dipole moment gives a nonzero *transition moment*, $\int \Psi_j^* \boldsymbol{\alpha} \Psi_i \, d\tau \neq 0$, which tells us that an electric dipole causes transitions between harmonic oscillator states (for certain symmetries of the states Ψ_j and Ψ_i).

The statement that the Hamiltonian does not mix eigenvectors belonging to different eigenvalues is equivalent to saying that there will not be spontaneous changes from one energy state to another brought about by the internal conditions of the system, as expressed by the Hamiltonian. A molecule cannot, by itself, jump from one harmonic oscillator state to another (unless there is an internal interaction of electronic or rotational states that make the oscillator wave functions nonorthogonal). That is why the eigenstates are called *normal modes* (in the mathematical sense of orthogonality) or *stationary states*.

A.4 Correspondence of Notations

An approximate correspondence of symbols is shown in Table A.1. The function, Ψ, and a column matrix, [ ], may each be represented by a ket, $| >$, and the

Table A.1 Correspondence of Quantum Operator Notations

Differential Operators	Bra-Ket Notation	Matrices
$\Psi = \sum a_i \phi_i$	$\vert\rangle = \sum a_i \vert i\rangle$	$\begin{bmatrix} \\ \end{bmatrix}$ (column)
Ψ^*	$\langle\vert$	$[\quad]$ (row)
$\boldsymbol{\alpha}$	$\boldsymbol{\alpha}$	$\begin{bmatrix} \alpha_{ij} \end{bmatrix}$
$\overline{\boldsymbol{\alpha}} = \int \Psi^* \boldsymbol{\alpha} \Psi \, d\tau$	$\overline{\boldsymbol{\alpha}} = \langle \vert \boldsymbol{\alpha} \vert \rangle$	$\overline{\boldsymbol{\alpha}} = \sum a_i^* a_j \alpha_{ij}$
$= \sum a_i^* \boldsymbol{\alpha}_{ij} a_j$		$= [\; a_i^* \;] \begin{bmatrix} \alpha_{ij} \end{bmatrix} \begin{bmatrix} a_j \end{bmatrix}$
$\alpha_{ij} \equiv \int \phi_i^* \boldsymbol{\alpha} \phi_j$		

operators, $\boldsymbol{\alpha}$, by a square matrix. However, the full correspondence appears only when the state Ψ is represented by a *basis set,* $\{\Phi_i\}$, which can be any *complete* set, that is, any set that satisfies the condition that any function may be expressed precisely as a linear sum of members of the set.

$$\Psi = \sum_i a_i \Phi_i \tag{A-23}$$

Then the average value, or expectation value, for the operator $\boldsymbol{\alpha}$ is

$$\overline{\boldsymbol{\alpha}} = \sum_{i,j} \int \Phi_i^* a_i^* \boldsymbol{\alpha} a_j \Phi_j \, d\tau = \sum_{i,j} a_i^* a_J \int \Phi_i^* \boldsymbol{\alpha} \Phi_j \, d\tau \tag{A-24}$$

The integral

$$\int \Phi_i^* \boldsymbol{\alpha} \Phi_j \, d\tau \equiv \alpha_{ij} \tag{A-25}$$

is called a *matrix element,* even when differential operator notation is being used. The numbers that appear in the row and column matrix vectors will be recognized as the coefficients, a_i, of the basis functions, Φ_i, and the elements of the square matrix operator, $\boldsymbol{\alpha}$, are the matrix elements, α_{ij}, as just defined.

Bibliography

A. Classical Thermodynamics

There are many books on classical thermodynamics, including many titled *Physical Chemistry* that emphasize thermodynamics. Only a few, of particular importance, are listed.

1. G. N. Lewis and M. Randall, *Thermodynamics and the Free Energy of Chemical Substances*. New York: McGraw-Hill Book Co., 1923; 623 pp.

A carefully written exposition of classical thermodynamics that has heavily influenced its successors.

A second edition, by K. Pitzer and L. Brewer (1961), includes new material, but with lower reliability than the original.

2. F. W. Sears and G. L. Salinger, *Thermodynamics, Kinetic Theory, and Statistical Thermodynamics,* 3rd ed. Reading, Mass.: Addison-Wesley Publishing Co., Inc., 1975.

A carefully written, clear exposition of classical thermodynamics, followed by brief discussions of kinetic theory and statistical mechanics.

3. M. Zemansky, *Heat and Thermodynamics*. New York: McGraw-Hill Book Co.

Through a series of revisions, from 1937 to the present (with revision by R. H. Dittman), this has been a popular choice for classical thermodynamics for physicists.

4. P. W. Bridgman, *The Nature of Thermodynamics*. Cambridge, Mass.: Harvard University Press, 1941.

Probing insights into some aspects of thermodynamics by a major worker in the field.

5. H. A. Bent, *The Second Law: An Introduction to Classical and Statistical Thermodynamics*. New York: Oxford University Press, 1965.

Contains significant information, with loose terminology and notation.

6. H. B. Callen, *Thermodynamics and an Introduction to Thermostatistics,* 2nd ed. New York: John Wiley & Sons, Inc., 1985; 483 pp.

An intensive analysis, suitable for a second course, at the senior or graduate level. The second edition has added some statistical mechanics but retains the classical base for understanding of thermodynamics.

7. R. P. Bauman, *An Introduction to Equilibrium Thermodynamics*. Englewood Cliffs, N.J.: Prentice Hall, Inc., 1966; 115 pp.
An elementary approach, intended for freshmen.

B. Statistical Mechanics

8. G. Rushbrooke, *Introduction to Statistical Mechanics*. Oxford: Clarendon Press, 1949.
A good, easy introduction to the field.

9. F. Reif, *Fundamentals of Statistical and Thermal Physics*. New York: McGraw-Hill Book Co., 1965; 651 pp.
An undergraduate (junior–senior) thermodynamics course developed primarily from the standpoint of statistical mechanics.

10. F. Reif, *Statistical Physics; Berkeley Physics Course,* Vol. 5. New York: McGraw-Hill Book Co., 1967; 398 pp.
A briefer, sophomore-level version with the same philosophy as *Fundamentals of Statistical and Thermal Physics*.

11. J. W. Gibbs, *Elementary Principles in Statistical Mechanics*. New Haven, Conn.: Yale University Press, 1902; New York: Dover Publications, Inc., 1960; 207 pp.
The original exposition of what is now called statistical mechanics, introducing the canonical and grand canonical ensembles.

12. R. C. Tolman, *The Principles of Statistical Mechanics*. Oxford: Oxford University Press, 1938; 661 pp.
The definitive early development of classical statistical mechanics and its modification for quantum mechanics. Minimal discussion of applications.

13. L. D. Landau and E. M. Lifshitz, *Statistical Physics*. Elmsford, N.Y.: Pergamon Press, Inc./Reading, Mass.: Addison-Wesley Publishing Co., Inc., 1958; 448 pp.
Like others in the same series (*A Course in Theoretical Physics*), this volume is a definitive modern exposition of the principles and applications.

14. R. Kubo, H. Ichimura, T. Usui, and N. Hashitsume, *Statistical Mechanics, An Advanced Course with Problems and Solutions*. Amsterdam: North-Holland/Elsevier, 1965; 425 pp.
A compressed discussion of the theory is followed, in each of the six chapters, with problems, with solutions given. The problems are suggestive of comprehensive examination questions.

15. D. ter Haar, *Statistical Mechanics*. New York: Rinehart & Company, Inc., 1954; 466 pp.; 2nd ed., 1966.
Written for graduate students, to combine basic theory and applications. Solid presentation of the topics with excellent historical notes.

16. Keith Stowe, *Introduction to Statistical Mechanics and Thermodynamics*. New York: John Wiley & Sons, Inc., 1984; 534 pp.
An introduction for physics students, beginning with statistical mechanics.

17. J. F. Lee, F. W. Sears, and D. L. Turcotte, *Statistical Thermodynamics*. Reading, Mass.: Addison-Wesley Publishing Co., Inc., 1963; 374 pp.
A one-semester course to follow thermodynamics for engineers.

18. A. Sommerfeld, *Thermodynamics and Statistical Mechanics; Lectures on Theoretical Physics,* Vol. V. New York: Academic Press, Inc., 1964; 401 pp.
A survey, with considerable depth, of classical and statistical thermodynamics, emphasizing historical development over modern quantum theory.

19. C. Kittel, *Elementary Statistical Physics.* New York: John Wiley & Sons, Inc., 1958; 228 pp.
A condensed discussion for first-year graduate students.

20. C. Kittel, *Thermal Physics.* New York: John Wiley & Sons, Inc., 1969; 410 pp.
Intended for undergraduate students. Introduces thermodynamics by means of statistical mechanics.

21. T. L. Hill, *An Introduction to Statistical Thermodynamics.* Reading, Mass.: Addison-Wesley Publishing Co., Inc., 1960; 508 pp.
Emphasis is on applications such as chemical equilibrium, polymers, electrolytes, and liquids and solutions.

22. H. Eyring, D. Henderson, B. J. Stover, and E. M. Eyring, *Statistical Mechanics and Dynamics.* New York: John Wiley & Sons, Inc., 1954; 508 pp.; 2nd ed., 1982.
A first-year graduate course for chemists, with emphasis on rates.

23. L. E. Reichl, *A Modern Course in Statistical Physics.* Austin, Tex.: University of Texas Press, 1980; 691 pp.
A graduate-level survey emphasizing applications.

A series of books by Fowler and by Guggenheim are significant.

24. E. A. Guggenheim, *Modern Thermodynamics by the Methods of Willard Gibbs.* London: Methuen & Company Ltd., 1933.
An early attempt to approach thermodynamics from statistical mechanics.

25. R. H. Fowler, *Statistical Mechanics,* 2nd ed. Cambridge: Cambridge University Press, 1936.
Good, early, graduate-level survey.

26. R. H. Fowler and E. A. Guggenheim, *Statistical Thermodynamics.* Cambridge: Cambridge University Press, 1939.
Comprehensive treatment, containing information not usually found in other books.

27. E. A. Guggenheim, *Thermodynamics, An Advanced Treatise for Chemists and Physicists.* Amsterdam: North-Holland, 1949; 4th ed., 1959.
Thermodynamics and statistical mechanics for chemists and physicists.

C. Probability and Error Theory Related to Scientific Experimentation

28. E. Whittaker and G. Robinson, *The Calculus of Observations; An Introduction to Numerical Analysis,* 4th ed., 1944. New York: Dover Publications, Inc., 1967; 597 pp.
A broad, rigorous treatment of numerical methods, such as interpolation, evaluation of determinants, numerical integration, statistics, least squares, Fourier analysis, and the solution of systems of linear equations, algebraic and transcendental equations, and differential equations.

29. H. D. Young, *Statistical Treatment of Experimental Data.* New York: McGraw-Hill Book Co., 1962; 172 pp.

Brief, standard discussion of statistical treatment of experimental errors, with derivations of equations.

30. J. R. Taylor, *An Introduction to Error Analysis*. Mill Valley, Calif.: University Science Books, 1982; 270 pp.

A good exposition of the meaning of uncertainty in experiments and how they can be estimated, including statistical analysis.

31. N. C. Barford, *Experimental Measurements: Precision, Error, and Truth*. Reading, Mass.: Addison-Wesley Publishing Co., Inc., 1967; 143 pp.

Interpretation of experimental results by means of elementary statistical analysis.

32. D. C. Baird, *Experimentation: An Introduction to Measurement Theory and Experiment Design*. Englewood Cliffs, N.J.: Prentice Hall, Inc., 1962; 198 pp.

Interpretation of experimental results by means of elementary statistical analysis.

33. R. Langley, *Practical Statistics, Simply Explained*. New York: Dover Publications, Inc., 1971; 399 pp.

A lucid presentation for nonmathematical people that covers the fundamental distributions and tests of significance, emphasizing common mistakes of experiment design and interpretation.

D. Information Theory Applied to Statistical Mechanics

34. L. Brillouin, *Science and Information Theory*. New York: Academic Press, Inc., 1962; 351 pp.

An introduction to information theory and its applications. Very readable and important.

35. L. Brillouin, *Scientific Uncertainty, and Information*. New York: Academic Press, Inc., 1964; 164 pp.

A sequel, emphasizing questions of interpretation and unsolved challenges. Considers philosophy of classical and quantum uncertainties, the meaning of scientific laws, unidirectionality of time, and so on.

36. Myron Tribus, *Thermostatics and Thermodynamics*. Princeton, N.J.: D. Van Nostrand Company, 1961; 649 pp.

An early attempt at building a course on thermodynamics and statistical mechanics on the foundation of information theory. Slanted for undergraduate engineering students.

37. C. E. Shannon, "A Mathematical Theory of Communication," *Bell System Technical Journal* **27,** 379–423, 623–656 (1948).

These papers were instrumental in the resurgence of interest in information theory in recent decades.

38. E. T. Jaynes, "Information Theory and Statistical Mechanics," *Physical Review* **106,** 620–630; **108,** 171–190 (1957).

Important papers contributing to the development of the theory.

E. Kinetic Theory of Gases

The kinetic theory typically refers to theory that emphasizes the behavior of individual molecules, averaged over a gas (as contrasted with the properties of a system or subsystem of many molecules, averaged). Examples of important reference works include:

39. S. Chapman and T. G. Cowling, with D. Burnett, *The Mathematical Theory of Non-Uniform Gases,* 3rd ed. Cambridge: Cambridge University Press, 1970.
Important reference for kinetic theory of gases.

40. J. H. Jeans, *The Dynamical Theory of Gases,* 4th ed., 1925. New York: Dover Publications, Inc., 1954.
The second edition was published in 1916 with quantum mechanics added in 1920. Historically important development of primarily classical theory.

41. J. H. Jeans, *An Introduction to the Kinetic Theory of Gases*. Cambridge: Cambridge University Press, 1940.
"Simpler and more physical" presentation than Jeans' earlier work.

42. E. H. Kennard, *Kinetic Theory of Gases*. New York: McGraw-Hill Book Co., 1938.
Comprehensive coverage of basic theory, with some statistical mechanics.

43. S. G. Brush, *The Kind of Motion We Call Heat; A History of the Kinetic Theory of Gases in the 19th Century.* Book 1: *Physics and the Atomists;* Book 2: *Statistical Physics and Irreversible Processes*. Amsterdam: North-Holland, 1976.
An important treatment of the history of our understanding of elementary thermodynamic concepts.

F. Molecular Structure and Molecular Spectra

44. G. Herzberg, *Molecular Spectra and Molecular Structure*. I. *Spectra of Diatomic Molecules,* 2nd ed., 1950; II. *Infrared and Raman Spectra of Polyatomic Molecules,* 1945; III. *Electronic Spectra of Polyatomic Molecules,* 1960. Princeton, N.J.: D. Van Nostrand Co., Inc.
This series (and its predecessor on atomic spectra) has been the standard reference for molecular spectroscopy since its appearance.

45. R. P. Bauman, *Absorption Spectroscopy.* New York: John Wiley & Sons, Inc., 1962.
Written as an introduction to the field, with discussions of both theoretical and analytical applications and techniques. (Much of the instrumentation discussion is now outdated.)

46. E. B. Wilson, Jr., J. C. Decius, and P. C. Cross, *Molecular Vibrations,* New York: McGraw-Hill Book Co., Inc, 1955; Dover, 1980.
A concise treatment of the theoretical analysis of vibrational spectra, including the F and G matrix techniques introduced by Wilson.

List of Tables

Answers to Selected Problems

2.1. $\beta = \frac{1}{T}$; $\beta = \frac{1}{V}\frac{\partial V}{\partial T} = \frac{1}{V}\frac{\partial}{\partial T}\left(\frac{nRT}{P}\right) = \frac{nR}{PV}$.

a. 1.29×10^{-3} K^{-1} = 1/773 K.

2.2. **b.** Error = 1.875×10^{-4} cm^3. The fractional error is (very nearly) $3(\alpha\,\Delta T)^2/3\alpha\,\Delta T = \alpha\,\Delta T = 25 \times 10^{-5} = 0.025\%$.

3.1. 10 kJ = 2.4 kcal. 40 g × 1 cal/g · K × 60 K = 2400 cal.

3.4. 100° C; 29.(17) g of steam condenses. If it is assumed that all the steam condenses, the calculation gives t = 145° C; therefore, the assumption is invalid, steam and water remain, and t = 100° C.

3.7. 7.8(4) kJ.

3.9. **a.** 0.27 J.

4.1. $\Delta T = 0$, so: **a.** $Q = 11$ kJ $= -W$; **b.** $W = -10.(96)$ kJ $= -nRT \ln(V_2/V_1) = -4.0 \times 8.314 \times 300 \times \ln 3$.

c. $\Delta E = 0$ **d.** $\Delta H = 0$

4.5. **a.** $Q = 1.56$ kJ $= \Delta H = nc_p\,\Delta T = 0.75$ mol $\times \frac{5}{2}R \times 100$K.

b. $W = -624$ J $= -P\Delta V = -\Delta(nRT) = -0.75 \times R \times 100$K.

c. $\Delta E = 935$ J $= nc_v\,\Delta T = 0.75 \times \frac{3}{2}R \times 100$K.

d. $\Delta H = 1.56$ kJ $= Q$

Check: $\Delta E = Q + W$.

5.1. **a.** $\Delta H = -289$ kJ; **b.** $\Delta H = -464$ kJ.

5.5. $\Delta H_R = -518.(6)$ kJ.

6.2. $c_v = 0.23(4)$ J/g · K $= 3R \times 1$ mol/106.4 g $= 0.056(0)$ cal/g · K.

6.5. **a.** $\overline{\overline{\theta}} = 6.4 \times 10^{-4}$ rad $= \sqrt{kT/\kappa}$; $\frac{1}{2}\kappa\overline{\theta^2} = \frac{1}{2}kT$.

b. $\overline{\overline{x}} = 3.8$ mm $= 2r\overline{\overline{\theta}} = 6\overline{\overline{\theta}}$.

c. $\overline{\overline{\omega}} = 6.4 \times 10^{-7}$ rad/s $= \sqrt{kT/I}$; $\frac{1}{2}I\overline{\omega^2} = \frac{1}{2}kT$.

d. $\overline{\overline{v}} = 3.8$ μm/s $= 6\overline{\overline{\omega}}$ (not significant).

7.1. **a.** 2.94.

7.3. O_2: 20.0%.

7.8. $d = 0.308$ Å.

7.11. **a.** κ(He) $= 4.9 \times 10^{-2}$ W/(m · K) (experimental value 15.5×10^{-2}).

b. κ(Ar) $= 7.3 \times 10^{-3}$ W/(m · K) (vs. 1.8×10^{-2}).

Calculated values agree with experiment to about a factor of 3.

8.1. $T_2 = 226$ K $= -47°$ C.
8.5. $V = 32.77 \times 10^{-3}$ m^3; Evaluate V in correction term from $V = nRT/P = 32.82$ L.
8.6. **b.** $A = 22.3$ m^2; $f = \Delta PA = 50 \times 10^3 \times 9.8$. [Note, however, that the assumption of constant speeds on shaped wing surfaces is a very rough approximation.]
9.4. **b.** $\Delta S = 69.4$ J/K.
9.6. **a.** $\Delta S = 42.6$ J/K.
10.3. **a.** $\varepsilon = 6.4\%$; **b.** $Q_H = 1.56$ kJ; **d.** $-Q_H = 1.56$ kJ.
10.4. **a.** M; values depend on size of stroke, n, and ideality.
d. T; $Q(\mathrm{A}) = nRT \ln(V_2/V_1)$ only for an ideal gas.
10.7. $\varepsilon_{\mathrm{CA}} = 14.4\%$ vs. 26.8%.
11.3. $P = 2.9 \times 10^8$ Pa.
11.5. **c.** $\Delta G_{A+B} = \Delta G_A + \Delta G_B = -25.9$ kJ $= -6.24$ kcal
12.1. $p_{(-5)} = 3.0(24)$ torr.
12.2. $p_{(-5)} = 3.1(79)$ torr.
12.10. $p_{25} = 1.4 \times 10^{-2}$atm $= 11$ torr.
12.11. **a.** Triple point $= 182.(37)$ K $= -90.(95)°$ C.
13.4. **a.** If the overall composition and temperature correspond to point A (15% aniline), the phases actually present will be pure phenol and a solution of aniline (approximately 20%) in phenol.
13.7. The lowest horizontal line, on the right, is the eutectic temperature for the Ag–Hg melt, occurring very near the melting point of pure mercury and very near the composition of pure mercury.
14.2. 9.63% decomposed.
14.6. $K_{\mathrm{sp}} = 8.1 \times 10^{-17}$.
14.8. $K_{\mathrm{eq}} = 4.0(6)$. $\mathcal{E}° = 0.789 + (-0.771) = 0.018$ V.
15.6. $P_{30}(25) = 5\%$. This should be compared with $P_{30}(30) = 0.0726$. The probability of getting 25 counts is 70% as great as the probability of getting 30 counts.
15.8. **b.** $P = 0.057$.
16.1. **b.** 4.58 bits.
18.8. The definition of a determinant, of order n, is the sum of $n!$ products, of n numbers each.

$$|a_{ij}| = \sum_j \mathrm{sign}(j_1, j_2, \cdots j_n) a_{1j_1} a_{2j_2} a_{3j_3} \cdots a_{nj_n}$$

where $(j_1, j_2 \cdots j_n)$ is the permutation of $\{j_i\}$,which is even or odd as there have been an even or odd number of transpositions from standard order. From this follows the property that interchanging any two rows or columns of a determinant changes the sign of the determinant.
20.6. **b.** $n_4/N = 5.668 \times 10^{-3}$.

The Bridgman relations (see back endpapers) provide a quick derivation of many thermodynamic equations. The symbols given are combined, as fractions, to give partial derivatives [P. W. Bridgman, *Phys. Rev.* 3, 273 (1914)]. For example,

$$\left(\frac{\partial \mathrm{E}}{\partial \mathrm{V}}\right)_{\mathrm{T}} = \frac{TV\beta - PV\kappa}{\kappa V} = T\frac{\beta}{\kappa} - P$$

Index

*Major entries are set in boldface type.

Bridgman Relations

$$\beta \equiv \frac{1}{V}\left(\frac{\partial V}{\partial T}\right)_P \qquad \kappa = -\frac{1}{V}\left(\frac{\partial V}{\partial P}\right)_T \qquad C_p = \left(\frac{\partial H}{\partial T}\right)_P$$

$$(\partial T)_P = -(\partial P)_T = 1$$

$$(\partial V)_P = -(\partial P)_V = \beta V$$

$$(\partial S)_P = -(\partial P)_S = \frac{C_p}{T}$$

$$(\partial E)_P = -(\partial P)_E = C_p - PV\beta$$

$$(\partial H)_P = -(\partial P)_H = C_p$$

$$(\partial G)_P = -(\partial P)_G = -S$$

$$(\partial F)_P = -(\partial P)_F = -(S + PV\beta)$$

$$(\partial V)_T = -(\partial T)_V = \kappa V$$

$$(\partial S)_T = -(\partial T)_S = \beta V$$

$$(\partial E)_T = -(\partial T)_E = TV\beta - PV\kappa$$

$$(\partial H)_T = -(\partial T)_H = -V + TV\beta$$

$$(\partial G)_T = -(\partial T)_G = -V$$

$$(\partial F)_T = -(\partial T)_F = -PV\kappa$$

$$(\partial S)_V = -(\partial V)_S = \frac{V}{T}(-C_p\kappa + TV\beta^2)$$

$$(\partial E)_V = -(\partial V)_E = -C_pV\kappa + TV^2\beta^2$$